Nanostructures and Nanomaterials for Batteries

Yu-Guo Guo

Nanostructures and Nanomaterials for Batteries

Principles and Applications

Springer

Yu-Guo Guo
Institute of Chemistry
Chinese Academy of Sciences
Beijing, China

ISBN 978-981-13-6232-3 ISBN 978-981-13-6233-0 (eBook)
https://doi.org/10.1007/978-981-13-6233-0

This Springer imprint is published by the registered company Springer Nature Singapore Pte Ltd.
The registered company address is: 152 Beach Road, #21-01/04 Gateway East, Singapore 189721, Singapore

Foreword

It is generally recognized that the dependence of modern society on the energy stored in a fossil fuel is not sustainable. The most sustainable energy source is the sun; its energy comes to planet Earth continually in the forms of sunlight, rainwater, and wind energy. However, these energy sources are all diffuse and arrive on a timescale completely different from that of the demands of modern society. Therefore, the sun's energy must be condensed, stored, and harvested for use when and where needed. Nature condenses and transforms sunlight and rainwater into stored chemical energy in plants and trees that are either harvested directly by animal life or, over time, are converted into a fossil fuel. Chemical energy can be burned directly to supply heat or mechanical energy as in an engine; it can also be stored and/or converted into electric power in an electrochemical cell. Rainwater is also condensed and transported by rivers. Flowing water is hydropower that can be stored in dams for subsequent conversion in turbines of its mechanical energy into electric power. Wind energy and sunlight can also be converted into electric power by, respectively, windmills and photovoltaic solar panels.

The advantage of electric power is that it can be transported inexpensively long distances to distributed collection sites and stored in a rechargeable battery. A battery may consist of a single cell or of multiple cells connected in series and/or in parallel for a required output power. Large-scale batteries may consist of several thousand cells. Rechargeable batteries can be either stationary or portable with different requirements for different applications.

This book is intended to be an introduction to electric-power storage in rechargeable batteries and the use of nanoparticles in their fabrication, particularly in the positive electrode. The hand-held Li-ion rechargeable battery has enabled the cell telephones, tablets, and laptop computers that have revolutionized modern life; but it has problems with too long a charging time, safety, cost, and charge/discharge cycle life. The development of rechargeable batteries that go beyond the Li-ion battery is an ongoing activity being extensively addressed worldwide in research laboratories.

Austin, USA

Dr. John B. Goodenough
Department of Mechanical Engineering
Cockrell School of Engineering
The University of Texas at Austin

FOREWORD

It is generally recognized that the dependence of modern society on the energy stored in a fossil fuel is not sustainable. The most sustainable energy source is the sun; its energy comes to planet Earth continually in the form of sunlight, rainwater, and wind energy. However, these energy sources are all diffuse and arrive on a timescale completely different from that of the demands of modern society. Therefore, the sun's energy must be condensed, stored, and harvested for use when and where needed. Nature condenses and transforms sunlight and rainwater into stored chemical energy in plants and trees that are either harvested directly by animal life or, over time, are converted into a fossil fuel. Chemical energy can be burned directly to supply heat or mechanical energy as in an engine; it can also be stored and converted into electric power in an electrochemical cell. Rain water is also condensed and transported by rivers. Flowing water is hydropower that can be stored in dams for subsequent conversion in turbines of its mechanical energy into electric power. Wind energy and sunlight can also be converted into electric power by, respectively, windmills and photovoltaic solar panels.

The advantage of electric power is that it can be transported inexpensively long distances to distributed collection sites and stored in a rechargeable battery. A battery may consist of a single cell or of multiple cells connected in series and/or in parallel for a required output power. Large-scale batteries may consist of several thousand cells. Rechargeable batteries can be either stationary or portable with different requirements for different applications.

This book is intended to be an introduction to electric-power storage in rechargeable batteries and the use of nanoparticles in their fabrication, particularly in the positive electrode. The hand-held Li-ion rechargeable battery has enabled the cell telephones, tablets, and laptop computers that have revolutionized modern life; but it has problems with too long a charging time, safety, cost, and charge/discharge cycle life. The development of rechargeable batteries that go beyond the Li-ion battery is an ongoing activity being extensively addressed worldwide in research laboratories.

John B Goodenough
Centennial Professor of Materials Science
Cockrell School of Engineering
The University of Texas at Austin.

Preface

As the worldwide demand for energy continues to increase, it is essential to develop advanced energy storage systems for efficient utilization of renewable energy resources. Various electrochemical energy storage (EES) systems, e.g., lead-acid battery, nickel-cadmium and nickel-metal hydride batteries, lithium-ion battery, and supercapacitor, have witnessed their tremendous success and penetrated into almost every aspect of modern civilizations. Specifically, rechargeable batteries with high energy density, high power density, low cost, long-term cycle stability, high safety, and abundant resources are highly desired. All the above parameters are largely dependent on the thermodynamic and kinetic properties of the key materials in the EES system. Nanomaterials show admirable physical and chemical properties relative to the bulk materials, such as high surface to volume ratio, favorable transport kinetics, and nanopore confinement, which offer new opportunities to address the challenges of advanced EES systems. As a result, the application of nanostructured materials in rechargeable batteries has shown great potential to introduce new active storage sites, shorten ion transport path, reduce specific surface current rate, and improve electrochemical stability.

In this book, we presented a comprehensive discussion on fundamental knowledge about nanostructures and nanomaterials toward their application in rechargeable batteries. The advantages and challenges accompanied by use of nanostructured electrode materials are presented, along with the rational design at material/structures level to reach optimal battery performance. In addition, we summarized recent advancement of the use of nanostructures and nanomaterials in the next-generation rechargeable battery systems, including high-energy lithium-ion batteries, lithium-metal batteries, sodium-ion batteries, all-solid-state batteries, and other types of rechargeable batteries; and give perspectives on the opportunities of nanomaterials in these advanced EES technologies.

Chapter 1 intends to present a broad picture of world energy utilization and introduce various types of energy storage technologies, current research about these technologies and their sustainability for the future. In order to be linked to the next chapter, Chap. 1 will also discuss basic working principles of crucial EES devices (e.g., primary battery, rechargeable battery, pseudocapacitor, and fuel cell) as well as key components/materials for these devices.

Chapter 2 is focusing on the fundamentals about the charge transfer and storage processes in various types of electrochemical cells (e.g., zinc-based primary cells, lead-acid cells, nickel-metal hydride cells, rechargeable lithium cells), and discusses effects of using nanostructured electrode materials on thermodynamic and kinetic performance of an electrochemical cell. With the discussions, it is aimed at providing insights into design principles for "kinetically stable" nanostructured electrode materials.

Chapters 3 and 4 intend to discuss the recent progress in the development of advanced electrode materials (both cathode and anode materials) for state-of-the-art Li-ion batteries and next-generation Li-metal batteries. Particularly, the two chapters focus on the creation of nanostructures and use of nanoengineering to improve the electrochemical storage performance of advanced electrode materials that not only operate based on traditional intercalation reaction, but also on new EES reactions that could enable realization of rechargeable Li batteries with boosted energy density, such as conversion reactions (S, O_2, transition metal oxides/sulfides/nitrides/fluorides), alloying reactions (Si, Ge, Sn, and their derivatives), and plating-stripping reactions (Li metal and Na metal).

Chapter 5 introduces the history, general knowledge, and development at the present stage of solid-state electrolytes and rechargeable all-solid-state cells that promise improved energy density and safety, extended shelf life, and competitive packaging cost for future storage market. Highlights were placed on the use of nanostructures and nanoengineering techniques to address the thermodynamic stability issue at electrode-electrolyte interface, and to improve the charge transfer kinetics in solid-state cell. Other factors that influence the overall storage performance of the cell, such as temperature and stress, were also discussed.

Chapter 6 elaborates the operation principle and fundamental electrochemistry of room temperature sodium-ion battery, a substitutive storage technology with improved resource sustainability for today's Li-ion battery. The electrochemical storage processes occurred in the above two types of batteries will be compared to understand the differences and similarities between them, as well as the benefits and challenges accompanied by the use of the battery. The chapter also discusses the recent research advancements of nanostructured electrode materials that trigger high-performance sodium-ion batteries, and the prospects for their future development.

Chapter 7 summarizes the recent progress on other crucial non-storage components of rechargeable Li cells, such as conductive additives, separators, and current collectors. The nanostructures and nanomaterials in these components, as well as their influences on the electrochemical storage properties of the cell, are highlighted.

The last chapter, Chap. 8, summarizes the advantages and challenges of emerging rechargeable batteries and their key materials that enable short-term and long-term goals of EES achievements, and more importantly, give perspectives on the opportunities and potential benefits of nanostructures and nanomaterials for these emerging storage technologies.

The potential readership of this book may include, but is not limited to scientists, undergraduate, and graduate students of related majors in energy, electrochemistry, materials science, as well as nanoscience and nanotechnology, and other professionals working in the related fields from both science and industry. Overall, we have taken this opportunity to polish the text thoroughly, making the information even more accessible, helpful, and up to date. We sincerely hope that the readers will contact us with valuable suggestions for its continued improvement.

Finally, I would like to thank all the contributors to this book. In particular, I appreciate the contribution of Foreword from Prof. John B. Goodenough at the University of Texas at Austin, one of the inventors of modern Li-ion battery technology and also a respectful scientist who still leads the frontier research of EES at his age of ninety seven. I would also thank his family for their continuous support during the career. To Dr. Sen Xin I am also indebted for his great help throughout the preparation of this book. Many thanks to them.

Beijing, China
March 2019

Yu-Guo Guo, Ph.D.
Professor of Chemistry and Materials Science

CAS Key Laboratory of Molecular
Nanostructure and Nanotechnology
Beijing National Laboratory
for Molecular Sciences (BNLMS)
Institute of Chemistry
Chinese Academy of Sciences (CAS)

University of Chinese Academy of Sciences

Contents

Contributors

Fei-Fei Cao Huazhong Agricultural University, Wuhan, Hubei, People's Republic of China

Hongcai Gao University of Texas at Austin, Austin, TX, USA

Jin-Zhi Guo Northeast Normal University, Changchun, Jilin, People's Republic of China

Yu-Guo Guo Institute of Chemistry, Chinese Academy of Sciences, Beijing, People's Republic of China

Yutao Li University of Texas at Austin, Austin, TX, USA

Yu-Bin Niu Institute of Chemistry, Chinese Academy of Sciences, Beijing, People's Republic of China

Peng-Fei Wang University of Maryland at College Park, College Park, MD, USA

Shu-Hua Wang Shandong University, Jinan, Shandong, People's Republic of China

Xing-Long Wu Northeast Normal University, Changchun, Jilin, People's Republic of China

Sen Xin University of Texas at Austin, Austin, TX, USA

Chun-Peng Yang University of Maryland, College Park, MD, USA

Huan Ye Huazhong Agricultural University, Wuhan, Hubei, People's Republic of China

Ya You University of Texas at Austin, Austin, TX, USA

Junpei Yue Institute of Chemistry, Chinese Academy of Sciences, Beijing, People's Republic of China

Chapter 1
Introduction to Electrochemical Energy Storage

Sen Xin, Hongcai Gao, Yutao Li and Yu-Guo Guo

Abstract Facing the challenge from a fast growth in global primary energy consumption during the last two decades, energy conversion and storage with high efficiency and sustainability is demanded. This chapter intends to discuss the broad picture of world energy utilization, and introduce various types of energy storage technologies, their advantages/disadvantages, research at the present stage and sustainability for the future. Specifically, this chapter will introduce the basic working principles of crucial electrochemical energy storage devices (e.g., primary batteries, rechargeable batteries, pseudocapacitors and fuel cells), and key components/materials for these devices.

Abbreviations

EDLC	Electrical double-layer capacitor
SMES	Superconducting magnetic energy storage
NEC	Nippon Electric Company
DC	Direct current
EES	Electrochemical energy storage
TES	Thermal energy storage
NiCd	Nickel-cadmium
NiMH	Nickel-metal hydride
SHE	Standard hydrogen electrode
VB	Valence band
CB	Conduction band
LUMO	Lowest unoccupied molecular orbital
HOMO	Highest occupied molecular orbital

S. Xin · H. Gao · Y. Li
University of Texas at Austin, Austin, TX 78712, USA

Y.-G. Guo (✉)
Institute of Chemistry, Chinese Academy of Sciences, Beijing 100190, People's Republic of China
e-mail: ygguo@iccas.ac.cn

Y.-G. Guo, *Nanostructures and Nanomaterials for Batteries*,
https://doi.org/10.1007/978-981-13-6233-0_1

1.1 Sustainable Energy Conversion and Storage

Sustainable energy conversion and storage based on renewable energy sources (e.g., solar energy, wind power, hydropower, tidal energy, geothermal energy and nuclear energy, etc.) are becoming important due to the increasing energy demand for economic and social development. During the last decades, the world has witnessed a fast growth in global primary energy consumption, in which the share of renewables kept ascending while the share of traditional energy sources, such as coal, continued to decline (Fig. 1.1) [1]. To realize a sustainable utilization of energy, especially the renewables, efficient energy conversion and storage techniques are required.

Though frequently used together, energy conversion and energy storage are conceptually different. Energy conversion, also termed as energy transformation, is the process of changing energy from one form into another. For example, in a heat machine, thermal energy is converted into mechanical energy so that the machine can do external work. Energy storage, on the other hand, is to capture or harvest energy produced at one time for use later. As a common phenomenon occurred in nature, plants harvest the solar energy under direct sunlight exposure and stores it in organic matters through photosynthesis. One should note that, both the energy conversion and storage processes follow the law of energy conservation, i.e., the total energy of an isolated system always remains constant. It means that energy can neither be created nor destroyed, but can only be transformed or transferred from one form to another.

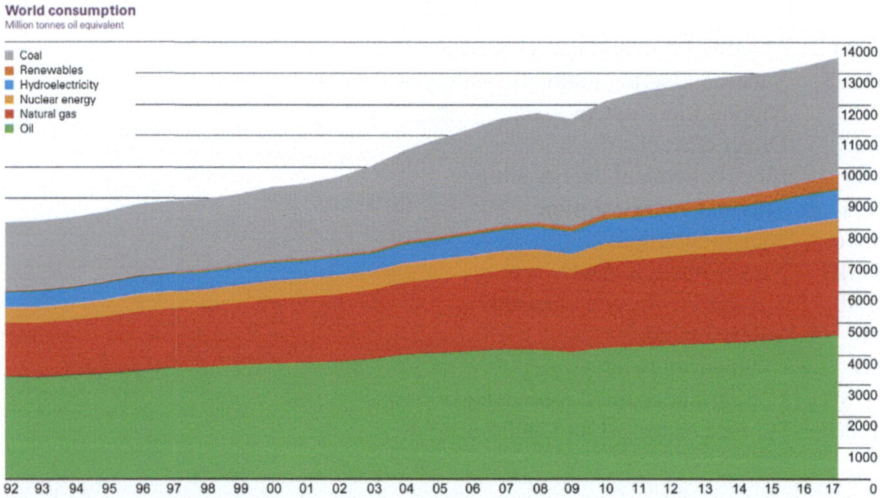

Fig. 1.1 World primary energy consumption from the year 1992 to 2017. All fuels except coal and hydroelectricity grew at above-average rates. Natural gas provided the largest increment to energy consumption at 83 million tonnes of oil equivalent (mtoe), followed by renewable power (69 mtoe) and oil (65 mtoe). Reproduced with the permission from Ref. [1], Copyright © BP p.l.c. 2018

As energy comes in multiple forms including radiation, chemistry, electricity, heat, and even gravity, energy storage devices that involve conversion of energy from forms that are difficult to store to storable forms are necessary for the efficient utilization of energy. Battery, for example, is a typical energy storage device, which converts and stores electrical energy through chemical reaction. In the following section, we will make a brief introduction to various techniques and devices for energy storage.

1.2 A Brief Introduction to Energy Storage Technology

Energy storage has existed since pre-history of human beings, though it was often not explicitly recognized as such. For example, by growing crops and storing dried wood, people were able to collect energy from nature and stored it for future needs. To power the modern civilization, many techniques have been developed to store energy in form of heat or electricity. Generally, these techniques can be classified into seven types according to the method/medium of implementation (refer to the following Table 1.1).

1.2.1 Fossil Fuels

A fossil fuel is a fuel that contains energy stored during ancient photosynthesis. The fossil fuels are usually formed by natural processes, such as anaerobic decomposition of buried dead organisms [2]. Coal, oil and nature gas represent typical fossil fuels that are used mostly around the world (Fig. 1.1). The extraction and utilization of energy from fossil fuels is mainly achieved through combustion reaction, which turns chemical energy directly into thermal energy for collection and transformation into other forms (e.g., mechanical or electrical energy).

Despite efforts to switch to clean sources of energy, the cost and resource advantages make the fossil fuels indispensable to our lives at least for the next few decades. For example, coal will remain a major source of energy in satisfying the world's enormous appetite for electricity in the upcoming years. However, the utilization of energy from fossil fuels has raised concerns, from both science and industry communities, about the sustainability of resources and the adverse effects on the environment. All the fossil fuels are non-renewable, which means predicted depletion of reserves in the near future as global fuel demands continue to raise. Besides, burning of fossil fuels leads to elevated total carbon emissions (mainly in form of greenhouse gases such as carbon dioxide) into the atmosphere, which is the potential reason for global warming. Combustion of fossil fuels also produces other air pollutants, such as nitrogen/sulfur oxides, sulfur dioxide, volatile organic compounds and heavy metals, which are believed threatens to the ecological system. To achieve better economic and environmental sustainability, there are always needs to reduce the dependence

Table 1.1 Various types of energy storage techniques classified according to the method/medium of implementation

Types of energy storage technology	Types of method/medium of implementation
Fossil fuels	Oil Coal Nature gas
Mechanical storage	Pumped storage of hydropower Compressed air energy storage Flywheel energy storage Hydraulic accumulator
Electrical/electromagnetic storage	Dielectric capacitor Electrical double-layer capacitor Superconducting magnetic energy storage
Electrochemical energy storage	Primary battery Rechargeable battery Faradaic supercapacitor (pseudocapacitor) Flow battery
Biological storage	Glycogen Starch
Thermal storage	Sensible heat storage Latent heat storage Chemical heat storage
Chemical storage	Biofuels Hydrated salts Hydrogen storage Power to gas

on fossil fuels, and the use of fossil fuels should proceed under strict administrative regulations.

1.2.2 Mechanical Storage

Mechanical storage refers to storage of excessive mechanical or electrical energy in a medium as kinetic energy, potential energy or other energy forms. Pumped storage in a hydropower plant, compressed air energy storage and flywheel energy storage are the three major methods of mechanical storage [3]. However, only for the flywheel the supplied and consumed energies are in mechanical form; the other two important applications, namely pumped hydro energy storage and compressed air energy storage, are for electrical power production at peak demand.

The storage of electricity in pumped hydropower plants are the mostly-used and largest storage systems for electricity [4, 5]. Designed specifically towards the peak-load shifting application in a grid, a pumped hydropower plant is operated based on the conversion between electrical energy and gravitational potential energy of water [5]. During off-peak times, water is electrically pumped from a lower reservoir to a higher reservoir at a lower cost of electricity, during which the electrical energy is transformed into gravitational potential of water. During the peak times, water is released back into the lower reservoir to spin a propeller-like device named turbine, which in turn rotates a generator to generate electricity. In another form of hydropower plant, water is collected by and stored in an impoundment, or a dam, and is dropped by its own gravity through a penstock pipe to spin the turbine that is linked to a generator and output electricity. The conversion process inevitably leads to loss of a certain amount of energy, however, the pumped storage systems can achieve an energy efficiency of up to 80% [6]. Meanwhile, the hydropower plants have very high reliability, with an expected service life of more than a century. However, the utilization of hydropower has its own limitations. The construction of a hydropower plant is costly and relies highly on suitable geological conditions, and may have potential hazardous effects on the ecological system around the site. Once the plant is built, it will require additional labors and costs to maintain operation of the facility.

To reduce the geological dependence and construction cost of mechanical storage at grid scale, compressed air energy storage systems were proposed [7, 8]. A compressed air energy storage power plant functions in a way similar to a hydropower plant, yet the storage medium is changed from water to compressed air. At utility scale, cheap electricity generated during off-peak periods is utilized to compress air into an underground cavern at a pressure 40–80 bar, which is then released next day to run gas turbines so that the generated electricity can meet the power demand during peak periods [7, 8]. Compared with the pumped storage of hydropower, the compressed air storage requires less on the geological conditions and has a high flexibility to meet application requirements at different scales. However, this storage technology has safety concerns, mainly catastrophic tank rupture. Safety regulations should be made to eliminate the rupture incident.

Flywheels are intended for use in medium and small-scale mechanical energy storage. The system works by accelerating a rotor (flywheel) to a very high speed and maintaining the energy in the system as rotational energy [9, 10]. During the operation, adding energy to the system increases the rotational speed of the flywheel, and extraction of energy from the system reduces the flywheel speed. With continuous efforts on materials and design, the systems are expected to deliver a higher energy/power density in the future.

1.2.3 Electrical/Electromagnetic Storage

Electromagnetic energy can be stored in the form of an electric field or a magnetic field. Conventional electrostatic capacitors, electrical double-layer capacitors (EDLCs) and superconducting magnetic energy storage (SMES) are most common storage techniques [11–13].

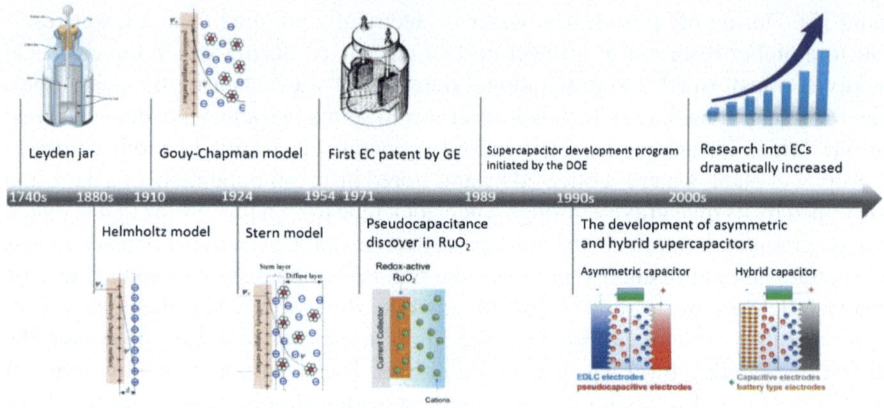

Fig. 1.2 History for the development of capacitors (upper panel) and schematics of various models proposed for the charge-storage mechanism inside a capacitor. Reproduced with the permission from Ref. [15], Copyright © 2018 American Chemical Society

The demonstration of the first capacitor can date back to the middle of the 18th century. The first capacitor, named a "Leyden jar", was invented separately by a German cleric Ewald Georg von Kleist in 1745 and a Dutch scientist Pieter van Musschenbroek in 1746 [14]. It consisted of two pieces of metal foils, water, and a conductive chain inside a glass jar, and static electricity could be generated by rotating the glass jar [14, 15]. Today's capacitors, though provided with much more complicated structures, still have the same basic components and energy storage mechanism as the Leyden jar (Fig. 1.2).

A conventional electrostatic capacitor contains at least two electrical conductors in forms of metal foil or thin film, or an electrolyte, separated by a non-conducting dielectric medium, which can be glass, ceramic, plastic film, or a layer of metal oxide. Ceramic and electrolytic capacitors are two commonly-used dielectric capacitors [16, 17]. Though varied in configurations and materials, the capacitors are operated on similar energy storage mechanisms. When a capacitor is attached across a battery, the two conductors experience a potential difference, and an electric field develops across the dielectric, so that a net positive charge accumulates on one plate while a net negative charge accumulates on the other. Capacitance is a key parameter for performance evaluation of a capacitor, which is defined as the ratio of the electric charge on each conductor to the potential difference between them according to the following equation [18]:

$$C = Q/V \tag{1.1}$$

where C denotes the capacitance (unit: Farad), Q denotes the charge quantity (unit: Coulomb), and V denotes the potential difference (unit: Volt). In the simplest parallel-plate model, the capacitance of a capacitor is proportional to the surface area of the conductors (A, unit: m^2) and inversely related to the gap inbetween (d, which is also the thickness of the dielectric, unit: m), which gives another form of the above equation:

$$C = \varepsilon_0 \times \varepsilon_r \times A/d \tag{1.2}$$

where ε_0 is the permittivity of air (8.84×10^{-12} F m^{-1}), and ε_r is the relative permittivity of the dielectric medium between the two plates.

Another performance parameter of capacitor is the energy stored in the electrostatic field (E, unit: Joule), which is equal to the total work (W, unit: Joule) done in establishing the electric field from an uncharged state and is calculated according to the following equation [19]:

$$E = W = (C \times V^2)/2 \tag{1.3}$$

Therefore, a capacitor with larger capacitance or higher voltage stores more energy. To improve the capacitance, the capacitor is expected to have a large surface area of conductor, a smaller distance of two plates and a higher permittivity of the dielectric. By applying a higher voltage, the capacitor may also be expected to achieve a higher energy storage, yet the voltage applied should not exceed the breakdown voltage (V_{bd}), which is calculated by multiplying the dielectric strength (E_{ds}) by the distance between the two plates (d) according to the following equation:

$$V_{bd} = E_{ds} \times d \tag{1.4}$$

Clearly, a capacitor with a higher dielectric strength can store more energy.

To further improve the energy stored on a capacitor, other forms of capacitors, such as EDLCs, were developed. After the invention of the Leyden jar, people realized that the storage of static electricity can occur at the interface between a solid electrode and a liquid electrolyte, and this knowledge helped to establish the original concept of an electric double layer. Compared with the conventional electrostatic capacitors, von Helmholz first studied the electrical charge-storage mechanism in capacitors and built the first electric double-layer model in 1853 [20]. After that, some pioneering electrochemists, including Gouy, Chapman, Stern, and Grahame, developed the modern theory of electric double-layer capacitance for EDLC [21–24]. In 1954, H. I. Becker at General Electric patented the first EDLC, which employed porous carbon electrodes immersed in an aqueous electrolyte (which functions as dielectric) and stored electric energy at the interfacial electric double layer [25]. Benefiting from the several orders of magnitude higher specific surface area of porous carbon electrode (A) and the small charge-separation distance at nanoscale (dielectric thickness, d), the EDLCs are able to store much more electric energy than conventional dielectric capacitors. The first nonaqueous-electrolyte-based EDLC was patented by Robert A. Rightmire at the Standard Oil of Ohio and later commercialized by Nippon Electric Company (NEC) in 1978 [26]. The nonaqueous EDLC could provide much higher operating voltage (3.4–4.0 V) and energy densities than Becker's aqueous EDLC due to a wider electrochemical window of nonaqueous electrolyte. Based on the technology, the NEC successfully developed the electrochemical capacitor market for back-up power inside key computer electronic components such as clock chips and complementary metal-oxide-semiconductor memories, which are still one of the main applications for current supercapacitors.

SMES systems store energy in the magnetic field created by direct current flow in a superconducting coil, which has been cryogenically cooled to a temperature below its superconducting critical temperature [12, 13, 27–29]. The general structure of SMES consists of three components: (1) a low-temperature/high-temperature superconducting coil, (2) a cryogenically cooled refrigerator and (3) a cooling agent (helium/nitrogen liquid). During the operation, a direct current (DC) voltage is applied across the terminals of a SMES coil to generate energy in a magnetic field [13]. After removing the voltage source, the current continues to flow as the coil attains a negligible resistance at its superconducting state, so that the current will not decay and the magnetic energy can be stored indefinitely. Upon discharging, the energy is released by a discharging coil, and the SMES can quickly transit between its fully charged state to fully discharged state due to its high efficiency. After discharging, the released energy goes through a power conditioning system, which employs an inverter/rectifier to transform alternating current power to DC power and vice versa, and is then fed back into the network. As the inverter/rectifier accounts for ca. 2–3% energy loss in each direction, the SMES system usually shows a round-trip efficiency of > 95% [12], making it an appealing choice for the future storage market.

1.2.4 Electrochemical Energy Storage

Although some historians claim that the first battery was invented hundreds of years before Christ, the traceable story of electrochemistry began with Alessandro Volta, who announced his invention of the voltaic pile, the first modern electrical battery, in the year 1800 [30]. During the next two centuries, electrochemical energy storage (EES) gradually became one of the most powerful storage techniques and penetrated into almost every aspect of modern civilization. With the invention of rechargeable lithium battery and supercapacitors in the past century, the EES devices have again witnessed its tremendous success in emerging fields such as consumer electronics and electricity-powered automobiles and show promises towards future applications.

One of the fundamental differences between EES and electrostatic storage lies in the occurrence of electrochemical redox reactions (or the faradaic process) during the storage process. For example, in the development of "supercapacitors" towards powering modern consumer electronics during 1950s–70s, there were two types of capacitors that came into being: EDLCs and pseudocapacitors. In 1971, the pseudocapacitance was first discovered and on a capacitor based on ruthenium oxide (RuO_2) thin film electrode [31]. While the RuO_2 electrode exhibited a typical capacitive behavior (the cyclic voltammograms showed a rectangular shape), the energy storage on the electrode went through a faradaic process [32].

Unlike the non-faradaic process occurred in EDLCs, charge carriers (e.g., electrons) are transferred across the electrode|electrolyte interface during the storage process of pseudocapacitors, causing oxidation or reduction reactions to readily occur [33, 34]. Since such reactions are governed by the Faraday's law (i.e., the amount of chemical reaction caused by the current flow is proportional to the amount of electricity passed), the process is named a faradaic process, and the capacitance contributed by the faradaic process is called pseudocapacitance. As both the faradaic and the non-faradaic processes could occur when the electrode reactions took place, the discovery of pseudocapacitance opened up a new approach to enhance the charge-storage capability of electrochemical capacitors [15]. From then on, a large variety of supercapacitors have become commercially available, which include EDLCs, pseudocapacitors, and asymmetric supercapacitors. Each type of the capacitors has their own features and target applications, such as backup energy for portable electronics or uninterrupted power supplies, and high-power energy sources for energy load leveling system in electric vehicles. For example, on the basis of this discovery, the Pinnacle Research Institute in Austin, Texas and the U.S. Department of Energy separately started research projects in the 1980s to develop high-performance supercapacitors targeted at military use and civil applications in electric and hybrid vehicles. In 1989, Department of Energy of the United States initiated an ultracapacitor development program and contracted it to Maxwell Technologies. The program aimed at development of capacitors suitable for battery or fuel cell levelling in an electric vehicle, with both short-term goal defined for 1998–2003 and long-term goal for thereafter [15]. After 2000s, the amount of studies related to supercapacitors boosted due to an ever-increasing demand for high-power, high-reliability, and high-safety energy storage devices, and these studies have led to many new physical and electrochemical findings for the capacitor system. Meanwhile, companies from all over the world such as Nesscap (Korea), ELTON (Russia), and Nippon Chemicon, NCC (Japan) emerged with a variety of ECs that offers energy densities ranging from 1 to 10 Wh kg^{-1} and power densities up to 20 kW kg^{-1} [35].

Batteries form another key type of EES application. Conceptually, an electric battery is actually a device consisting of one or more electrochemical cells connected in a certain pattern (series, parallel or series-and-parallel) to power external appliances (Fig. 1.3) [36–38]. An electrochemical cell is a device able to either generate electrical energy from electrochemical redox reactions or utilize the reactions for storage of electrical energy. The cell usually consists of two electrodes, namely, the anode and the cathode, which are separated by an electronically insulative yet ionically conductive substance named an electrolyte (Fig. 1.3a) [39]. In practical configurations, the two solid electrodes are usually kept apart by an electrolyte-permeable separator [37]. There are two types of charge carriers working in the electrochemical cell: electrons and ions. During the cell operation, the working ions migrate towards a certain direction in the electrolyte while the electrons transfer through external circuit [37]. The total amount of energy produced by a battery depends on both the difference between the electrochemical potentials of both electrodes and the total

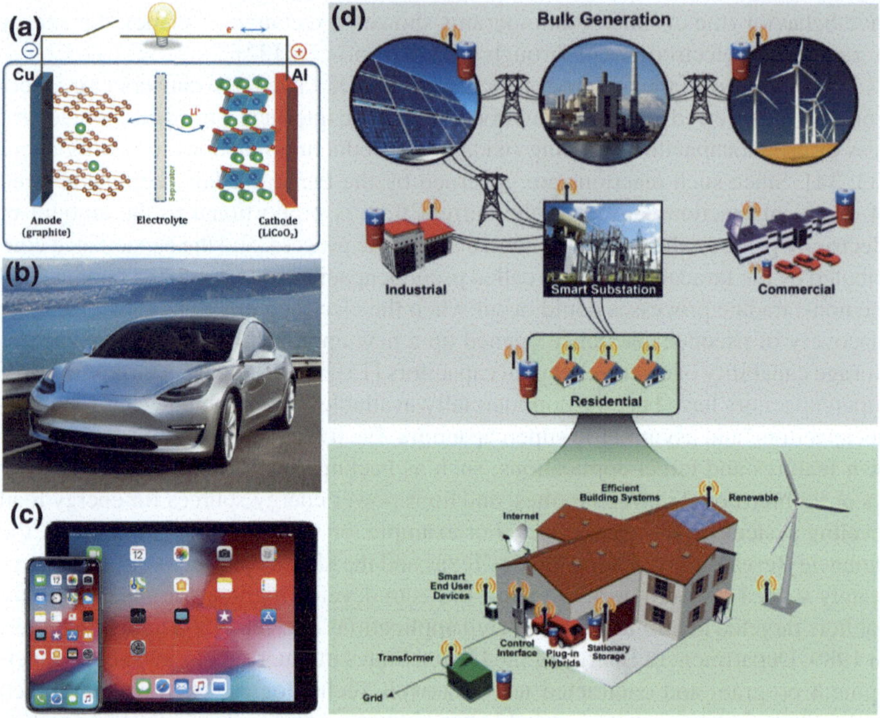

Fig. 1.3 **a** Schematic illustration of the modern Li-ion battery. **b–d** Application of batteries in **b** automobiles, **c** consumer electronics and **d** future grid-scale storage. Wherein, **a** was reproduced with permission from Ref. [87], Copyright © 2013 American Chemical Society. **d** was reproduced with permission from Ref. [36], Copyright © 2011 American Chemical Society. **b** and **c** were obtained from website of Tesla, Inc. and Apple, Inc.

number of electrons that are transferred during the electrochemical reaction, and in the simplest case, can be calculated by the following equation:

$$E = W = V \times Q \tag{1.5}$$

wherein, W is the electrical work (unit: Joule), V is the potential difference between two electrodes (unit: Volt), and Q is the amount of electrons transferred during electrochemical reaction (unit: Coulomb). Apparently, a battery that outputs a higher voltage and stores a larger quantity of charges can generate more electrical energy. The development of modern battery is a history of human pursuit for advanced energy storage devices that are able to store more energy. From the invention of voltaic pile in 1800, to the first rechargeable lead-acid battery in 1859 and the first nickel-cadmium battery in 1899, and finally to the first commercial Li-ion battery in 1991, the energy output of modern battery systems kept raising. However, the development of batteries still fell far behind the development in other fields of electronics. As a result, they were often regarded as the heaviest and costliest components in an electronic device [39]. The lack of high-energy and low-cost batteries slowed down the progress of emerging storage fields such as electric cars, wearable electronics and grid-scale storage [4, 40, 41]. To improve the storage ability of batteries at reduced costs, it is critical to develop new materials and new battery systems. By replacing conventional lithium cobalt oxide ($LiCoO_2$) cathode with high-nickel layered oxide cathodes or replacing graphite with silicon (or silicon oxide) in the anode, Li-ion batteries with higher energy output are to be realized [42–47]. In addition, new electrochemical systems, such as lithium-sulfur batteries, lithium-oxygen batteries, sodium-sulfur batteries, sodium-ion batteries, potassium-ion batteries, magnesium-ion batteries, are now under development to meet future storage applications [48–70].

1.2.5 Thermal Energy Storage

Even before the human history, thermal energy was readily available in nature. Freely available solar thermal energy from sun helps to maintain the ambient thermal condition required for sustaining our lives on earth. The use of fire, a direct source of thermal energy, is regarded as the most important milestone in the human evolution. Nowadays, our lifestyle has created plenty of new applications for thermal energy. The supply source of heat is diversified and can include solar thermal energy, geothermal energy, fossil-fuel power plants, nuclear power plants, industrial waste heat and biomass, and these sources together generate a huge amount of thermal energy [71–76]. On the other hand, the demand for thermal energy is continuously increasing year by year. However, there are gaps between demand and supply of thermal energy, which include (a) time difference between generation and consumption of thermal energy, (b) difference in cost of thermal energy between peak and off-peak hours use and (c) distance between source of thermal energy and place of consumption [77]. In case of improper management of the gaps, the generated thermal energy will dissipate into the environment and result in wastage. To improve the utilization efficiency of thermal energy, various thermal energy storage (TES) technologies have been developed, which enables excessive thermal energy to be stored and used hours, days, or months later, at scales ranging from individual process, building, district, town, or even region. A wide variety of materials are used for TES application. A TES material should possess suitable thermo-physical properties such as appropriate melting point, high thermal/chemical stability, high latent/specific heat and fast thermal conduction, while is also required to have a low cost, low volume change (during the storage process), favorable eco-friendliness and low toxicity/flammability [77, 78]. Based on the type of material employed for heat or cold storage, the TES systems can be generally classified into three types [77], including,

(1) Sensible heat storage system, which commonly serves large size and high temperature TES applications. With the help of the low-cost TES materials such as water and molten salts, the sensible heat storage system can store a very large amount of heat over a large temperature range, and its storage capacity can be significantly improved by using a TES material with solid-liquid phase change temperature between minimum and maximum temperatures of operation.
(2) Latent heat storage system, which mainly serves low-temperature human thermal comfort applications. In the system, the TES materials store thermal energy in their latent heat during a constant temperature process like phase change. Usually solid-liquid phase change is used due to its high specific latent heat, yet solid-solid phase changes are also used as they have predominant advantages such as no leakage and no need for encapsulation. As latent heat is 50–100 times larger than sensible heat, the energy storage density of latent heat storage materials near the phase change temperature could be very high, which enables realization of compact TES systems. Also, as the phase change material maintains a constant temperature during its operation, the outlet temperature of the heat transfer fluid is steady during discharge. The main technical barrier for the latent heat storage system is poor thermal conductivity of phase change material (especially the solid phase), which could adversely affect the thermal charging and discharging rates of the system. As such, thermal conductivity enhancement of the TES material has become a major research area.
(3) Chemical heat storage system, which uses reversible reactions that involve heat absorption and release to store thermal energy. One example of an experimental storage system based on chemical reaction energy is the salt hydrate technology, which uses the reaction energy created when salts are hydrated or dehydrated. As the dried salt can be stored at room temperature for prolonged times, the system is especially advantageous for seasonal thermal energy storage. Storage of energy in molecular bonds is also investigated, which could show an ultrahigh energy density equivalent to lithium-ion batteries [79]. However, as the development of thermal–chemical heat storage materials are in the laboratory stage, the systems currently have very limited practical applications.

1.3 Fundamentals About Electrochemical Energy Storage

In this section, we will introduce the operation principles of typical EES devices (e.g., primary batteries, rechargeable batteries, pseudocapacitors and fuel cells) from the electrochemical points of view, as well as their key components and materials.

1.3.1 How Is Energy Stored and Released Through Electrochemical Reactions?

In a heat engine (e.g., an internal combustion engine), chemical energy that stored in fossil fuels is converted to thermal energy through redox reaction, which is then utilized to do mechanical work. The energy conversion process in an EES device undergoes in a quite similar way: the electrochemical redox reaction on the electrode helps to transform the chemical energy stored in the device into electric energy to drive the external equipments during the discharge process, and in some cases, convert the electric energy back into the chemical energy for storage in the EES device during the reverse charge process. According to the fundamentals of chemistry, an oxidation is the removal of electrons from a species, a reduction is the addition of electrons to a species, and the net effect is a redox reaction, in which there is a transfer of electrons from one species (electron donor, or reductant) to another (electron acceptor, or oxidant) [80]. Conceptually, any redox reaction can be expressed as the difference of two reduction half-reactions, and the reduced and oxidized species in a half-reaction form a redox couple. The electron transfer may be accompanied by other events, such as atom or ion transfer. In general, one could write a couple as Ox/Red and the corresponding half reaction as follows [80]:

$$\mathrm{Ox} + \nu \mathrm{e}^- \rightarrow \mathrm{Red}$$

The reduction and oxidation processes of an electrochemical reaction are separated in the cell, with oxidation occurring at one electrode and reduction at the other. As the reaction proceeds, the electrons released via the oxidation reaction ($\mathrm{Red1} \rightarrow \mathrm{Ox1} + \nu \mathrm{e}^-$) at one electrode travel through the external circuit to reach the other electrode, at which they trigger the reduction reaction ($\mathrm{Ox2} + \nu \mathrm{e}^- \rightarrow \mathrm{Red2}$). While the electrode at which oxidation occurs is called the anode, the one at which reduction occurs is called the cathode [80]. The reduction process withdraws electrons from the cathode and leaves a relatively positive charge on it (corresponding to a high potential). Meanwhile, the oxidation process feeds electrons to the anode, leaving it a relatively negative charge (corresponding to a low potential). As a result, the cathode has a higher potential than the anode. In practical batteries, the potential difference between the two electrodes drives the electrons through an external circuit to do electrical work when the cell reaction has not reached its chemical equilibrium state. When the potential difference is large, a given number of electrons travelling between the electrodes can do a large amount of electrical work. A cell in which the overall reaction is at equilibrium does not do any work, and the potential difference is zero.

The maximum non-expansion work a closed system can do equals to the Gibbs energy change of the system ($W_{e,\mathrm{max}} = \Delta G$). In case of the electrochemical cell, the non-expansion work is identified as the electrical work, and ΔG is the Gibbs energy change during the advancement of a given cell reaction ($\Delta_r G$). In this way, we can correlate the Gibbs energy change ($\Delta_r G$) of cell reaction and the cell potential (E_{cell}) as follows [80]:

$$\Delta_r G = -vFE_{cell} \tag{1.6}$$

where F is the Faraday's constant (96,485 C mol^{-1}), and v is the number of the electrons that participate in the half reactions. This equation bridges the electrochemical and thermodynamic parameters of a galvanic cell. By measuring the potential difference between the two electrodes, one is able to calculate the Gibbs energy change and changes of other thermodynamic parameters (such as internal energy, enthalpy and entropy) of a given electrochemical system using the above equation and other basic thermodynamic equations. In this way, the maximum work outputted by a cell is obtained. Note that the maximum work is outputted when the change of the system is thermodynamically reversible, the electrochemical reaction that occurs in the cell should be controlled so that it undergoes in a quasi-static manner, or in other words, the current that passes through the cell is negligible (no current) and the cell is operated at constant compositions of electrode and electrolyte. In effect, the cell reaction is poised for change, but not actually changing. In practical cases, there is measurable current flow through the cell and the cell reaction does not proceed in a quasi-static way. In that case, the energy (E) available from a single cell at a given discharge current (I_d), a given voltage (V_d), and a given time duration (Δt_d) is,

$$E = \int_0^{\Delta t_d} I_d V_d(q)\mathrm{d}t \tag{1.7}$$

where $I_d = \mathrm{d}q/\mathrm{d}t$ and q is the state of charge of the cell [37]. As a battery consists of one or more interconnected cells, the output current I_d and/or time t to depletion of the stored energy can be increased by connecting the cells in parallel, and the voltage V_d can be increased by connecting the cells in series. Both routes lead to improved power ($P = I_d \times V_d$) and energy output of batteries so that they can be used for large-scale storage purposes, for example, the electric vehicles and grids.

1.3.2 Energy Storage Devices Operated by Electrochemical Reactions

There are many types of EES devices, each of them targets at specific storage applications. In the following sections, we will introduce four types of EES devices that are most commonly used in our daily lives.

1.3.2.1 Primary Cells

A primary cell is designed for one-time use and discarded. The electrochemical reaction occurring in the cell is not reversible, hence, the cell is not rechargeable. As a primary cell is used, chemical reactions in the cell use up the power-generating chemicals, and the cell stops producing electricity and becomes useless [81, 82]. Typical examples of primary cells include zinc-carbon cells and alkaline cells [83–85]. A zinc-carbon cell was the first commercial dry battery developed from the wet Leclanché cell patented in 1866. It delivers about 1.5 V of direct current from the electrochemical reaction between zinc and manganese dioxide [83]. A dry Leclanché cell is usually made of a zinc can (serving as the anode), and an inert carbon rod cathode that collects the current from the manganese dioxide electrode [83]. The cell usually employs an aqueous paste of ammonium chloride (possibly mixed with some zinc chloride) as electrolyte [83]. An alkaline cell is an upgraded version of the zinc-carbon cell, which employs an alkaline electrolyte of potassium hydroxide to replace the acidic ammonium chloride electrolyte [85, 86]. In this way, the cell obtains a higher energy density and longer shelf-life. During World War II and the postwar period, significant advancements were made with new types of cells, and the energy density of cell was improved from less than 50 Wh kg^{-1} with the early zinc-carbon cells to more than 500 Wh kg^{-1} currently obtained with lithium primary and zinc/air cells [82].

The primary cell is a convenient source of power for portable electronic devices, lighting, watches, toys, household appliances and a wide variety of other applications [81, 82]. The cells are low in cost, convenient and easy to use, simple in structure, and require little maintenance, and can be sized and shaped to fit various applications. Other advantages of primary cells include good shelf life, reasonable energy and power density, and good reliability [81, 82]. The primary cells have dominated the battery market for a long period of time. After the early twenty-first century, they began losing market share to secondary cells as the relative costs declined for the latter. However, the discovery of new energy storage chemistry and efforts in improving the existing characteristics have opened up many new opportunities for the primary cells, making them an indispensable part of today's energy storage market.

1.3.2.2 Secondary Cells

A secondary cell (rechargeable cell) is a type of electrochemical cell that, after being discharged into a load, can be recharged many times to restore its energy for the next-time use. The concept of the secondary cell is opposed to that of the primary (disposable) cell, which is supplied fully charged and discarded after use. Several electrochemical systems were proposed and commercialized, including lead-acid, nickel-cadmium (NiCd), nickel-metal hydride (NiMH) and today's lithium-ion cells [37, 87–89]. Secondary batteries are produced in many different shapes and sizes, ranging from button cells for portable electronic applications to megawatt systems for grid-scale storage. The advantages of rechargeable cells over the primary counterparts are obvious. Though having higher initial costs than the disposable cells, the rechargeable cells have much lower total costs of ownership as they can be recharged inexpensively many times before replacement. On the other hand, since the secondary cells can be recharged and reused for hundreds of times before being discarded, they also produce less waste that could pollute the environment. Because of these distinct advantages, there has been a shift from the use of primary cells to secondary cells as power supply during the past years. After the invention and successful marketing of commercial Li-ion batteries, the rechargeable cells soon dominated the consumer electronics field, and are now actively expanding their applications in emerging markets such as automobiles and grids.

The realization of secondary cells relies on reversible electrochemical reactions. One typical example is the Li-ion battery, which enables reversible Li insertion and extraction from the layered cathode compounds (e.g., $LiCoO_2$) and graphite anode to complete its discharge and charge process, so that it can be repeatedly cycled for hundreds of times before depleting its service life [87]. In the following chapters, we will discuss in details the electrochemical fundamentals of various types of secondary cells, and the approaches to optimize the storage performance of each type of cell.

1.3.2.3 Pseudocapacitors

As discussed in Sect. 1.2.4, a pseudocapacitor store electrical energy by Faradaic electron charge transfer between its electrodes and electrolyte. The energy storage process is usually accomplished by electrosorption, redox reactions and intercalation processes, termed pseudocapacitance [15, 33]. The redox reaction in a pseudocapacitor usually occurs at the surface or near surface of active materials, which is accompanied by a valence state change of electrode material as a result of electron transfer. The active centers that contribute to the pseudocapacitance locate near the surface of metal oxide particles, at a distance, $\ll(2Dt)^{1/2}$, where D is the diffusion coefficient for charge-compensating ions ($cm^2\ s^{-1}$) and t is the diffusion time range (s) [90]. Energy storage involving pseudocapacitance shows an intermediate electrochemical behavior between pure electrostatic EDLC and solid-state diffusion in bulk battery-type materials [1]. For example, ruthenium dioxide (RuO_2), which was the first electrode material reported with pseudocapacitive electrochemical behavior, shows a typical capacitive feature as demonstrated by rectangular-shape cyclic voltammetry profile (i.e., the capacitance arises from the linear relationship between the extent of charge (ΔQ) and the potential change (ΔU)), though the charge storage is actually enabled by rapid and reversible charge-transfer Faradaic reactions across the electrode|electrolyte interface [31]. The redox reaction significantly improves the energy storage ability of the capacitor. The maximum capacitance available from such pseudocapacitor is ~5000 F cm^{-3}, which is significantly larger than the maximum value (~825 F cm^{-3}) achieved with a double-layer capacitor from 1 cm^3 of a compact high surface area carbon [91, 92].

Pseudocapacitance also occurs in the case of ion insertion into layered compound materials [93, 94]. To maintain electric neutrality, the intercalation is accompanied by a change of metal valence. Materials with both "intrinsic" or "extrinsic" pseudocapacitive properties, can trigger intercalation charge storage [15, 31]. This specific storage process usually provides high capacitance, wide potential window and favorable rate capability.

1.3.2.4 Fuel Cells

A fuel cell is an electrochemical cell that converts the energy from chemical form into electrical form through the reaction between hydrogen with oxygen or oxidizing agents and produces water as its byproduct [95]. It is considered as one of the most promising sources of renewable and clean energy because of its low emission of sulfur and nitrogen oxides and low noise-level operation [95]. Different from the conventional electrochemical cells where the energy comes from preserved chemicals, the fuel cell requires a continuous source of fuel hydrogen and oxygen to maintain its operation. As long as fuel and oxygen are supplied, the generation of electricity from the cell sustains.

The fuel cells were invented in 1838 and were then equipped to power satellites and space capsules in NASA space programs. Since then, they have been used as primary and backup power supplies for commercial and residential buildings and in remote areas and were also used to power automobiles for military and civil use.

Fuel cells come in varieties. However, these cells share the same working principles. Each of the fuel cells consists of an electrolyte layer (usually a polymer membrane) in contact with two electrodes on either side [96, 97]. During the operation, protons flow from the anode to the cathode through the electrolyte to trigger two chemical reactions at the electrode|electrolyte interfaces and give rise to fuel consumption and generation of water or depleted oxidant, while the electrons move from the anode to the cathode through an external circuit to generate electricity. The chemical reactions involved in the anode and cathode and its overall reactions are given as,

$$\text{Anode reaction: } H_2 \rightarrow 2H^+ + 2e^-$$
$$\text{Cathode reaction: } 1/2O_2 + 2H^+ + 2e^- \rightarrow H_2O$$
$$\text{Overall reaction: } H_2 + 1/2O_2 \rightarrow H_2O$$

Individual fuel cells output relatively small electrical potentials of ~ 0.7 V. Hence, the cells are usually "stacked" or connected in series to provide sufficient voltage for requirements of practical applications. To facilitate the electrochemical reaction, catalysts were employed at the anode side [98]. Besides, a proton exchange membrane was also employed to facilitate the permeation of generated protons from the anode to the cathode. To function, the membrane must be conductive to protons yet insulative to electrons to prevent short circuit of the cell. The membrane must also not allow either gas to pass to the other side of the cell, a problem known as gas crossover. Although fundamental research focus on the electrode materials, catalysts (as well as their supports), membranes and other key materials, a practical fuel cell requires more complex configuration which further includes thermal and water management system, and management systems for single cell and battery pack. The fuel cell technology has already shown its promises to compete with Li-ion cells for the next-generation vehicle storage, and its development requires more efforts from the field.

1.3.3 Key Components and Materials for EES Application

All EES devices, no matter how practically complicated they are, are composed of two key components: electrodes and electrolytes [39]. In the following section, we will briefly introduce the fundamentals about these components and materials.

1.3.3.1 Electrodes

In an EES device, an electrode is referred to as either an anode or a cathode. As discussed in Sect. 1.3.1, the anode is defined as the electrode at which electrons leave the cell and oxidation occurs, whereas the cathode is defined as the electrode at which electrons enter the cell so that reduction takes place [80]. Although it is not possible to obtain the absolute potential value of a single electrode, it is feasible to measure the potential difference between cathode and anode in an electrochemical cell (that is, the cell potential E_{cell}). If we define the potential of one electrode as zero, we could assign value to the other. To compare the potential of different electrodes, a special electrode, standard hydrogen electrode (SHE), was selected as the reference. The electrode is operated based on the following half reaction [80]:

$$2H^+(aq) + 2e^- \rightarrow H_2(g)$$

The redox reaction occurs at a platinized platinum electrode, which is dipped in an acidic solution with pure hydrogen gas bubbled through. In this way, the SHE is expressed as follows [80]:

$$Pt(s)|H_2(g)|H^+(aq)$$

And the potential of SHE is set to be zero ($E^\ominus = 0$). To achieve the standard conditions, the proton activity in the solution must be 1 (that is, pH = 0) and the pressure of the hydrogen gas must be 1 bar (100 kPa).

For a given electrode "X", its standard potential electrode ($E^\ominus$(X)) is then obtained by pairing the electrode (as the cathode) and the SHE (as the anode) to assemble a cell as follows [80]:

Table 1.2 Standard electrode potentials of some typical redox couples at 298 K

Redox couple	$E^{\ominus}$/V
$2H^{+}(aq) + 2e^{-} \rightarrow H_2(g)$	0
$Cu^{2+}(aq) + 2e^{-} \rightarrow Cu(s)$	0.34
$Zn^{2+}(aq) + 2e^{-} \rightarrow Zn(s)$	−0.76
$Mg^{2+}(aq) + 2e^{-} \rightarrow Mg(s)$	−2.37
$K^{+}(aq) + e^{-} \rightarrow K(s)$	−2.93
$Na^{+}(aq) + e^{-} \rightarrow Na(s)$	−2.71
$Li^{+}(aq) + e^{-} \rightarrow Li(s)$	−3.04

$$\mathrm{Pt(s)|H_2(g)|H^+(aq)||X};\ E^{\ominus}(\mathrm{X}) = E_{\mathrm{cell}}$$

Table 1.2 lists standard potentials of some typical redox couples at 298 K.

According to Eq. 1.6 and the second law of thermodynamics, the electrochemical reaction in a cell is spontaneous if,

$$\Delta_r G = -\nu F E_{\mathrm{cell}} < 0 \tag{1.8}$$

On the other hand,

$$E_{cell} = E_c - E_a \tag{1.9}$$

wherein, E_c and E_a stand for electrode potentials of cathode and anode, respectively).

According to Eqs. 1.8 and 1.9, $E_{\mathrm{cell}} > 0$ or $E_c > E_a$. Therefore, to forward the cell reaction, the anode usually holds a lower electrode potential than the cathode. The above criteria also apply to the standard conditions, where $\Delta_r G^{\ominus}$, $E_{cell}^{\ominus}$, $E_c^{\ominus}$ and $E_a^{\ominus}$ are used. In this way, by comparing the standard electrode potentials versus SHE of the known electrodes, one is able to tell the cathode from the anode and calculate the theoretical voltage output of the assembled cell.

Another important thermodynamic parameter is the Fermi level (denoted as ε_F) of electrode [80]. Fermi level, or electrochemical potential (denoted as μ), is a term used to describe the top of the collection of electron energy levels at absolute zero temperature (0 K) [99, 100]. In a metal electrode, the closely packed atoms have strong overlaps between one another. All electrons are on the average attracted to the same extent by the lattice of positive charge, but they interact with each other. Electrons are fermions, by the Pauli principle only two electrons, each with opposite spins, can have the same energy. Therefore, the electronic levels in a metal solid are not discrete like in an atom of the same material. Instead, a continuum of levels is established with the available electrons filling the states from the bottom upwards to form a "Fermi sea" of electron energy states (Fig. 1.4a). The Fermi level corresponds to the energy at which the 'top' electrons sit, or in other words, the surface of the Fermi sea at absolute zero where no electrons have enough energy to escape. At finite temperatures, electrons can be thermally excited to levels above the Fermi level, leaving an unoccupied state (or a hole). However, the distribution of electrons and holes is restricted to a certain region in the vicinity of Fermi level, forming a "transition state" with probability of occupancy ranging from 0 to 1 (Fig. 1.4a) [101]. In this way, the Fermi level also refers to the energy when the electron occupation probability equals to 0.5 in the electronic energy level that follows the Fermi-Dirac distribution at any temperature [102]. In a perfect semiconductor electrode, the electronic states are delocalized just as in metals, which also leads to formation of bands of allowed electronic energies. However, the current-carrying bands in semiconductors do not overlap as they do in metals; they are separated by a band gap, and the Fermi level lies within the gap (Fig. 1.4b). The band below the Fermi level is completely filled at $T = 0$ K and is named a valence band (VB). The band above is empty at $T = 0$ K and is named a conduction band (CB) [101]. In a pure or intrinsic semiconductor, the Fermi level locates close to the center of the band gap (Fig. 1.4b). At finite temperature, a few electrons have energy above the Fermi level and they may be excited from VB into CB to contribute to electric current (Fig. 1.4b). The introduction of dopants into the semiconductor helps to alter the position of Fermi level [101]. For example, donors that have localized electronic states with energies immediately below the CB of the semiconductor can donate their electrons to the CB, leading to elevation of Fermi level towards the lower edge of CB (n-type semiconductor). On the other hand, acceptors that have empty states just above the VB can accept an electron from the VB and as a result, shift the Fermi level towards the upper edge of the VB (p-type semiconductor) [101].

A thorough understanding about the Fermi level and its relationship with electronic band structure of solid materials and with voltage and energy output of electrode is essential to learning solid-state physics and electrochemical properties of electrode materials for EES application. For the metal electrode, the electrons at the Fermi level hold the highest kinetic energy and are the least bound with the metal solid (at $T = 0$ K). As a consequence, they are the first to leave the metal in electrochemical reactions. Also, the Fermi level is the first to receive electrons that are injected into the metal from the environment. Since the transfer of electrons is linked to occurrence of redox reaction, one can bridge the Fermi level (or electrochemical potential) of solid material with the electrochemical redox potential of electrode composed of that material.

It is known that the unit of voltage (volt, V) can be expressed by the following equation:

$$\mathrm{V} = \mathrm{Joule/Coulomb} \tag{1.10}$$

Therefore, a volt is simply the energy (J) required to move a charge (C). Since electrons are charged, an applied voltage can alter the "energy" of the electrons in a metal electrode. Hence, electrochemists are therefore able to alter the energy of the Fermi level by applying a voltage to the electrode to enable occurrence or avoidance of the electrochemical reactions.

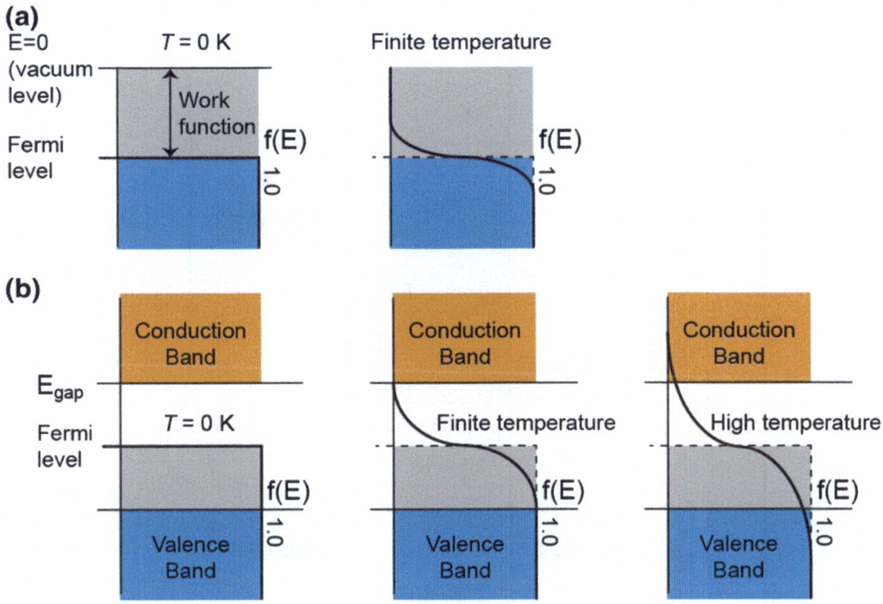

Fig. 1.4 Schematic energy diagrams of **a** a pure metal and **b** an intrinsic semiconductor material at different temperatures. Wherein, the work function in **a** denotes the minimum energy required to remove electron from surface of the metal. The closer an electron is to the vacuum level, the weaker it is bound to the solid. In **b**, no electrons can exist above the VB at 0 K, since none have energy above the Fermi level and there are no available energy states in the band gap. At finite temperatures, some electrons have energy above the Fermi level. At elevated temperatures, some electrons can reach the CB and contribute to electric current

The energy storage process occurred in an electrode material involves transfer and storage of charges. In addition to the intrinsic electrochemical properties of the materials, the dimensions and structures of the materials may also influence the energy storage process in an EES device [103, 104]. More details about the size effect on charge storage of electrode materials will be presented in the next chapter.

1.3.3.2 Electrolytes

In an electrochemical cell, the anode is the reductant, the cathode is the oxidant, and the two electrodes are separated by an ionically conductive yet electronically isolated matter named an electrolyte [39]. On the schematic of relative electron energies of the cell, the energy separation E_g of the lowest unoccupied molecular orbital (LUMO) and the highest occupied molecular orbital (HOMO) of the electrolyte is the "electrochemical window" of the electrolyte (Fig. 1.5) [38]. An anode with an electrochemical potential μ_A (or Fermi energy ε_{FA}) above the LUMO will reduce the electrolyte unless a passivation layer that forms at the anode|electrolyte interface

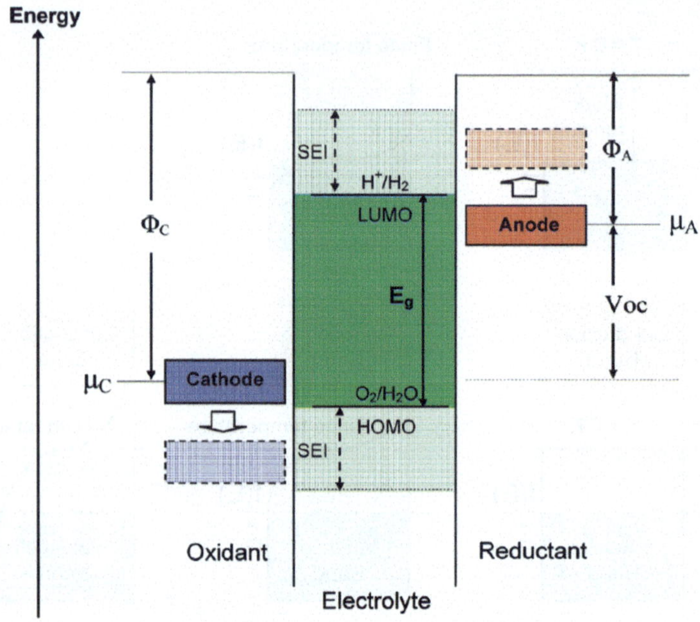

Fig. 1.5 Schematic energy diagram of an electrochemical cell at its open-circuit voltage. Φ_A and Φ_C are the work functions of both electrodes. E_g is the "thermodynamically stable" window of the electrolyte. A μ_A > LUMO and/or a μ_C < HOMO requires a kinetic stability by the formation of a solid electrolyte interface. Reproduced with the permission from Ref. [37], Copyright © 2010 American Chemical Society

hinders electron transfer from the anode to the electrolyte LUMO (Fig. 1.5). A cathode with an electrochemical potential μ_C (or Fermi energy ε_{FC}) below the HOMO will oxidize the electrolyte unless a passivation layer blocks electron transfer from the electrolyte HOMO to the cathode (Fig. 1.5) [38]. Therefore, a thermodynamically stable cell requires locating the electrochemical potentials μ_A and μ_C of both electrodes within the window of the electrolyte. In this way, the open-circuit voltage V_{OC} of a cell is constrained to meet the following relationship,

$eV_{OC} = \mu_A - \mu_C \leq E_g$, where e is the magnitude of the electron charge [38].

The energy density of a secondary cell can output is ΛV_{OC}. Wherein, Λ is the capacity per unit weight of reversible charge transfer (Ah g^{-1}) between both electrodes. In order to obtain a cell with higher energy density, it is required to improve the mass-based capacity Λ or the open-circuit voltage V_{OC}, or both. Since V_{OC} is limited by E_g, an electrolyte with wider electrochemical window will promise a cell with higher energy output [38]. The cells that adopt conventional aqueous electrolyte ($E_g = \sim 1.3$ eV) shows a limited V_{OC}. In order to obtain a cell with a higher V_{OC} and therefore a higher energy density, it is required to turn to a nonaqueous electrolyte with a larger E_g. This finding, in turn, has shifted the research towards Li-ion battery since the electrolyte offers a wider electrochemical window E_g [38]. In addition to the large energy separation, the electrolyte should also meet several additional requirements for battery use, including,

(1) a sufficiently high conductivity of working ion (e.g., $\sigma_{\mathrm{Li}} > 10^{-4}\,\mathrm{S\,cm^{-1}}$ for Li-ion batteries over their temperature range of operation);
(2) a sufficiently low electronic conductivity ($\sigma_e < 10^{-10}\,\mathrm{S\,cm^{-1}}$);
(3) a transference number σ_w/σ_{total} (where σ_w is the conductivity of the working ion, and σ_{total} includes conductivities by the working ion and other ions, as well as σ_e);
(4) high chemical stability over ambient temperature ranges and against temperature change during cell operation;
(5) high chemical stability with respect to the electrodes, including the retention of electrode|electrolyte interface upon drastic volume change of electrode particles during cycling, and the ability to passivate the electrode surface in case the electrode potential lies outside the electrolyte window;
(6) last, a good electrolyte should be safe, environmentally friendly and have a low cost.

An electrolyte can be prepared into solid, liquid or even (liquified) gas form [38, 105, 106]. Each of them has its unique electrochemistry in bulk and at interface in contact with the electrode, which may significantly influence the energy storage properties of the EES device. In the following chapters, we will present more discussions about the thermodynamic and kinetic aspects of charge transfer and storage in an electrolyte-containing EES device.

1.3.3.3 Other Crucial Components and Materials

In addition to electrodes and electrolytes, an EES device may also consists of other components such as current collector, separator, conductive additive and binder. Each of these components plays an important role in the energy storage process and may affect the practical performance of the EES device. For example, a current collector is a conductive solid that is connected to support the electrode and collect electrical energy from it [52]. During the storage process, a highly efficient transport of electrons provided by the current collector is a guarantee of excellent electrode performance.

A separator is another critical component in liquid electrolyte-based EES devices, which is placed between the cathode and the anode to prevent electrical contact of the electrodes while enable free ion transport inbetween [107]. To provide ion conduction pathways, the separator is, in most cases, a nanoporous film consisting of either polymer or non-woven fabrics. Besides, the separator must be chemically and electrochemically stable towards the electrolyte and electrode materials, and mechanically robust to bear the pressure and tension during the cell assembly process [107]. Refer to Chap. 7 for further information on these crucial components and materials for use in EES devices.

1.4 Summary

In this chapter, we made a brief introduction to various types of sustainable energy conversion and storage technologies that have shown potentials to meet the rapidly advanced global energy demand. Especially, we focused on the electrochemical energy storage technology and typical EES devices including batteries and supercapacitors. Operational fundamentals, and key components and materials of these devices were also discussed. In the next chapter, we will present more details on the fundamental electrochemistry of typical devices and crucial materials toward EES applications.

References

1. Anonymous. (2018). *BP statistical review of world energy* (pp. 1–56). London, UK: BP p.l.c.
2. Sato, M. (1990). Thermochemistry of the formation of fossil fuels. In R. J. Spencer & I.-M. Chou (Eds.), *Fluid–mineral interactions: A tribute to H. P. Eugster* (pp. 271–283). Geochemical Society.
3. Gőğűş, Y. A. (2009). Mechanical energy storage. In Y. Gogus (Ed.), *Energy storage systems: Encyclopedia of life support systems* (pp. 190–226). EOLSS Publishers/UNESCO.
4. Dunn, B., Kamath, H., & Tarascon, J.-M. (2011). Electrical energy storage for the grid: A battery of choices. *Science, 334,* 928–935.
5. Rehman, S., Al-Hadhrami, L. M., & Alam, M. M. (2015). Pumped hydro energy storage system: A technological review. *Renewable and Sustainable Energy Reviews, 44,* 586–598.
6. Yang, C.-J. (2016). Pumped hydroelectric storage. In T. M. Letcher (Ed.), *Storing energy: With special reference to renewable energy sources* (pp. 25–38). Elsevier.
7. Venkataramani, G., Parankusam, P., Ramalingam, V., et al. (2016). A review on compressed air energy storage—A pathway for smart grid and polygeneration. *Renewable and Sustainable Energy Reviews, 62,* 895–907.
8. Luo, X., Wang, J., Dooner, M., et al. (2014). Overview of current development in compressed air energy storage technology. *Energy Procedia, 62,* 603–611.
9. Amiryar, M. E., & Pullen, K. R. (2017). A review of flywheel energy storage system technologies and their applications. *Applied Sciences, 7,* 286.
10. Mousavi, G. S. M., Faraji, F., Majazi, A., et al. (2017). A comprehensive review of Flywheel Energy Storage System technology. *Renewable and Sustainable Energy Reviews, 67,* 477–490.
11. Burt, R., Birkett, G., & Zhao, X. S. (2014). A review of molecular modelling of electric double layer capacitors. *Physical Chemistry Chemical Physics, 16,* 6519–6538.
12. Tixador, P. (2008). *Superconducting magnetic energy storage: Status and perspective*. Paper presented at: IEEE/CSC & ESAS European Superconductivity News Forum (ESNF).
13. Vulusala, G. V. S., & Madichetty, S. (2018). Application of superconducting magnetic energy storage in electrical power and energy systems: A review. *International Journal of Energy Research, 42,* 358–368.
14. Pandolfo, T., Ruiz, V., Sivakkumar, S., et al. (2013). General properties of electrochemical capacitors. In F. Beguin & E. Frackowiak (Eds.), *Supercapacitors: Materials, systems, and applications* (pp. 69–109). Wiley.
15. Shao, Y., El-Kady, M. F., Sun, J., et al. (2018). Design and mechanisms of asymmetric supercapacitors. *Chemical Reviews, 118,* 9233–9280.
16. Ho, J., Jow, T. R., & Boggs, S. (2010). Historical introduction to capacitor technology. *IEEE Electrical Insulation Magazine, 26,* 20–25.

17. Both, J. (2015). Electrolytic capacitors, 1890 to 1925: Early history and basic principle. *IEEE Electrical Insulation Magazine, 31,* 22–29.
18. Ulaby, F. T., & Ravaioli, U. (2015). *Fundamentals of applied electromagnetics* (7th ed.). Pearson Education Dorling Kindersley.
19. Simon, P., & Gogotsi, Y. (2008). Materials for electrochemical capacitors. *Nature Materials, 7,* 845–854.
20. Helmholtz, H. V. (1853). On the conservation of force. *Annals of Physics, 165,* 211.
21. Gouy, G. (1910). Constitution of the electric charge at the surface of an electrolyte. *Comptes Rendus, 149,* 654–657.
22. Chapman, D. L. (1913). A contribution to the theory of electrocapillarity. *Philosophical Magazine, 25,* 475–481.
23. Stern, O. Z. (1924). Theory of the electrical double layer. *Zeitschrift für Elektrochemie und Angewandte Physikalische Chemie, 30,* 508–516.
24. Grahame, D. C. (1947). The electrical double layer and the theory of electrocapillarity. *Chemical Reviews, 41,* 441–501.
25. Becker, H. I. (1957). *Low voltage electrolytic capacitor*. U.S. Patent No. 2,800,616. United States: General Electric Co.
26. Rightmire, R. A. (1966). *Electrical energy storage apparatus*. U.S. Patent No. 3,288,641. United States: Standard Oil Co.
27. File, J., & Mills, R. G. (1963). Observation of persistent current in a superconducting solenoid. *Physical Review Letters, 10,* 93–96.
28. Buckles, W., & Hassenzahl, W. V. (2000). Superconducting magnetic energy storage. *IEEE Power Engineering Review, 20,* 16–20.
29. Luongo, C. A. (1996). Superconducting storage systems: An overview. *IEEE Transactions on Magnetics, 32,* 2214–2223.
30. Scrosati, B. (2011). History of lithium batteries. *Journal of Solid State Electrochemistry, 15,* 1623–1630.
31. Trasatti, S., & Buzzanca, G. (1971). Ruthenium dioxide: A new interesting electrode material. Solid state structure and electrochemical behaviour. *Journal of Electroanalytical Chemistry and Interfacial Electrochemistry, 29,* A1–A5.
32. Conway, B. E. (1991). Transition from "supercapacitor" to "battery" behavior in electrochemical energy storage. *Journal of the Electrochemical Society, 138,* 1539–1548.
33. Augustyn, V., Simon, P., & Dunn, B. (2014). Pseudocapacitive oxide materials for high-rate electrochemical energy storage. *Energy & Environmental Science, 7,* 1597–1614.
34. Conway, B. E., Birss, V., & Wojtowicz, J. (1997). The role and utilization of pseudocapacitance for energy storage by supercapacitors. *Journal of Power Sources, 66,* 1–14.
35. Miller, J. R. (2013). Market and applications of electrochemical capacitors. In F. Béguin & E. Frąckowiak (Eds.), *Supercapacitors: Materials, systems, and applications* (pp. 509–526). Wiley.
36. Yang, Z., Zhang, J., Kintner-Meyer, M. C. W., et al. (2011). Electrochemical energy storage for green grid. *Chemical Reviews, 111,* 3577–3613.
37. Goodenough, J. B. (2013). Evolution of strategies for modern rechargeable batteries. *Accounts of Chemical Research, 46,* 1053–1061.
38. Goodenough, J. B., & Kim, Y. (2010). Challenges for rechargeable Li batteries. *Chemistry of Materials, 22,* 587–603.
39. Armand, M., & Tarascon, J. M. (2008). Building better batteries. *Nature, 451,* 652–657.
40. Manzetti, S., & Mariasiu, F. (2015). Electric vehicle battery technologies: From present state to future systems. *Renewable and Sustainable Energy Reviews, 51,* 1004–1012.
41. Kim, J., Kumar, R., Bandodkar, A. J., et al. (2017). Advanced materials for printed wearable electrochemical devices: A review. *Advanced Electronic Materials, 3,* 1600260.
42. You, Y., Celio, H., Li, J., et al. (2018). Modified high-nickel cathodes with stable surface chemistry against ambient air for lithium-ion batteries. *Angewandte Chemie International Edition, 57,* 6480–6485.

43. You, Y., & Manthiram, A. (2018). Progress in high-voltage cathode materials for rechargeable sodium-ion batteries. *Advanced Energy Materials, 8,* 1701785.
44. Zhang, Y.-C., You, Y., Xin, S., et al. (2016). Rice husk-derived hierarchical silicon/nitrogen-doped carbon/carbon nanotube spheres as low-cost and high-capacity anodes for lithium-ion batteries. *Nano Energy, 25,* 120–127.
45. Yin, Y.-X., Xin, S., Wan, L.-J., et al. (2011). Electrospray synthesis of silicon/carbon nanoporous microspheres as improved anode materials for lithium-ion batteries. *The Journal of Physical Chemistry C, 115,* 14148–14154.
46. Xu, R., Wang, G., Zhou, T., et al. (2017). Rational design of Si@carbon with robust hierarchically porous custard-apple-like structure to boost lithium storage. *Nano Energy, 39,* 253–261.
47. Xu, Q., Sun, J.-K., Yu, Z.-L., et al. (2018). SiO_x encapsulated in graphene bubble film: An ultrastable Li-ion battery anode. *Advanced Materials, 30,* 1707430.
48. Yin, Y.-X., Xin, S., Guo, Y.-G., et al. (2013). Lithium-sulfur batteries: Electrochemistry, materials, and prospects. *Angewandte Chemie International Edition, 52,* 13186–13200.
49. Xin, S., Gu, L., Zhao, N.-H., et al. (2012). Smaller sulfur molecules promise better lithium-sulfur batteries. *Journal of the American Chemical Society, 134,* 18510–18513.
50. Xin, S., Yin, Y.-X., Guo, Y.-G., et al. (2014). A high-energy room-temperature sodium-sulfur battery. *Advanced Materials, 26,* 1261–1265.
51. You, Y., Yao, H.-R., Xin, S., et al. (2016). Subzero-temperature cathode for a sodium-ion battery. *Advanced Materials, 28,* 7243–7248.
52. Jin, S., Xin, S., Wang, L., et al. (2016). Covalently connected carbon nanostructures for current collectors in both the cathode and anode of Li–S batteries. *Advanced Materials, 28,* 9094–9102.
53. Xin, S., Yin, Y.-X., Wan, L.-J., et al. (2013). Encapsulation of sulfur in a hollow porous carbon substrate for superior Li–S batteries with long lifespan. *Particle & Particle Systems Characterization, 30,* 321–325.
54. Xin, S., Yu, L., You, Y., et al. (2016). The electrochemistry with lithium versus sodium of selenium confined to slit micropores in carbon. *Nano Letters, 16,* 4560–4568.
55. Xin, S., Guo, Y.-G., Chang, Z., et al. (2016). Progress of rechargeable lithium metal batteries based on conversion reactions. *National Science Review, 4,* 54–70.
56. Wang, Z., You, Y., Yuan, J., et al. (2016). Nickel-doped $La_{0.8}Sr_{0.2}Mn_{1-x}Ni_xO_3$ nanoparticles containing abundant oxygen vacancies as an optimized bifunctional catalyst for oxygen cathode in rechargeable lithium-air batteries. *ACS Applied Materials & Interfaces, 8,* 6520–6528.
57. Xue, L., Gao, H., Zhou, W., et al. (2016). Liquid K-Na alloy anode enables dendrite-free potassium batteries. *Advanced Materials, 28,* 9608–9612.
58. Gao, H., Xue, L., Xin, S., et al. (2017). A plastic-crystal electrolyte interphase for all-solid-state sodium batteries. *Angewandte Chemie International Edition, 56,* 5541–5545.
59. Zhou, W., Xue, L., Lü, X., et al. (2016). $Na_xMV(PO4)_3$ (M = Mn, Fe, Ni) structure and properties for sodium extraction. *Nano Letters, 16,* 7836–7841.
60. Du, X.-L., You, Y., Yan, Y., et al. (2016). Conductive carbon network inside a sulfur-impregnated carbon sponge: A bioinspired high-performance cathode for Li–S battery. *ACS Applied Materials & Interfaces, 8,* 22261–22269.
61. Chen, Z.-H., Du, X.-L., He, J.-B., et al. (2017). Porous coconut shell carbon offering high retention and deep lithiation of sulfur for lithium-sulfur batteries. *ACS Applied Materials & Interfaces, 9,* 33855–33862.
62. Xin, S., You, Y., Li, H.-Q., et al. (2016). Graphene sandwiched by sulfur-confined mesoporous carbon nanosheets: A kinetically stable cathode for Li–S batteries. *ACS Applied Materials & Interfaces, 8,* 33704–33711.
63. Xu, D.-W., Xin, S., You, Y., et al. (2016). Built-in carbon nanotube network inside a biomass-derived hierarchically porous carbon to enhance the performance of the sulfur cathode in a Li–S battery. *ChemNanoMat, 2,* 712–718.
64. Gao, H., Xin, S., Xue, L., et al. (2018). Stabilizing a high-energy-density rechargeable sodium battery with a solid electrolyte. *Chem, 4,* 833–844.

65. Gao, H., Xue, L., Xin, S., et al. (2018). A high-energy-density potassium battery with a polymer-gel electrolyte and a polyaniline cathode. *Angewandte Chemie International Edition, 57,* 5449–5453.
66. Zhang, J., Zhang, C., Li, W., et al. (2018). Nitrogen-doped perovskite as a bifunctional cathode catalyst for rechargeable lithium-oxygen batteries. *ACS Applied Materials & Interfaces, 10,* 5543–5550.
67. Xue, L., Zhou, W., Xin, S., et al. (2018). Room-temperature liquid Na–K anode membranes. *Angewandte Chemie International Edition, 57,* 14184–14187.
68. Gao, H., Seymour, I. D., Xin, S., et al. (2018). $Na_3MnZr(PO_4)_3$: A high-voltage cathode for sodium batteries. *Journal of the American Chemical Society, 140,* 18192–18199.
69. Wu, N., Yang, Z.-Z., Yao, H.-R., et al. (2015). Improving the electrochemical performance of the $Li_4Ti_5O_{12}$ electrode in a rechargeable magnesium battery by lithium-magnesium co-intercalation. *Angewandte Chemie International Edition, 54,* 5757–5761.
70. Hu, X.-C., Shi, Y., Lang, S.-Y., et al. (2018). Direct insights into the electrochemical processes at anode/electrolyte interfaces in magnesium-sulfur batteries. *Nano Energy, 49,* 453–459.
71. Denholm, P., King, J. C., Kutcher, C. F., et al. (2012). Decarbonizing the electric sector: Combining renewable and nuclear energy using thermal storage. *Energy Policy, 44,* 301–311.
72. O'Sullivan, M., Yeh, A., & Mannington, W. (2010). Renewability of geothermal resources. *Geothermics, 39,* 314–320.
73. Rezaie, B., Reddy, B. V., & Rosen, M. A. (2017). Assessment of the thermal energy storage in Friedrichshafen district energy systems. *Energy Procedia, 116,* 91–105.
74. Edwards, J., Bindra, H., & Sabharwall, P. (2016). Exergy analysis of thermal energy storage options with nuclear power plants. *Annals of Nuclear Energy, 96,* 104–111.
75. Miró, L., Brückner, S., & Cabeza, L. F. (2015). Mapping and discussing Industrial Waste Heat (IWH) potentials for different countries. *Renewable and Sustainable Energy Reviews, 51,* 847–855.
76. Pantaleo, A. M., Camporeale, S. M., Miliozzi, A., et al. (2017). Thermo-economic assessment of an externally fired hybrid CSP/biomass gas turbine and organic rankine combined cycle. *Energy Procedia, 105,* 174–181.
77. Alva, G., Lin, Y., & Fang, G. (2018). An overview of thermal energy storage systems. *Energy, 144,* 341–378.
78. Cabeza, L. F., Martorell, I., Miró, L., et al. (2015). Introduction to thermal energy storage (TES) systems. In L. F. Cabeza (Ed.), *Advances in thermal energy storage systems* (pp. 1–28). Woodhead Publishing.
79. Kolpak, A. M., & Grossman, J. C. (2011). Azobenzene-functionalized carbon nanotubes as high-energy density solar thermal fuels. *Nano Letters, 11,* 3156–3162.
80. Atkins, P., & Paula, J. D. (2014). *Physical chemistry: Thermodynamics, structure, and change* (10th ed.). New York: W. H. Freeman and Company.
81. Linden, D. (2001). Primary batteries—Introduction. In D. Linden & T. B. Reddy (Eds.), *Handbook of batteries* (pp. 1–21). McGraw-Hill Education.
82. Linden, D., & Reddy, T. B. (2010). An introduction to primary batteries. In T. B. Reddy (Ed.), *Linden's handbook of batteries* (pp. 1–18). McGraw-Hill Education.
83. Brooke Schumm, J. (2010). Zinc-carbon batteries—Leclanché and zinc chloride cell systems. In T. B. Reddy (Ed.), *Linden's handbook of batteries* (pp. 1–41). McGraw-Hill Education.
84. Sayilgan, E., Kukrer, T., Civelekoglu, G., et al. (2009). A review of technologies for the recovery of metals from spent alkaline and zinc–carbon batteries. *Hydrometallurgy, 97,* 158–166.
85. Scarr, R. F., Hunter, J. C., & Slezak, P. J. (2001). Alkaline-manganese dioxide batteries. In D. Linden & T. B. Reddy (Eds.), *Handbook of batteries* (pp. 1–32). McGraw-Hill Education.
86. Kozawa, A., & Powers, R. A. (1972). Electrochemical reactions in batteries. Emphasizing the MnO_2 cathode of dry cells. *Journal of Chemical Education, 49,* 587–591.
87. Goodenough, J. B., & Park, K.-S. (2013). The Li-ion rechargeable battery: A perspective. *Journal of the American Chemical Society, 135,* 1167–1176.
88. May, G. J., Davidson, A., & Monahov, B. (2018). Lead batteries for utility energy storage: A review. *Journal of Energy Storage, 15,* 145–157.

89. Bergstrom, S. (1952). Fiftieth anniversary: Anniversary issue on storage batteries: Nickel-cadmium batteries—Pocket type. *Journal of the Electrochemical Society, 99,* 248C–250C.
90. Gogotsi, Y., & Penner, R. M. (2018). Energy storage in nanomaterials—Capacitive, pseudocapacitive, or battery-like? *ACS Nano, 12,* 2081–2083.
91. El-Kady, M. F., Shao, Y., & Kaner, R. B. (2016). Graphene for batteries, supercapacitors and beyond. *Nature Reviews Materials, 1,* 16033.
92. Conway, B. E. (1999). Electrochemical capacitors based on pseudocapacitance. In B. E. Conway (Ed.), *Electrochemical supercapacitors: Scientific fundamentals and technological applications* (pp. 221–257). Springer.
93. Kim, J. W., Augustyn, V., & Dunn, B. (2012). The effect of crystallinity on the rapid pseudocapacitive response of Nb_2O_5. *Advanced Energy Materials, 2,* 141–148.
94. Augustyn, V., Come, J., Lowe, M. A., et al. (2013). High-rate electrochemical energy storage through Li^+ intercalation pseudocapacitance. *Nature Materials, 12,* 518–522.
95. Kirubakaran, A., Jain, S., & Nema, R. K. (2009). A review on fuel cell technologies and power electronic interface. *Renewable and Sustainable Energy Reviews, 13,* 2430–2440.
96. Smitha, B., Sridhar, S., & Khan, A. A. (2005). Solid polymer electrolyte membranes for fuel cell applications—A review. *Journal of Membrane Science, 259,* 10–26.
97. Peighambardoust, S. J., Rowshanzamir, S., & Amjadi, M. (2010). Review of the proton exchange membranes for fuel cell applications. *International Journal of Hydrogen Energy, 35,* 9349–9384.
98. Liu, H., Song, C., Zhang, L., et al. (2006). A review of anode catalysis in the direct methanol fuel cell. *Journal of Power Sources, 155,* 95–110.
99. Khan, S. U. M., Kainthla, R. C., & Bockris, J. O. M. (1987). The redox potential and the Fermi level in solution. *The Journal of Physical Chemistry, 91,* 5974–5977.
100. Kittel, C. (2004). *Introduction to solid state physics* (8th ed.). Wiley.
101. Schmickler, W., & Santos, E. (2010). Metal and semiconductor electrodes. In *Interfacial electrochemistry* (pp. 9–18). Springer.
102. Gao, J., Shi, S.-Q., & Li, H. (2016). Brief overview of electrochemical potential in lithium ion batteries. *Chinese Physics B, 25,* 018210.
103. Guo, Y.-G., Hu, J.-S., & Wan, L.-J. (2008). Nanostructured materials for electrochemical energy conversion and storage devices. *Advanced Materials, 20,* 2878–2887.
104. Maier, J. (2005). Nanoionics: Ion transport and electrochemical storage in confined systems. *Nature Materials, 4,* 805–815.
105. Xin, S., You, Y., Wang, S., et al. (2017). Solid-state lithium metal batteries promoted by nanotechnology: Progress and prospects. *ACS Energy Letters, 2,* 1385–1394.
106. Rustomji, C. S., Yang, Y., Kim, T. K., et al. (2017). Liquefied gas electrolytes for electrochemical energy storage devices. *Science, 356*, eaal4263.
107. Lee, H., Yanilmaz, M., Toprakci, O., et al. (2014). A review of recent developments in membrane separators for rechargeable lithium-ion batteries. *Energy & Environmental Science, 7,* 3857–3886.

Chapter 2
Charge Transfer and Storage of an Electrochemical Cell and Its Nano Effects

Sen Xin, Hongcai Gao and Yu-Guo Guo

Abstract Efficient charger transfer and storage forms the precondition for stable operation of an electrochemical energy storage device. Nanomaterials, due to their admirable structure properties such as reduced particle dimensions and high surface to volume ratio, have shown promises in facilitating storage kinetics and enabling novel storage chemistry of electrode materials. In this chapter, we will introduce the fundamentals about the charge transfer and storage processes in various types electrochemical cells (e.g., zinc-based primary cells, lead-acid cells, nickel-metal hydride cells, rechargeable Li cells), and discuss the effects of using nanostructured electrode materials on the thermodynamic and kinetic properties of the charge storage/transfer process in an electrochemical cell. With the discussions, we aim to provide insights into design principles for "kinetically stable" nanostructured electrode materials towards their practical applications in future electrochemical cells.

Abbreviations

EES	Electrochemical energy storage
LIB	Li-ion battery
VRLA	Valve-regulated lead-acid
NiCd	Nickel-cadmium
NiMH	Nickel-metal hydride
MH	Metal hydride
RLC	Rechargeable lithium cell
NCA	$LiNi_{1-x-y}Co_xAl_yO_2$
NCM	$LiNi_{1-x-y}Co_xMn_yO_2$
NHE	Normal hydrogen electrode
SEI	Solid electrolyte interphase

S. Xin · H. Gao
University of Texas at Austin, Austin, TX 78712, USA

Y.-G. Guo (✉)
Institute of Chemistry, Chinese Academy of Sciences, Beijing 100190, People's Republic of China
e-mail: ygguo@iccas.ac.cn

Y.-G. Guo, *Nanostructures and Nanomaterials for Batteries*,
https://doi.org/10.1007/978-981-13-6233-0_2

VB	Valence band
CB	Conduction band
SSE	Solid-state electrolyte
SCE	Solid crystalline electrolyte
LiBOB	Lithium bis(oxalato)borate
VC	Vinylene carbonate
DMAc	Dimethylacetamide
LiTFSI	Lithium bis(trifluoromethane)sulfonimide
LiFSI	Lithium bis(fluorosulfonyl) imide
γ-BL	γ-butyrolactone
PC	Propylene carbonate
THF	Tetrahydrofuran
DMC	Dimethyl carbonate
EC	Ethylene carbonate
DEC	Diethyl carbonate
DME	Dimethoxyethane
DOL	Dioxolane

2.1 Storage of Charges in an Energy Storage Device

The operation of an electrochemical energy storage (EES) device relies on storage (release) of positive/negative charges in (from) the electrode materials. Upon discharging the device, the prestored charges are released from the electrode materials and migrate through the electrolyte, while the electrons move along the external circuit to do electrical work. In this way, the difference between the electrochemical potentials of cathode and anode is minimized by the end of discharge. For rechargeable batteries and pseudocapacitors, the reverse reaction is enabled through the charge process, so that the released charges can be restored back into the electrode for another discharge/charge cycle. As discussed in the previous chapter, the energy output of an EES device depends on both the potential difference between the two electrodes and the total number of charges that are transferred from one electrode to another during the electrochemical reaction. Therefore, an EES system with a larger potential difference or ability to transfer more charges during its operation is expected to deliver higher energy. All the above parameters are closely related to the thermodynamic and kinetic properties of key materials in the EES system. Nanostructured materials have been extensively studied for energy-related applications such as batteries, supercapacitors, catalysts, thermoelectrics and hydrogen storage due to their admirable properties such as high surface to volume ratio, favorable transport kinetics, altered physical properties and nanopore confinement [1–18]. Specifically, for battery applications, these characteristics make it possible to introduce new active storage sites, shorten ion transport path and include electron conductive path, reduce specific surface current rate, and improve electrochemical stability [2, 3, 12, 19]. Nevertheless, the use of nanostructures and nanomaterials in electrochemical cells involves its own limitations and challenges. To further enhance the storage capability of nanostructured materials, new mechanisms and optimized structures are required; but before that, it is necessary to understand the fundamental electrochemistry of charge storage and transfer in various types of electrochemical cells, as will be detailed below.

2.1.1 Storage of Charges in a Primary Cell

Primary batteries have existed for over 100 years, but up to 1940, zinc-carbon battery was the only one in wide use. Many of the significant advancements were made during 1970s–90s, under the stimulation of concurrent development of portable electronics and demand for space exploitation and military use [20, 21]. During the period, alkaline cell began to dominate the market share of primary cells, while new types of primary cells, such as lithium cells and zinc-air cells that were able to deliver specific energy at least twice that of most conventional aqueous primary cells, were developed and marketed. Although a number of anode-cathode combinations have been proposed for and used in primary cell systems (Table 2.1), only a few have achieved commercial success. By far, zinc has been the most successful anode material for primary cells because of its low cost, easy availability, high electrochemical equivalence and electrolyte compatibility, and long shelf life [20, 21]. Magnesium has also been successfully used in an active primary battery, particularly for military applications (although the commercial interests have been limited), because of its low cost, high energy density and good shelf life [20, 21]. Currently, there is an increasing focus on lithium, which outputs the highest gravimetric energy density and lowest standard potential among the metals [20]. The primary cells, which use lithium as anode and nonaqueous electrolytes that are electrochemically stable against lithium to pair with different cathode materials, offer the opportunity for obtaining higher energy density and other advantages in the performance characteristics of primary systems.

2.1.1.1 Zinc-Carbon Cells and Alkaline Cells

As introduced in Sect. 1.3.2.1, zinc-carbon cells and alkaline cells form the two major types of primary cells that are commercially available at present. Zinc-carbon cells consist of two major types: Leclanché and zinc chloride systems [22]. Both types are the most widely used of all the primary systems worldwide. By powering various kinds of electric appliances for daily-life use (flashlights, alarms, clocks, radios, remote controller, personal care devices etc.), the zinc-carbon cells are characterized as having low cost, ready availability, and acceptable performance [22]. A zinc-carbon (Leclanché) cell for general-purpose applications employs a zinc can as anode, which contains a layer of ammonium chloride (NH_4Cl) or zinc chloride ($ZnCl_2$) aqueous paste that separates the zinc can from a mixture of powdered carbon (usually

Table 2.1 Major characteristics and applications of common types of primary batteries.

Cell type	Characteristics	Applications
Zinc-carbon (Zinc/MnO_2)	Common, low-cost primary battery	Flashlight, radios, toys, etc.
Alkaline (Zn/alkaline/MnO_2)	Most popular general-purpose battery; low-temperature and high-rate operation; low cost	Most popular primary battery; used in a variety of battery-operated instruments
Mercury (Zn/HgO)	Highest volumetric capacity among conventional types; flat discharge; long shelf life	Medical devices (e.g., hearing aids and pacemakers), and military equipment; the use is now limited due to hazardous impact of mercury on the environment
Zinc/air (Zn/O_2)	Highest energy density; low cost; rich resource abundance	Special applications, medical devices, military electronics
Magnesium (Mg/MnO_2)	High-capacity primary battery; long shelf life	Use for military purpose and in aircrafts
Lithium/soluble cathode	High energy density; long shelf life; stable storage performance over a wide temperature range	A variety of applications requiring high energy density and long shelf life, e.g., from utility meters to military electronics applications
Lithium/solid cathode	High energy density; good rate capability and low-temperature performance; long shelf life; competitive cost	Replacement for conventional button and cylindrical cell applications, such as digital cameras
Lithium/solid electrolyte	Extremely long shelf life; low-power battery	Medical electronics

Data obtained from Ref. [20]

graphite powder) and manganese oxide (MnO_2) that packed around a carbon rod as cathode (Fig. 2.1a) [22, 23]. As the cell is discharged, zinc is oxidized and MnO_2 is reduced. A simplified overall cell reaction is,

$$Zn + 2MnO_2 \rightarrow ZnO + Mn_2O_3$$

The chemical reactions that occur on both electrodes of the Leclanché cell are so much complicated that controversy over the details continues from its appearance [22]. Recuperation of chemicals may occur simultaneously with the discharge reactions, resulting in several intermediate states that confuse the reaction chemistry. The non-stoichiometric nature of the cathode material (MnO_x, $x > 1.9$) makes the reaction mechanism even more complex. Reactions that occur in a general-purpose Leclanché cell (with ammonium chloride as the primary electrolyte) can be described as follows [22]:

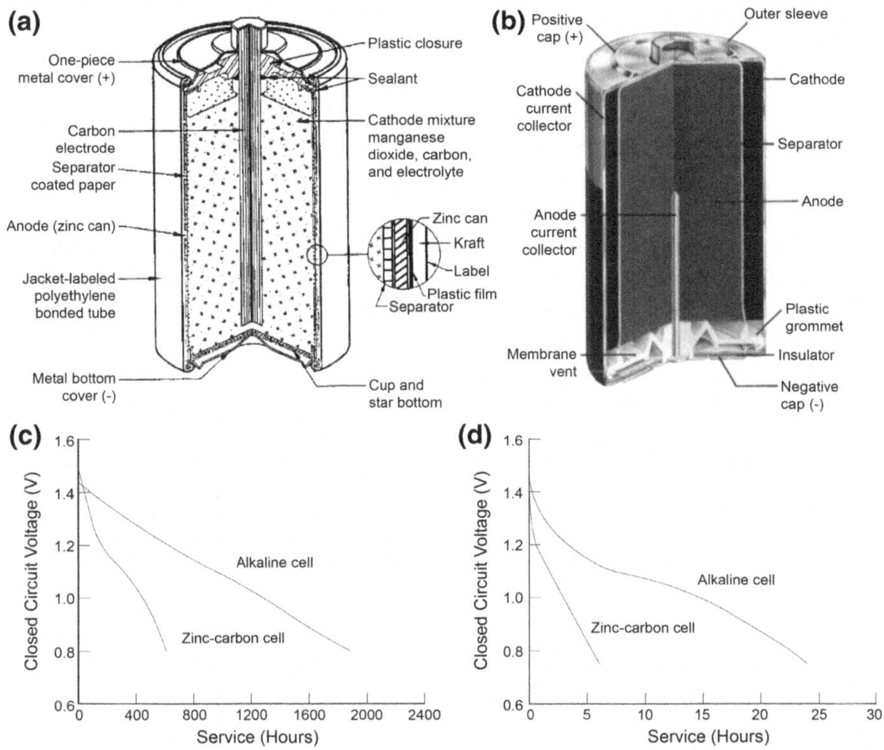

Fig. 2.1 Configurations of **a** cylindrical zinc-carbon cell and **b** alkaline cells, and performance comparison between the two cells: **c** typical light-drain discharge test (30 mA continuous test at 293 K), and **d** typical heavy-drain discharge test (500 mA continuous test at 293 K). **a** was reproduced with the permission from Ref. [22], Copyright © 2011, 2002, 1995, 1984 by The McGraw-Hill Companies, Inc.; **b–d** were reproduced with the permission from Ref. [25], Copyright © 2002, 1999, 1994, 1972 by The McGraw-Hill Companies, Inc

$$\text{Light discharge: } Zn + 2MnO_2 + 2NH_4Cl \rightarrow 2MnOOH + Zn(NH_3)_2Cl_2$$
$$\text{Deep discharge: } Zn + 2MnO_2 + NH_4Cl + H_2O \rightarrow 2MnOOH + NH_3 + Zn(OH)Cl$$
$$\text{Prolonged discharge: } Zn + 6MnOOH \rightarrow 2Mn_3O_4 + ZnO + 3H_2O$$

For heavy-duty applications, zinc chloride is usually substituted for ammonium chloride as the primary electrolyte, and the upgraded zinc chloride cells offer about twice the service life of general-purpose zinc-carbon cells, or up to four times in continuous-use or high-drain applications. In that case, the overall cell reactions are [22]:

$$\text{Light/deep discharge: } 4Zn + 8MnO_2 + 9H_2O + ZnCl_2 \rightarrow 8MnOOH + ZnCl_2 + 4ZnO + 5H_2O$$
$$\text{Prolonged discharge: } Zn + 6MnOOH + 2Zn(OH)Cl \rightarrow 2Mn_3O_4 + ZnCl_2 + 2ZnO + 4H_2O$$

The open-circuit voltage of zinc-carbon cell is derived from the potentials of anode (zinc) and cathode (MnO_2) materials, respectively. As most zinc-carbon cells share similar anode alloys, the open-circuit voltage usually depends on the type or mixture of MnO_2 used in the cathode, and the composition and pH value of the electrolyte system. Normally, the cell has an open-circuit voltage of around 1.5 V [22].

A further upgraded version of zinc-carbon cell is an alkaline cell, otherwise known as an alkaline-manganese dioxide or zinc-manganese dioxide battery ($Zn/KOH/MnO_2$), which offers up to eight times the battery life of zinc chloride cells [24, 25]. The active materials in the alkaline cell are electrolytically produced MnO_2, an aqueous KOH electrolyte, and powdered zinc metal (Fig. 2.1b). The electrolyte usually contains 35–52% of KOH, which affords greater conductivity and a reduced hydrogen gassing rate. The use of powdered zinc as the anode aims at providing a large surface area for high-rate capability and distributing solid and liquid phases more homogeneously.

Upon discharging, the MnO_2 cathode first undergoes a one-electron reduction to yield an oxyhydroxide [25],

$$MnO_2 + H_2O + e^- \rightarrow MnOOH + OH^-$$

The MnOOH product forms a solid solution with the reactant, leading to 17% volume expansion in the cathode [26]. At a lower voltage, MnOOH can be further discharged through the following reaction [23, 25, 27]:

$$3MnOOH + e^- \rightarrow Mn_3O_4 + OH^- + H_2O$$

The second step of reaction offers one-third of capacity of the first reaction based on MnO_2, and no additional volume change occurs during the reaction. Also, the reaction proceeds slower than the first reaction and therefore, is useful only under low-rate discharge conditions [25].

During the discharge process, the anode reaction of Zn in highly caustic electrolyte first produces soluble zincate ions [25],

$$Zn + 4OH^- \rightarrow Zn(OH)_4^{2-} + 2e^-$$

As the reaction proceeds, the electrolyte is gradually saturated with zincate. At this moment, zinc hydroxide becomes the main product of anode reaction, which then slowly dehydrates to ZnO in a water-starved environment [25],

$$Zn + 2OH^- \rightarrow Zn(OH)_2^{2-} + 2e^-$$
$$Zn(OH)_2 \rightarrow ZnO + H_2O$$

These oxidized products of zinc, though vary in their forms, share equivalent oxidation states with negligible difference in potential. As such, it is usually hard to tell the transition from one form to another in the discharge profile. Generally, the anode reaction is kinetically favorable and contribute little to the cell impedance [25]. However, under certain circumstances (e.g., high-rate or low-temperature operation, low KOH or high zincate concentration), the formed discharge products are too densely attached to the surface of electrode, giving rise to passivation of zinc. The use of Zn powder with large surface area could alleviate anode passivation and lead to improved anode performance [24].

With the above discussion, the total reactions of the alkaline cell at different depth of discharge are described as follows [25]:

$$2MnO_2 + Zn + 2H_2O \rightarrow 2MnOOH + Zn(OH)_2 (\text{one electron per mole of } MnO_2)$$
$$3MnO_2 + 2Zn \rightarrow Mn_3O_4 + 2ZnO (4/3 \text{ electrons per mole of } MnO_2)$$

The initial open-circuit voltage of the alkaline cell is *ca.* 1.5–1.65 V, depending on the purity of the cathode material and the ZnO content of the anode. The average cell voltage during discharge to the end voltage (0.75 V) is about 1.2 V [25].

The side reactions in alkaline cells include the reactions occurred on MnOOH and zinc. Through its equilibrium with soluble Mn(III), MnOOH can react with the zincate ion to form a complex compound hetaerolite ($ZnMn_2O_4$), which is less electroactive than MnOOH and may raise the cell impedance [25]. Besides, the MnOOH/MnO_2 solid solution may recrystallize into a less active form, resulting in notable loss in cell voltage [24]. In addition to the passivation of zinc, the anode may also trigger undesirable gas-generating reactions [24, 25]. Zinc metal is electrochemically active and can reduce water to produce hydrogen gas. The gassing rate on pure zinc is low. Heavy-metal impurities of trace amount introduced in the anode, however, could act as cathodic sites for hydrogen evolution and promote the gassing reaction. It is noted that the gassing reaction may occur during shelf life or operation period of alkaline cells, which results in buildup of gas pressure and causes dimensional distortion and leakage of the cell. Besides, the gassing reaction is usually accompanied by corrosion of zinc, which accounts for loss of anode capacity and chemical self-discharge of cathode by hydrogen gas [25]. To reduce gassing, inhibitors, such as metals (e.g., lead or indium), metal oxides (e.g., ZnO) or organics (e.g., end-substituted polyethylene oxide compounds) are added into the anode to reduce the driving force of zinc by mass action. Reduction of impurity level in cell components has also been proven effective [25].

Alkaline cells are provided with theoretical capacities considerably higher than that of zinc-carbon cells of the same size, and the reasons are as follows. First, the alkaline cell uses MnO_2 of much higher purity than that is used in most zinc-carbon cells, which ensures a high electrochemical activity of cathode [25]. Besides, the cathode contains only a small amount of electrolyte (the cathode has a very dense structure), and the space occupied by other cell components (separator, current collectors, etc.) is minimized [25]. In addition to having a high capacity, the alkaline cells also show improved rate capabilities. On one hand, the KOH electrolyte has a very high conductivity; on the other hand, the powdered zinc anode shows a large surface area in contact with the electrolyte (compared to the low-area zinc can used in the zinc-carbon cell). Consequently, the cell shows a low internal resistance before and during its entire service life, which contributes to improved kinetics of electrochemical reaction [25].

The alkaline cell outperforms the zinc-carbon cell under a wide range of operational conditions. In a typical light-drain discharge test (Fig. 2.1c), the alkaline cell demonstrates a much longer service life owing to its higher theoretical capacity [25]. Under a heavy-drain continuous discharge condition (Fig. 2.1d), the difference between the performance of the two types of cells becomes even significant, which is largely ascribed to the better rate performance of the alkaline cell.

2.1.1.2 Lithium Primary Cells

The development of primary cells with high energy density was started in the 1960s with focus on systems using nonaqueous electrolyte and Li metal anode [28]. In the early 1970s, Li primary cells sought their opportunities first in military applications. With continuous efforts made to improve the cell design, formulation and safety, these cells have attracted increasing attention in a variety of applications including digital cameras, memory backup circuits, watches and even standby power for bulk storage [28]. Compared with the conventional primary cells, the Li primary cells have many advantages including high operation voltage, high specific energy and energy density (up to 870 Wh kg^{-1} or 1180 Wh L^{-1}, which is 2–5 times higher than that of conventional zinc-based cells), wide temperature range of operation, flat discharge characteristics, high power density, and ultralong shelf life (up to 10 years at room temperature and 1 year at 70 °C has been demonstrated) [28]. Only the zinc-based systems (Zn/O_2, Zn/HgO, and Zn/Ag_2O cells) noted for their high energy densities approach the energy output of the lithium systems at ambient temperature. Nevertheless, the operation of Zn/O_2 cell relies notably on the atmospheric condition, while the others suffer from either low specific energy or poor performance at lower temperatures.

Li primary cells are designed using a number of different chemistries. Therefore, the term "lithium battery" applies to many different types of cells, each using lithium as the anode and pairing with different cathodes/electrolytes to provide them with different working chemistries, designs, configurations, dimensions, and other physical and mechanical features [28]. Based on the type of electrolyte and cathode material involved, the Li primary cells can be classified into three main categories as shown in the following Table 2.2.

There are plenty of choices on the cathode material for primary Li cells. The general requirements on cathode materials include high voltage versus Li anode, high energy density and favorable compatibility with the electrolyte [28]. In order to achieve a high power density of cell, the cathode material should be intrinsically conductive or combined with a host material with high conductivity (e.g., carbon). In case the cathode reaction yields a metal and a soluble Li salt, the cathode may also exhibit improved conductivity as the discharge proceeds. Other desirable fea-

Table 2.2 Classification of lithium primary cells

Cell type	Soluble cathode (liquid or gas)	Solid cathode	Solid electrolyte
Electrolyte	Organic or inorganic	Organic	Solid state
Power capability	Moderate to high, W	Low to moderate power, mW-W	Very low power, μW
Size (Ah)	0.5–10,000	0.03–1,200	0.003–2.4
Operating range (°C)	−80 to 70	−40 to 50	0–100
Shelf life (years)	5–20	5–8	10–25
Typical cathodes	SO_2; $SOCl_2$; SO_2Cl_2	V_2O_5; $AgV_2O_{5.5}$; MnO_2; CF_x; CuS; FeS_2; FeS	PbI_2/PbS/Pb; I_2(P2VP)
Key characteristics	High energy output, high power output, low-temperature	High energy output for moderate power requirements, nonpressurized cells	Excellent shelf life, leakage proof, long-term microampere discharge

Data obtained from Ref. [28]

tures of the cathode material shall include low cost, rich abundance of elemental resource, nontoxicity and nonflammability. Table 2.3 summarizes some materials that have been studied for cathode candidates in primary Li batteries, along with the crucial electrochemical parameters such as cell reactions, theoretical cell voltages and capacities.

Primary cells are designed for one-time use and should be discarded after its service. Technically, the electrochemical reactions in many primary cells are reversible, and some primary cells can be recharged for several cycles under careful management of charging conditions and at low current rate of charge. However, even if successful, the charged cells may not deliver full capacity and may have poor capacity retention after being charged. In most cases recharging a primary cells is impractical and could be dangerous since the cells are usually tightly sealed without any adequate mechanism to permit the release of gases that form during charging [20]. The gassing could cause leakage, rupture, or explosion of cells, resulting in personal injury or other hazards. Therefore, recharging a primary cell is a practice that should be avoided.

2.1.2 *Storage of Charges in a Rechargeable Cell*

Since the invention of lead-acid cell in 1859, rechargeable cells (or secondary cells) have been in existence for 160 years. Now they are widely used in many applications including electric tools, toys, lightening, photographic, consumer electronics, and more recently, rechargeable cells are being developed as a power source for electric and hybrid electric vehicles, and grid storage [29]. The rechargeable cells are oper-

Table 2.3 Classification of lithium primary cells

Cathode material	Valence change	Density ($g\,cm^{-3}$)	Cell reaction	Theoretical values			
				Capacity (mAh g^{-1})	Capacity (mAh cm^{-3})	Cell voltage (V)	Energy (Wh kg^{-1})
SO_2	1	1.37	$2Li + 2SO_2 \rightarrow 2Li_2S_2O_4$	419	–	3.1	1170
$SOCl_2$	2	1.63	$4Li + 2SOCl_2 \rightarrow 4LiCl + S + SO_2$	450	–	3.65	1470
SO_2Cl_2	2	1.66	$2Li + SO_2Cl_2 \rightarrow 2LiCl + SO_2$	397	–	3.91	1405
Bi_2O_3	6	8.5	$6Li + Bi_2O_3 \rightarrow 3Li_2O + 2Bi$	350	2970	2	640
$Bi_2Pb_2O_5$	10	9	$10Li + Bi_2Pb_2O_5 \rightarrow 5Li_2O + 2Bi + 2Pb$	290	2640	2	544
$(CF)_x$	1	2.7	$xLi + (CF)_x \rightarrow xLiF + xC$	860	2320	3.1	2180
$CuCl_2$	2	3.1	$2Li + CuCl_2 \rightarrow 2LiCl + Cu$	400	1220	3.1	1125
CuF_2	2	2.9	$2Li + CuF_2 \rightarrow 2LiF + Cu$	530	1520	3.54	1650
CuO	2	6.4	$2Li + CuO \rightarrow Li_2O + Cu$	670	4260	2.24	1280
CuS	2	4.6	$2Li + CuS \rightarrow Li_2S + Cu$	560	2570	2.15	1050
FeS	2	4.8	$2Li + FeS \rightarrow Li_2S + Fe$	610	2950	1.75	920
FeS_2	4	4.9	$4Li + FeS_2 \rightarrow 2Li_2S + Fe$	890	4350	1.8	1304
MnO_2	1	5	$Li + Mn^{IV}O_2 \rightarrow Mn^{III}O_2(Li^+)$	310	1540	3.5	1005
MoO_3	1	4.5	$2Li + MoO_3 \rightarrow Li_2O + Mo_2O_5$	190	840	2.9	525
$AgCl$	1	5.6	$Li + AgCl \rightarrow LiCl + Ag$	190	1040	3.267	583
Ag_2CrO_4	2	5.6	$2Li + Ag_2CrO_4 \rightarrow Li_2CrO_4 + 2Ag$	160	900	3.35	515
V_2O_5	1	3.6	$Li + V_2O_5 \rightarrow LiV_2O_5$	150	530	3.4	490

Data obtained from Ref. [28]

ated by various types of electrochemistry, among which the Li-ion batteries (LIBs) based on Li intercalation chemistry are the most successful and are now accounting for a major share of rechargeable cell market. In the following sections, we will briefly introduce charge storage electrochemistry of typical rechargeable cells including lead-acid cells, nickel-cadmium and nickel-metal hydride cells (pre-lithium rechargeable cells), rechargeable lithium cells and some post-lithium rechargeable cells based on earth-abundant alkali metal anodes (sodium, potassium) and alkali earth metal anodes (magnesium, calcium and aluminum).

2.1.2.1 Storage of Charges in a Pre-lithium Rechargeable Cell

Lead-acid batteries are now provided by a well-established, worldwide industrial supplier base and hold the largest market share for rechargeable cells [30–32]. They are used extensively in communication devices, data networks, power tools, security and emergency power supplies, mining and material-handling equipment, and electric and hybrid vehicles (including off-road vehicles) in addition to its traditional usage for automotive engine ignition and vehicle lighting [31, 33, 34]. Despite of a low energy-to-weight/volume ratio of the cell, there are good reasons for its popularity on the storage market, which include dependable chemistry, low cost per watt and ease of manufacture and recycle. Besides, the cells are provided with a relatively large power-to-weight ratio, so that they are able to supply high surge current to meet the high-power requirements when starting vehicle motor.

The history of lead-acid cells began with the invention of Raymond Gaston Planté in 1859, although batteries containing sulfuric acid or lead components were discussed earlier [31]. Subsequent to Planté's achievements, numerous efforts were done on improving utilization and production of active materials, design of critical components, and cell configuration, which enabled the cell with much improved performance and durability.

The lead-acid cell employs lead dioxide (PbO_2) as the cathode material and metallic lead, in a high-surface-area porous structure, as the anode material. Two crystalline forms of PbO_2 (alpha or orthorhombic PbO_2 and beta or tetragonal PbO_2) were contained on the charged cathode [32]. The α-PbO_2 is less electrochemically active and slightly lower in capacity per unit weight than the β-PbO_2, yet it does promote longer cycle life of the cell [31]. The cathode material is a major factor that influences the performance and life of the lead-acid battery, while the anode controls cold-temperature performance (for example, engine starting). The electrolyte is a sulfuric acid solution, and the acid concentration is typically about 37 wt% in a fully charged condition. In practical cell configuration, active materials of both electrodes are supported on lead alloy grids which are stacked together and serve as current collectors, and interleaved with a porous, electrically insulating separator [32]. Both flooded and sealed valve-regulated lead-acid (VRLA) cells are made, and the grids may be in the form of flat pasted plates or tubular plates [30]. To improve the cell response to pulse power requirements and avoid the performance decay of the anode during high-rate partial-state-of-charge cycling, more advanced config-

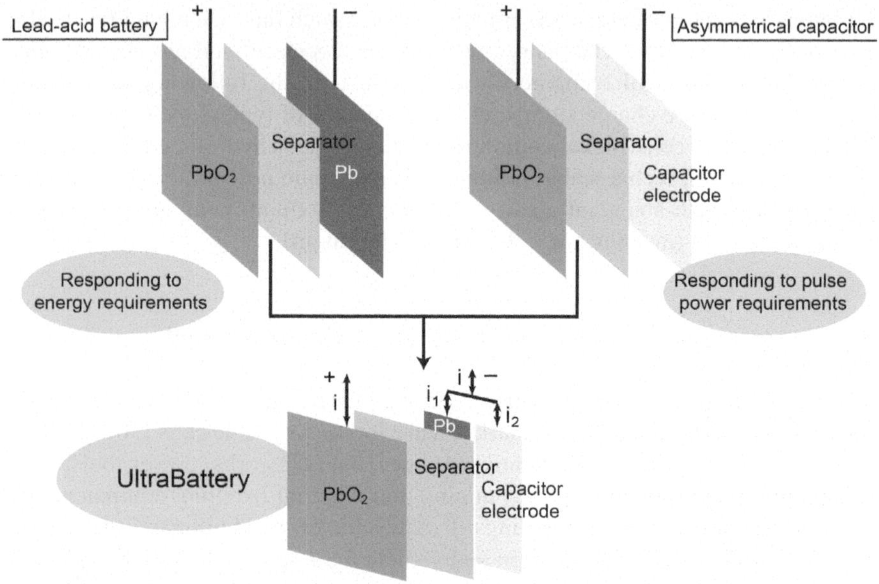

Fig. 2.2 Schematic illustration showing the UltraBattery concept which combines the electrochemistry of lead-acid cell asymmetrical capacitor. Reproduced with the permission from Ref. [31], Copyright © 2011, 2002, 1995, 1984 by The McGraw-Hill Companies, Inc

urations (e.g., UltraBattery invented by Australia's Commonwealth Scientific and Industrial Research Organization) adopt the combination of a lead-acid cell and a capacitor in one case (Fig. 2.2) [31, 34, 35]. The anode is vertically divided in halves, with a lead electrode on one half and a carbon double layer capacitor electrode on the other. The two components of the negative electrode are connected in parallel and the capacitor half acts as a buffer to share current with the anode and reduce the rate of charge/discharge [34, 35]. The new design offers important advantages over both conventional lead-acid batteries and asymmetric lead-based supercapacitors, and accounts for improved high-rate charge acceptance and self-balancing of the cell [34, 35].

During the discharge of the cell, both electrodes are converted to lead sulfate and the process reverses on charge [31]:

$$\text{Anode: } Pb \rightleftarrows Pb^{2+} + 2e^-; Pb^{2+} + SO_4^{2-} \rightleftarrows PbSO_4$$

$$\text{Cathode: } PbO_2 + 4H^+ + 2e^- \rightleftarrows Pb^{2+} + 2H_2O; Pb^{2+} + SO_4^{2-} \rightleftarrows PbSO_4$$

$$\text{Overall: } Pb + PbO_2 + 2H_2SO_4 \rightleftarrows 2PbSO_4 + 2H_2O$$

The nominal cell voltage is around 2.05 V, which represents the highest value among all aqueous-electrolyte cells. The voltage on open circuit is a direct function of the electrolyte concentration, ranging from 2.125 V for a cell with 1.28 specific gravity (or relative density) of electrolyte to 2.05 V with 1.21 specific gravity. The selection of specific gravity of electrolyte depends on the service requirements of battery. The electrolyte concentration should be sufficiently high to facilitate ion conduction, but not so high as to cause corrosion of other components of the cell, as it would shorten operation life and increase risk of self-discharge [31].

The basic electrode processes in positive and negative electrodes involve a dissolution-precipitation process rather than solid-state ion transport or film formation mechanism. The discharge/charge chemistry was discovered by Gladstone and Tribs in 1882 and was known as a double-sulfate reaction [36]. Due to consumption of sulfuric acid during discharge, the electrolyte may also be treated as an "active" material and in certain battery designs can be a capacity-limiting factor.

As cells approach top-of-charge and the majority of the $PbSO_4$ has been converted to Pb or PbO_2, the cell voltage on charge becomes greater than the gassing voltage (~2.39 V). As a result, the overcharge reactions begin, which result in the production of hydrogen and oxygen and also loss of water [31, 34].

$$\begin{aligned}
&\text{Anode: } 2H^+ + 2e^- \rightarrow H_2\\
&\text{Cathode: } H_2O - 2e^- \rightarrow (1/2)O_2 + 2H^+\\
&\text{Overall: } H_2O \rightarrow H_2 + (1/2)O_2
\end{aligned}$$

Fortunately, the overpotential at the occurrence of reactions is sufficiently high so that water loss is manageable by controlling the charging voltage [31, 34]. For flooded batteries, correct selection of the grid materials and charging parameters minimizes water loss so that battery maintenance by supplement of water only needs to be performed occasionally [32]. For sealed-type cells, electrolyte is either immobilized in an absorptive glass mat separator or gelled with finely dispersed silica, so that gas channels are formed between both electrode plates to enable fast migration of oxygen from the cathode where it is generated to the anode where it reacts with lead as described by the following reactions:

$$\begin{aligned}
&\text{Anode: } H_2O \rightarrow 2H^+ + (1/2)O_2 + 2e^-\\
&\text{Cathode: } Pb + (1/2)O_2 + H_2SO_4 \rightarrow PbSO_4 + H_2O
\end{aligned}$$

In this way, oxygen is recombined with Pb to produce lead sulfate which is the normal discharge product and can be recharged back to lead as in the normal charge process [32]. In addition to oxygen evolution, hydrogen evolution at the anode side also requires attentions. Grid alloys with a high hydrogen overpotential are usually selected to reduce hydrogen evolution at the anode. Besides, one-way valves are usually equipped (on the VRLA-type cell) to enable release of small amount of hydrogen while preventing entrance of air into the cell [32].

The nickel-cadmium cell (NiCd cell) is another type of commercial rechargeable cell that are widely used. Invented by Waldemar Jungner in 1899 and later patented by Thomas Edison in 1902, the NiCd cell served many industrial applications, for example, railroad service, military purpose and aerospace [37–40]. The cell shows a higher energy density than the lead-acid cell, besides, it also shows many merits such as low maintenance, long cycle/shelf life, wide temperature range of operation (−20 °C to 70 °C), and favorable high-rate performance [41]. The cell contains a nickel hydroxide cathode, a metallic cadmium anode, a separator and an alkaline electrolyte. The cell usually has a metal case equipped with a self-sealing safety valve, and both electrodes are isolated from each other by the separator and are rolled inside the case [42]. The NiCd cells can be prepared in many shapes and configurations including vented pocket plate, sintered plate, fiber and plastic-bonded plate, and even sealed types, yet they all share the same basic electrochemistry (Fig. 2.3a) [37–39, 43]. On discharge, trivalent nickel oxy-hydroxide is reduced to divalent nickel hydroxide with consumption of water, while metallic cadmium is oxidized to form cadmium hydroxide. On charge, the opposite reactions take place. The reactions are expressed as follows (forward reactions occur during the discharge process) [37–40]:

$$\text{Cathode: } 2\text{NiOOH} + 2\text{H}_2\text{O} + \text{e}^- \rightleftarrows 2\text{Ni(OH)}_2 + 2\text{OH}^-$$
$$\text{Anode: } \text{Cd} + 2\text{OH}^- \rightleftarrows \text{Cd(OH)}_2 + 2\text{e}^-$$
$$\text{Net reaction: } \text{Cd} + 2\text{NiOOH} + 2\text{H}_2\text{O} \rightleftarrows \text{Cd(OH)}_2 + 2\text{Ni(OH)}_2$$

The electromotive force of the cell is 1.29 V. Different from the lead-acid cell, the specific gravity (which is normally 1.2 g mL^{-1}) and composition of alkaline (KOH) electrolyte does not change significantly during the operation.

Water electrolysis is the major competitive reaction (overcharge reaction) to the charging process of aqueous-nickel cells. Toward the end of the charging cycle, hydrogen and oxygen are separately evolved at the anode and the cathode. In an open cell, the gases are allowed to vent. The overcharge reactions in a vented cell are as follows [39]:

$$\text{Cathode: } 4\text{OH}^- \rightarrow 2\text{H}_2\text{O} + \text{O}_2 + 4\text{e}^-$$
$$\text{Anode: } 4\text{H}_2\text{O} + 4\text{e}^- \rightarrow 2\text{H}_2 + 4\text{OH}^-$$
$$\text{Net reaction: } 2\text{H}_2\text{O} \rightarrow 2\text{H}_2 + \text{O}_2$$

With continuous consumption of water during the cell overcharge, there is a need for addition of water in the vented cell configuration.

In a sealed cell, the anode usually has a higher effective capacity than the cathode, so that the cathode reaches fully charged state before the anode begins to evolve hydrogen. The evolved oxygen then migrates to the anode, where it reacts with and discharges the cadmium to yield cadmium hydroxide. The reactions are expressed as follows [37]:

$$\text{Cathode: } 4\text{OH}^- \rightarrow 2\text{H}_2\text{O} + \text{O}_2 + 4\text{e}^-$$
$$\text{Anode: } 2\text{Cd(OH)}_2 + 4\text{e}^- \rightarrow 2\text{Cd} + 4\text{OH}^-$$
$$\text{Recombination reaction on the anode: } 2\text{Cd} + \text{O}_2 + 2\text{H}_2\text{O} \rightarrow 2\text{Cd(OH)}_2$$

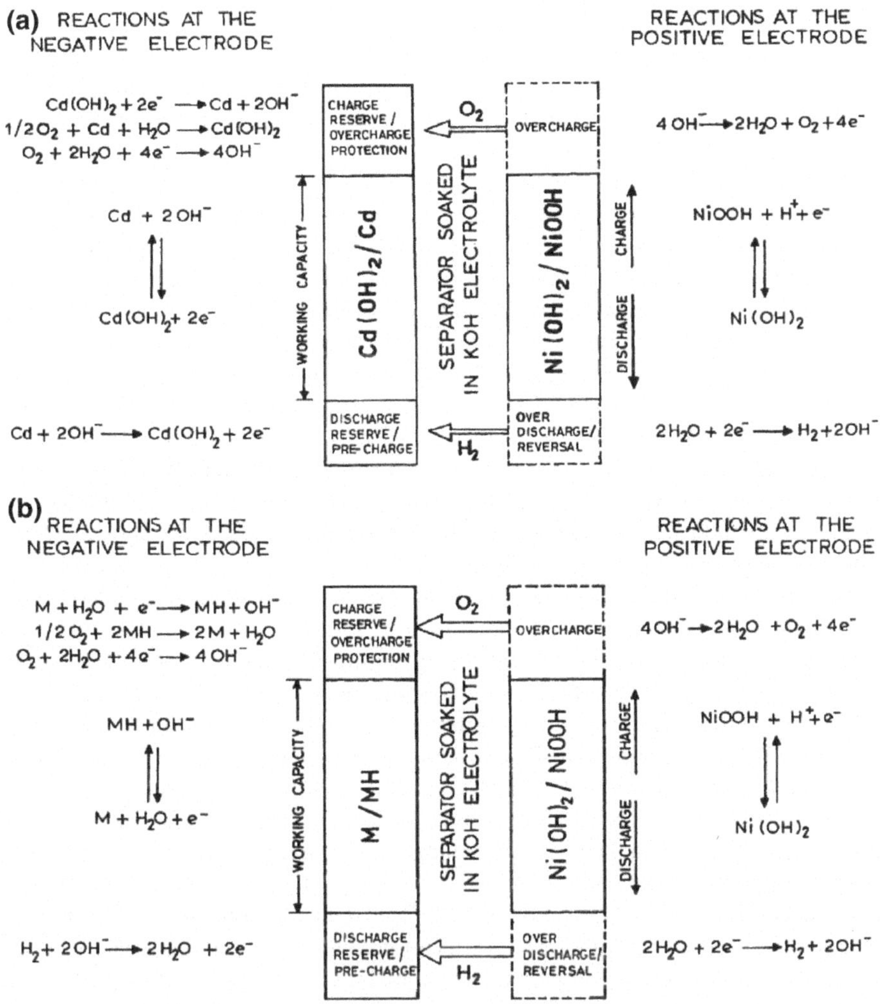

Fig. 2.3 Schematic illustrations showing the design principles of **a** a sealed NiCd cell and **b** a sealed NiMH cell. Reproduced with the permission from Ref. [40], Copyright © 2001 Elsevier Science B.V

The overcharge reaction is exothermic, particularly the chemical recombination reaction [38]. As a result, electricity in a sealed cell is converted into heat without any net chemical change [37]. As the overcharge reaction on the cathode starts before the cell is fully charged, certain oxygen evolution during the charge process is unavoidable. At elevated temperatures, the oxygen evolution starts at a lower voltage, resulting in low charging efficiency and charge imbalance on both electrode plates that accounts for reduced cell capacity [37, 40]. On the other hand, the recombination reaction helps to consume the evolved oxygen and prevent the buildup of pressure inside the cell [27, 32]. In practical cell design, a separator that is permeable to oxygen is employed so that oxygen can pass through to the anode. A limited amount of electrolyte is also used to facilitate the oxygen transfer. Besides, a safety venting mechanism is used in the sealed design to prevent rupture in case of excessive pressure buildup due to a malfunction, high charge rate, or abuse [37].

Another issue related to the use of NiCd cells is voltage depression (memory effect). A sealed NiCd cell may suffer a reversible loss of capacity when it is cycled repetitively with low depth of discharge (discharge process is terminated before reaching its full capacity). This condition is known as "voltage depression" or referred to as "memory effect" as the battery appears to "remember" the lower capacity of the shallow discharge. Operation of cells at higher temperatures accelerates this type of loss. The voltage depression occurs at shallow discharge because the active materials that have not participated in the electrochemical reaction, particularly the cadmium electrode, may undergo a physical change that leads to an increase in resistance. The effect has also been ascribed to structural changes in the nickel electrode [44]. Fortunately, the loss of capacity due to memory effect is reversible and the battery can restore its full capacity with a few full discharge/charge reconditioning cycles.

In addition to overcharging and memory effect, the NiCd cells also suffer from a higher material cost and environmental concerns with the anode use of cadmium. To address these problems, nickel-metal hydride (NiMH) cell was proposed in the 1960s as a derivative of NiCd cell [45]. Being equipped with the same cathode material (NiOOH) as that in the NiCd cell, the anode material of the NiMH cell uses a hydrogen-absorbing alloy instead of cadmium (Fig. 2.3b) [40]. With an upgraded cell design, the NiMH cell shows distinctive advantages including higher energy density compared to lead-acid and nickel-cadmium cells, good high-temperature and high-rate capabilities, high charge retention, long cycle/shelf life, improved safety against high-voltage operation and abusive overcharge, sealed maintenance-free design, and lower cost/pollution with cadmium-free anode [40, 45]. It soon became a commercially successful rechargeable cell for vehicle propulsion applications, while also competed against rechargeable lithium technologies for the market share of consumer electronics (including laptops and cell phones) in the early 1990s [45].

Figure 2.3b shows the operation principle of the NiMH cell. During the discharge process, nickel oxyhydroxide is reduced to nickel hydroxide,

$$\mathrm{NiOOH} + \mathrm{H_2O} + e^- \rightarrow \mathrm{Ni(OH)_2} + \mathrm{OH^-}$$

Meanwhile, metal hydride (MH) is oxidized to the metal alloy (M),

$$\mathrm{MH} + \mathrm{OH^-} \rightarrow \mathrm{M} + \mathrm{H_2O} + e^-$$

The overall reaction on discharge is,

$$\mathrm{MH} + \mathrm{NiOOH} \rightarrow \mathrm{M} + \mathrm{Ni(OH)_2}$$

The above reaction normally shows a voltage output of 1.35 V and is completely reversible during the charge process [40, 45].

The metal hydride (MH) anode is the key material that enables the NiMH cell, which is made up of engineered intermetallic alloys capable of storing hydrogen in the solid form. A large variety of alloys have been used/proposed as the active materials in the anode of NiMH cell, such as A_2B [46], AB [47], AB_2 [48], AB_3 [49], A_2B_7 [50], A_5B_{19} [51], AB_5 [52], body-centered-cubic solid solution [53] and their combinations. Among these alloys, the AB_5-type materials are more common and are provided with other advantages such as low raw-material cost, easy activation and formation and high discharge-rate capability [45]. On the other hand, there has been significant ongoing development to improve the properties of A_2B_7 and AB_2-type materials due to their inherently higher hydrogen storage capacities (380 mA h g^{-1} and 440 mA h g^{-1}, respectively, compared to 320 mA h g^{-1} reported for the AB_5-type material), which potentially lead to higher energy and bring cost reduction [45].

Similar to NiCd cell, the NiMH cell also involves the generation of oxygen gas on the fully charged cathode toward the end of the charge and in overcharge. In a sealed NiMH cell, an oxygen-recombination mechanism is employed to prevent the buildup of pressure inside the cell. A starved electrolyte design and an appropriate porous separator is employed to facilitate the diffusion of oxygen from the cathode to the anode, triggering reaction of oxygen with the hydrogen electrode to produce water [45].

$$4MH + O_2 \rightarrow 4M + 2H_2O$$

With the recombination reaction, the anode is not fully charged, which prevents the evolution of hydrogen at the early stage. To further prevent buildup of gas pressure, the charge current is controlled at the end of charge and during overcharge to limit the generation of oxygen to below the rate of recombination, so that the cell realizes stable cycling free of maintenance.

A sealed NiMH cell may also experience a reversible voltage depression and loss of capacity when it is partially discharged and charged repetitively without the benefit of a full discharge [45]. The phenomenon is known as its "memory effect" and is ascribed to a raised internal resistance due to reversible changes in physical characteristics of uncycled active materials on the electrode during partial discharging. However, the memory effect is reversible, and the battery can restore its full capacity within a few full discharge/charge cycles. Also, as the extent of voltage depression and capacity loss depends on the depth of discharge, the memory effect can be avoided or minimized by discharging to an appropriate end voltage. For example, the effect is largely suppressed if the cutoff voltage of discharge is set between 1.15 and 1.10 V per cell compared with those provided with higher terminating discharge voltages (e.g., 1.2 V per cell).

Since the market launch of portable NiMH cell in 1987, the technology has been continuously developed and great strides have been made in terms of increasing specific energy (from 52 to 80–110 Wh kg^{-1}, so that the energy density achieves 430 Wh L^{-1}), specific power (from 180 to 850–2000 W kg^{-1}), cycle life (from 300 to > 1000) and charge retention (from 70% loss in one month to < 15% in one year) [45]. With significantly higher energy output than lead-acid cell (30–40 Wh kg^{-1}) and NiCd cell (40–60 Wh kg^{-1}), the target use of NiMH cell covers consumer electronics and (hybrid) electric vehicles, while is continuously expanding into markets dominated by other battery chemistries.

2.1.2.2 Storage of Charges in a Rechargeable Lithium Cell

Rechargeable lithium cells (RLCs) form another type of major competitors to NiMH cells. The history of RLCs began in the 1960s, when chemists in Europe were exploring the chemistry of reversible insertion of lithium into layered transition metal sulfides [54]. In the 1970s, M. Stanley Whittingham from Exxon Mobil Corporation proposed an RLC that was able to achieve reversible Li insertion chemistry by using a layered titanium disulfide (TiS_2) cathode [55]. However, the Whittingham's cell adopted Li metal as the anode, during the repeated charging process, plating of Li on the anode surface resulted in the formation of Li dendrites that grew across the electrolyte to the cathode to create an internal short circuit and ignite the flammable electrolyte, so that the cells easily caught fire [54]. On the other hand, TiS_2 as cathode material is provided with a high cost and unstable chemistry during air exposure. In the search of more practical electrode materials, Samar Basu from the University of Pennsylvania demonstrated intercalation of Li in graphite [56, 57]. This finding led to the development of an operable Li-intercalated graphite electrode by Rachid Yazami to provide an alternative to the Li metal anode [57]. In 1979–1980, a research group led by John B. Goodenough at Oxford University, reported reversible Li insertion and extraction in a layered lithium cobalt oxide ($LiCoO_2$) compound at ~4 V versus Li^+/Li [58]. Compared with TiS_2, $LiCoO_2$ is easily prepared, and can provide good electrochemical performance and improved safety properties, and is less sensitive to process variation and air moisture. This innovation provided the first cathode material that enabled intercalation chemistry in commercial RLCs. After that, Akira Yoshino, at the Asahi Kasei Corporation, made the first rechargeable LIB by combining the $LiCoO_2$ cathode with a graphitic-carbon anode, which is commercialized by Sony Corporation in 1991 to power the first portable phone [56]. Compared with other types of rechargeable cells, the LIBs are provided with distinctive merits such as maintenance-free sealed design, broad temperature range of operation, long cycle/shelf life, low self-discharge rate, no memory effect, high-rate and high-power operation capabilities, high coulombic and energy efficiency, and most importantly, ultrahigh specific energy (> 265 Wh kg^{-1}) and energy density (> 700 Wh L^{-1}) that are several times of their competitors [59]. After more than two decades of development, the LIBs have dominated the consumer electronics market and are now actively expanding their applications into emerging markets including electric vehicles and grid storage [2, 3, 60–63].

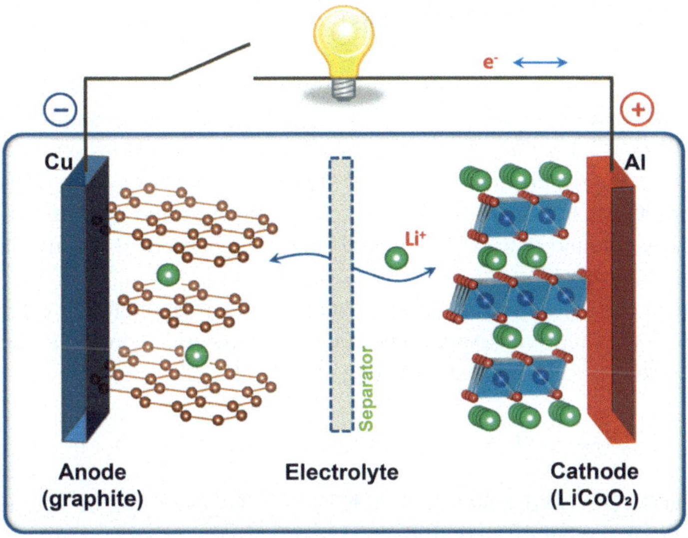

Fig. 2.4 Schematic illustration showing the operation principle of the first LIB ($LiCoO_2/Li^+$ electrolyte/graphite). Reproduced with the permission from Ref. [60], Copyright © 2013 American Chemical Society

The first LIB, assembled with a discharged graphitic carbon anode and a discharged $LiCoO_2$ cathode, was operated by a reversible reaction where Li ions (guests) are reversibly extracted or inserted into the layered host to form a sandwich-like structure without significant structural change to the host (Fig. 2.4) [60, 64, 65]. Typical reactions were expressed as follows (forward reactions are charge reactions) [60, 66]:

$$\text{Cathode: } LiCOO_2 \rightleftarrows Li_{i_{kx}}COO_2 + xe^- + xLi^+$$
$$\text{Anode: } C + ye^- + yLi^+ \rightleftarrows Li_yC(0 < y < 0.167)$$
$$\text{Net reaction: } LiCoO_2 + (x/y)C \rightleftarrows (x/y)Li_yC + Li_{1-x}COO_2$$

While the cathode reaction shows a theoretical electrode potential (E_c) of ~4.0 V versus Li^+/Li, the anode shows a theoretical potential of ~0.2 V (E_a) versus Li^+/Li [60, 64–66]. In this way, the full cell can be operated at a voltage of ~3.8 V. The theoretical specific capacity of electrode active material in a Li-ion cell is calculated by the Faraday's law:

$$Q_{tren} = (n \times F)/(3600 \times M_w)\left(\text{unit: Ah g}^{-1}\right), \quad (2.1)$$

where n is the number of charge carriers, F is the Faraday constant (96,485 C mol^{-1}) and M_w is the molecular weight of active material used in the electrode.

For the $LiCoO_2$ cathode, in case a complete extraction of Li occurs during charge, the theoretical specific capacity of $LiCoO_2$ is calculated as follows:

$$Q_{\text{theo}} = (1 \times 96485)/(3600 \times 97.87) = 0.274\text{Ah g}^{-1}.$$

However, the electrochemical Li storage of $LiCoO_2$ is enabled by the Co(IV)/Co(III) couple, which is pinned near the top of the O-2p bands. As a result, the original $Li_{1-x}CoO_2$ cathode evolves oxygen or inserts protons on removing Li^+ beyond x = 0.55, which gives a specific capacity of $LiCoO_2$ of ~150 mAh g^{-1} [60]. On the other hand, by the end of the charge process, the graphitic carbon will form a LiC_6 compound (y = 0.167), which leads to a theoretical specific capacity of 372 mA h g^{-1} [66].

In practical case, the specific capacity of electrode in an operating cell is usually different from the theoretical value, and can be calculated according to the voltage-time profile from the galvanostatic cycling test:

$$Q_{\text{prac.}} = (I \times A \times t_{\text{co}})/(3600 \times M_{\text{w}})\left(\text{unit: Ah g}^{-1}\right), \tag{2.2}$$

wherein, I is the current density (unit: A m^{-2}), A is the area (unit: m^2), t_{co} is the time (unit: s) to reach the cut off potential (V_{co}).

The theoretical electrode potential (E) and specific capacity (Q_{theo}) of the electrode material are important parameters since they determine the theoretical energy that can be stored in an electrochemical cell according to Eq. 1.5 in Chapter 1. The theoretical electrode potential of the anode/cathode material directly links to its chemical potential (μ), which further determines the open-circuit voltage (V_{oc}) of the cell according to the following equation:

$$V_{\text{oc}} = E_{\text{c}} - E_{\text{a}} = (\mu_{\text{a}} - \mu_{\text{c}})/e \quad (\text{unit: V}), \tag{2.3}$$

where e is the magnitude of the electron charge [64, 65].

According to Eq. 1.6 in Chapter 1, V_{oc} (or E_{cell}) determines the Gibbs free energy change ($\Delta_r G$) of a given cell reaction (that is, the maximum work an electrochemical cell can do), and the direction of spontaneous reaction (forward or backward).

On the other hand, the theoretical specific capacity of an electrochemical cell (denoted as Q_{cell}), which is also the maximum total charge transfer that can be enabled in a cell on discharge or charge, is determined by the theoretical specific capacities of both electrodes and can be calculated as follows:

$$Q_{\text{cell}} = Q_{\text{theo-a}} \times Q_{\text{theo-c}}/(Q_{\text{theo-a}} + Q_{\text{theo-c}})\left(\text{unit: Ah g}^{-1}\right) \tag{2.4}$$

where $Q_{theo\text{-}a}$ denotes the theoretical specific capacity of the anode while $Q_{theo\text{-}c}$ denotes the theoretical specific capacity of the cathode.

The theoretical specific energy (unit: Wh kg^{-1}) of an electrochemical cell can be calculated by multiplying its theoretical specific capacity (Q_{cell}) and theoretical voltage output (V_{oc}) of the cell.

In practical cells, the density of stored energy in a fully charged cell may be obtained by measuring the time $\Delta t = \Delta t\ (I_{\text{dis}})$ for its complete discharge at a constant current $I_{\text{dis}} = \mathrm{d}q/\mathrm{d}t$ (where q represents the state of charge):

$$\text{Energy density} = \int_0^{\Delta t} IV(t)dt/wt = \int_0^{Q} V(q)dq/wt, \tag{2.5}$$

where wt is the weight of the cell [64].

From the above discussion, it is clear that an electrochemical cell can reach higher gravimetric specific energy (or volumetric energy density) by two approaches, increasing the cell voltage, or increasing the cell capacity, or both. When it comes to each electrode, it requires to use a cathode material with higher electrode potential versus Li^+/Li (or lower chemical potential μ_c) or an anode with lower electrode potential (or higher chemical potential μ_a), or at least one electrode is provided with a high specific capacity (or volumetric capacity density). To enable cell application at industrial level, other parameters, such as material/manufacturing cost, tap/packing density of the electrode material should also be taken into consideration. In the search of advanced electrode materials for high energy LIBs, cathode materials based on traditional intercalation chemistry (such as layered oxides and polyanionic compounds) continued to receive intensive attentions due to their stable cycling performance, high tap density and high working potential, which are beneficial for realizing practical LIBs with high energy output and long lifespan (Table 2.4) [60, 61, 64–66]. Especially, transition metal layered oxides that belong to the high-nickel-content $LiNi_{1-x-y}Co_xAl_yO_2$ (NCA) or $LiNi_{1-x-y}Co_xMn_yO_2$ (NCM) family are the most promising candidates to enable boosting on specific energy at cell level of rechargeable Li batteries to higher than 350 Wh kg^{-1} and up to 500 Wh kg^{-1} if a Li metal anode is paired, which is a desirable choice of the automotive industry for "near-future" electric vehicle generation [61]. Significant advancements in energy density at cell level and reductions in cost are also expected to come from new storage chemistries, such as conversion-type cathodes (e.g., NiF_2, FeF_3 and other transition metal fluorides) and Group VIA cathodes (O_2, S, Se and Te) (Table 2.4) [67–90]. For example, through a two-electron electrochemical redox process, S can react with Li in a stepwise fashion to generate lithium sulfide (Li_2S) by the end of discharge [72, 78]. A sulfur cathode with a typical starting composition of cyclo-S_8 molecules shows two plateaus on its discharge/charge profile, with a higher voltage (usually starts at ~2.35 V vs. Li^+/Li) corresponding to the conversion between S_8 and Li_2S_4 intermediate, and a lower one (usually starts at ~2.0 V vs. Li^+/Li) corresponding to the conversion between Li_2S_4 and Li_2S [32, 76]. In this way, S is provided with a significantly higher theoretical specific capacity (1675 mA h g^{-1}) than the conventional layered oxide cathodes based on the intercalation chemistry, which thus offsets the deficiency of low average working potential (*ca.* 2.1 V vs. Li^+/Li) and contributes to an attractively high specific energy (*ca.* 2600 Wh kg^{-1}) at cell level if paired with a Li metal anode [51, 82].

Table 2.4 Electrochemical-storage-related features of some typical cathode materials

Material	Specific Capacity (mA h g^{-1})	Operation Voltage (V vs. Li^+/Li)	Comments
$LiCoO_2$	155	3.9[a]	The most common cathode material, though limited by the potential biohazard and high price of Co
$LiMn_2O_4$	120	4.05[a]	Low price, safer electrochemistry, poor high temperature stability
$LiNi_{0.5}Mn_{1.5}O_4$	130	4.6[a]	Requires working with an electrolyte that is stable at a high voltage
$LiFePO_4$	160	3.45[a]	Low price and safe cathode material, but low in volumetric energy
$LiNi_{1-x-y}Co_xMn_yO_2$	200 ($LiNi_{0.8}Co_{0.1}Mn_{0.1}O_2$)	3.8 ($LiNi_{0.8}Co_{0.1}Mn_{0.1}O_2$)[a]	Strong competitors for $LiCoO_2$. Possess an equivalent energy density to $LiCoO_2$, yet are much safer and less expensive
$LiNi_{1-x-y}Co_xAl_yO_2$	200 ($LiNi_{0.8}Co_{0.15}Al_{0.05}O_2$)	3.73 ($LiNi_{0.8}Co_{0.15}Al_{0.05}O_2$)[a]	
CuF_2	528[b]	3.55[b]	Conversion-type cathodes. High in capacity but poor in electrochemical stability
NiF_2	554[b]	2.96[b]	
FeF_3	712[b]	2.75	
O_2	3350[b] (in the case of generating Li_2O)	2.9	Very low price. Rich resource abundance. Extremely high specific capacity. Poor reversibility of electrochemical reaction vs. Li (poor capacity retention and large voltage polarization)
S	1675[b]	2.2	Low price and rich abundance. High in specific capacity and volumetric capacity density (3461 mAh cm^{-3}) but poor in electrochemical stability

(continued)

Table 2.4 (continued)

Material	Specific Capacity (mA h g^{-1})	Operation Voltage (V vs. Li^+/Li)	Comments
Se	679[b]	2.1	High in volumetric capacity density (3265 mAh cm^{-3}). Poor electrochemical stability. Toxic
Te	420[b]	1.8	High in volumetric capacity density (2622 mAh cm^{-3}). Poor electrochemical stability. Improved conductivity. Toxic
I_2	211[b]	3.6	
Br_2	335[b]	3.9	Liquid cathode. High voltage. Toxic

Data obtained from Refs. [66, 71]
Note [a]Midpoint voltage obtained at a rate of 1/20 C
[b]Theoretical value

On the anode side, the intercalation chemistry has been dominated by graphite and its derivatives, and opportunities are found with the materials operated by new storage mechanism, such as interfacial/nanopore-storage-type porous carbons, alloy-type Si, Sn, Ge, P, Al and their derivatives, conversion-type transition metal oxides and sulfides, and the plating-/stripping-type Li metal [3, 17, 75, 91–120]. For example, silicon (Si) has shown its promises in the past several years as a high-capacity anode material for the next-generation high-energy LIBs [121]. By forming a $Li_{4.4}Si$ alloy with Li, each Si atom is able to host up to 4.4 Li ions by the end of discharge, which thus leads to a high theoretical specific capacity (*ca.* 4200 mA h g^{-1}) that is 10 times higher than that of commercial graphite (372 mA h g^{-1}). In addition, Si shows a suitable (de)lithiation potential (*ca.* 0.4 V vs. Li^+/Li) that is compatible with graphitized carbon and a high element abundance (second richest in earth crust) to ensure its resource sustainability for energy storage [2, 12, 121, 122]. Therefore, substitution (or partial substitution) of graphite by silicon or its derivatives (e.g., silicon monoxide) in the anode is an effective approach to increase the energy output of LIBs [107, 123, 124].

Another option for the “beyond-lithium” technology is the use of Li metal anode. Operated based on simple plating/stripping electrochemistry, Li is able to provide the highest theoretical specific capacity (3860 mA h g^{-1}) and the lowest electrode potential (−3.04 V versus NHE, normal hydrogen electrode) among the metal anodes [101–103, 105, 106, 120]. With the distinct advantages, Li metal holds the highest possible energy density as an anode for rechargeable Li batteries, and has thereby been regarded as the “holy grail” material for pushing the cell-level energy output of rechargeable Li battery to > 500 Wh kg^{-1} in the near future [59, 61, 123]. It is also the ultimate anode choice for the next-generation solid-state Li metal batteries, Li-S batteries and Li-air batteries [71, 78, 104, 109, 118, 125].

Electrode materials realized by the new storage mechanisms are promising. However, the practical use of non-intercalation electrode materials in RLCs is still premature, and many challenges, such as slow diffusion of charge carriers (Li^+/e^-), drastic volume variation during (de)lithiation, unstable evolution of bulk structure upon cycling, and poor thermodynamic stability at electrode|electrolyte interface, remain to be addressed. In the following chapters, we will discuss in details about the fundamental chemistry of the "beyond-lithium" storage systems and strategies to stabilize the chemistry.

2.1.2.3 Storage of Charges in a Post-lithium Rechargeable Cell

Although RLCs have proven their tremendous success in the industry, the future of Li-based storage technology is limited by its low reserve (1.6×10^7 tons, 20 ppm) and inhomogeneous distribution in the earth crust. Besides, the demand of Li will continue to raise due to an increasing global consumption, which inevitably raises the cost of Li-containing products and brings about economical and geopolitical concerns. The situation gets worse with the anode use of Li metal as the cost of Li foil is largely dependent on its thickness [126, 127]. Although much efforts have been made on optimizing the utilization of Li, researchers are still looking for alternative elements that are abundant in the earth crust [128].

Sodium, potassium, magnesium, calcium, as well as aluminum, are among the highest abundant elements in the earth crust (Fig. 2.5). For instance, the amount of Na (23,000 ppm) in the earth crust is 1150 times more than Li, which leads to a significantly reduced price (by about 50 times) in material. Aluminum is the fifth most abundant element in the earth crust (82,000 ppm) and is widely distributed, and the price of Al metal is only 1/150 of Li metal [128]. In addition to high resource abundance, these metals also have impressive electrochemical parameters to support their anode use in rechargeable cells (Fig. 2.5) [128]. For example, Na metal has been reported to be a promising anode material with a much larger specific capacity of 1166 mA h g^{-1} and a low redox potential of −2.71 V versus NHE (*ca.* 0.3 V higher than that of the Li anode), so that sodium-based rechargeable cells may show competitive energy output to that of the RLCs [117, 127–140]. Potassium metal also has a large theoretical specific capacity of 685 mA h g^{-1} and an even lower redox potential of −2.931 V versus NHE that approaches the potential of the Li anode, making it another appealing anode choice for building high-energy rechargeable cells [141–144]. The electrochemical stripping and plating of magnesium is a two-electron process, which results in a high theoretical volumetric capacity density of 3832 mA h cm^{-3} (or a specific capacity of 2205 mA h g^{-1}) to make it potentially useful in storage devices with a compact design [145–148]. In addition, magnesium involves with less safety problems (a dendrite-free anode) and is relatively stable in air [147]. Calcium shows many similarities to magnesium. Both of them are abundant in the earth crust and relatively stable in air and both involve with two-

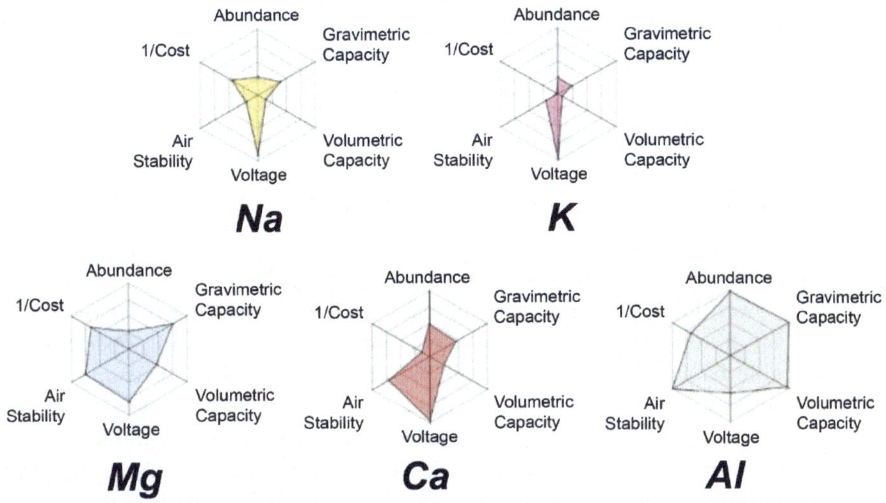

Fig. 2.5 Comparison of storage-related properties among the earth-abundant metal anodes. Reproduced with the permission from Ref. [128], Copyright © 2018 John Wiley & Sons, Inc

electron transfer electrochemistry. Due to a larger atomic weight, calcium has a lower specific capacity (1337 mA h g^{-1}) compared to magnesium, yet its much lower redox potential (−2.87 V vs. NHE, which is close to the redox potentials of Li and K) than magnesium (−2.37 V vs. NHE) offsets the deficiency in capacity and make it an promising choice to pair with low-voltage and high-capacity cathode materials such as sulfur and oxygen [149–151]. Aluminum, through a three-electron redox process, is able to deliver a high specific capacity (2978 mA h g^{-1}) comparable to that of Li, and a volumetric capacity density (8040 mA h cm^{-3}) four times larger than that of Li (2061 mA h cm^{-3}) due to its higher density [152]. Admittedly, the relatively high redox potential of Al (−1.662 V vs. NHE) makes it hard to operate an Al-based full cell at high voltage (for example, the theoretical voltage output of S-cathode-based rechargeable cells drops from 2.63 V to 1.27 V when replacing the Li metal by Al metal in the anode, the ultrahigh volumetric capacity density, ultralow price and air/moisture-stable feature of Al metal still offer opportunities to realize a high-energy-density, low-cost, and safe energy storage system [153, 154].

Although rechargeable cells based on these earth-abundant metal anodes show prospects in the future-oriented energy storage, the use of these metal anodes is accompanied with various problems. The relevant research is too preliminary to support realization of practically operable devices at present, and only a few reports on K, Ca, and Mg metal anodes can demonstrate satisfactory performance in laboratory-level experiments [128]. The problems on alkali metal anodes (Na, K) include hazardous dendrite formation/growth that could lead to electrochemically "dead" metal on the anode or short circuit of the cells, and formation of unstable solid electrolyte interphase (SEI) that accounts for low Coulombic efficiency and continuous electrolyte consumption. For alkali earth metal anode (Ca, Mg, Al), the common difficulty lies in surface passivation due to formation of cation-insulating SEI upon use of carbonate electrolyte, which results in fast performance fading of these anodes [128].

2.2 Transfer of Charges in a Rechargeable Cell

Electrochemical energy can be stored in a rechargeable battery and released in term of electric power through electrochemical reactions. The charged particles (ions and electrons) are transported between the anode and the cathode during operation of a rechargeable battery in repeated charge and discharge cycles. For example, the discharge process in a LIB includes the following steps: (i) simultaneous separation of Li ions and electrons in the anode and solid-state diffusion of them towards opposite directions; (ii) motion of Li ions to the anode|electrolyte interface and diffusion of Li ions through the electrolyte; (iii) migration of electrons from the anode to the current collector and further to external circuit to power a device; (iv) simultaneous solid-state diffusion and combination of Li ions and electrons in the cathode (Fig. 2.6). Because the electrochemical reactions in a rechargeable battery proceed through changing the oxidation states of the anode and the cathode, the feasibility to transfer electrons and ions in the electrodes and electrolytes determines the performance of the battery. The transport mechanisms of ions and electrons in battery systems will be discussed in the following sections.

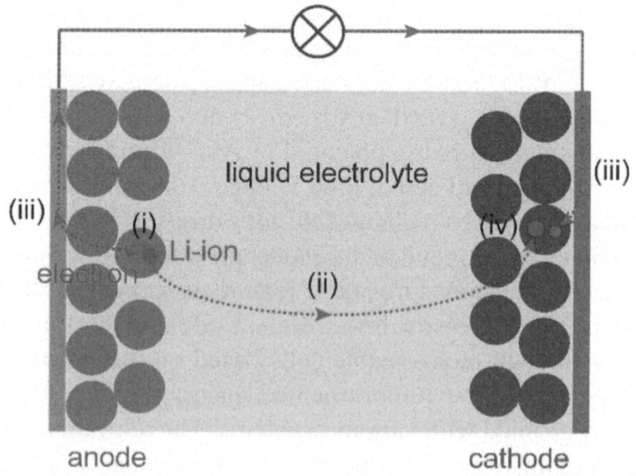

Fig. 2.6 Transport of Li ions and electrons during the discharge process of a LIB. Reproduced with permission from Ref. [155]. Copyright © 2015 Royal Society of Chemistry

2.2.1 Transfer of Charges in Solid Electrode Material

The specific capacity and energy density of a rechargeable battery are largely determined by the cathode materials that have framework structures to allow reversible insertion and extraction of guest ions. Li ion is the guest ion in a rechargeable LIB and the typical cathode hosts include layered (e.g., $LiCoO_2$) and spinel (e.g., $LiMn_2O_4$) transition-metal oxides and polyanionic compounds (e.g., $LiFePO_4$). The migration of guest ions in insertion cathodes proceeds through a solid-state diffusion process, and ion diffusivity (D_i) is a parameter that characterizes the ability of ion motion in the electrodes. A strong dependence of ion diffusivity on temperature is generally observed in solid electrode materials according to the Arrhenius equation [156]:

$$D_i = D_0 \exp\left(-\frac{\Delta G}{k_B T}\right), \tag{2.6}$$

where ΔG is the energy barrier for ion diffusion; k_B is the Boltzmann constant; and D_0 is the empirical prefactor. The diffusion time (τ) of guest ions in solid electrode materials can be expressed as follows [6, 157],

$$\tau = \lambda^2 / D_i \tag{2.7}$$

Therefore, the diffusion time can be shortened by increasing the ion diffusivity (D_i) in the electrodes, and/or reducing the diffusion length (λ) through designing small or nanosized electrode materials. The ion diffusivity is exponentially inversely proportional to the energy barrier ΔG, and consequently, a slight decrease of energy barrier can significantly increase the ion diffusivity. Although elevating the temperature (T) will also increase the ion diffusivity, operation of a battery at high temperature poses safety concerns.

The solid-state diffusion of guest ions in the electrode during charge-discharge cycles is accompanied by migration of electrons. Electronic conductivity (σ_e) is a parameter to characterize the ability of electron motion in the electrodes. Most of the cathode materials in LIBs exhibit a semiconducting feature with a gap between the valence band (VB) and the conduction band (CB), and the electronic conductivity is originated from the band structures. Both electrons and holes can contribute to electronic conduction, and the electronic conductivity can be expressed as follows [158],

$$\sigma_e = n_i e \mu_e + p_i e \mu_h, \tag{2.8}$$

where n_i and p_i are the concentrations of electrons and holes, respectively; μ_e and μ_h are the mobilities of electrons and holes, respectively. For the intrinsic semiconducting electrode materials, excitation of electrons in the VB near the Fermi level into the CB accounts for the electronic conductivity. The excitation of every single electron generates a hole in the VB, and therefore, the concentration of electrons is equal to that of holes in the intrinsic semiconducting electrode materials. For extrinsic semiconducting electrode materials, doping of the materials introduces different concentrations of holes and electrons, and movement of electrons or holes contribute to electronic conduction in n-type or p-type electrode materials, respectively.

Selection of electrode materials with high electronic conductivity is essential to improve the rate performances of LIBs. Different material design strategies, such as doping and addition of conductive components, are also important to increase the electronic conductivity of the electrode materials. For example, the introduction of extrinsic charge carriers through doping can increase dramatically the electronic conductivity, and the addition of conductive components, such as carbon coating, can promote significantly the diffusion of electrons in the electrodes [140]. Moreover, the ionic conductivity of electrode materials should also be considered because both the electrons and the guest ions participate in the electrochemical reactions of two electrodes during charge and discharge cycles. For example, graphite anode in LIBs with zero bandgap exhibits high electronic conductivity, but its high-rate performance is limited by a low ionic conductivity (10^{-10}–10^{-8} S cm^{-1}) [159].

2.2.2 *Transfer of Charges in a Liquid Electrolyte*

The main function of an electrolyte in a rechargeable battery is to transport guest ions and to insulate electrons between the anode and the cathode during charge and discharge cycles. The electrolyte in a LIB belongs to the electrochemically inactive component; however, its ion conductivity and its chemical reactivity with the electrodes with the formation of SEI has a significant influence on the rate capability, cycling stability, and safety property of LIB. [160] Electrolytes for LIBs typically include inorganic solid electrolytes, polymer solid electrolytes, organic liquid electrolytes, aqueous liquid electrolytes, ionic liquid electrolytes, and composite electrolytes. Organic liquid electrolytes comprising lithium salts dissolved in a mixture of organic solvents are commonly used in commercial LIBs at present.

The fundamental requirement of an electrolyte in a LIB is to have a high Li^+ conductivity (σ_{Li}) to facilitate the transportation of Li ions and a low electronic conductivity to insulate the electrons. A conductivity at the order of 10^{-3} S cm^{-1} at room temperature is generally required for a LIB electrolyte to achieve an acceptable rate performance. The Li^+ conductivity in a liquid electrolyte can be expressed as follows [161]:

$$\sigma_{\mathrm{Li}} = \frac{q^2 c}{k_B T} D_L, \tag{2.9}$$

where D_L is the diffusivity of Li ion in liquid electrolyte, q is the charge of Li ion, and c is the concentration of Li ion. The diffusivity is related to the mobility of Li ion in a liquid electrolyte according to the Stokes-Einstein equation [162]:

$$D_L = uk_BT = \frac{k_BT}{6\pi\mu R_0}, \tag{2.10}$$

where u is the mobility of Li ion, μ is the viscosity of the solution, and R_0 is the radius of Li ion. Therefore, the selection of electrolyte solvents with high solubility and low viscosity is preferred in order to achieve a high conductivity of Li ions. Profited from these merits, high conductivity is generally obtained in liquid electrolytes with organic carbonates as the solvents (10^{-1} S cm^{-1}) and polymer electrolytes gelled with organic plasticizers (10^{-2} S cm^{-1}).

The total ionic conductivity of a liquid electrolyte can be easily measured; however, the contribution of a specific species of interest (e.g., Li ions in the LIB) to the total ionic conductivity is not readily to obtain. The Li^+ transference number (t_{Li}) that quantifies the contribution of Li^+ conductivity to the total ionic conductivity in a liquid electrolyte is defined as [163]:

$$t_{Li} = \frac{u_{Li}}{\Sigma u_i} \tag{2.11}$$

In a liquid electrolyte, a more stable and larger solvation shell can be formed around the Li ions, whereas anions with smaller charge/radius ratios are weakly solvated and exhibit higher mobility than the Li ions. Therefore, organic liquid electrolytes usually exhibit a smaller Li^+ transference number (0.2–0.4) than the anion transference number (0.6–0.8); and the anions are usually more efficiently transported in the organic liquid electrolytes to induce concentration polarization of the battery especially at high charge/discharge current densities. Because of the larger ionic radius, the weaker solvation shell of Na ion might be an advantage to have a higher Na^+ transference number than that of the Li ion.

2.2.3 Transfer of Charges in a Solid Electrolyte

The transport mechanism of mobile ions in solid-state electrolytes (SSEs), including polymer electrolytes and inorganic solid electrolytes, is fundamentally different from that in liquid electrolytes. In liquid electrolytes, the transport of ions is accompanied by the movement of solvation shells in solvent medium; and the nature of the solvents, the salt and its concentration determine the ionic conductivity of a liquid electrolyte. On the other hand, ion transport in SSEs is related to the crystalline structure or local environment of the framework; and the electrostatic force and other interactions with the lattice are needed to overcome. The ionic conductivity of an SSE is typically much lower than that of a liquid electrolyte at room temperature. For example, the ionic conductivity of a polymer electrolyte based on poly(ethylene oxide) is ranging from 10^{-6} to 10^{-4} S cm^{-1} at room temperature, and the ionic conductivity of an inorganic solid electrolyte is usually lower than 10^{-3} S cm^{-1} [164].

2.2.3.1 Transfer of Charges in a Solid Crystalline Electrolyte

Solid crystalline electrolytes (SCEs) are normally composed of coordinated polyhedrons and a spatial distribution of mobile ions. A framework structure is formed by the coordinated polyhedrons in a solid electrolyte with structural defects. In SCEs, the ionic conduction usually depends on the amounts of mobile ions and structural defects, and more structural defects are therefore required on the equivalent (or nearly equivalent) sites for mobile ions to occupy than the number of mobile ions in the framework structure [165]. The structural defects commonly include interstitial ions and vacancies (missing ions) that can be generated by substitution (Fig. 2.7). The diffusion of mobile ions in a SCE can proceed through random hopping between vacancies, direct hopping between interstitial sites, and exchange between interstitial sites. The ionic conductivity of an inorganic SCE typically follows the Arrhenius equation [166]:

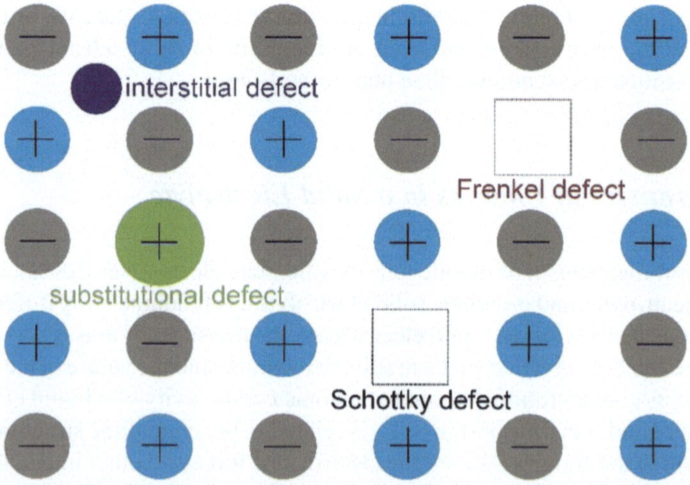

Fig. 2.7 Schematic illustration showing the structural defects in a solid crystalline electrolyte. Reproduced with permission from Ref. [167]. Copyright © 2018 John Wiley & Sons, Inc

$$\sigma = \sigma_0 \exp\left(-\frac{E_a}{k_B T}\right) \tag{2.12}$$

where σ is the ionic conductivity, σ_0 is the pre-exponential factor, E_a is the activation energy of ion diffusion, k_B is the Boltzmann constant, and T is the absolute temperature. The activation energy corresponds to the energy needed for the mobile ions to transport across the bottlenecks in the framework structure; and the temperature and pre-exponential factor also have a significant influence on the ionic conductivity.

Because of strong electrostatic interactions between the mobile cations and the framework skeleton formed by polyanions, mobile ions with higher valence (e.g., Mg^{2+}, Ca^{2+} and Al^{3+}) usually exhibit a lower ionic conductivity and diffusivity than the monovalent ions (e.g., Li^+, Na^+ and K^+) [168]. The general approaches to increase the ionic conductivity include increasing the concentration of mobile ions, reducing the steric restrictions, and constructing continuous diffusion channels in the framework structure. The activation energy can be reduced by increasing the lattice volume of the crystalline framework, which leads to an increased ionic conductivity [169]. Substitution by different cations has proven to be an effective approach to tune the lattice volume of crystal structure and to modify the bottleneck size for diffusion of mobile ions; and the ionic conductivity of a solid crystalline electrolyte can be increased by substitution with aliovalent cations.

2.2.3.2 Transfer of Charges in an Amorphous Glass Electrolyte

Different from the long-range ordered structure of solid crystalline electrolytes, amorphous glass electrolytes have short-range or medium-range ordered structure to contribute to the ionic conductivity. The excitation of ordered lattice mobile ions at local sites to neighboring sites and the collective diffusion of mobile ions induce transport of mobile ions in glass electrolytes [170]. The presence of abundant defects in amorphous glass electrolytes and the interaction between mobile ions and structural skeleton determines the transport capability of mobile ions. The following criteria should be generally considered in designing a glass electrolyte with high conductivity: (i) the availability of suitable transport channels for mobile ions in the disordered sub-lattices; (ii) the presence of interconnected vacancies and interstitial sites with low activation energy to allow continuous diffusion of mobile ions; and (iii) the weak interactions between mobile ions and skeleton of framework.

Increasing attention has been recently focused on the amorphous glass electrolytes, because of their advantages over solid crystalline electrolytes, including high ionic conductivity, low grain-boundary resistance, and feasibility to prepare them into thin-film. Oxides and sulfides are the two main categories of glass electrolytes that conduct Li ions. Most of the oxide glass electrolytes have a low Li^+ conductivity at room temperature for practical applications, usually in the range of 10^{-8}–10^{-6} S cm^{-1}; however, some oxide glass electrolytes with a high Li^+ concentration such as lithium ortho-oxosalts exhibit a relatively high ion conductivity (10^{-6} S cm^{-1}) [171]. The same level of ion conductivity is also realized for the amorphous lithium phosphorous oxynitride in thin-film batteries where the resistance is minimized because of the reduced thickness of the solid electrolyte [172]. Because of the high polarizability of sulfur ions, the sulfide glass electrolytes possess high Li^+ conductivities of 10^{-5}–10^{-3} S cm^{-1} at room temperature, such as Li_2S-SiS_2 and Li_2S-P_2S_5, and the exploration of solid-state lithium batteries with sulfide electrolytes is actively undertaken at present [173].

2.2.3.3 Transfer of Charges in a Polymer Electrolyte

In solid polymer electrolytes, the mobile ions coordinate with the polar groups (e.g., -O-, -O-CO-, -F, and -CN) of polymer chains. The segment movement of polymer chains promotes the continuous transportation of mobile ions from one coordination site to another. Most of the solid polymer electrolytes exhibit a low ionic conductivity at room temperature because of the restricted movement of polymer segments. The complexity of solid polymer electrolyte systems and the lack of simple correlations between structure and properties prevent the generation of physical models that comprehensively describe ion transport in such systems. Nevertheless, the ionic conductivity of solid polymer electrolytes can be experimentally modeled by the Arrhenius equation or the Vogel-Tammann-Fulcher (VTF) equation or a combination of them. In most cases, a solid polymer electrolyte is more suitably fitted by the VTF equation [174]:

$$\sigma_i = \sigma_0 T^{-1/2} \exp\left(-\frac{B}{T - T_0}\right), \tag{2.13}$$

where B is the pseudo-activation energy of ion conductivity (expressed in units of E_a/k_B), and T_0 is the reference temperature which is usually 10–50 K below the experimental (kinetic) glass transition temperature (T_g) of the solid polymer electrolyte. The VTF behavior is correlated with the transport of mobile ions and long-range segment movements of the polymer chains. The parameters in the VTF equation could be obtained by fitting the ionic conductivity against temperature according to the linearized equation [158]:

$$\log_{10}\left(\sigma_i T^{1/2}\right) = \log_{10}\sigma_0 - 0.43\frac{E_a}{k_B(T - T_0)} \tag{2.14}$$

The ionic conductivity of a solid polymer electrolyte can be improved by decreasing the crystallinity of polymer and increasing the concentration of dissociated mobile ions. The construction of solid polymer electrolytes with nanomaterials is also effective to increase the ionic conductivity. Distribution of nanometer-sized particles into the polymer matrix is a general approach to obtain composite polymer electrolytes. The additives of nanoparticles can be classified as active or passive in terms of their intrinsic ionic conductivity. The active additives (e.g., $Li_{1.5}Al_{0.5}Ti_{1.5}(PO_4)_3$ and $Na_3Zr_2(SiO_4)(PO_4)$) exhibit intrinsic ionic conductivity, leading to complex ion conduction behaviors at different weight ratios [175]. The addition of passive nanoparticles (e.g., Al_2O_3, SiO_2, and TiO_2) can also increase the ionic conductivity by one order of magnitude because of the Lewis acid-base interaction between the surface of nanoparticles and the lithium salt or polymer segments [176]. The addition of active or passive nanoparticles may also reduce the crystallinity and glass-transition temperature of solid polymer electrolyte, coherently contributing to increased ionic conductivity.

2.2.4 Transfer of Charges Across Interface Between Electrode and Electrolyte

2.2.4.1 Transfer of Charges Across a Solid-Liquid Interface

Liquid electrolytes consisting of organic solvents and Li salts are widely used in commercial LIBs. The electrochemical performance of a LIB is influenced by the (electro)chemical interface between solid electrodes and liquid electrolytes. In principle, both electrodes can form SEI during their contact with liquid electrolyte. To maintain stable operation of the cell, the SEI should have an electrochemically stable structure to conduct Li ions and insulate electrons.

The interfacial stability between graphite anode and organic liquid electrolyte can be tuned by solvents, additives, and lithium salts. The reactivity between a liquid electrolyte and a discharged graphite anode determines the minimum temperature where exothermic reaction occurs; therefore, the thermal stability of a new electrolyte component would be evaluated by its ability to postpone the onset temperature. The replacement of conventional lithium salt $LiPF_6$ by lithium bis(oxalato)borate (LiBOB) can delay the onset temperature of electrolyte decomposition by 90 °C [177]. The interface on the graphite anode is also affected by the chemical structure and concentration of electrolyte additives. For example, the interface with $LiPF_6$-based electrolyte can be stabilized by the addition of vinylene carbonate (VC) and a low concentration (1%) of dimethylacetamide (DMAc); however, a higher concentration of DMAc (10%) results in a more serious decomposition of the electrolyte [178, 179]. The electrolyte additives exhibit different working mechanism: a polymeric interface could be formed with the VC additive to provide a good protection of anode surface, while the aggressive attack of PF_6^- anion and its degradation products by carbonate solvents can be neutralized by Lewis-base additive of DMAc.

The chemical reactivity between a cathode and an organic liquid electrolyte is usually the critical source that leads to heat generation and thermal runaway in a LIB. The interfacial reactions on cathode side has proven to be more challenging to mitigate through developing new electrolyte components than the corresponding reactions at anode side, since the change of electrolyte compositions and the interfacial species would hardly prevent the electrochemical oxidation of liquid electrolyte and other components in the cathode, such as polymer binder, especially at a high voltage. For example, the thermal stability of graphite anode can be enhanced by using the LiBOB salt, while the reaction of LiBOB with the 4.0-V cathode materials, such as $LiCoO_2$ and $LiNi_{1-x-y}Co_xMn_yO_2$, deteriorated the electrochemical performance in comparison with the baseline electrolyte having $LiPF_6$ as the lithium salt [180]. Another example is that the dissolution of Mn from cathode could be hardly prevented by using electrolyte additives such as VC and fluoroethylene carbonate, but the deposition of Mn on the anode surface is minimized with the electrolyte additive of VC [181]. However, an exception is that the Lewis-base additive of DMAc that was proven to be effective on graphite anode could also protect the surface of cathode materials from corrosion, for example, $LiCoO_2$, $LiMn_2O_4$, and $LiNi_{0.8}Co_{0.2}O_2$ [182].

The optimization of electrolyte is much more effective to improve the stability of LIBs with cathodes having lower voltage outputs (and lower energy densities) as evidenced in the works about $LiFePO_4$ cathode. For example, the cycling stability of $LiFePO_4$ cells at elevated temperatures could be improved by adding tris(pentafluorophenyl)borane into the electrolyte [183], and the thermal stability of LIBs with $LiFePO_4$ cathode could also be increased by adding LiBOB [184]. However, the intrinsic safety and slow degradation of the $LiFePO_4$ cathode with a lower working voltage makes the role of electrolyte optimization and interface modification in stabilizing the $LiFePO_4$ cathode less significant.

2.2.4.2 Transfer of Charges Across a Solid-Solid Interface

Solid electrolytes are actively explored in Li-metal batteries with the aim of increasing the energy density through using metallic Li as the anode. However, because of the repeated volume change of Li metal in plating and stripping cycles, a stable interface could be hardly maintained between a solid electrolyte and a lithium-metal anode. An ideal interface between a solid electrolyte and a Li-metal anode should be highly ionically conductive to transfer Li ions at a low impedance and electronically insulating to prevent further side reactions. If a mixed conducting interface is formed that can still conduct electrons, the reaction between Li-metal anode and solid electrolyte will continuously proceed and eventually lead to short-circuit of the battery. If the interface is only ionically conductive, the electron transfer is prevented and future side reactions can be stopped to maintain a stable interface between the Li-metal anode and the solid electrolyte [185]. Therefore, the interfacial reactions between a Li-metal anode and a solid electrolyte are acceptable if the reactions lead to formation of an interface that is ionically conductive and electronically insulating to prevent the further side reactions.

The reactions between Li-metal anode and highly conductive sulfide solid electrolytes, such as $Li_{10}GeP_2S_{12}$, $Li_{10}SiP_2S_{12}$ and $Li_{10}Si_{0.3}Sn_{0.7}P_2S_{12}$, generally lead to the formation of interfacial film with mixed ionic and electronic conduction ability [186]. A continuous decomposition of sulfide solid electrolyte accompanied by proliferation of interfacial film as a result of crossover of electrons through the mixed conducting interface would increase the interfacial resistance of battery during long-term cycling. Interfacial coating and element doping are promising strategies to relieve the continuous interfacial reactions of sulfide solid electrolyte against a Li-metal anode [187]. Some oxide solid electrolytes, such as the garnet-type solid electrolytes, are relatively stable against Li-metal; however, the formation of lithium dendrites is observed in the holes, grain boundaries, and defect sites of the solid-electrolyte pellet [188]. In contrast, the growth of lithium dendrites is suppressed by the glass solid-electrolyte of Li_2O-Al_2O_3-P_2O_5-TiO_2-GeO_2, as a result of the formation of a stable interface to provide a soft contact between the solid electrolyte and Li-metal. In contrast, the formation of a rigid interface, as in the case of $Li_{6.1}Ga_{0.3}La_3Zr_2O_{12}$, will increase the interfacial resistance and promote the growth of lithium dendrites to terminate charge/discharge cycling of the battery.

In addition to the large interfacial impedance and short cycling life, potential safety concerns also arise as a result of the unstable interface between Li-metal anode and some oxide solid electrolytes. Molten Li-metal reacts drastically at 200 °C with a sintered solid electrolyte pellet of $Li_{1.5}Al_{0.5}Ge_{1.5}(PO_4)_3$ in an argon-filled glove box with H_2O and O_2 contents below 0.1 ppm, and the rapid reaction at high temperatures leads to uncontrolled degradation of the interface with rapid emission of O_2 gas and strong thermal runaway [189]. The formation of a mixed conducting interface could account for the continuous reactions between solid electrolyte and Li-metal, and safety hazards arise as a result of the rigorous thermal runaway in solid-state Li-metal batteries. Therefore, the release of oxygen from solid electrolytes under some extreme circumstances (such as thermal abuse) would accelerate the interfacial reactions and lead to explosion. The stability of oxygen with the solid electrolytes and their reactions on battery safety should be evaluated for the design of solid electrolytes.

On the cathode side, a space charge layer is often formed against the solid electrolyte; and the space charge layer shows a great impact on the properties of cathode|electrolyte interface and plays an important role in determining the cycle life of the battery. The interface between a $LiCoO_2$ cathode and a sulfide solid electrolyte β-Li_3PS_4 has a disordered structure and adsorption sites to attract Li ions. The Li ions in the subsurface region of the solid electrolyte begin to transfer even at an under-potential condition, leading to increment of interfacial resistance [190]. The space charge layer can be modified at atomic scale to improve the performance of the battery. The introduction of a $LiNbO_3$ buffer layer between the $LiCoO_2$ cathode and the β-Li_3PS_4 electrolyte suppresses the formation of space charge layer and provides a smooth transport pathway for Li ions to eliminate the Li^+ adsorption sites for both the cathode and the electrolyte.

2.3 Fundamentals About Use of Nanostructures and Nanomaterials in a Rechargeable Cell

The shift from bulk-sized electrode materials to nano-sized counterparts has witnessed tremendous progress in the development of lithium-based rechargeable cells in recent decades. Nanostructured material has emerged as a promising solution to tackle fundamental problems in conventional battery materials, while also enabled discoveries of new electrode materials and new storage mechanisms that lead to substantially improved battery performance. However, the use of nanomaterials in rechargeable cells also raises problems and challenges. In this section, we will discuss the fundamentals of "nano" effects, both positive and negative, on the charge storage and transfer properties of a rechargeable cell. Insights about general design principles for "kinetically stable" nanostructured electrode materials towards their practical applications in future electrochemical cells will also be presented.

2.3.1 *"Nano" Effect on the General Storage Properties of a Rechargeable Cell*

The specific energy of an electrochemical cell is defined as $Q_{\text{cell}} \times V_{\text{oc}}$, where Q_{cell} is the capacity of reversible charge transfer per unit weight (unit: mA h g^{-1}) between the two electrodes, and V_{oc} is the open-circuit voltage. In practical operation, Q_{cell} is determined by the ionic current density in the cell, which is further controlled by the rate of ion migration in bulk electrode/electrolyte and ion transfer across the electrode|electrolyte interface [65]. Compared to electron transfer in the external circuit, the ion migrations in the bulk electrode and across the electrode|electrolyte interface are usually much slower [65]. Therefore, when the cell is operated at high current densities, the ion motion within the electrode and across the interface is too slow for the charge distribution to reach its equilibrium state, which accounts for reversible capacity decay upon increasing the current density and capacity loss recovery upon reducing the discharge/charge rate [65]. To prevent the reversible capacity loss during high-power operation of electrochemical cells, the electrodes and electrolyte are usually provided with large surface area and small thickness.

Take rechargeable Li-ion cell as an example. The first-generation LIBs have electrodes that are composed of powders with particle dimensions at millimeter scale, and the electrolyte is trapped among the active particles and the pores of a separator [8]. Although the cell is provided with a high energy density, the high-power operation of the device is limited by the intrinsic diffusivity of Li^+ in the intercalation hosts (*ca.* 10^{-8} cm^2 s^{-1} or lower), which limits the rate of intercalation/deintercalation reaction and therefore, leads to low specific capacity at high current densities [6, 8]. In order to be used in electric vehicles, the LIBs should be able to work at higher discharge/charge rates. Nanostructured materials including nanostructures and nanomaterials, with their physical and chemical properties that are significantly different from the bulk materials, have shown their potential to influence the practical performance of the LIBs [2, 3, 6, 8, 12]. The positive effects of using nanostructured materials in LIBs are summarized as follows.

(1) Nanomaterials are usually provided small particle size in at least one dimension, which enables fast ion diffusion within the particles. As a result, nanomaterials may trigger much shortened time duration for ion intercalation into the bulk material (the time for intercalation decreases with the square of particle size, for example, a reduction of λ from 10 μm (typical size of commercial electrode materials) to 100 nm reduces, for a material with $D_i = 10^{-10}$ cm^2 s^{-1}, the value of τ from 5000 to 0.5 s according to Eq. 2.7) [6, 191]. The fast ion migration thus forms the main precondition that enables high-power operation of electrochemical cell. Moreover, it may enable the occurrence of electrode reactions that cannot take place on materials composed of micrometer-sized particles; for example, reversible Li intercalation into mesoporous β-MnO_2 without destroying of the rutile structure [192]. For very small particles, the chemical potentials for Li^+ and e^- may be altered, which could result in a change of electrode potential (thermodynamics of the reaction) [193].

(2) Nanomaterials usually show a high specific surface area and porous structure, which permits a high contact area with electrolyte and deep penetration of solvated working ions into bulk material, and hence, a high Li^+ flux across the interface to favor high-rate (dis)charge. The unique structural characteristics may also trigger new storage mechanisms such as storage through interfacial Faradaic reactions and storage in nanopores. For example, in transition metal oxides/fluorides/sulfides/nitrides that can undergo conversion reaction with Li, an extra Li storage capacity is observed in nanometer-sized particles at low potential which can be explained by an interfacial Li storage model [194, 195]. According to the model, Li ions are preferably stored on the ionic conducting side of the interface (Li_yX, where X = O, F, S, Se, N), while electrons (e^-) are localized on the metallic side (M = Fe, Co, Ni, Cr, Mn, Cu, etc.) to result in rapid charge separation. Provided with a large interface area of Li_yX/M, the storage mechanism may form a bridge between a supercapacitor and a battery electrode and offer a reasonable compromise between rate and capacity. In addition to interfacial storage, surface storage, which is energetically more favorable for nanometer-sized particles than for bulk insertion, may also contribute to the overall capacity of electrode materials [196]. For nanomaterials with a porous structure, there may be additional capacity contribution from the ion storage in nanopores. For example, it has been widely reported that Li can be reversibly stored in the sub-nano pores of pyrolytic carbons that are derived from organic matters including carbohydrates and phenolic resin [17, 114, 117, 127, 197].

(3) Nanostructured electrode materials may help to stabilize the electrode chemistry during the ion intercalation process [1, 198]. Electrode materials operated based on new storage mechanisms other than insertion chemistry, such as alloy-type anodes (e.g., Si, Sn, P, and Al), plating-/stripping-type Li metal anode, conversion-type anodes, and S (Li-S batteries) and O_2 (Li-air batteries) cath-

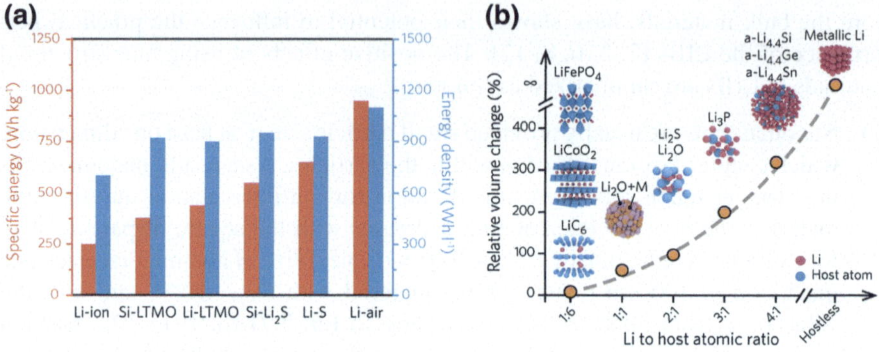

Fig. 2.8 **a** Practical specific energies and energy densities of state-of-the-art LIBs (Li-ion), and future-oriented RLCs based on new storage chemistries such as Si/Li metal anode vs. Li transition-metal oxide layered cathode, Si-Li_2S, Li-S, and Li-air. Casings, separators, and electrolytes are all considered, and Li metal cells are calculated on the basis of 100% excess Li. **b** Increased specific capacity is accompanied by significantly increased volume change, which could bring challenges to the practical applications of new battery chemistries. Reproduced with the permission from Ref. [2], Copyright © 2017 American Chemical Society

odes, show their promises to power the next-generation RLCs due to their ability to offer much higher Li storage capacity (Fig. 2.8a) [2, 3, 6, 12, 78, 103]. However, the increased capacity is often obtained at a cost of drastic lattice/volume changes and significant bond breaking of electrode materials, which undermine the practical electrochemical performance [2, 6, 12, 199]. Figure 2.8b plots the relative volume expansion versus the Li to host atomic ratio. Conventional graphite anode shows less than 10% volume change accompanied with a small Li host ratio (1:6) [2]. In contrast, the Si anode is able to accommodate up to 4.4 Li ions per single atom, yielding one of the highest known theoretical charge capacities (4200 mA h g^{-1}) yet also leading to > 300% expansion in volume by the end of lithiation [2, 12, 92, 93, 124, 199]. This drastic volume variation would result in cracking and pulverization of the electrode, and loss of electric contact of active material particles from the current collector, and finally lead to severe capacity fade and short battery lifetimes. Fortunately, the use of Si nanostructures helps to alleviate these issues. According to the work by Huggins and Nix, there is a critical particle size below which fracture of materials will not propagate further [200]. The critical size can be calculated as follows:

$$h_C = (23/\pi)(3k_{IC}/B\varepsilon_T)^2, \tag{2.15}$$

where h_C is the critical crack size, k_{IC} is the fracture roughness (unit: MPa $m^{0.5}$), B is the biaxial Young's modulus of the material, and ε_T is the total strain. For silicon, the fracture toughness is 10 MPa $m^{0.5}$, and the strain associated with the full uptake of Li by Si ($Li_{4.4}Si$) is approximately 400%. Using the above equation, the critical particle size to maintain the Si structure is estimated to be 75 nm [200]. Also, alloy particles within the nano-regime often display higher plasticity and deformability to accommodate a larger degree of volume variation. Experimentally, by tailoring the dimensions of Si particles to nanoscale, many works have demonstrated high reversible capacity and improved cyclability of Si anode [94]. The success in the fundamental storage chemistry of nano-Si composites paves the way for its commercial use in the next-generation LIBs [2, 12].

(4) By using the nanotechnology, it is feasible to control exposure of particular crystal facets of electrode material that is energetically preferred for Li storage [1]. These improved attributes can be ascribed to the unique surface properties of the exposed facets, which include the surface energy and diffusion barrier for Li^+. For example, Li insertion occurs faster on the (001) facet of anatase TiO_2 than on the (101) facet because of a higher charge transfer rate constant (10^{-8} vs. 2×10^{-9} cm s^{-1}) and chemical diffusion coefficient (7×10^{-14} vs. 2×10^{-13} cm^2 s^{-1}) for Li^+ insertion [201]. Therefore, exposure of the (001) facet can result in a lower energy barrier for Li^+ intercalation, leading to improved storage performance of TiO_2 anode. As a successful demonstration, Chen et al. synthesized hierarchically-structured spheres from ultrathin anatase TiO_2 nanosheets with nearly 100% exposed (001) facets [202, 203]. With the distinct structural merits, the TiO_2 anode shows favorable rate performance, as proved by a high reversible capacity of 95 mA h g^{-1} delivered at an ultrahigh discharge/charge rate of 20 C.

The nanotechnology is also a powerful method to generate surface-coating and functionalization layers on electrode materials, leading to improved interfacial chemistry. For example, by forming a conductive carbon coating (e.g., through pyrolysis of carbohydrates or graphene wrapping) on the surface of active particles, the interparticle electron transfer is favored and the active material is protected from unfavorable side reactions in the battery [17, 107, 108, 113, 204, 205]. Finally, nanotechnology gives people flexibility to engineer every single component within a battery (separator, current collector, etc.), bringing novel functions that are unachievable by the conventional methods [2].

However, nanomaterials are certainly not a panacea, and the battery use of nanomaterials has its own limitations. First, the nanoparticles may be more difficult to synthesize, which raises the production cost of materials and challenges on its mass production [8, 12].

Second, as the nanomaterials are provided with high specific surface area, an extra contribution form surface free energy should be considered for the chemical potential, which can be approximately given by,

$$\mu^0(r) = \mu^0(r = \infty) + 2(\gamma/r)V, \tag{2.16}$$

where γ is the effective surface tension, V is the partial molar volume, and r is the effective grain radius [6]. The effect of the excessive surface free energy $(2(\gamma/r)V)$

on the Li storage process of electrode materials is two sided. While it contributes to a high electroactivity toward Li storage, the excessive surface free energy also leads to disadvantages. The nanoparticles tend to form agglomerates, and therefore, it is difficult to disperse them and mix them with conductive additive and binder to prepare electrodes [6]. As a result, the electrode prepared barely with nanomaterials may show a significantly raised contact resistance, which reasonably explains the frequent occurrence of performance failure. During cycling of battery, electrochemical agglomeration of nanomaterials may happen, leading to aggravated performance fade [6]. The high surface area of nanoparticles also raises the risk of parasitic reactions between electrode and electrolyte, which leads to formation of thick SEI layer by consuming a massive amount of Li ions supplied from the cathode. The thick SEI may result in decreased conductivity at the interface, and the consumption of Li ions causes a high level of irreversibility of electrode reaction and thus, a poor cycling performance. Moreover, for transition metal oxide anodes that undergo conversion reaction, the SEI layered formed on the nanoparticles during Li uptake may disappear completely due to catalysis by transition metals upon Li extraction, which may bring about capacity fade and safety problem [6]. In the next section, we will discuss in details the formation chemistry of SEI and the strategies for its stabilization in rechargeable cells especially those involved with nanomaterials with high storage capacity.

Finally, a decrease of particle dimension from micrometer scale to nanometer scale usually results in a powders of low tap density and consequently a low volumetric capacity of an electrode [6, 8, 12]. It would inevitably lower the energy density of practical batteries.

2.3.2 "Nano" Effect on the Charge Transfer Across Electrode|Electrolyte Interface

In the rechargeable Li batteries, the chemical reactions at the electrode|electrolyte interface result in irreversible formation of a passivating layer on the surface of electrode during initial charge of cell fabricated in a discharged state. In 1979, Peled firstly realized the electrically insulating and ionically conductive SEI [206]. The SEI layer is usually provided with a thickness of several to tens of nanometers and consists of various organic and inorganic components. During the formation of SEI, both electrode and electrolyte are consumed, leading to irreversible capacity loss and accounting for a low Coulombic efficiency especially in the first cycle. On the other hand, the SEI physically isolates the electrode from the components of the electrolyte and prevents further reactions between the electrode and the electrolyte when the chemical potential of electrode material falls out of the electrolyte window [65]. For example, if an anode has a μ_a above the lowest unoccupied molecular orbital of liquid electrolyte (or the conductive band of solid electrolyte), it will reduce the electrolyte unless the reaction is blocked by formation of SEI; similarly, a cathode with a μ_c located below the highest occupied molecular orbital oxidizes the electrolyte unless the reaction is stopped by an SEI layer [65]. In case of using metal Li as the anode, a Li^+ permeable SEI not only makes Li dynamically stable in certain organic solvents, but adjusts the distribution of Li ions from the bulk electrolyte to the anode as well through the following steps [120]: (1) the solvated Li^+ gets rid of its solvent molecules and access to the internal Schottky vacancy of the SEI; (2) the Li^+, by relaying itself in Schottky vacancies, continuously migrates through the bulk of the SEI; (3) the Li^+ that passes through the SEI finally reaches the anode surface and accepts an electron from the current collector before being deposited as Li metal. As a result, the SEI contributes to an even deposition of Li ions, which helps to suppress the dendrite nucleation/growth and improve the plating/stripping efficiency of Li anode [206]. Therefore, the SEI formation on the electrode is critical for stable operation of the cell.

In routine LIBs, the SEI on the graphite anode can be extremely stable during several thousands of cycles in the carbonate electrolyte, which paves the way for the commercial success of the LIBs [12]. However, when it comes to high-capacity anode materials, such as conversion-type transition metal oxides/sulfides/fluorides/phosphides/nitrides, alloy-type Si, Sn, P and Al, plating-stripping-type Li metal, intake/release of a large amounts of Li by/from the host will induce massive volume changes during repetitive charge/discharge process [207]. For example, the alloy-type anodes could experience volume expansions as high as 420% for Si, 260% for Ge and Sn, and 300% for P, all significantly larger than 10% for traditional graphite anodes [2]. As a result, cracking, fracture and pulverization of bulk-sized active materials may occur, which leads to continuous evolution and significant change of the electrode|electrolyte interface and makes it challengeable to maintain a stable SEI for high-capacity electrode materials. For example, solid Si structures expand towards the electrolyte during lithiation and contract during the reverse process [208]. The large volume change breaks the SEI and expose new electrode surface to the electrolyte to induce growth of new SEI. The gradual buildup of SEI over cycling consumes electrolyte and Li, extends Li^+ diffusion distance and causes material degradation due to mechanical stress from a thick SEI [208]. The result is a continuously decayed storage performance of Si.

As discussed in the above section, tailoring the particle size of bulk electrode materials to nanoscale helps to alleviate the stress during (de)lithiation of electrode and therefore, prevent the cracking and fracture of materials, which could be beneficial for stable formation of SEI. Further, with advanced nanoengineering techniques, it is now feasible to encapsulate active nanoparticles into appropriate nanostructures with an electrolyte blocking layer for stabilization of the SEI after formation cycles and a pre-defined void space to accommodate volume variation of electrode materials [98, 100, 208–210]. However, compared with electrodes composed of micrometer-scale materials, the nanostructured electrode usually shows a significantly higher surface area that is in contact with the electrolyte, which raises the risk of side reactions involving irreversible electrolyte decomposition and lithium consumption. The results are low initial Coulombic efficiency of nanostructured electrode materials and reduced overall capacity and energy density of LIBs. Therefore, control of the specific

SEI (electrode|electrolyte interface area) plays a critical role in boosting the practical storage performance of nanochemistry-based LIBs. Although high surface area is an inherent feature of nanomaterials, the surface area at electrode|electrolyte interface is tunable by engineering the secondary structures. Figure 2.9a shows the calculation of specific SEI area for a micrometer-sized secondary particle that consists of assembly of primary particles at nanoscale. Provided with the same diameter of primary particles and gap size inbetween, significantly reduced specific SEI area is observed with increment in number of primary nanoparticles in the assembly. Hence, compared with primary nanoparticles, the use of micron-scale secondary particles helps to reduce deleterious side reactions at the interface, so that the electrode material could achieve higher initial Coulombic efficiency and reversible capacity upon cycling. For example, inspired by the pomegranate grown in nature, Liu et al. reported synthesis of micrometer-sized Si secondary assembly composed of primary Si nanoparticles encapsulated in a carbon shell with preserved space (Fig. 2.9b) [209]. Further, a carbon layer is formed on the secondary assembly and functions as an electrolyte-blocking layer (Fig. 2.9b). At the nanoscale, the preserved void space enables free expansion of lithiated Si particles and fast migration of Li ions; at the micron-scale, the formation of secondary assembly and the outer carbon shell helps to reduce specific SEI and provide fast electron transport pathways via interconnected carbon framework [209]. With the above structural benefits, the Si composites show high efficiency, high storage capacity and stable cycling performance when applied as the anode material for LIBs. As a bonus merit, the space-efficient packing of these primary particles also enables high tap density and volumetric mass loading (Fig. 2.9b), which are important parameters for practical evaluation of electrode [209].

The Li metal anode is a case that requires specific attention. The Fermi energy of Li is so high that it can react with most of the electrolyte (both liquid and solid) and passivates its surface (Table 2.5). The parasitic reaction between Li and electrolyte species occurs in a time duration of milliseconds or less when fresh Li surface is exposed to solution and a current applied, and the insoluble products deposit on the metallic anode surface and form the SEI [211]. Several mechanisms have been proposed for the SEI formation chemistry on the Li metal anode (Fig. 2.10). For example, the Peled model (Fig. 2.10a), a primary mechanism that describes the formation of SEI via a stepwise and preferential reduction reaction of certain electrolyte components [206, 212]. The formed SEI shows an integral structure except for several Schottky defects for Li^+ migration. (2) The mosaic model (Fig. 2.10b), of which the formation relates to simultaneous proceeding of several reductive decompositions on the negatively charged anode surface, triggering deposition of a mixture of insoluble products on the anode [213, 214]. The formed SEI shows a mosaic morphology, which allows the Li ions to migrate through the boundary of multiphase products. (3) The Coulombic interaction mechanism (Fig. 2.10c), in which the decomposition products are lined up with positively charged Li^+ as a "head" and the partially positively charged carbons as a "foot", such a unique double electric layer allows strong adhesion of the products onto the existing film [215, 216]. In addition to these models, other models of SEI layers were also proposed, including polymer electrolyte interphase model, solid polymer layer model, and compact stratified layer model [217, 218]. Although

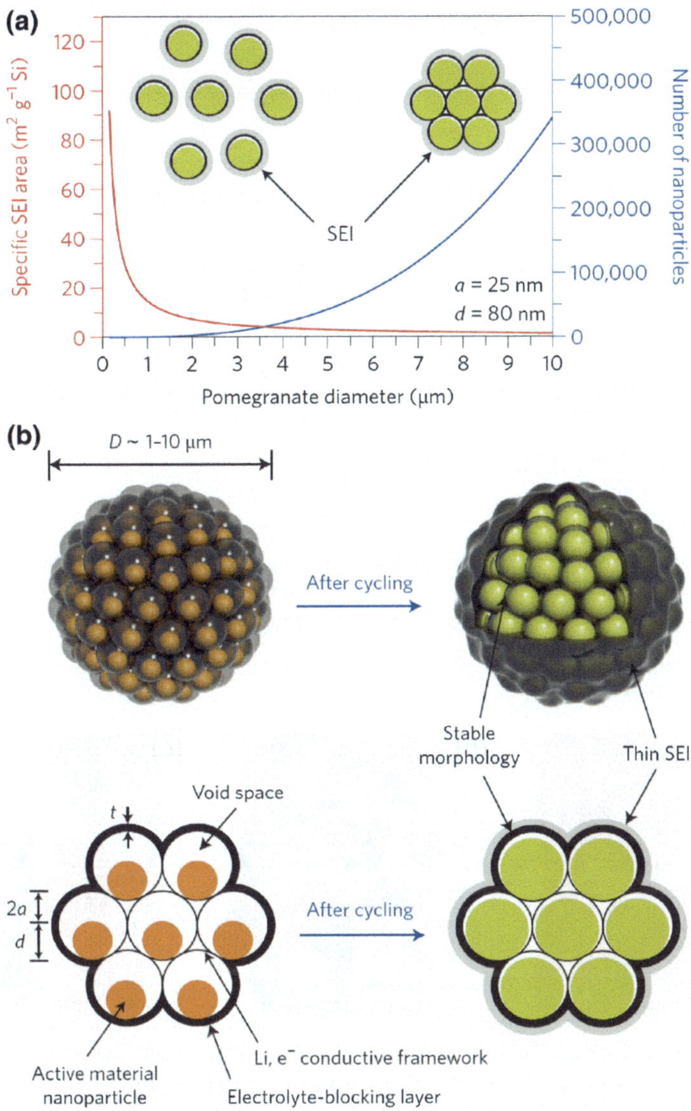

Fig. 2.9 **a** Calculated surface area in contact with the electrolyte (specific SEI area) for secondary particles composed of primary nanoparticles, and the number of primary nanoparticles in one secondary particle versus its diameter. Wherein, *a* is the gap size between primary nanoparticles and their shells, *d* is the diameter of primary nanoparticles. **b** Schematic of a pomegranate microparticle before and after electrochemical Li uptake. The micrometer size of the secondary particle helps to increase the tap density and decrease the surface area in contact with the electrolyte. *D* is the diameter of the secondary particle and *t* is the thickness of the outer shell. Reproduced with the permission from Ref. [209], Copyright © 2014 Springer Nature Publishing AG

Table 2.5 Typical contents of the SEI forming on the Li metal anode as reported in the literatures

Electrolyte	Component
Fresh Li plate	Outer surface: LiOH or Li_2CO_3, hydrocarbon, carbonate; Inner part: Li_2O, carbide
$LiBF_4$ (1.0 M) in PC for 3 days	LiF, LiOH or Li_2CO_3, hydrocarbon, carbonate
$LiBF_4$ (1.0 M) in r-BL for 3 days	LiF, LiOH or Li_2CO_3, hydrocarbon
$LiBF_4$ (1.0 M) in THF for 3 days	LiF, hydrocarbon
$LiClO_4$ (1.0 M) and HF (5.0 M) in PC	Outer part: LiF, LiOH (Li_2CO_3 or $LiOCO_2R$); Inner part: Li_2O
$LiAsF_6$ in DMC, EC-DMC (1:1), and EC-DEC (1:1, v/v)	$ROCO_2Li$, ROLi, and Li_2CO_3
$LiPF_6$ (1 M), vinylene carbonate (2 vol.%), $LiNO_3$ (0.1 M), EC/DMC (1:1, v/v)	$ROCO_2Li$, $(–CH_2CH_2O–)_n$, Li_2CO_3, Li_3N, $LiNO_2$, LiF, C–F
$LiNO_3$ in DOL/DME	LiN_xO_y
Li_2S_6 in DOL/DME	Li_2S, Li_2S_2
$LiNO_3$ and Li_2S_6 in DOL/DME	$Li_2S_2O_3$, Li_2SO_4, LiN_xO_y, Li_2S, Li_2S_2
LiTFSI (0.8 M), Li_2S_6 (0.2 M) in DIOX/DME (1:1, v/v)	Top layer: $Li_2S_2O_3$, Li_3N, $–NSO_2CF_3$; Bottom layer: Li_2S, Li_2S_2
LiTFSI (0.5 M), LiFSI (0.5 M) in DOL/DME (2:1, v/v)	$Li_2NSO_2CF_3$, $Li_yC_2F_x$, LiF, $Li_2S_2O_4$, Li_2S, etc.

Data obtained from Ref. [120]

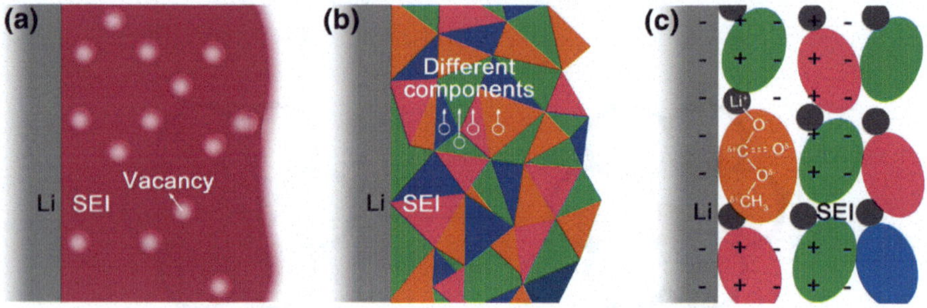

Fig. 2.10 Schemes of these mechanisms for SEI formation. **a** The Peled's model, **b** Mosaic model and **c** Coulombic interaction mechanism. Reproduced with permission from Ref. [120], Copyright © The Authors

these mechanisms vary from each other, they do provide insights into the practical SEI formation on Li metal and indicate that the electrolyte components (solvents, salts and additives) account for the main factor that influences the SEI formation.

On the other hand, the Li metal has a virtually infinite relative volume change during its stripping/plating process owing to its "hostless" nature. Unfortunately, although the formed SEI is able to homogenize the Li^+ flux from the electrolyte, its low mechanical strength makes it unable to prevent the drastic volume change during Li plating/stripping process [206]. Consequently, dendrites form and grow on repeated charging, and finally penetrate through the separator to give an internal short-circuit of cell with incendiary, even explosive outcomes [60]. On the other hand, the formation and growth of Li dendrites leads to periodic breaking and reformation of SEI during Li plating/stripping, which leads to rapid consumption of electrolyte, irreversible loss of lithium and low Coulombic efficiency of the anode [219]. Thus, the challenge of stabilizing the SEI of the Li metal anode is even greater than that of the Si anode. In view of the nucleation and growth chemistry of Li dendrites, other factors that could potentially influence the SEI formation on Li metal shall include temperature, current density and homogeneity/microstructure of anode, and pressure for assembling and operation of cell [120]. To suppress dendrite growth and maintain stable operation of the Li metal anode, the physical and chemical properties of SEI must be improved. As discussed above, the reduction products of conventional electrolyte constitute the main components of the SEI. Therefore, the modulation of electrolyte is the most efficient approach for SEI stabilization. For example, Shkrob et al. found that Li salts with the bis(fluorosulfonyl)imide (FSI^-) anion can form a robust SEI protection layer to prevent the parasitic reaction between liquid electrolyte and Li metal, which leads to dendrite-free cycling of Li anode and promises commercial electrolytes with relatively low viscosity and stable chemistry for rechargeable Li batteries [220, 221]. Other salts, such as lithium difluoro(oxalate)borate, were also reported with SEI-forming ability that protected the anode from suffering dendrite issue in a working cell [222]. Electrolyte additives with higher reduction voltages than solvents and salts may also have a positive effect on forming a stable SEI on the Li metal [223]. For example, halogenated salt additives in the electrolyte provide an on-electrode SEI layer consisting mainly of fluorine-rich products (most likely LiF), LiOH and Li_2CO_3 due to in situ reactions between electrolyte and Li metal. The components in the SEI could act as a catalyst to reduce the activation energy during the migration of Li^+ across the SEI layer and its deposition on the anode, which results in to reduced Li^+ diffusion resistance and much improved operation stability of anode [221, 224–226]. The addition of hydrogen fluoride and even trace amount of water into conventional carbonate-based electrolytes could give similar results [213, 227, 228]. Lithium nitrate ($LiNO_3$) is another important additive. Due to its strong oxidation effect, the use of $LiNO_3$ to assist the SEI formation yields a surface film consisting of both inorganic species (such as LiN_xO_y) and organic species (such as ROLi and $ROCO_2Li$) [229]. Other additives, such as toluene and bis(2,2,2-trifluoroethyl) ether, also exhibited superior results in improving the SEI stability and cycling performance of rechargeable Li metal batteries [230, 231]. The addition of Cs^+ ions at low concentrations to repel Li^+ from sharp nanotips were also reported effective in preventing dendrite growth [232].

Nanoengineering also shows promises in tackling the unstable SEI issue of Li metal anode. Encapsulation of Li metal into various types of current collectors with conductive nanostructured scaffolds (e.g., 3D porous Cu skeleton, carbon nanotube arrays, graphitized carbon fibers, spherical carbon granules) were found able to render low and homogeneous local current density to suppress the growth of Li dendrites and facilitates transport of ions/electrons, so that they enabled high-performance Li metal anode [75, 101, 105, 106, 233, 234]. Another promising strategy is to engineer a chemically stable and mechanically robust nanostructured interfacial layer between Li metal anode and electrolyte, for example, a layer of interconnected hollow carbon nanospheres or ultrathin BN/graphene, which triggers stable formation of SEI while protects the anode surface from dendrite formation [235, 236]. Construction of an artificial SEI on the Li anode has been proved as a feasible strategy for stabilizing the Li electrochemistry. For example, through in situ formation of a uniform Li_3PO_4 layer with high Young's modulus (~10–11 GPa) on the Li metal anode, the artificial SEI can restrain the unfavorable reactions between Li metal and electrolyte and suppress the growth of Li dendrites [102].

The formation chemistry and composition of the SEI at the interface between Li anode and SSE differs significantly from those generated from the liquid electrolytes. Although the SSE is expected to be chemically compatible with the Li metal and able to suppress dendrite formation, the reality is far from satisfactory. The major problem that hinders the anode use of Li in solid-state cells lies in an unstable electrochemical interface between anode and SSE. Although most of the SSEs, including oxides and sulfides, can react with Li, many of them show poor contact with the Li anode due to their "lithiophobic" nature [118, 237–243]. As a result, the SSE shows a poor wettability against Li, which accounts for a high impedance at the interface, and moreover, induces an inhomogeneous distribution of Li^+ flux that leads to dendrite formation in the grain boundaries. Using the nanoengineering technique, it is now feasible to artificially introduce inbetween the Li anode and the SSE a Li^+ conductive interlayer that wets both components. The interlayer improves the Li^+ conduction and homogenizes the Li^+ flux at the interface, so that the interfacial resistance is minimized, and the dendrite nucleation/formation is reduced [238, 241–243]. A detailed discussion regarding the formation chemistry of SEI in solid-state cells and its stabilization strategies will be given in Chapter 5.

The formation chemistry of SEI in other rechargeable storage systems, including rechargeable Na, K, Ca, Mg and Al cells, involves with their own characteristics. For example, the SEI formed by conventional electrolyte (which consists of $NaPF_6$ and carbonate solvents) on Na anode is usually not stable and tends to dissolve during the operation to give self-discharge, which leads to continuous corrosion of metal and growth of moss-like dendrites [160, 244, 245]. The anode use of K metal in rechargeable cells faces similar problems, and artificial SEI coating, optimized current collector design and electrolyte modification may help to alleviate/solve these problems. As opposed to alkali-metal anodes, which have SEI with a cation transfer number (t_+) close to unity, alkaline-earth metals such as calcium, are covered by an SEI that consists of CaF_2 or $CaCl_2$ which are simply anionic conductors (t_- close to unity) [206]. Therefore, in calcium and magnesium cells the anode can be discharged at high rates but cannot be charged even at very high voltage [246]. At present, the relevant studies about optimization of SEI components in these new storage systems are still preliminary. Future efforts on understanding the fundamentals about SEI formation in these systems will be desired.

2.3.3 *Design Principles for "Kinetically Stable" Nanostructured Electrode Materials*

Referring to Sect. 2.3.1, the use of nanostructured electrode materials in an EES device involves its own benefits and challenges. A rational design of nanomaterials towards battery application should maintain the benefits while compensating the disadvantages. In realizing a "kinetically stable" rechargeable cell, electrode materials with nano/micro hierarchical structures, which is formed by "bottom-up" self-assembly of nanometer-sized building blocks into a micron- or submicron-sized secondary structure, have been demonstrated to be an optimal choice [1, 6, 12–14]. On one hand, the nanosized building blocks enable the composites electrode fast mass diffusion and reaction kinetics, high reaction depth and large storage capacity, and accommodation of volume change during cycling. On the other hand, the micron-sized assembly helps to lower the surface energy of nanomaterials and prevent the undesired side reactions at electrode|electrolyte interface, thereby guarantees favorable thermodynamic stability as the electrode reaction proceeds. Such a heterogeneous structure may bring additional benefits such as high tap density, low manufacturing cost and reduced risk of nano-toxicity associated with nanoparticle synthesis and processing. The design concept has been widely adopted to improve the storage performance of various cathode and anode materials of rechargeable cells [98, 136, 137, 247–252].

Based on the nano/micro hierarchically structured electrode materials, a further optimized nanostructure design of electrode materials for RLCs is shown to be the introduction of hierarchically 3D mixed conducting networks on both nanoscale and microscale levels [3, 248]. Typically, such a network is constructed from micron-scaled building blocks, which further consists of nanosized subunits (Fig. 2.11). Materials with high electronic conductivities, such as RuO_2, copper and carbon, have been employed to construct the conducting networks, among which carbon is an appealing choice due to its low cost, light weight, high electronic conductivity, and favorable thermodynamic stability) [3, 73, 91, 107, 131, 132, 204, 248, 253–255]. Nanocarbon networks of different building blocks (e.g., carbon nanoparticles, carbon nanotubes, graphene, and nanoporous carbon), are usually employed to integrate with nanostructured electrode materials, and the nanoscale network is superimposed by a similar net formed by conductive carbon black on the microscale. While the nanoscopic network offers negligible diffusion times, enhanced local conductivities (especially at the interface), additional ion storage sites (synergistic storage effect), and flexible structure buffer, and hence is the key to the excellent power performance of electrode material, the microscopic network guarantees high absolute capacity (high tap density), stable SEI formation, and easy fabrication [3, 6]. Recently, the design concept of mixed conducting networks has been extended to high-capacity electrode materials such as porous carbon, alloy metals, transition metal oxides/sulfides, Li metal, and Group VI cathodes, putting new insights for development of the next-generation high-energy rechargeable electrochemical cells toward future use in 5G electronics, vehicles and grid storage [17, 73, 75–77, 79–84, 86, 87, 92, 93, 95, 107, 108, 111, 114, 117, 119, 129, 131, 205, 256].

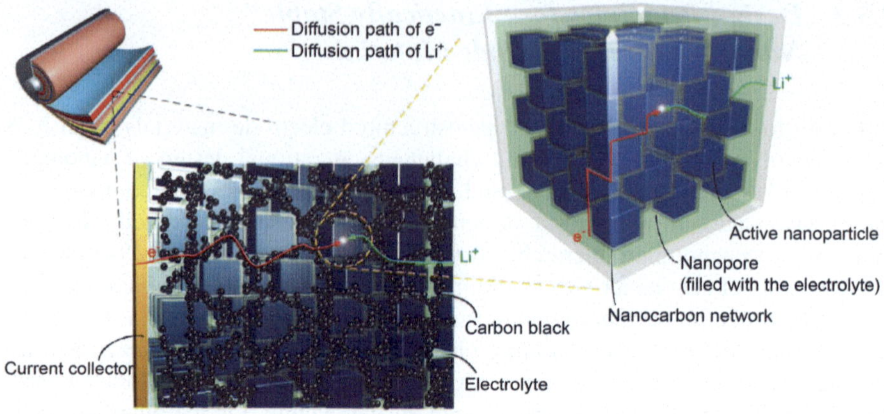

Fig. 2.11 Design scheme of the hierarchically mixed conducting network in a LIB (left), in which the middle drawing gives the general view of the microscale conducting network, and the right drawing gives the fine structure of an enlarged "building block" consisting of a nanocarbon network and active nanoparticles. Reproduced with permission from Ref. [3], Copyright © 2012 American Chemical Society

References

1. Zhang, Q., Uchaker, E., Candelaria, S. L., et al. (2013). Nanomaterials for energy conversion and storage. *Chemical Society Reviews, 42,* 3127–3171.
2. Liu, Y., Zhou, G., Liu, K., et al. (2017). Design of complex nanomaterials for energy storage: Past success and future opportunity. *Accounts of Chemical Research, 50,* 2895–2905.
3. Xin, S., Guo, Y.-G., & Wan, L.-J. (2012). Nanocarbon networks for advanced rechargeable lithium batteries. *Accounts of Chemical Research, 45,* 1759–1769.
4. Pan, Y.-X., Cong, H.-P., Men, Y.-L., et al. (2015). Peptide self-assembled biofilm with unique electron transfer flexibility for highly efficient visible-light-driven photocatalysis. *ACS Nano, 9,* 11258–11265.
5. Pan, Y.-X., Peng, J.-B., Xin, S., et al. (2017). Enhanced visible-light-driven photocatalytic H_2 evolution from water on noble-metal-free CdS-nanoparticle-dispersed Mo_2C@C nanospheres. *ACS Sustainable Chemistry & Engineering, 5,* 5449–5456.
6. Guo, Y.-G., Hu, J.-S., & Wan, L.-J. (2008). Nanostructured materials for electrochemical energy conversion and storage devices. *Advanced Materials, 20,* 2878–2887.
7. Xue, D.-J., Wang, J.-J., Wang, Y.-Q., et al. (2011). Facile synthesis of germanium nanocrystals and their application in organic-inorganic hybrid photodetectors. *Advanced Materials, 23,* 3704–3707.
8. Bruce, P. G., Scrosati, B., & Tarascon, J.-M. (2008). Nanomaterials for rechargeable lithium batteries. *Angewandte Chemie International Edition, 47,* 2930–2946.
9. Men, Y.-L., You, Y., Pan, Y.-X., et al. (2018). Selective CO evolution from photoreduction of CO_2 on a metal-carbide-based composite catalyst. *Journal of the American Chemical Society, 140,* 13071–13077.
10. Pan, Y.-X., You, Y., Xin, S., et al. (2017). Photocatalytic CO_2 reduction by carbon-coated indium-oxide nanobelts. *Journal of the American Chemical Society, 139,* 4123–4129.
11. Pan, Y.-X., Sun, Z.-Q., Cong, H.-P., et al. (2016). Photocatalytic CO_2 reduction highly enhanced by oxygen vacancies on Pt-nanoparticle-dispersed gallium oxide. *Nano Research, 9,* 1689–1700.

12. Sun, Y., Liu, N., & Cui, Y. (2016). Promises and challenges of nanomaterials for lithium-based rechargeable batteries. *Nature Energy, 1,* 16071.
13. Yin, Y.-X., Xin, S., & Guo, Y.-G. (2013). Nanoparticles engineering for lithium-ion batteries. *Particle & Particle Systems Characterization, 30,* 737–753.
14. Li, N.-W., Yin, Y.-X., Xin, S., et al. (2017). Methods for the stabilization of nanostructured electrode materials for advanced rechargeable batteries. *Small Methods, 1,* 1700094.
15. Li, X., Zhang, C., Xin, S., et al. (2016). Facile synthesis of MoS_2/reduced graphene Oxide@Polyaniline for high-performance supercapacitors. *ACS Applied Materials & Interfaces, 8,* 21373–21380.
16. Gao, H., Xin, S., & Goodenough, J. B. (2017). The origin of superior performance of $Co(OH)_2$ in hybrid supercapacitors. *Chem, 3,* 26–28.
17. Song, L., Xin, S., Xu, D.-W., et al. (2016). Graphene-wrapped graphitic carbon hollow spheres: Bioinspired synthesis and applications in batteries and supercapacitors. *ChemNanoMat, 2,* 540–546.
18. Guo, W., Xin, S., Ji, M., et al. (2011). Supercapacitor-battery hybrid energy storage devices from an aqueous nitroxide radical active material. *Chinese Science Bulletin, 56,* 2433–2436.
19. Jiang, C., Hosono, E., & Zhou, H. (2006). Nanomaterials for lithium ion batteries. *Nano Today, 1,* 28–33.
20. Linden, D., & Reddy, T. B. (2010). An introduction to primary batteries. In T. B. Reddy (Ed.), *Linden's handbook of batteries* (pp. 8.1–8.18). McGraw-Hill Education.
21. Linden, D. (2001). Primary batteries—Introduction. In D. Linden & T. B. Reddy (Eds.), *Handbook of batteries* (pp. 7.1–7.21). McGraw-Hill Education.
22. Brooke Schumm, J. (2010). Zinc-carbon batteries—Leclanché and zinc chloride cell systems. In T. B. Reddy (Ed.), *Linden's handbook of batteries* (pp. 9.1–9.41). McGraw-Hill Education.
23. Sayilgan, E., Kukrer, T., Civelekoglu, G., et al. (2009). A review of technologies for the recovery of metals from spent alkaline and zinc–carbon batteries. *Hydrometallurgy, 97,* 158–166.
24. Nardi, J. C., & Brodd, R. J. (2010). Alkaline-manganese dioxide batteries. In T. B. Reddy (Ed.), *Linden's handbook of batteries* (pp. 11.11–11.17). McGraw-Hill Education.
25. Scarr, R. F., Hunter, J. C., & Slezak, P. J. (2001). Alkaline-manganese dioxide batteries. In D. Linden & T. B. Reddy (Eds.), *Handbook of batteries* (pp. 10.11–10.32). McGraw-Hill Education.
26. Kozawa, A., & Powers, R. A. (1972). Electrochemical reactions in batteries. Emphasizing the MnO_2 cathode of dry cells. *Journal of Chemical Education, 49,* 587.
27. Almeida, M. F., Xará, S. M., Delgado, J., et al. (2006). Characterization of spent AA household alkaline batteries. *Waste Management, 26,* 466–476.
28. Reddy, T. B. (2010). Lithium primary batteries. In T. B. Reddy (Ed.), *Linden's handbook of batteries* (pp. 14.11–14.90). McGraw-Hill Education.
29. Reddy, T. B. (2010). An introduction to secondary batteries. In T. B. Reddy (Ed.), *Linden's handbook of batteries* (pp. 15.11–15.20). McGraw-Hill Education.
30. Parker, C. D., & Garche, J. (2004). Battery energy-storage systems for power-supply networks. In D. A. J. Rand, J. Garche, P. T. Moseley, et al. (Eds.), *Valve-regulated lead-acid batteries* (pp. 295–326). Amsterdam: Elsevier.
31. Salkind, A., & Zguris, G. (2010). Lead-acid batteries. In T. B. Reddy (Ed.), *Linden's handbook of batteries* (pp. 16.11–16.87). McGraw-Hill Education.
32. Ruetschi, P. (1977). Review on the lead-acid battery science and technology. *Journal of Power Sources, 2,* 3–120.
33. Bullock, K. R., & Salkind, A. J. (2010). Valve regulated lead-acid batteries. In T. B. Reddy (Ed.), *Linden's handbook of batteries* (pp. 17.11–17.39). McGraw-Hill Education.
34. May, G. J., Davidson, A., & Monahov, B. (2018). Lead batteries for utility energy storage: A review. *Journal of Energy Storage, 15,* 145–157.
35. Cooper, A., Furakawa, J., Lam, L., et al. (2009). The ultrabattery—A new battery design for a new beginning in hybrid electric vehicle energy storage. *Journal of Power Sources, 188,* 642–649.
36. Mantell, C. L. (1983). *Batteries and energy systems* (2nd ed.). New York: McGraw-Hill.

37. Carcone, J. A. (2010). Portable sealed nickel-cadmium batteries. In T. B. Reddy (Ed.), *Linden's handbook of batteries* (pp. 21.21–21.33). McGraw-Hill Education.
38. Erbacher, J. K. (2010). Industrial and aerospace nickel-cadmium batteries. In T. B. Reddy (Ed.), *Linden's handbook of batteries* (pp. 19.11–19.23). McGraw-Hill Education.
39. Lucero, R. D. (2010). Vented sintered-plate nickel-cadmium batteries. In T. B. Reddy (Ed.), *Linden's handbook of batteries* (pp. 20.21–20.28). McGraw-Hill Education.
40. Shukla, A. K., Venugopalan, S., & Hariprakash, B. (2001). Nickel-based rechargeable batteries. *Journal of Power Sources, 100,* 125–148.
41. Halpert, G. (1984). Past developments and the future of nickel electrode cell technology. *Journal of Power Sources, 12,* 177–192.
42. Chen, H., Cong, T. N., Yang, W., et al. (2009). Progress in electrical energy storage system: A critical review. *Progress in Natural Science, 19,* 291–312.
43. Fleischer, A. (1948). Sintered plates for nickel-cadmium batteries. *Journal of the Electrochemical Society, 94,* 289–299.
44. Sato, Y., Ito, K., Arakawa, T., et al. (1996). Possible cause of the memory effect observed in nickel-cadmium secondary batteries. *Journal of the Electrochemical Society, 143,* L225–L228.
45. Fetcenko, M., & Koch, J. (2010). Nickel-metal hydride batteries. In T. B. Reddy (Ed.), *Linden's handbook of batteries* (pp. 22.21–22.51). McGraw-Hill Education.
46. Goo, N. H., Woo, J. H., & Lee, K. S. (1999). Mechanism of rapid degradation of nanostructured Mg_2Ni hydrogen storage alloy electrode synthesized by mechanical alloying and the effect of mechanically coating with nickel. *Journal of Alloys and Compounds, 288,* 286–293.
47. Zhang, Q. A., Lei, Y. Q., Wang, C. S., et al. (1998). Structure of the secondary phase and its effects on hydrogen-storage properties in a $Ti_{0.7}Zr_{0.2}V_{0.1}Ni$ alloy. *Journal of Power Sources, 75,* 288–291.
48. Züttel, A., Meli, F., & Schlapbach, L. (1994). Electrochemical and surface properties of $Zr(V_xNi_{1-x})_2$ alloys as hydrogen-absorbing electrodes in alkaline electrolyte. *Journal of Alloys and Compounds, 203,* 235–241.
49. Liao, B., Lei, Y. Q., Chen, L. X., et al. (2004). A study on the structure and electrochemical properties of $La_2Mg(Ni_{0.95}M_{0.05})_9$ (M = Co, Mn, Fe, Al, Cu, Sn) hydrogen storage electrode alloys. *Journal of Alloys and Compounds, 376,* 186–195.
50. Yasuoka, S., Magari, Y., Murata, T., et al. (2006). Development of high-capacity nickel-metal hydride batteries using superlattice hydrogen-absorbing alloys. *Journal of Power Sources, 156,* 662–666.
51. Zhao, Y., Zhang, L., Ding, Y., et al. (2017). Comparative study on the capacity degradation behavior of Pr_5Co_{19}-type single-phase Pr_4MgNi_{19} and La_4MgNi_{19} alloys. *Journal of Alloys and Compounds, 694,* 1089–1097.
52. Willems, J. J. G., & Buschow, K. H. J. (1987). From permanent magnets to rechargeable hydride electrodes. *Journal of the Less Common Metals, 129,* 13–30.
53. Young, K., Nei, J., Wong, D. F., et al. (2014). Structural, hydrogen storage, and electrochemical properties of Laves phase-related body-centered-cubic solid solution metal hydride alloys. *International Journal of Hydrogen Energy, 39,* 21489–21499.
54. Goodenough, J. B. (2018). How we made the Li-ion rechargeable battery. *Nature Electronics, 1,* 204.
55. Whittingham, M. S. (1976). Electrical energy storage and intercalation chemistry. *Science, 192,* 1126–1127.
56. Zanini, M., Basu, S., & Fischer, J. E. (1978). Alternate synthesis and reflectivity spectrum of stage 1 lithium—graphite intercalation compound. *Carbon, 16,* 211–212.
57. Basu, S., Zeller, C., Flanders, P. J., et al. (1979). Synthesis and properties of lithium-graphite intercalation compounds. *Materials Science and Engineering, 38,* 275–283.
58. Mizushima, K., Jones, P. C., Wiseman, P. J., et al. (1980). Li_xCoO_2 $(0<x<-1)$: A new cathode material for batteries of high energy density. *Materials Research Bulletin, 15,* 783–789.
59. Liu, J., Bao, Z., Cui, Y., et al. (2019). Pathways for practical high-energy long-cycling lithium metal batteries. *Nature Energy, 4,* 180–186.

60. Goodenough, J. B., & Park, K.-S. (2013). The Li-ion rechargeable battery: A perspective. *Journal of the American Chemical Society, 135,* 1167–1176.
61. Myung, S.-T., Maglia, F., Park, K.-J., et al. (2017). Nickel-rich layered cathode materials for automotive lithium-ion batteries: Achievements and perspectives. *ACS Energy Letters, 2,* 196–223.
62. Yang, Z., Zhang, J., Kintner-Meyer, M. C. W., et al. (2011). Electrochemical energy storage for green grid. *Chemical Reviews, 111,* 3577–3613.
63. Dunn, B., Kamath, H., & Tarascon, J.-M. (2011). Electrical energy storage for the grid: A battery of choices. *Science, 334,* 928–935.
64. Goodenough, J. B. (2013). Evolution of strategies for modern rechargeable batteries. *Accounts of Chemical Research, 46,* 1053–1061.
65. Goodenough, J. B., & Kim, Y. (2010). Challenges for rechargeable Li batteries. *Chemistry of Materials, 22,* 587–603.
66. Dahn, J., & Ehrlich, G. M. (2010). Lithium-ion batteries. In T. B. Reddy (Ed.), *Linden's Handbook of Batteries* (pp. 26.21–26.79). McGraw-Hill Education.
67. Villa, C., Kim, S., Lu, Y., et al. (2019). Cu-substituted NiF_2 as a cathode material for Li-Ion batteries. *ACS Applied Materials & Interfaces, 11,* 647–654.
68. Li, T., Li, L., Cao, Y. L., et al. (2010). Reversible three-electron redox behaviors of FeF_3 nanocrystals as high-capacity cathode-active materials for Li-ion batteries. *The Journal of Physical Chemistry C, 114,* 3190–3195.
69. Fan, X., Hu, E., Ji, X., et al. (2018). High energy-density and reversibility of iron fluoride cathode enabled via an intercalation-extrusion reaction. *Nature Communications, 9,* 2324.
70. Jung, S.-K., Kim, H., Cho, M. G., et al. (2017). Lithium-free transition metal monoxides for positive electrodes in lithium-ion batteries. *Nature Energy, 2,* 16208.
71. Xu, J., Ma, J., Fan, Q., et al. (2017). Recent progress in the design of advanced cathode materials and battery models for high-performance lithium-X (X = O_2, S, Se, Te, I_2, Br_2) batteries. *Advanced Materials, 29,* 1606454.
72. Yin, Y.-X., Xin, S., Guo, Y.-G., et al. (2013). Lithium-sulfur batteries: Electrochemistry, materials, and prospects. *Angewandte Chemie International Edition, 52,* 13186–13200.
73. Xin, S., Gu, L., Zhao, N.-H., et al. (2012). Smaller sulfur molecules promise better lithium-sulfur batteries. *Journal of the American Chemical Society, 134,* 18510–18513.
74. Yang, C.-P., Xin, S., Yin, Y.-X., et al. (2013). An advanced selenium-carbon cathode for rechargeable lithium-selenium batteries. *Angewandte Chemie International Edition, 52,* 8363–8367.
75. Jin, S., Xin, S., Wang, L., et al. (2016). Covalently connected carbon nanostructures for current collectors in both the cathode and anode of Li–S batteries. *Advanced Materials, 28,* 9094–9102.
76. Xin, S., Yin, Y.-X., Wan, L.-J., et al. (2013). Encapsulation of sulfur in a hollow porous carbon substrate for superior Li-S batteries with long lifespan. *Particle & Particle Systems Characterization, 30,* 321–325.
77. Xin, S., Yu, L., You, Y., et al. (2016). The electrochemistry with lithium versus sodium of selenium confined to slit micropores in carbon. *Nano Letters, 16,* 4560–4568.
78. Xin, S., Guo, Y.-G., Chang, Z., et al. (2016). Progress of rechargeable lithium metal batteries based on conversion reactions. *National Science Review, 4,* 54–70.
79. Wang, Z., You, Y., Yuan, J., et al. (2016). Nickel-doped $La_{0.8}Sr_{0.2}Mn_{1-x}Ni_xO_3$ nanoparticles containing abundant oxygen vacancies as an optimized bifunctional catalyst for oxygen cathode in rechargeable lithium-air batteries. *ACS Applied Materials & Interfaces, 8,* 6520–6528.
80. Du, X.-L., You, Y., Yan, Y., et al. (2016). Conductive carbon network inside a sulfur-impregnated carbon sponge: A bioinspired high-performance cathode for Li–S battery. *ACS Applied Materials & Interfaces, 8,* 22261–22269.
81. Xin, S., You, Y., Li, H.-Q., et al. (2016). Graphene sandwiched by sulfur-confined mesoporous carbon nanosheets: A kinetically stable cathode for Li–S batteries. *ACS Applied Materials & Interfaces, 8,* 33704–33711.

82. Xu, D.-W., Xin, S., You, Y., et al. (2016). Built-in carbon nanotube network inside a biomass-derived hierarchically porous carbon to enhance the performance of the sulfur cathode in a Li-S battery. *ChemNanoMat, 2,* 712–718.
83. Chen, Z.-H., Du, X.-L., He, J.-B., et al. (2017). Porous coconut shell carbon offering high retention and deep lithiation of sulfur for lithium-sulfur batteries. *ACS Applied Materials & Interfaces, 9,* 33855–33862.
84. Xu, J., Xin, S., Liu, J.-W., et al. (2016). Elastic carbon nanotube aerogel meets tellurium nanowires: A Binder- and collector-free electrode for Li-Te batteries. *Advanced Functional Materials, 26,* 3580–3588.
85. Du, Z., Guo, C., Wang, L., et al. (2017). Atom-thick interlayer made of CVD-grown graphene film on separator for advanced lithium-sulfur batteries. *ACS Applied Materials & Interfaces, 9,* 43696–43703.
86. Zhang, S.-F., Wang, W.-P., Xin, S., et al. (2017). Graphitic nanocarbon-selenium cathode with favorable rate capability for Li–Se batteries. *ACS Applied Materials & Interfaces, 9,* 8759–8765.
87. Zhang, J., Zhang, C., Li, W., et al. (2018). Nitrogen-doped perovskite as a bifunctional cathode catalyst for rechargeable lithium-oxygen batteries. *ACS Applied Materials & Interfaces, 10,* 5543–5550.
88. Wu, Y., Wang, T., Zhang, Y., et al. (2016). Electrocatalytic performances of g-C_3N_4-$LaNiO_3$ composite as bi-functional catalysts for lithium-oxygen batteries. *Scientific Reports, 6,* 24314.
89. Yang, C.-P., Yin, Y.-X., Guo, Y.-G., et al. (2015). Electrochemical (De)lithiation of 1D sulfur chains in Li–S batteries: A model system study. *Journal of the American Chemical Society, 137,* 2215–2218.
90. Yang, C.-P., Yin, Y.-X., & Guo, Y.-G. (2015). Elemental selenium for electrochemical energy storage. *The Journal of Physical Chemistry Letters, 6,* 256–266.
91. Cao, F.-F., Deng, J.-W., Xin, S., et al. (2011). Cu-Si nanocable arrays as high-rate anode materials for lithium-ion batteries. *Advanced Materials, 23,* 4415–4420.
92. Zhang, Y.-C., You, Y., Xin, S., et al. (2016). Rice husk-derived hierarchical silicon/nitrogen-doped carbon/carbon nanotube spheres as low-cost and high-capacity anodes for lithium-ion batteries. *Nano Energy, 25,* 120–127.
93. Yin, Y.-X., Xin, S., Wan, L.-J., et al. (2011). Electrospray synthesis of silicon/carbon nanoporous microspheres as improved anode materials for lithium-ion batteries. *The Journal of Physical Chemistry C, 115,* 14148–14154.
94. Yan, Y., Yin, Y.-X., Xin, S., et al. (2013). High-safety lithium-sulfur battery with prelithiated Si/C anode and ionic liquid electrolyte. *Electrochimica Acta, 91,* 58–61.
95. Cong, H.-P., Xin, S., & Yu, S.-H. (2015). Flexible nitrogen-doped graphene/SnO_2 foams promise kinetically stable lithium storage. *Nano Energy, 13,* 482–490.
96. Yin, Y., Xin, S., Wan, L., et al. (2012). SnO_2 hollow spheres: Polymer bead-templated hydrothermal synthesis and their electrochemical properties for lithium storage. *Science China Chemistry, 55,* 1314–1318.
97. Yin, Y.-X., Xin, S., Wan, L.-J., et al. (2012). Synthesis of nanostructured SnO_2/C microfibers with improved performances as anode material for Li-ion batteries. *Journal of Nanoscience and Nanotechnology, 12,* 2581–2585.
98. Zhang, W.-M., Hu, J.-S., Guo, Y.-G., et al. (2008). Tin-nanoparticles encapsulated in elastic hollow carbon spheres for high-performance anode material in lithium-ion batteries. *Advanced Materials, 20,* 1160–1165.
99. Wu, N., Yao, H.-R., Yin, Y.-X., et al. (2015). Improving the electrochemical properties of the red P anode in Na-ion batteries via the space confinement of carbon nanopores. *Journal of Materials Chemistry A, 3,* 24221–24225.
100. Li, S., Niu, J., Zhao, Y. C., et al. (2015). High-rate aluminium yolk-shell nanoparticle anode for Li-ion battery with long cycle life and ultrahigh capacity. *Nature Communications, 6,* 7872.
101. Yang, C.-P., Yin, Y.-X., Zhang, S.-F., et al. (2015). Accommodating lithium into 3D current collectors with a submicron skeleton towards long-life lithium metal anodes. *Nature Communications, 6,* 8058.

102. Li, N.-W., Yin, Y.-X., Yang, C.-P., et al. (2016). An artificial solid electrolyte interphase layer for stable lithium metal anodes. *Advanced Materials, 28,* 1853–1858.
103. Zhang, R., Li, N.-W., Cheng, X.-B., et al. (2017). Advanced micro/nanostructures for lithium metal anodes. *Advanced Science, 4,* 1600445.
104. Zeng, X.-X., Yin, Y.-X., Li, N.-W., et al. (2016). Reshaping lithium plating/stripping behavior via bifunctional polymer electrolyte for room-temperature solid Li metal batteries. *Journal of the American Chemical Society, 138,* 15825–15828.
105. Zuo, T.-T., Wu, X.-W., Yang, C.-P., et al. (2017). Graphitized carbon fibers as multifunctional 3D current collectors for high areal capacity Li anodes. *Advanced Materials, 29,* 1700389.
106. Zuo, T.-T., Yin, Y.-X., Wang, S.-H., et al. (2018). Trapping lithium into hollow silica microspheres with a carbon nanotube core for dendrite-free lithium metal anodes. *Nano Letters, 18,* 297–301.
107. Xu, Q., Sun, J.-K., Yu, Z.-L., et al. (2018). SiO_x encapsulated in graphene bubble film: An ultrastable Li-ion battery anode. *Advanced Materials, 30,* 1707430.
108. Xue, D.-J., Xin, S., Yan, Y., et al. (2012). Improving the electrode performance of Ge through Ge@C core-shell nanoparticles and graphene networks. *Journal of the American Chemical Society, 134,* 2512–2515.
109. Xin, S., You, Y., Wang, S., et al. (2017). Solid-state lithium metal batteries promoted by nanotechnology: Progress and prospects. *ACS Energy Letters, 2,* 1385–1394.
110. Ye, H., Xin, S., Yin, Y.-X., et al. (2017). Advanced porous carbon materials for high-efficient lithium metal anodes. *Advanced Energy Materials, 7,* 1700530.
111. Shan, T.-T., Xin, S., You, Y., et al. (2016). Combining nitrogen-doped graphene sheets and MoS_2: A unique film–foam–film structure for enhanced lithium storage. *Angewandte Chemie International Edition, 55,* 12783–12788.
112. Zhou, F., Xin, S., Liang, H.-W., et al. (2014). Carbon nanofibers decorated with molybdenum disulfide nanosheets: Synergistic lithium storage and enhanced electrochemical performance. *Angewandte Chemie International Edition, 53,* 11552–11556.
113. Zhang, B.-L., Xin, S., Qin, H., et al. (2018). Stable lithium storage in nitrogen-doped carbon-coated ferric oxide yolk-shell nanospindles with preserved hollow space. *ChemPlusChem, 83,* 99–107.
114. Wu, X.-L., Chen, L.-L., Xin, S., et al. (2010). Preparation and Li storage properties of hierarchical porous carbon fibers derived from alginic acid. *Chemsuschem, 3,* 703–707.
115. Li, W., Yin, Y.-X., Xin, S., et al. (2012). Low-cost and large-scale synthesis of alkaline earth metal germanate nanowires as a new class of lithium ion battery anode material. *Energy & Environmental Science, 5,* 8007–8013.
116. Chen, Z., Yan, Y., Xin, S., et al. (2013). Copper germanate nanowire/reduced graphene oxide anode materials for high energy lithium-ion batteries. *Journal of Materials Chemistry A, 1,* 11404–11409.
117. Yu, Z.-L., Xin, S., You, Y., et al. (2016). Ion-catalyzed synthesis of microporous hard carbon embedded with expanded nanographite for enhanced lithium/sodium storage. *Journal of the American Chemical Society, 138,* 14915–14922.
118. Zhou, W., Wang, S., Li, Y., et al. (2016). Plating a dendrite-free lithium anode with a polymer/ceramic/polymer sandwich electrolyte. *Journal of the American Chemical Society, 138,* 9385–9388.
119. Xu, R., Wang, G., Zhou, T., et al. (2017). Rational design of Si@carbon with robust hierarchically porous custard-apple-like structure to boost lithium storage. *Nano Energy, 39,* 253–261.
120. Cheng, X.-B., Zhang, R., Zhao, C.-Z., et al. (2016). A review of solid electrolyte interphases on lithium metal anode. *Advanced Science, 3,* 1500213.
121. Li, J.-Y., Xu, Q., Li, G., et al. (2017). Research progress regarding Si-based anode materials towards practical application in high energy density Li-ion batteries. *Materials Chemistry Frontiers, 1,* 1691–1708.
122. Hassan, F. M., Chabot, V., Elsayed, A. R., et al. (2014). Engineered Si electrode nanoarchitecture: A scalable postfabrication treatment for the production of next-generation li-ion batteries. *Nano Letters, 14,* 277–283.

123. Cao, Y., Li, M., Lu, J., et al. (2019). Bridging the academic and industrial metrics for next-generation practical batteries. *Nature Nanotechnology, 14,* 200–207.
124. Ko, M., Chae, S., Ma, J., et al. (2016). Scalable synthesis of silicon-nanolayer-embedded graphite for high-energy lithium-ion batteries. *Nature Energy, 1,* 16113.
125. Bruce, P. G., Freunberger, S. A., Hardwick, L. J., et al. (2011). Li–O_2 and Li–S batteries with high energy storage. *Nature Materials, 11,* 19–29.
126. Pan, H., Hu, Y.-S., & Chen, L. (2013). Room-temperature stationary sodium-ion batteries for large-scale electric energy storage. *Energy & Environmental Science, 6,* 2338–2360.
127. You, Y., & Manthiram, A. (2018). Progress in high-voltage cathode materials for rechargeable sodium-ion batteries. *Advanced Energy Materials, 8,* 1701785.
128. Zhao, H., Xu, J., Yin, D., et al. (2018). Electrolytes for batteries with earth-abundant metal anodes. *Chemistry—A European Journal, 24,* 18220–18234.
129. Xu, Q., Sun, J.-K., Yue, F.-S., et al. (2018). Stable sodium storage of red phosphorus anode enabled by a dual-protection strategy. *ACS Applied Materials & Interfaces, 10,* 30479–30486.
130. Zhou, W., Li, Y., Xin, S., et al. (2017). Rechargeable sodium all-solid-state battery. *ACS Central Science, 3,* 52–57.
131. Xin, S., Yin, Y.-X., Guo, Y.-G., et al. (2014). A high-energy room-temperature sodium-sulfur battery. *Advanced Materials, 26,* 1261–1265.
132. You, Y., Yao, H.-R., Xin, S., et al. (2016). Subzero-temperature cathode for a sodium-ion battery. *Advanced Materials, 28,* 7243–7248.
133. Gao, H., Xue, L., Xin, S., et al. (2017). A plastic-crystal electrolyte interphase for all-solid-state sodium batteries. *Angewandte Chemie International Edition, 56,* 5541–5545.
134. Wang, P.-F., You, Y., Yin, Y.-X., et al. (2016). Suppressing the P2–O2 phase transition of $Na_{0.67}Mn_{0.67}Ni_{0.33}O_2$ by magnesium substitution for improved sodium-ion batteries. *Angewandte Chemie International Edition, 55,* 7445–7449.
135. Gao, H., Xin, S., Xue, L., et al. (2018). Stabilizing a high-energy-density rechargeable sodium battery with a solid electrolyte. *Chem, 4,* 833–844.
136. You, Y., Xin, S., Asl, H. Y., et al. (2018). Insights into the improved high-voltage performance of Li-incorporated layered oxide cathodes for sodium-ion batteries. *Chem, 4,* 2124–2139.
137. You, Y., Dolocan, A., Li, W., et al. (2019). Understanding the air-exposure degradation chemistry at a nanoscale of layered oxide cathodes for sodium-ion batteries. *Nano Letters, 19,* 182–188.
138. Wang, P.-F., Yao, H.-R., You, Y., et al. (2018). Understanding the structural evolution and Na+ kinetics in honeycomb-ordered O′3-$Na_3Ni_2SbO_6$ cathodes. *Nano Research, 11,* 3258–3271.
139. You, Y., Yu, X., Yin, Y., et al. (2015). Sodium iron hexacyanoferrate with high Na content as a Na-rich cathode material for Na-ion batteries. *Nano Research, 8,* 117–128.
140. Gao, H., Seymour, I. D., Xin, S., et al. (2018). $Na_3MnZr(PO_4)_3$: A high-voltage cathode for sodium batteries. *Journal of the American Chemical Society, 140,* 18192–18199.
141. Xue, L., Li, Y., Gao, H., et al. (2017). Low-cost high-energy potassium cathode. *Journal of the American Chemical Society, 139,* 2164–2167.
142. Xue, L., Gao, H., Zhou, W., et al. (2016). Liquid K-Na alloy anode enables dendrite-free potassium batteries. *Advanced Materials, 28,* 9608–9612.
143. Gao, H., Xue, L., Xin, S., et al. (2018). A high-energy-density potassium battery with a polymer-gel electrolyte and a polyaniline cathode. *Angewandte Chemie International Edition, 57,* 5449–5453.
144. Xue, L., Zhou, W., Xin, S., et al. (2018). Room-temperature liquid Na–K anode membranes. *Angewandte Chemie International Edition, 57,* 14184–14187.
145. Wu, N., Yang, Z.-Z., Yao, H.-R., et al. (2015). Improving the electrochemical performance of the Li4Ti5O12 electrode in a rechargeable magnesium battery by lithium-magnesium Co-intercalation. *Angewandte Chemie International Edition, 54,* 5757–5761.
146. Wu, N., Yin, Y.-X., & Guo, Y.-G. (2014). Size-dependent electrochemical magnesium storage performance of spinel lithium titanate. *Chemistry—An Asian Journal, 9,* 2099–2102.
147. Shterenberg, I., Salama, M., Gofer, Y., et al. (2014). The challenge of developing rechargeable magnesium batteries. *MRS Bulletin, 39,* 453–460.

148. Wu, N., Lyu, Y.-C., Xiao, R.-J., et al. (2014). A highly reversible, low-strain Mg-ion insertion anode material for rechargeable Mg-ion batteries. *NPG Asia Materials, 6,* e120.
149. Yu, X., Boyer, M. J., Hwang, G. S., et al. (2019). Toward a reversible calcium-sulfur battery with a lithium-ion mediation approach. *Advanced Energy Materials, 9,* 1803794.
150. Reinsberg, P., Bondue, C. J., & Baltruschat, H. (2016). Calcium-oxygen batteries as a promising alternative to sodium-oxygen batteries. *The Journal of Physical Chemistry C, 120,* 22179–22185.
151. Ouchi, T., Kim, H., Spatocco, B. L., et al. (2016). Calcium-based multi-element chemistry for grid-scale electrochemical energy storage. *Nature Communications, 7,* 10999.
152. Muldoon, J., Bucur, C. B., & Gregory, T. (2014). Quest for nonaqueous multivalent secondary batteries: Magnesium and beyond. *Chemical Reviews, 114,* 11683–11720.
153. Gao, T., Li, X., Wang, X., et al. (2016). A rechargeable Al/S battery with an ionic-liquid electrolyte. *Angewandte Chemie International Edition, 55,* 9898–9901.
154. Yu, X., Boyer, M. J., Hwang, G. S., et al. (2018). Room-temperature aluminum-sulfur batteries with a lithium-ion-mediated ionic liquid electrolyte. *Chem, 4,* 586–598.
155. Tang, Y., Zhang, Y., Li, W., et al. (2015). Rational material design for ultrafast rechargeable lithium-ion batteries. *Chemical Society Reviews, 44,* 5926–5940.
156. Galwey, A. K., & Brown, M. E. (2002). Application of the Arrhenius equation to solid state kinetics: can this be justified? *Thermochimica Acta, 386,* 91–98.
157. Zhu, C., Usiskin, R. E., Yu, Y., et al. (2017). The nanoscale circuitry of battery electrodes. *Science, 358,* eaao2808.
158. Sun, C., Liu, J., Gong, Y., et al. (2017). Recent advances in all-solid-state rechargeable lithium batteries. *Nano Energy, 33,* 363–386.
159. Buqa, H., Goers, D., Holzapfel, M., et al. (2005). High rate capability of graphite negative electrodes for lithium-ion batteries. *Journal of the Electrochemical Society, 152,* A474–A481.
160. Peled, E., & Menkin, S. (2017). Review—SEI: Past, present and future. *Journal of the Electrochemical Society, 164,* A1703–A1719.
161. Hayamizu, K. (2017). Direct relations between ion diffusion constants and ionic conductivity for lithium electrolyte solutions. *Electrochimica Acta, 254,* 101–111.
162. Orädd, G., Edman, L., & Ferry, A. (2002). Diffusion: A comparison between liquid and solid polymer LiTFSI electrolytes. *Solid State Ionics, 152–153,* 131–136.
163. Wohde, F., Balabajew, M., & Roling, B. (2016). Li^+ transference numbers in liquid electrolytes obtained by very-low-frequency impedance spectroscopy at variable electrode distances. *Journal of the Electrochemical Society, 163,* A714–A721.
164. Koksbang, R., Olsen, I. I., & Shackle, D. (1994). Review of hybrid polymer electrolytes and rechargeable lithium batteries. *Solid State Ionics, 69,* 320–335.
165. Bachman, J. C., Muy, S., Grimaud, A., et al. (2016). Inorganic solid-state electrolytes for lithium batteries: Mechanisms and properties governing ion conduction. *Chemical Reviews, 116,* 140–162.
166. Li, W., Wu, G., Araújo, C. M., et al. (2010). Li^+ ion conductivity and diffusion mechanism in α-Li_3N and β-Li_3N. *Energy & Environmental Science, 3,* 1524–1530.
167. Gao, Z., Sun, H., Fu, L., et al. (2018). Promises, challenges, and recent progress of inorganic solid-state electrolytes for all-solid-state lithium batteries. *Advanced Materials, 30,* 1705702.
168. Saha, P., Datta, M. K., Velikokhatnyi, O. I., et al. (2014). Rechargeable magnesium battery: Current status and key challenges for the future. *Progress in Materials Science, 66,* 1–86.
169. Zeier, W. G., Zhou, S., Lopez-Bermudez, B., et al. (2014). Dependence of the Li-ion conductivity and activation energies on the crystal structure and ionic radii in $Li_6MLa_2Ta_2O_{12}$. *ACS Applied Materials & Interfaces, 6,* 10900–10907.
170. Tatsumisago, M., Nagao, M., & Hayashi, A. (2013). Recent development of sulfide solid electrolytes and interfacial modification for all-solid-state rechargeable lithium batteries. *Journal of Asian Ceramic Societies, 1,* 17–25.
171. Kanehori, K., Matsumoto, K., Miyauchi, K., et al. (1983). Thin film solid electrolyte and its application to secondary lithium cell. *Solid State Ionics, 9–10,* 1445–1448.

172. Bates, J. B., Dudney, N. J., Neudecker, B., et al. (2000). Thin-film lithium and lithium-ion batteries. *Solid State Ionics, 135,* 33–45.
173. Cao, C., Li, Z.-B., Wang, X.-L., et al. (2014). Recent advances in inorganic solid electrolytes for lithium batteries. *Frontiers in Energy Research, 2.*
174. Druger, S. D. (1994). Ionic transport in polymer electrolytes based on renewing environments. *The Journal of Chemical Physics, 100,* 3979–3984.
175. Leo, C. J., Subba Rao, G. V., & Chowdari, B. V. R. (2002). Studies on plasticized PEO–lithium triflate–ceramic filler composite electrolyte system. *Solid State Ionics, 148,* 159–171.
176. Scrosati, B., Croce, F., & Persi, L. (2000). Impedance spectroscopy study of PEO-based nanocomposite polymer electrolytes. *Journal of the Electrochemical Society, 147,* 1718–1721.
177. Jiang, J., & Dahn, J. R. (2003). Comparison of the thermal stability of lithiated graphite in LiBOB EC/DEC and in $LiPF_6$ EC/DEC. *Electrochemical and Solid-State Letters, 6,* A180–A182.
178. Li, W., Xiao, A., Lucht, B. L., et al. (2008). Surface analysis of electrodes from cells containing electrolytes with stabilizing additives exposed to high temperature. *Journal of the Electrochemical Society, 155,* A648–A657.
179. Smart, M. C., Lucht, B. L., & Ratnakumar, B. V. (2008). Electrochemical characteristics of MCMB and $LiNi_xCo_{1-x}O_2$ electrodes in electrolytes with stabilizing additives. *Journal of the Electrochemical Society, 155,* A557–A568.
180. Jiang, J., & Dahn, J. R. (2004). ARC studies of the thermal stability of three different cathode materials: $LiCoO_2$; $Li[Ni_{0.1}Co_{0.8}Mn_{0.1}]O_2$; and $LiFePO_4$, in $LiPF_6$ and LiBoB EC/DEC electrolytes. *Electrochemistry Communications, 6,* 39–43.
181. Cho, I. H., Kim, S.-S., Shin, S. C., et al. (2010). Effect of SEI on capacity losses of spinel lithium manganese oxide/graphite batteries stored at 60 °C. *Electrochemical and Solid-State Letters, 13,* A168–A172.
182. Li, W., & Lucht, B. L. (2007). Inhibition of solid electrolyte interface formation on cathode particles for lithium-ion batteries. *Journal of Power Sources, 168,* 258–264.
183. Chang, C.-C., Chen, T.-K., Her, L.-J., et al. (2009). Tris(pentafluorophenyl)borane as an electrolyte additive to improve the high temperature cycling performance of $LiFePO_4$ cathode. *Journal of the Electrochemical Society, 156,* A828–A832.
184. Ping, P., Wang, Q., Sun, J., et al. (2010). Thermal stabilities of some lithium salts and their electrolyte solutions with and without contact to a $LiFePO_4$ electrode. *Journal of the Electrochemical Society, 157,* A1170–A1176.
185. Cheng, X.-B., Zhang, R., Zhao, C.-Z., et al. (2017). Toward safe lithium metal anode in rechargeable batteries: A review. *Chemical Reviews, 117,* 10403–10473.
186. Bron, P., Roling, B., & Dehnen, S. (2017). Impedance characterization reveals mixed conducting interphases between sulfidic superionic conductors and lithium metal electrodes. *Journal of Power Sources, 352,* 127–134.
187. Xie, D., Chen, S., Zhang, Z., et al. (2018). High ion conductive Sb_2O_5-doped β-Li_3PS_4 with excellent stability against Li for all-solid-state lithium batteries. *Journal of Power Sources, 389,* 140–147.
188. Wu, B., Wang, S., Lochala, J., et al. (2018). The role of the solid electrolyte interphase layer in preventing Li dendrite growth in solid-state batteries. *Energy & Environmental Science, 11,* 1803–1810.
189. Chung, H., & Kang, B. (2017). Mechanical and thermal failure induced by contact between a $Li_{1.5}Al_{0.5}Ge_{1.5}(PO_4)_3$ solid electrolyte and Li metal in an all solid-state Li cell. *Chemistry of Materials, 29,* 8611–8619.
190. Haruyama, J., Sodeyama, K., Han, L., et al. (2014). Space-charge layer effect at interface between oxide cathode and sulfide electrolyte in all-solid-state lithium-ion battery. *Chemistry of Materials, 26,* 4248–4255.
191. Aricò, A. S., Bruce, P., Scrosati, B., et al. (2005). Nanostructured materials for advanced energy conversion and storage devices. *Nature Materials, 4,* 366–377.
192. Jiao, F., & Bruce, P. G. (2007). Mesoporous crystalline β-MnO_2—A reversible positive electrode for rechargeable lithium batteries. *Advanced Materials, 19,* 657–660.

193. Balaya, P., Bhattacharyya, A. J., Jamnik, J., et al. (2006). Nano-ionics in the context of lithium batteries. *Journal of Power Sources, 159,* 171–178.
194. Maier, J. (2005). Nanoionics: ion transport and electrochemical storage in confined systems. *Nature Materials, 4,* 805–815.
195. Jamnik, J., & Maier, J. (2003). Nanocrystallinity effects in lithium battery materials Aspects of nano-ionics. Part IV. *Physical Chemistry Chemical Physics, 5,* 5215–5220.
196. Amatucci, G. G., Badway, F., Du Pasquier, A., et al. (2001). An asymmetric hybrid nonaqueous energy storage cell. *Journal of the Electrochemical Society, 148,* A930–A939.
197. Wang, Q., Li, H., Chen, L., et al. (2001). Monodispersed hard carbon spherules with uniform nanopores. *Carbon, 39,* 2211–2214.
198. Meethong, N., Huang, H.-Y. S., Carter, W. C., et al. (2007). Size-dependent lithium miscibility gap in nanoscale $Li_{1-x}FePO_4$. *Electrochemical and Solid-State Letters, 10,* A134–A138.
199. Choi, J. W., & Aurbach, D. (2016). Promise and reality of post-lithium-ion batteries with high energy densities. *Nature Reviews Materials, 1,* 16013.
200. Huggins, R. A., & Nix, W. D. (2000). Decrepitation model for capacity loss during cycling of alloys in rechargeable electrochemical systems. *Ionics, 6,* 57–63.
201. Fang, W. Q., Gong, X.-Q., & Yang, H. G. (2011). On the unusual properties of anatase TiO_2 exposed by highly reactive facets. *The Journal of Physical Chemistry Letters, 2,* 725–734.
202. Ding, S., Chen, J. S., Wang, Z., et al. (2011). TiO_2 hollow spheres with large amount of exposed (001) facets for fast reversible lithium storage. *Journal of Materials Chemistry, 21,* 1677–1680.
203. Chen, J. S., Tan, Y. L., Li, C. M., et al. (2010). Constructing hierarchical spheres from large ultrathin anatase TiO_2 nanosheets with nearly 100% exposed (001) facets for fast reversible lithium storage. *Journal of the American Chemical Society, 132,* 6124–6130.
204. Cao, F.-F., Wu, X.-L., Xin, S., et al. (2010). Facile synthesis of mesoporous TiO_2-C nanosphere as an improved anode material for superior high rate 1.5 V rechargeable Li ion batteries containing $LiFePO_4-C$ cathode. *The Journal of Physical Chemistry C, 114,* 10308–10313.
205. Jiang, K.-C., Xin, S., Lee, J.-S., et al. (2012). Improved kinetics of $LiNi_{1/3}Mn_{1/3}Co_{1/3}O_2$ cathode material through reduced graphene oxide networks. *Physical Chemistry Chemical Physics, 14,* 2934–2939.
206. Peled, E. (1979). The electrochemical behavior of alkali and alkaline earth metals in nonaqueous battery systems—The solid electrolyte interphase model. *Journal of the Electrochemical Society, 126,* 2047–2051.
207. Beaulieu, L. Y., Eberman, K. W., Turner, R. L., et al. (2001). Colossal reversible volume changes in lithium alloys. *Electrochemical and Solid-State Letters, 4,* A137–A140.
208. Wu, H., Chan, G., Choi, J. W., et al. (2012). Stable cycling of double-walled silicon nanotube battery anodes through solid–electrolyte interphase control. *Nature Nanotechnology, 7,* 310–315.
209. Liu, N., Lu, Z., Zhao, J., et al. (2014). A pomegranate-inspired nanoscale design for large-volume-change lithium battery anodes. *Nature Nanotechnology, 9,* 187–192.
210. Zhang, H., Zhou, L., Noonan, O., et al. (2014). Tailoring the void size of iron Oxide@Carbon yolk-shell structure for optimized lithium storage. *Advanced Functional Materials, 24,* 4337–4342.
211. Odziemkowski, M., & Irish, D. E. (1992). An electrochemical study of the reactivity at the lithium electrolyte/bare lithium metal interface: I. Purified electrolytes. *Journal of the Electrochemical Society, 139,* 3063–3074.
212. Peled, E., & Straze, H. (1977). The kinetics of the magnesium electrode in thionyl chloride solutions. *Journal of the Electrochemical Society, 124,* 1030–1035.
213. Kanamura, K., Shiraishi, S., & Takehara, Z. I. (1996). Electrochemical deposition of very smooth lithium using nonaqueous electrolytes containing HF. *Journal of the Electrochemical Society, 143,* 2187–2197.
214. Kanamura, K., Tamura, H., Shiraishi, S., et al. (1995). XPS analysis of lithium surfaces following immersion in various solvents containing $LiBF_4$. *Journal of the Electrochemical Society, 142,* 340–347.

215. Ein-Eli, Y. (1999). A new perspective on the formation and structure of the solid electrolyte interface at the graphite anode of Li-ion cells. *Electrochemical and Solid-State Letters, 2,* 212–214.
216. Ein-Eli, Y., McDevitt, S. F., & Laura, R. (1998). The superiority of asymmetric alkyl methyl carbonates. *Journal of the Electrochemical Society, 145,* L1–L3.
217. Garreau, M. (1987). Cyclability of the lithium electrode. *Journal of Power Sources, 20,* 9–17.
218. Thevenin, J. G., & Muller, R. H. (1987). Impedance of lithium electrodes in a propylene carbonate electrolyte. *Journal of the Electrochemical Society, 134,* 273–280.
219. Xu, K. (2004). Nonaqueous liquid electrolytes for lithium-based rechargeable batteries. *Chemical Reviews, 104,* 4303–4418.
220. Hu, J. J., Long, G. K., Liu, S., et al. (2014). A LiFSI–LiTFSI binary-salt electrolyte to achieve high capacity and cycle stability for a Li–S battery. *Chemical Communications, 50,* 14647–14650.
221. Shkrob, I. A., Marin, T. W., Zhu, Y., et al. (2014). Why bis(fluorosulfonyl)imide is a "magic anion" for electrochemistry. *The Journal of Physical Chemistry C, 118,* 19661–19671.
222. Wu, F., Zhu, Q., Chen, R., et al. (2015). Ionic liquid-based electrolyte with binary lithium salts for high performance lithium–sulfur batteries. *Journal of Power Sources, 296,* 10–17.
223. Heine, J., Hilbig, P., Qi, X., et al. (2015). Fluoroethylene carbonate as electrolyte additive in tetraethylene glycol dimethyl ether based electrolytes for application in lithium ion and lithium metal batteries. *Journal of the Electrochemical Society, 162,* A1094–A1101.
224. Lu, Y., Tu, Z., Shu, J., et al. (2015). Stable lithium electrodeposition in salt-reinforced electrolytes. *Journal of Power Sources, 279,* 413–418.
225. Miao, R., Yang, J., Feng, X., et al. (2014). Novel dual-salts electrolyte solution for dendrite-free lithium-metal based rechargeable batteries with high cycle reversibility. *Journal of Power Sources, 271,* 291–297.
226. Lu, Y., Tu, Z., & Archer, L. A. (2014). Stable lithium electrodeposition in liquid and nanoporous solid electrolytes. *Nature Materials, 13,* 961–969.
227. Togasaki, N., Momma, T., & Osaka, T. (2014). Enhancement effect of trace H_2O on the charge–discharge cycling performance of a Li metal anode. *Journal of Power Sources, 261,* 23–27.
228. Qian, J., Xu, W., Bhattacharya, P., et al. (2015). Dendrite-free Li deposition using trace-amounts of water as an electrolyte additive. *Nano Energy, 15,* 135–144.
229. Aurbach, D., Pollak, E., Elazari, R., et al. (2009). On the surface chemical aspects of very high energy density, rechargeable Li–sulfur batteries. *Journal of the Electrochemical Society, 156,* A694–A702.
230. Choi, J.-W., Cheruvally, G., Kim, D.-S., et al. (2008). Rechargeable lithium/sulfur battery with liquid electrolytes containing toluene as additive. *Journal of Power Sources, 183,* 441–445.
231. Gordin, M. L., Dai, F., Chen, S., et al. (2014). Bis(2,2,2-trifluoroethyl) ether as an electrolyte Co-solvent for mitigating self-discharge in lithium-sulfur batteries. *ACS Applied Materials & Interfaces, 6,* 8006–8010.
232. Ding, F., Xu, W., Graff, G. L., et al. (2013). Dendrite-free lithium deposition via self-healing electrostatic shield mechanism. *Journal of the American Chemical Society, 135,* 4450–4456.
233. Wang, S.-H., Yin, Y.-X., Zuo, T.-T., et al. (2017). Stable Li metal anodes via regulating lithium plating/stripping in vertically aligned microchannels. *Advanced Materials, 29,* 1703729.
234. Ye, H., Xin, S., Yin, Y.-X., et al. (2017). Stable Li plating/stripping electrochemistry realized by a hybrid Li reservoir in spherical carbon granules with 3D conducting skeletons. *Journal of the American Chemical Society, 139,* 5916–5922.
235. Yan, K., Lee, H.-W., Gao, T., et al. (2014). Ultrathin two-dimensional atomic crystals as stable interfacial layer for improvement of lithium metal anode. *Nano Letters, 14,* 6016–6022.
236. Zheng, G., Lee, S. W., Liang, Z., et al. (2014). Interconnected hollow carbon nanospheres for stable lithium metal anodes. *Nature Nanotechnology, 9,* 618–623.
237. Han, F., Zhu, Y., He, X., et al. (2016). Electrochemical stability of $Li_{10}GeP_2S_{12}$ and $Li_7La_3Zr_2O_{12}$ solid electrolytes. *Advanced Energy Materials, 6,* 1501590.

238. Li, Y., Xu, B., Xu, H., et al. (2017). Hybrid polymer/garnet electrolyte with a small interfacial resistance for lithium-ion batteries. *Angewandte Chemie International Edition, 56,* 753–756.
239. Li, Y., Zhou, W., Xin, S., et al. (2016). Fluorine-doped antiperovskite electrolyte for all-solid-state lithium-ion batteries. *Angewandte Chemie International Edition, 55,* 9965–9968.
240. Hartmann, P., Leichtweiss, T., Busche, M. R., et al. (2013). Degradation of NASICON-type materials in contact with lithium metal: Formation of mixed conducting interphases (MCI) on solid electrolytes. *The Journal of Physical Chemistry C, 117,* 21064–21074.
241. Luo, W., Gong, Y., Zhu, Y., et al. (2016). Transition from superlithiophobicity to superlithiophilicity of garnet solid-state electrolyte. *Journal of the American Chemical Society, 138,* 12258–12262.
242. Wang, C., Gong, Y., Liu, B., et al. (2017). Conformal, nanoscale ZnO surface modification of garnet-based solid-state electrolyte for lithium metal anodes. *Nano Letters, 17,* 565–571.
243. Han, X., Gong, Y., Fu, K., et al. (2016). Negating interfacial impedance in garnet-based solid-state Li metal batteries. *Nature Materials, 16,* 572–579.
244. Mogensen, R., Brandell, D., & Younesi, R. (2016). Solubility of the solid electrolyte interphase (SEI) in sodium ion batteries. *ACS Energy Letters, 1,* 1173–1178.
245. Sun, B., Li, P., Zhang, J., et al. (2018). Dendrite-free sodium-metal anodes for high-energy sodium-metal batteries. *Advanced Materials, 30,* 1801334.
246. Meitav, A., & Peled, E. (1982). Solid electrolyte interphase (SEI) electrode: Part III. Deposition-dissolution process of lithium in thionyl-chloride solution. *Journal of Electroanalytical Chemistry and Interfacial Electrochemistry, 134,* 49–63.
247. Derrien, G., Hassoun, J., Panero, S., et al. (2007). Nanostructured Sn–C composite as an advanced anode material in high-performance lithium-ion batteries. *Advanced Materials, 19,* 2336–2340.
248. Guo, Y.-G., Hu, Y.-S., Sigle, W., et al. (2007). Superior electrode performance of nanostructured mesoporous TiO_2 (Anatase) through efficient hierarchical mixed conducting networks. *Advanced Materials, 19,* 2087–2091.
249. Park, K.-S., Schougaard, S. B., & Goodenough, J. B. (2007). Conducting-polymer/iron-redox-couple composite cathodes for lithium secondary batteries. *Advanced Materials, 19,* 848–851.
250. Cao, A.-M., Hu, J.-S., Liang, H.-P., et al. (2005). Self-assembled vanadium pentoxide (V_2O_5) hollow microspheres from nanorods and their application in lithium-ion batteries. *Angewandte Chemie International Edition, 44,* 4391–4395.
251. You, Y., Celio, H., Li, J., et al. (2018). Modified high-nickel cathodes with stable surface chemistry against ambient air for lithium-ion batteries. *Angewandte Chemie International Edition, 57,* 6480–6485.
252. Cao, F.-F., Xin, S., Guo, Y.-G., et al. (2011). Wet chemical synthesis of Cu/TiO_2 nanocomposites with integrated nano-current-collectors as high-rate anode materials in lithium-ion batteries. *Physical Chemistry Chemical Physics, 13,* 2014–2020.
253. Hu, Y.-S., Guo, Y.-G., Dominko, R., et al. (2007). Improved electrode performance of porous $LiFePO_4$ using RuO_2 as an oxidic nanoscale interconnect. *Advanced Materials, 19,* 1963–1966.
254. Wu, X.-L., Jiang, L.-Y., Cao, F.-F., et al. (2009). $LiFePO_4$ nanoparticles embedded in a nanoporous carbon matrix: Superior cathode material for electrochemical energy-storage devices. *Advanced Materials, 21,* 2710–2714.
255. Li, X., Jin, H., Liu, S., et al. (2015). Carambola-shaped $LiFePO_4/C$ nanocomposites: directing synthesis and enhanced Li storage properties. *Journal of Materials Chemistry A, 3,* 116–120.
256. Kang, H., Liu, H., Li, C., et al. (2018). Polyanthraquinone-triazine—A promising anode material for high-energy lithium-ion batteries. *ACS Applied Materials & Interfaces, 10,* 37023–37030.

Chapter 3
Nanostrucutres and Nanomaterials for Lithium-Ion Batteries

Fei-Fei Cao, Huan Ye and Yu-Guo Guo

Abstract Lithium-ion batteries have shown great promise in portable electronics and electric vehicles due to their long lifespan and high safety. However, hurdles relating to the sluggish dynamics and poor cycling stability restrict the practical application. Nanostructured materials, due to their significantly decreased particles size, are thought to effectively address the above issues. In this section, we will concentrate on the recent progress in the development of advanced electrode materials including cathodes and anodes for lithium-ion batteries. Particularly, the utilization of nanostructured materials and techniques to decrease the particles size and elevate the diffusion coefficient of electrode materials, and thus shorten the diffusion distance for lithium ions and alleviate the large volume change of electrodes, ensuring enhanced rate capability and improved stability for lithium-ion batteries will be discussed in detail.

Abbreviations

LIB	Lithium-ion battery
3D	Three-dimensional
Li^+	Lithium ion
TM	Transition metal
J-T	Jahn-Teller
NCA	$LiNi_{1-x-y}Co_xAl_yO_2$
NCM	$LiNi_{1-x-y}Co_xMn_yO_2$
DOD	Depth of discharge
LLO	Lithium-rich layered oxide
CNT	Carbon nanotube
SWCNT	Single-walled carbon nanotube

F.-F. Cao · H. Ye
Huazhong Agricultural University, Wuhan 430070, Hubei, People's Republic of China

Y.-G. Guo (✉)
Institute of Chemistry, Chinese Academy of Sciences, Beijing 100190, People's Republic of China
e-mail: ygguo@iccas.ac.cn

Y.-G. Guo, *Nanostructures and Nanomaterials for Batteries*,
https://doi.org/10.1007/978-981-13-6233-0_3

MWCNT	Multi-walled carbon nanotube
LBL	Layer-by-layer
HPCF	Hierarchical porous carbon fiber
MOF	Metal-organic framework
SEI	Solid electrolyte interphase
NP	Nanoparticle
0D	Zero-dimensional
1D	One-dimensional
NW	Nanowire
NT	Nanotube
CVD	Chemical vapor deposition
SEM	Scanning electron microscope
TEM	Transmission electron microscope
NF	Nanofiber
PCNF	Porous carbon nanofiber
PAN	Polyacrylonitrile
DMF	N-dimethylformamide
GT	Graphitic tube
ALD	Atomic layer deposition
2D	Two-dimensional
GO	Graphene oxide
rGO	Reduced graphene oxide
GDY	Graphdiyne
PDA	Polydopamine
CNM	Carbon nanotube macrofilm
PDDA	Poly(diallydimethylammonium chloride)
NC	Nanocrystal
CNF	Carbon nanofiber
STEM	Scanning transmission electron microscopy
SMC	Submicrocube
HN	Hollow nanosphere
NDC	Nitrogen-doped carbon
LTO	$Li_4Ti_5O_{12}$
PCP	Porous carbon polyhedra
TNHC	Tin nanoparticles encapsulated elastic hollow carbon sphere
ZIF	Zeolitic imidazolate framework
NS	Nanosheet
ICE	Initial Coulombic efficiency

3.1 General Introduction of Lithium-Ion Batteries

The evolution of rechargeable lithium batteries involved into three stages including lithium metal secondary battery, rocking chair battery, and solid-state lithium battery. Before the late 1980s, researchers mainly concentrated on developing secondary batteries based on Li metal as anode on account of the fact that Li is the most electropositive (–3.04 V vs. standard hydrogen electrode), and Li has its highest specific capacity (3860 mA h g^{-1}), as well as Li demonstrates the lightest weight (0.53 g cm^{-3}), thus promoting the energy storage systems with high energy density [1]. One seminal work of developing the TiS_2 cathode combined with Li metal anode was conducted by Exxon in 1972 [2]. Despite great progress on this system, its practical application was soon impeded by its poor cycling performance related to dendritic Li growth and the formation of inhomogeneous solid electrolyte interphase (SEI) during each subsequent discharge/recharge cycle. Substituting Li metal for an alloy addressed the dendrite obsession but the use of alloy electrodes was still restricted by a limited number of cycles due to their serious volume expansion during operation [3]. In the meantime, intercalation materials of oxides with higher capacities and voltages were also proposed to be a new class of advanced battery materials. Later, Goodenough synthesized one remarkable material of Li_xMO_2 (where M refers to Co, Ni or Mn) [4, 5], which gradually developed into a widely used cathode in today's batteries. In 1990, a rechargeable lithium battery was assembled based on the carbonaceous material anode in combination with $LiCoO_2$ cathode, and first proposed the Li-ion battery concept. In 1991, Sony Corporation launched the first commercial C/$LiCoO_2$ Li-ion batteries [6]. Owing to their high discharge voltage exceeding 3.6 V and high gravimetric energy densities approaching to 120–300 W h kg^{-1}, they rapidly found applications as power sources for portable electronic devices, electric vehicles and hybrid electric vehicles.

Lithium-ion battery (LIB) mainly consists of three basic components including an anode (usually graphitic carbon), a cathode (lithium transition-metal oxides), and an electrolyte. LIB is a device that realizes the reversible transformation between chemical energy and electronic energy based on Li ion insertion and extraction. During charge, Li migrates from the intercalation cathode through the organic electrolyte to the graphitic carbon anode in the form of Li^+ ions. Simultaneously, electron is removed from the d-orbital of transition metal (TM) of intercalation cathode and is transferred through an external circuit, which in combination with incoming Li^+ ions to insert into the graphitic carbon to form LiC_6, realizing the conversion of electric energy into chemical energy. During discharge, Li^+ ions are extracted from the anode and inserted into the cathode.

For a material with a given chemical composition, its microstructure, crystal structure, and electronic structure would significantly affect its chemical, electrochemical, and physical properties. Therefore, numerous efforts have been devoted to developing high-performance electrode materials to cater to the kinetic and thermodynamic needs for advanced LIBs. Comparing with their micrometer-sized counterparts, nanostructures and nanomaterials show much improved thermodynamics and dynamics. To date, well-known advantages with nanostructured materials are illustrated: (i) the enhancement of the reversibility of Li^+ ion insertion and extraction reaction in nanostructured electrodes without destroying the crystal structure; (ii) the reduction in the diffusion or transport distance of the Li ions and electrons accompanying the Li-ion insertion and extraction reactions; and (iii) the change in the site energy of Li ions occupying the host lattice.

3.2 Cathode Materials for Lithium-Ion Batteries

As a key component of the cell, the cathodes have a remarkable effect on the electrochemical performance. The choice of the cathodes relates with the type of the assembled batteries. For rechargeable Li batteries, because the anode is metallic Li, the cathodes can be Li-containing or Li-free materials. In contrast, for LIBs, owing to the utilization of graphitic carbon as the anode, the cathode must serve as a source of Li, thus requiring a Li-containing intercalation compounds with high air stability to facilitate the cell assembly. A cathode must meet the following critical requirements to realize the goal of high energy density: (i) a high specific capacity and a high voltage; (ii) a high electrochemical reversibility; (iii) a good electronic and ionic conductivity; (iv) a superior structural stability; (v) low cost and environmentally benignity.

The cathode materials can be classified into two categories. One group comprises layered lithium compounds with a close-packed anion lattice, in which TM cations occupy alternate layers between the anion sheets and Li ions are intercalated into remaining empty layers (Fig. 3.1a, c). This category has the inherent advantage of higher specific energy and operating voltage owing to their more compact lattices and highly oxidizing redox-active couples. $LiTiS_2$, $LiCoO_2$, $LiNiO_2$, $LiMnO_2$, and their derivatives of $LiNi_{1-x}Co_xO_2$, and $LiNi_xMn_xCo_{1-2x}O_2$ belong to this group. The other group consists of materials with more open structures, such as vanadium oxides, molybdenum oxides, the tunnel or layered compounds of manganese oxides, and transition metal phosphates, such as the olivine $LiFePO_4$ (Fig. 3.1b, c). Although they demonstrate lower operating voltage, the reduced cost, improved safety and rate capability of these materials compared to the first group make them competitive cathode candidates.

3.2.1 *Trichalcogenides and Related Materials*

Sulfur-rich transition metal sulfides such as TiS [3, 8] MoS_3 [9], TiS_4 [10], and $NbSe_3$ [11] have been explored as promising cathodes due to their high electronic conductivity and low solubility in liquid electrolytes. For TiS_3, the polysulfide group in TiS(S-S) firstly reacted with two lithium, breaking the S–S bond to form Li_2TiS_3 by a two-phase reaction. Then the titanium was reduced from Ti^{4+} to Ti^{3+} by a single-

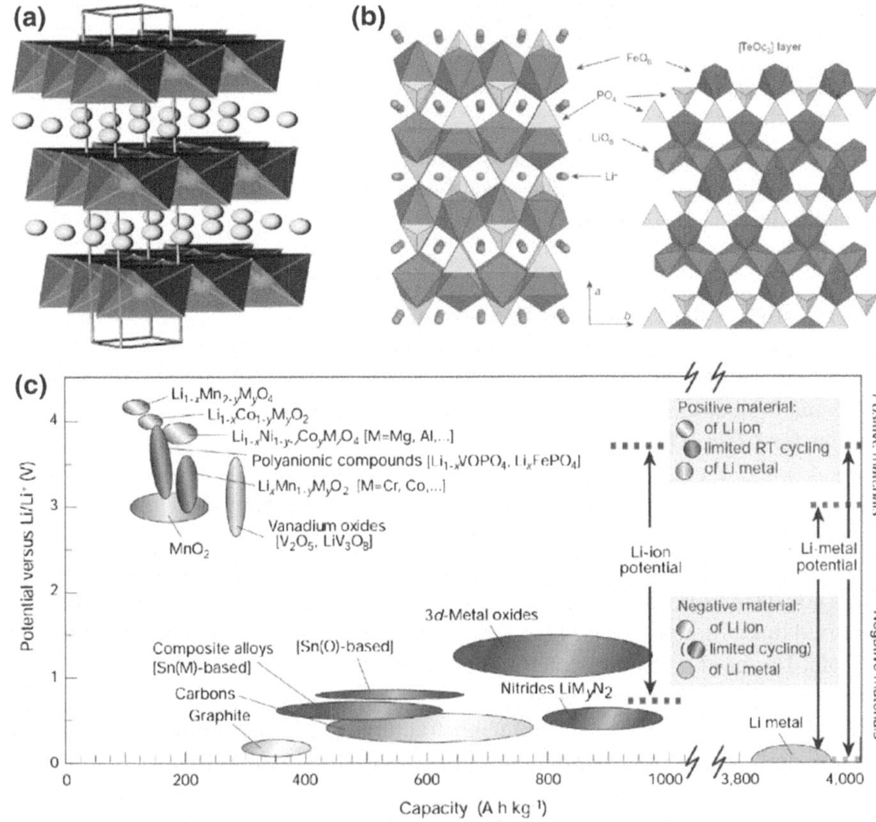

Fig. 3.1 **a** Crystal structure of layered $LiTiS_2$, $LiCoO_2$, $LiNiO_2$, and $LiNi_yMn_yCo_{1-2y}O_2$. Reproduced with the permission from Ref. [7], Copyright © 2004 American Chemical Society. **b** Crystal structure of olivine $LiFePO_4$ in projection along [001]. Reproduced with the permission from Ref. [1], Copyright © 2001 Nature. **c** Thermodynamic equilibrium voltage versus theoretical gravimetric capacity for various electrode materials. Reproduced with the permission from Ref. [1], Copyright © 2001 Nature

phase reaction which is similar to TiS_2 itself [8]. During the subsequent cycling, only the second step contributed the reversible capacity. On the contrary, $NbSe_3$ was found to be reversible with three lithium to form Li_3NbSe_3 by a single-phase reaction [11]. A lot of other high-rate high-conductivity materials such as TiS_2 and VSe_2 have been exploited to serve as the additives to compensate the chalcogenide-rich materials that demonstrate low reaction rates and conductivity.

3.2.2 *Nanostructured Vanadium Oxides*

Vanadium oxide is a typical intercalation compound as a result of its layered structure. For Li-ion intercalation applications, vanadium oxide offers the essential advantages of low cost, an abundant source, easy synthesis, and high energy densities. This section intends to discuss the categories of nanostructured vanadium oxides as cathode materials for LIBs (Fig. 3.2).

3.2.2.1 High-Valent Oxides of Vanadium

Because of the abundant sources as well as high theoretical capacity (294 mA h g^{-1}), vanadium oxide (V_2O_5) has attracted much attention as a well-known cathode material for LIBs. The high capacity originates from the insertion of two Li ions during 4.0–2.0 V (vs. Li^+/Li), which is much higher than $LiCoO_2$ (140 mA h g^{-1}) and $LiFePO_4$ (170 mA h g^{-1}). Even so, the poor structural stability, low electronic conductivity, and low diffusion coefficient of Li^+ prevent its large-scale application. To overcome these drawbacks, numerous efforts have been employed, including design novel structure, dope other elements, surface-coating techniques, or make self-supported three-dimensional (3D) porous V_2O_5 based carbon materials [12–15]. For example, Liu et al. [16] fabricated hollow porous V_2O_5 nanotubes (NTs) through electrospinning method. Wu et al. [17] reported a kind of hierarchical sisal-like V_2O_5 microstructure. These special nanostructures could shorten the Li^+ diffusing distance, and increase the electrochemical activity. Haris et al. [18] coated acetylene

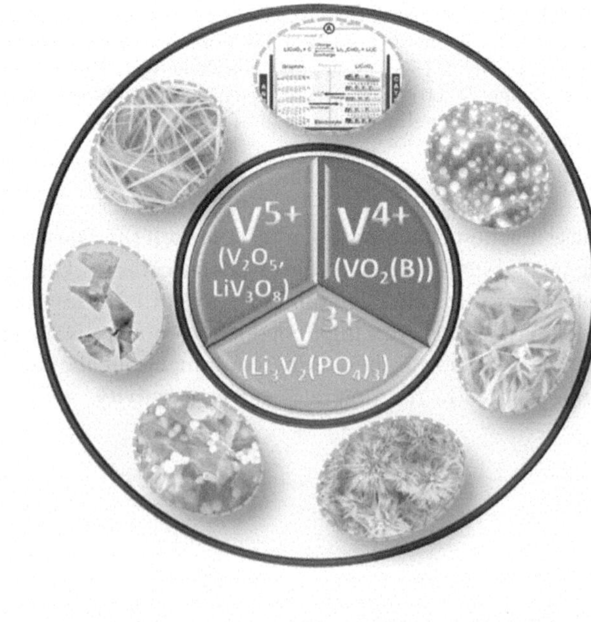

Fig. 3.2 Functional vanadium-based nanostructure materials ranging from 0D NPs to 3D interconnected networks for LIBs. Reproduced with the permission from Ref. [15], Copyright © 2015 The Royal Society of Chemistry

black onto the surface of V_2O_5. Attributing to the synergistic effect of nanosized V_2O_5 and conductive acetylene black, the corresponding nanocomposite exhibited enhanced electrochemical performance. To further enhance the energy density of LIBs, Chen et al. [19] reported lightweight binder-free positive electrodes by combing V_2O_5/nitrogen-doped carbon (NDC) core/shell arrays and graphene foam. This material delivered a noticeable initial capacity of 293 mA h g^{-1} at 1 C in the voltage range of 2.0–4.0 V and 223 mA h g^{-1} at 5 C after 1000 cycles.

3.2.2.2 Mixed-Valent Oxides of Vanadium

Amongst various vanadium oxides (e.g., V_2O_5, VO_2, V_2O_3, V_3O_7, V_6O_{13}, etc.), V_6O_{13} exhibits an outstanding capacity and energy density because of the large number of Li^+ that can be reversibly intercalated into the V lattice (up to 8 Li^+ per formula unit with all the V ions being reduced to V^{3+} oxidation state) [20]. It is reported that the theoretical specific capacity can reach 417 mA h g^{-1}. In practical, the limited number of electronic sites in V_6O_{13} bulk state decreases the certain lithium uptake, leading to the specific capacity around or below 310 mA h g^{-1}. Therefore, nanostructuring has become one of the most powerful means to improve electrochemical performance of V_6O_{13}. For example, Zou et al. [21] synthesized V_6O_{13} nanosheets (NSs) through a facile hydrothermal synthesis followed by an appropriate heat treatment. Ding et al. [20] further reported a 3D V_6O_{13} nanotextiles built from interconnected one-dimensional (1D) nanogrooves with diameter of 20–50 nm. However, their poor phase stability upon lithium insertion were not suitable for scale-up.

Monoclinic lithium trivanadate (LiV_3O_8) is another promising cathode candidate due to its low-cost, natural abundance, high energy density, and good safety [22]. This structure can be of benefit to reversibly accept multiple Li^+, leading to the high discharge specific capacity. According to the reported literatures, 3 Li^+ can be incorporated into the amorphous LiV_3O_8, contributing to the theoretical capacity of 280 mA h g^{-1} [23]. However, its electrochemical performance and practical applications as a cathode for LIBs is largely impeded by its poor electronic conductivity and low Li^+ diffusion coefficient. The structure damage during the insertion and desertion of lithium is another fatal issue before practical application. To address these issues, numerous strategies were employed, including nanosizing LiV_3O_8 structure to shorten Li^+ diffusion distance, surface coating to enhance the electronic conductivity, and doping of guest species to improve the structure stability [15].

3.2.2.3 Vanadium Phosphate Phases

Transition-metal phosphates, another class of Li intercalation material, have been widely regarded as the next generation cathode materials for LIBs due to the mobile Li ions and the redox-active metals within a rigid phosphate network. Among numerous phosphates, monoclinic lithium vanadium phosphate ($Li_3V_2(PO_4)_3$) is proposed as a highly prospective cathode owing to the following competitive advantages: (i) abundant natural resources; (ii) high theoretical capacity (197 mA h g^{-1}); (iii) superior structure stability [24]. Because of the two separated [VO_6] octahedral arrangement, the $Li_3V_2(PO_4)_3$ demonstrates inferior electronic conductivity, which prevent its widespread application.

Surface carbon coating is the most effective way to improve its electronic conductivity so that the cathode material can be utilized at high current rates [25]. Firstly, carbon coating can prevent the $Li_3V_2(PO_4)_3$ particles from growing up and aggregating during the calcination process at high temperature. Secondly, in the synthesis process, carbon acts as reducing agent to reduce V^{5+} to V^{3+}, which helps simplify the atmosphere requirement. Thirdly, the introduced carbon can improve its electronic conductivity. In fact, the effect of carbon coating on the $Li_3V_2(PO_4)_3$ particle is linked to the homogeneity, thickness, and graphitization degree of carbon. Although carbon coating can improve the conductivity of $Li_3V_2(PO_4)_3$, excess carbon can obstruct the diffusion of Li^+ and seriously decrease the tap density, and therefore, a delicate balance between carbon content and battery performance should be considered. For example, Rui et al. [26] studied $Li_3V_2(PO_4)_3$/C nanocomposites with different carbon contents ranging from 5.7, 9.6, 11.6 to 15.3 wt%. The electrochemical results demonstrated that the sample with 11.6 wt% carbon content shows the lowest charge transfer resistance and the best electrochemical properties, especially for the rate capability, viz., the capacity decreasing by only 7.2% from 125 mA h g^{-1} at 0.5 C to 116 mA h g^{-1} at 5 C between 3.0–4.3 V versus Li^+/Li. Liu et al. introduced graphene into $Li_3V_2(PO_4)_3$ to improve the rate performance and cycling stability of $Li_3V_2(PO_4)_3$/graphene nanocomposite [27].

Besides, ion-doping is also exploited to adjust the intrinsic electronic conductivity and the diffusion of Li^+. According to the literature survey, lots of multivalent cations, including Co^{2+}, Mn^{2+}, Ni^{2+}, Ca^{2+}, Fe^{3+}, Cr^{3+}, Al^{3+}, Ce^{3+}, Sn^{4+}, and Mo^{6+}, etc., have been introduced as dopants in the $Li_3V_2(PO_4)_3$ system [28]. Doping individual aliovalent cations at the V sites would cause the problem of charge imbalance, tending to introduce some impurities in the product, and therefore, co-doping is adopted. Besides, doping metal cations (e.g., Na^+, and Ca^{2+}) at the Li sites instead of V sites or anion doping (e.g., F^-, and Cl^-) was also applied to enhance the electrochemical performance of $Li_3V_2(PO_4)_3$.

3.2.3 Nanostructured Lithium Transition Metal Oxides

Lithium can be insertion/de-insertion from lithium transition metal oxides, thus delivering high specific capacity and energy density. This section intends to discuss the categories of nanostructured lithium transition metal oxides as cathode materials for LIBs.

3.2.3.1 Lithium Cobalt Oxide

$LiCoO_2$ is the most extensively used cathode material for commercial LIBs due to a high theoretical capacity (274 mA h g^{-1} based on 1.0 Li/Co), reasonable rate capability and good cycle life, as well as facile and scalable production [29]. However, due to structural limitations, only around 0.5 Li/Co can be reversibly cycled to yield a useable specific capacity of ~140 mA h g^{-1} [7]. In addition, $LiCoO_2$ cathode suffers from rapid capacity suppression in prolonging cycling associating with the following three main factors: (i) severe structure variation related to phase transition from hexagonal to monoclinic phase during overcharging process, (ii) the dissolution of Co into the electrolyte, (iii) the decomposition of electrolyte at high voltage associated with the release of oxygen from lattice.

Targeted at the above problems of $LiCoO_2$ cathode, tremendous efforts have been adopted, including doping of foreign atoms (Mg or Al) in bulk $LiCoO_2$ [30, 31], surface coatings with inorganic materials (metal-phosphates [32], oxides [33–38], and fluorides [39–41]) and electrochemically active materials, using additives in liquid electrolytes, and replacing the liquid electrolyte with solid electrolyte [42].

Doping is the simplest strategy for improving the structural and thermal stability. Various dopants such as Al, Mg, Ti, Nb, Mo and Na have been introduced into $LiCoO_2$. The doped $LiCoO_2$ has shown structural stability and kinetic advantage over the bare $LiCoO_2$ due to the suppression of the undesired phase transition and the promotion of the Li^+ transport. Among different dopants, the Al and Mg are considered as the most promising candidates due to their low cost. In 1998, Ceder et al. [43] reported that the introduction of Al into $LiCoO_2$ increased the intercalation voltage of $LiCoO_2$ (Fig. 3.3a, b). Soon afterwards, Jang et al. [30] found that Al-doped $LiCoO_2$ improved Li diffusivity and alleviated the structure variation. In addition, Tukamoto et al. [31] reported partial doping of Mg into $LiCoO_2$ may lead to improved long-term stability on cycling.

As for $LiCoO_2$, nanocoatings of various phosphates, oxides and fluorides have been studied and great improvements in capacity retention have been demonstrated. For example, Kim et al. made an extensive study on the effect of the MPO_4 (M = Al, Fe, and Ce) nanoparticle (NP) coatings on $LiCoO_2$ cathode materials (Fig. 3.3c–f) [32]. They found that the $LiCoO_2$ fully covered by thin $AlPO_4$ or $FePO_4$ NPs coating exhibits an intercalation capacity as high as of 230 mA h g^{-1} at a high cutoff voltage of 4.8 V (Fig. 3.3g). The improvement in the electrochemical performance in the coated $LiCoO_2$ is due to the suppression of cobalt dissolution and the refrainment of the non-uniform distribution of local strain by the coating layer. In addition to coating of phosphates, surface modification of $LiCoO_2$ by coating various oxides such as ZnO [44], ZrO_2 [36], or Al_2O_3 [33] has been widely explored. In the case of ZnO-coated $LiCoO_2$, the ZnO coating was supposed to suppress the cobalt dissolution and prevent the inorganic surface films such as LiF from covering the $LiCoO_2$ particles, and thus lead to the high-voltage cyclability [45]. ZrO-coated $LiCoO_2$ demonstrates a similar manner by protecting the $LiCoO_2$ cathode surface and reducing the electrolyte decomposition at high voltages [46, 47]. As for Al_2O_3-coated $LiCoO_2$, owing to the formation of a solid solution of $LiCo_{1-y}Al_yO_2$, the structural stability and the capacity

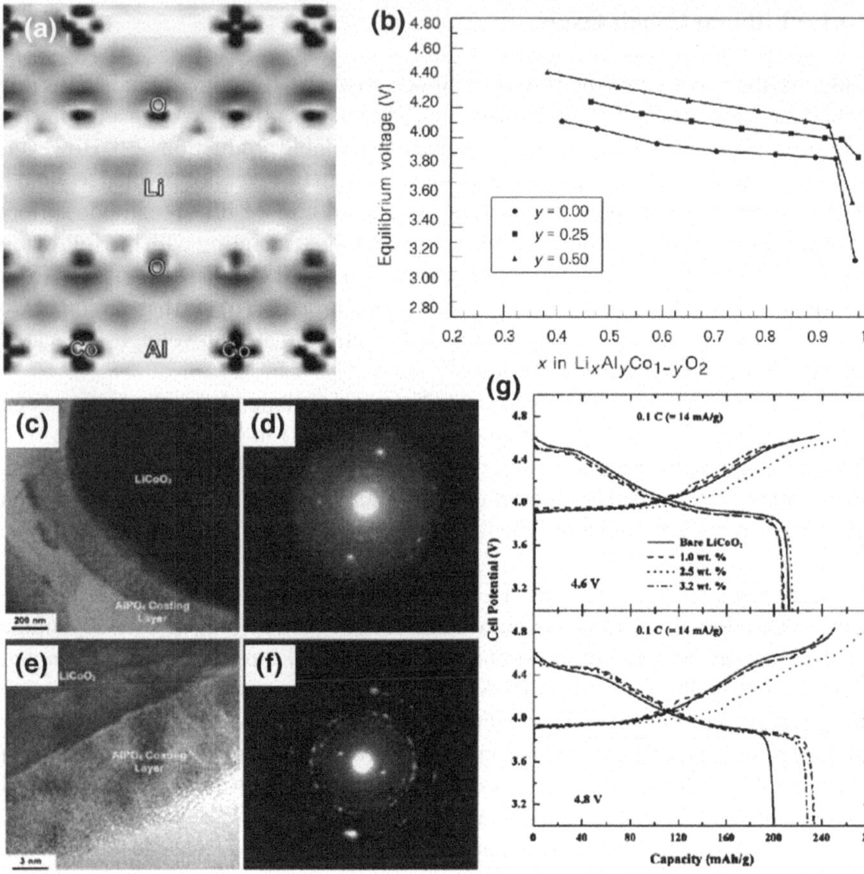

Fig. 3.3 **a** Electron density difference between $Li(Al_{0.33}Co_{0.67})O_2$ and $(Al_{0.33}Co_{0.67})O_2$ in a plane perpendicular to the direction of layering in the structure. **b** Equilibrium potential versus Al content in bare $LiCoO_2$ and in $Li_xAl_yCo_{1-y}O_2$. Reproduced with the permission from Ref. [43], Copyright © 1998 Nature. Structural characterization of the 3.2 wt% $AlPO_4$-coated $LiCoO_2$ (**c**, **d**), the 1.0 wt% $AlPO_4$-coated $LiCoO_2$ (**e**, **f**). **g** Charge-discharge curves comparison of the bare $LiCoO_2$ and 1.0, 2.5 3.2 wt% $AlPO_4$-coated $LiCoO_2$ at 0.1 C with charge-cutoff voltages of 4.6 and 4.8 V. Reproduced with the permission from Ref. [32], Copyright © 2004 The Electrochemical Society

retention get significantly improved even at a higher cut-off voltage of 4.4 V [48, 49]. In addition, various other metal oxide coatings such as MgO, SnO_2, SiO_2, and CuO etc., have also been employed for $LiCoO_2$ because they are helpful to suppress the release of oxygen, and therefore improve the stability of the interior structure and suppress the decomposition of the electrolyte.

Although metal oxide coatings lead to a certain improvement in electrochemical performances, the coated oxides are supposed to react with trace water in a $LiPF_6$-based electrolyte and finally convert into fluorides to stabilize the active surface of $LiCoO_2$. On the base of this theory, various metal fluoride coatings such as CeF_3,

AlF_3, MgF_2, and AlW_xF_y etc. coatings have been exploited to improve the capacity retention [39–41]. For instance, improved capacity retention at a cut-off of 4.5 V was realized through a uniform 10 nm coating of AlF_3 on $LiCoO_2$ [40].

Despite the $LiCoO_2$ cathode dominates the commercial rechargeable lithium battery market, the recent rise of the safety issues associated with the reaction between $LiCoO_2$ and liquid electrolyte in the fully charged state and the high toxicity of cobalt, as well as the high cost of $LiCoO_2$ due to the limited availability of cobalt have stimulated research efforts in battery systems using other cathode materials, such as $LiCo_{1-y}Ni_yO_2$ or $LiNiO_2$.

3.2.3.2 Lithium Nickel Oxide

As an alternative material to $LiCoO_2$, the relatively low cost of Ni promotes research on $LiNiO_2$, isostructural with $LiCoO_2$, as a promising cathode candidate for LIBs [50, 51]. $LiNiO_2$ exhibits a high theoretical capacity of ~275 mA h g^{-1}, which is close to that of $LiCoO_2$ cathode. Owing to the less overlap of the Ni 3d band with the top of O 2p band in its electronic structure, the actual reversible specific capacity of $LiNiO_2$ can be up to 200 mA h g^{-1} (Fig. 3.4a) [52, 53]. Figure 3.4b presents the typical valance electron configurations of d orbitals in TM cationic species. Mn^{3+} and Ni^{3+} ions have a lone electron in the e.g. level. The electron energy level in the nondegenerate orbit splits and engenders a Jahn–Teller distortion, which may lower the crystal structural reversibility of $LiNiO_2$ (Fig. 3.4c). However, it has been confirmed that formation of a stoichiometric $LiNiO_2$ with a Li/Ni ratio of 1:1 is difficult. The reasons for this phenomenon can be attributed to the similar ionic radii of Li^+ (76 ppm) and Ni^{2+} (69 ppm), which might result in the so-called "cation disorder", impeding Li ions transport in the layered framework [54, 55]. Comparing with $LiCoO_2$ cathode, which can be synthesized by sintering the mixture of Li, O, and Co sources under suitable temperature, the $LiNiO_2$ cathode requires to be calcined under rigorous oxidizing atmosphere (such as O_2). A seminal work of preparing $Li_{1-x}Ni_{1+x}O_2$ cathode was conducted by Ohzuku's group [56]. In the preparation process, the mixture of a slight excess of $LiNO_3$ and $Ni(OH)_2$ was sintered at 750 °C under the protection of O_2. The $Li_{1-x}Ni_{1+x}O_2$ cathode is proven to deliver a high reversible capacity of 150 mA h g^{-1} in the voltage range of 2.5–4.2 V. Arai et al. also proved that excess lithium is crucial to obtain high reversible capacity [57].

To prevent the cation disorder and suppress multiple phase transitions, and further improve the cyclability of $LiNiO_2$, a series of $LiNiO_2$ derivatives prepared by partial substitution of Ni with Co, Mn, Al, Mg, Fe, and Ti were put forward. It was shown that the incorporation of Co^{3+} is beneficial for the suppression of Ni^{2+}, reduction of cation mixing and stabilization of the layer structure during cycling, as well as the improvement of both electrical and ionic conductivity of the sample [58]. Therefore, a phase-pure $LiNi_{1-x}Co_xO_2$ ($x = 0 - 1$) and $LiNiO_2$–$LiCoO_2$ ($LiNi_{1-y}Co_yO_2$) solid solution were developed. Among these cathodes, $LiNi_{0.5}Co_{0.5}O_2$ exhibited good stability against air and a single-phase reaction, while $LiNi_{0.74}Co_{0.26}O_2$ demonstrated a superior cycling performance by delivering a reversible capacity of 200 mA h

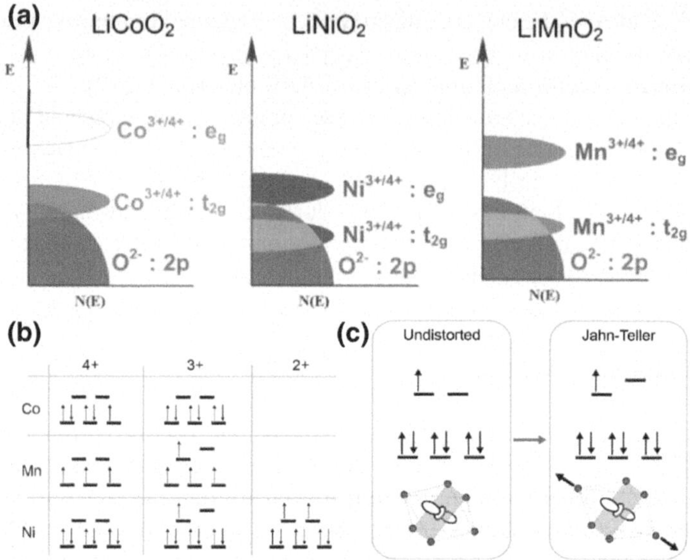

Fig. 3.4 **a** Comparison of electronic density of states in $LiCoO_2$, $LiNiO_2$ and $LiMnO_2$ [53]. Copyright © 2008 The Royal Society of Chemistry. **b** Typical valance electron configurations in layered oxides and **c** splitting of e.g. levels. Reproduced with the permission from Ref. [55], Copyright © 2017 Wiley-VCH

g^{-1} for 70 cycles without remarkable capacity fading [59]. Nevertheless, $LiNiO_2$ in $LiNi_{1-y}Co_yO_2$ was observed to convert to the rock-salt phase during the electrochemically or chemically delithiated process [60]. This transformation might lead to the release of oxygen from lattice, initiating a violent exothermic reaction to challenge the battery safety [57, 61]. To address this problem, a trace of Mn was proposed to introduce into $LiNiO_2$. The Mn^{4+} doping not only can suppress the surface irreversible structural reconstruction but also can improve thermal stability in in-depth delithiated state [62]. However, once the amount of Mn reaches relative high value, there exists an obsession for Mn-doped $LiNiO_2$ referring to the transformation of layered structure to spinel deriving from the reduction of Mn^{4+} to Mn^{3+}. In addition, this electrochemically inactive Mn^{4+} can promote more Ni^{2+} generation, deteriorating the electrochemical performances, which has been confirmed by Guilmard et al. [62]. Al^{3+} demonstrates a similar function to electrochemically inactive Mn^{4+} in the layered structure. Doping of $LiNiO_2$ with Al was proved to decrease the cationic disorder in the Li layers, inhibit phase transition, enhance the thermal and structural stability, which functions the same as that of Co^{3+} [63]. Besides, the introduction of Al^{3+} efficiently increases the operating voltage by strengthening the Al–O bond. But noteworthy, Al^{3+} is also electrochemically inactive and addition of more Al^{3+} will greatly decrease the capacity [51]. Therefore, a balanced design on the decreased capacity and increased operation voltage of Al-doped $LiNiO_2$ should be taken consideration. The electrochemical performance of other metallic elements (such as,

Mg, Ti, Fe) doped $LiNiO_2$ were also investigated, but the results were inferior to those Co, Mn or Al-doped $LiNiO_2$ [64–66].

3.2.3.3 Lithium Manganese Dioxide

Because monoclinic $LiMnO_2$ has a relatively high theoretical discharge capacity (285 mA h g^{-1}) and Mn is environmental benign and low cost, extensive researches have been concentrated on $LiMnO_2$ as an alternative choice [67]. However, $LiMnO_2$ suffers from severe capacity fading, which is mainly attributed to the dissolution of manganese upon cycling at elevated temperature, and structural instability due to Jahn-Teller (J-T) distortion [67, 68]. It was usually supposed that $LiMnO_2$ can irreversibly transforms to the thermodynamically stable spinel-like structure upon cycling. To stabilize the structure for elevated temperature performance, partial substitution of Mn ion with other metal ions has been investigated. Considering its non-toxic, cost-effectiveness, light and available in nature, substitution of Mn ion with Al ion has been proven to be an effective method. Thus, monoclinic $LiMn_{1-x}Al_xO_2$ was prepared. Though the Al substitution ($LiAl_{1-x}Mn_xO_2$) exhibited better capacity retention than the pure $LiMnO_2$ counterpart at 55 °C, $LiMn_{1-x}Al_xO_2$ oxides decreased the original capacity due to the introduction of electrochemically inactive element and did not completely inhibit J–T distortion [69, 70]. By substitution of Mn with rich-electronic Ni to make the electronic properties of Mn to be cobalt-like, the $LiNi_{0.5}Mn_{0.5}O_2$ compound demonstrated a high storage capacity and reversibility [71]. Later, improvements on capacity, reversibility and thermal stability through the co-doping of Ni and Co have been reported [72]. Another alternative method to improve the structural stability is coating the $LiMnO_2$ surface with metal oxides, which might alleviate the dissolution of Mn in the electrolyte and maintain its robust framework [70].

Because monoclinic $LiMnO_2$ is thermodynamically unstable and thus it cannot be obtained by the same approaches as applied for materials like $NaMnO_2$. In 1996, monoclinic $LiMnO_2$ is successfully synthesized by ion exchange from $NaMnO_2$ [73]. However, concerns associated with the purity, time-consuming and complicated process remain as its drawbacks. Therefore, intensive study has been focused to solid state methods, hydrothermal methods, sol gel methods, Pechini's route, and reverse microemulsion processes [74]. For example, Li et al. synthesized orthorhombic $LiMnO_2$ NPs and $LiMnO_2$ nanorods by a one-step hydrothermal method and a two-step hydrothermal method, respectively [74]. They found that the $LiMnO_2$ nanorods exhibited higher storage capacity and better cyclability than the $LiMnO_2$ NPs. This superior electrochemical performance can be attributed to two aspects: on one hand, the nanostructure shortened the pathway for Li^+ transport enabling superior rate capability; on the other hand, the nanostructured materials accommodated large strain thus improving the cycling stability.

3.2.3.4 Spinel-Type $LiMn_2O_4$

Spinel-type $LiMn_2O_4$ is particularly appealing due to its intrinsic advantages of low cost, high natural abundance, environmental benignity and high Mn^{4+}/Mn^{3+} redox potential (4 V vs. Li^+/Li) [75]. $LiMn_2O_4$ has a cubic spinel structure with space group symmetry Fd-3m, in which manganese and lithium atoms occupy the 16d octahedral and 8a tetrahedral sites of the cubic close-packed oxygen ions framework, respectively. The spinel $LiMn_2O_4$ can deliver a high theoretical reversible capacity of approaching 148 mA h g^{-1} by extracting all of the Li^+ in the tetrahedral sites, which shows extremely competitive activity in LIBs [76]. The 8a tetrahedral sites together with 16c octahedral sites form a 3D lithium diffusion pathway. Owing to its relatively low Li^+ diffusion coefficient ($10^{-9}-10^{-11}$ cm^2 s^{-1}), $LiMn_2O_4$ electrode usually demonstrated poor rate capability [76]. Decreasing particle size and creating nanopores inside the electrode material have been certified effectively to enhance electrocatalytic activity of the $LiMn_2O_4$ electrode by increasing the specific surface area of the material. Wang et al. reported improved capacity retention and rate capability of $LiMn_2O_4$ by preparing porous $LiMn_2O_4$ [77].

However, one main obstacle that impedes the realistic application of $LiMn_2O_4$ electrodes is the rapid capacity decay during long-term cycling, which is mainly caused by the dissolution of Mn into the liquid electrolyte by disproportionation reaction and structural transformation from cubic to tetragonal phase induced by Mn^{III} J-T distortion [78]. Targeted at the capacity decay issue, two most common approaches can be exploited. One is to dope suitable amounts of cations or anions into the bulk or surface spinel to improve the cycling stability. The commonly used cationic dopants mainly refer to Al, Cr, Co, Ga, Gd, La, Ce, Pr, Nd, Sm, Tb, Er, Sc, Y, B, Fe, Mg, Ni, Zn, Cu, Ti, Ru, Si, Sn, Li, and Na, etc., while the reported anions mainly include F^-, Cl^-, Br^-, S^{2-}, PO_3^{2-}, etc. Sometimes, single ion doping cannot effectively address all the issues challenging the electrochemical performance of $LiMn_2O_4$ [78]. Thus, dual or multiple ions co-doping has also been developed as an efficient method to improve its electrochemical performances.

As mentioned in the previous sections, Al has its intrinsic advantages of natural abundant, nontoxic, cost-effective and lighter than other TM dopants. Thus, Al has usually been considered as an effective dopant, which had been successfully confirmed by Myung et al. [79]. Their tests verified that the volume change of $LiAl_xMn_{2-x}O_4$ ($0 \leq x \leq 0.6$) during insertion and deinsertion of Li^+ was effectively suppressed, which may be attributed to the much stronger Al–O bonding in $LiAl_xMn_{2-x}O_4$ than that of Mn–O bonding. Sun et al. disclosed the relationship between structural degradation and capacity loss of the $LiAl_{0.2}Mn_{1.8}O_4$ electrode at elevated temperatures [80]. They reported that the Al-doped $LiMn_2O_4$ with smaller lattice parameter could reduce the dissolution of Mn into the electrolyte and stabilize the structural integrity. However, the assembled $Li/LiAl_{0.2}Mn_{1.8}O_4$ cell demonstrated relatively higher capacity loss at 80 °C than that of pure $LiMn_2O_4$ due to the formation of tetragonal $Li_2Mn_2O_4$ and rocksalt Li_2MnO_3 on the surface of the discharged electrode. Likewise, various preparation methods including wet-milling method, in situ polymerized methods, hydrothermal synthesis, solid state reaction, template method, and their combinations have been employed to prepare nanostructured Al-doped spinels to stabilize its cycling performance. Apart from Al atom, other trivalent cations such as Cr^{3+}, Co^{3+}, Ga^{3+}, Gd^{3+}, La^{3+}, and Tb^{3+} have also been utilized as a dopant not only to stabilize the framework of spinel structure but also to suppress the dissolution behavior of Mn. In addition, it should be pointed out that 0.25 Ni^{2+}-doping in $LiMn_2O_4$ lead to obvious increasement in redox potential of $LiMn_2O_4$. Recent study suggested, despite the fact that Sn^{4+}-doping accelerated the dissolution of Mn due to the increased Mn^{3+}/Mn^{4+} ratio, the crystal structure of Sn-doped spinels was stabilized because of the decrease of the lattice parameter [81]. Improved battery performance was also achieved in Li-rich $Li_{1.1}Mn_{1.9}O_4$ spinel [82].

Oxygen deficiency in the overheated $LiMn_2O_4$ will induce the Mn^{III} J-T distortion and the dissolution of Mn, and thus deteriorate the capacity retention. Xia et al. reported that Al-doped spinels could avoid the formation of oxygen deficiency at elevated temperatures and found that Al-doped spinels with stoichiometric oxygen exhibited one-phase reaction at 60 °C corresponding to free-oxygen deficiency, and the capacity retention of Al-doped spinels could reach as high as 100% at 60 °C [83]. Takahashi et al. reported that the oxygen deficiency can be significantly decreased with slight (merely 0.03) Mg^{2+}-doping [84]. Significantly, it is found that partial Mg^{2+} and Ti^{4+} co-doping in $LiMn_2O_4$ could increase its crystallinity and improve its storage performance at 60 °C. Wei et al. stabilized the crystal structure of $LiMn_2O_4$ by forming a concentration-gradient $LiMn_{1.87}Ni_{0.13}O_4$ spinel with a homogeneous $LiMn_2O_4$ core and continuously Ni-increasing shell [85]. Benefiting from the well suppression of the oxygen deficiency and the dissolution of Mn, the concentration-gradient $LiMn_{1.87}Ni_{0.13}O_4$ spinel showed considerably improved cycling performance with 90.2% of the reversible capacity retained after 200 cycles at 0.5 C rate at 55 °C. In contrast, the pristine $LiMn_2O_4$ electrode delivered merely 57.8% capacity retention after 200 cycles.

Although single ion-doped $LiMn_2O_4$ spinels have their own advantages, such as high redox potential of $LiMn_2O_4$, good capacity retention, and superior structural stability of $LiMn_2O_4$, their comprehensive electrochemical performance and high temperature stability are poor, which would impede their further practical application. The synergistic effect of binary or multiple-cation doped $LiMn_2O_4$ spinels have been investigated, particularly at elevated temperatures. Results showed that (Li, Al)-co-doping had synergetic effects on improving structural stability, suppressing J-T distortion, increasing average Mn valence, etc., which would result in not only larger initial discharge capacities but also much higher capacity retention [86]. At the meantime, other dual cation (such as Y–Co, Zn–Ti) and multi cation (such as Cu–Al–Ti, Li–Co–Gd, Li–Mg–Ca and Li–Mg–Ca–Ti) were employed to investigate the advantage of the synergetic effects on the improvement of high temperature performance [78]. In addition, partial substitution of O^{2-} ion with other anions (such as fluoride, chloride, sulfide, etc.) has been investigated to improve the crystallinity and structural stability of anion doped $LiMn_2O_4$ spinels, and thus enhance the cyclability at elevated temperatures.

Another effective method is to coat the surface with various protective layers to stabilize the $LiMn_2O_4$/electrolyte interface and inhibit the Mn dissolution of $LiMn_2O_4$

spinel. To the best of our knowledge, the coating materials mainly consist of various oxides, fluorides, phosphates, lithium compounds, carbon materials, carbonates, and polymers etc. For example, Al_2O_3-coated $LiMn_2O_4$ spinel is successfully fabricated by melting impregnation method [87]. The capacity retention and structural stability were obviously improved during cycling. Besides, enhanced electrochemical performance of $LiMn_2O_4$ spinels by other coating oxides or mixed oxides, such as ZnO, ZrO_2, MgO, Cr_2O_3, Mn_2O_3, La_2O_3, Co_3O_4, CeO_2, TiO_2, SiO_2, VO_x, Co–Al mixed oxide have also been reported [78].

3.2.3.5 Mixed Nickel-Cobalt Dioxide

As mentioned in the previous sections, $LiNiO_2$ has a favorable specific capacity compared with commercial $LiCoO_2$ cathode but exhibits serious safety hazards because of a violent exothermic reaction between the organic electrolyte and the collapsing delithiated Li_xNiO_2 structure. Delithiated Li_xCoO_2 was demonstrated to be more thermally stable than its Li_xNiO_2 counterpart. Thus, substituting a portion of Ni with Co is demonstrated effective to address the safety concerns surrounding $LiNiO_2$. It is commonly acknowledged that the Co cations in $LiNi_{1-y}Co_yO_2$ are all trivalent (Co^{3+}), which play a key role in decreasing the unfavorable cationic disordering, suppressing the irreversibly structural transformation, and enhancing the ionic/electronic conductivity. However, the evolution of oxygen related to the conversion of $Co^{3+/4+}$ impair the lithium storage capacity. The lithium storage capacity of $LiNi_{1-y}M_yO_2$ is determined by the Ni content: the more the Ni element is used, the higher capacity the $LiNi_{1-y}M_yO_2$ cathode delivers. Recent studies on the $LiNi_{1-y}Co_yO_2$ cathodes have demonstrated that, within the range $0.2 \leq y \leq 0.4$, there is a minimal variation in the unit cell volume upon cycling. Therefore, a balanced design on the improved structural stability and decreased capacity of $LiNi_{1-y}Co_yO_2$ should be taken consideration.

$LiNi_{0.8}Co_{0.2}O_2$ is particularly appealing because of the high storage capacity (due to the high Ni content) and superior cyclability (owing to the sufficient Co content). However, it seems necessary to emphasize that the electrochemical performance of $LiNi_{0.8}Co_{0.2}O_2$ varies with synthetic methods, morphologies and heat treatments of precursors. Wang et al. directly deposited $LiNi_{0.8}Co_{0.2}O_2$ thin films onto current collectors in a single step via a novel reactive spray deposition technology [88]. The $LiNi_{0.8}Co_{0.2}O_2$ thin film electrode demonstrated a superior battery performance, in terms of high initial discharge capacity and good cyclic stability (83% remained after 50 cycles).

Other cations, such as, Mg, Ti, Al, Nb and Ga, have been employed as efficient dopants to improve charge/discharge reversibility, slow down the capacity fading, and/or increase thermal stability in the fully delithiated state. Doping of Nb into the $LiNi_{0.8}Co_{0.2}O_2$ electrode has been demonstrated effectively to improve the thermal stability and electrochemical performance at the same time. Substitution of divalent nickel with redox-inactive element such as magnesium has been proven to be an effective strategy to avert the full extraction of all the lithium and thus minimize possible structural distortion [89, 90]. Significantly, it is found that partial aluminum substitution in $LiNi_{1-x-y}Co_xAl_yO_2$ (NCA) could prevent the cell from overcharge and maintain the single hexagonal phase, and thus, improve the safety and suppress the capacity fading [91]. Benefiting from these advantages, the cathode is widely commercialized, including in Tesla electric vehicles. Nevertheless, owing to the oxygen released during the delithiation process and the formation of NiO-like phase, the NCA cathode suffers from poor cyclic performance and poor thermal stability upon cycling at high temperatures and high current rates [92]. To address this problem, considerable efforts including surface modification, doping, core-shell, have been devoted to developing the NCA battery materials.

3.2.3.6 Mixed Manganese-Cobalt Dioxide

Layered $LiMnO_2$ oxide has once been developed as a promising cathode candidate through the ion-exchange reaction between α-$NaMnO_2$ and LiBr not only due to the low cost and low toxicity of manganese but also the considerable theoretical specific capacity (285 mA h g^{-1}). Although the monoclinic $LiMnO_2$ oxide delivered a large initial capacity of more than 270 mA h g^{-1}, a limited capacity of only 130 mA h g^{-1} can be recovered rendering this appealing cathode of less interest. The reasons for this large capacity fading can be attributed to structural transformation, the J-T distortion into the spinel manganese oxide, and the serious dissolution of Mn into liquid electrolyte.

Substituting Co ions for a slight of Mn^{3+} might stabilize the structure of layered $LiMnO_2$ and enhance its cyclic life at the same time. In 1998, Robert Armstrong and Gitzendanner [93] synthesized the layered $Li(Mn_{0.9}Co_{0.1})O_2$ through ion-exchange method at 150 °C. By replacing only 10% of the Mn by Co, the $Li(Mn_{0.9}Co_{0.1})O_2$ delivered an increased storage capacity of 200 mA h g^{-1}. In addition, the cooperative J-T distortion induced by localized high spin $Mn^{3+}(3d^4)$ in $LiMnO_2$ appeared to be suppressed for $Li_x(Mn_{0.9}Co_{0.1})O_2$. Soon after, Bruce et al. showed that merely 2.5% Co doped cathode would deliver a capacity of 200 mA h g^{-1} in a voltage range of 2.4–4.6 V at C/8 via ion-exchange in ethanol at 80 °C [94]. Other new lithiated mixed manganese cobalt oxides, $Li_{0.45}Mn_{0.85}Co_{0.15}O_2$, reported by Franger et al. exhibited superior cycling performance by delivering a high reversible capacity of 160 mA h g^{-1} for 40 cycles [95]. Recently, Xu et al. investigated the crystal structures, electronic structures, charge transfer, electronic configuration, valence states and magnetic moment of the $LiMn_{0.5}Co_{0.5}O_2$ oxide by using first-principles calculations based on the density functional theory [96]. Results showed that, owing to the doping of Co, the thermodynamic stability of $LiMn_{0.5}Co_{0.5}O_2$ got remarkably improved and the J-T distortion was effectively suppressed.

In addition, it is found that partial substitution of manganese ions by cobalt, iron, or nickel could increase the electronic conductivity of the $LiMnO_2$ oxide. For pure $LiMnO_2$, the electronic conductivity is around 10^{-5} S cm^{-1}, which is too low to allow rapid reaction. A combination of first-principles calculations and experimental data demonstrated that the cobalt doping could decrease the band gaps and thus enhance the electronic conductivity by almost two orders of magnitude.

3.2.3.7 Mixed Nickel-Manganese Dioxide

For Co doped manganese oxides, manganese is the redox-active ion while cobalt is the structure stabilizer. Doping cations other than cobalt, such as substitution by nickel results in a system where the manganese serves as the structure stabilizer and nickel as the electrochemically active ion. The resultant compounds might be considered as substituted nickel oxides in which the nickel participates in the redox reaction; the manganese helps in reducing the cost and stabilizing the lattice.

In 2001, a cathode material of $LiNi_{1/2}Mn_{1/2}O_2$ reported by Ohzuku et al. [97] was soon emerging as an attractive optional for $LiCoO_2$ in terms of the high operating voltage (2.5–4.5 V), high reversible capacity (150 mA h g^{-1}), superior cyclability, and improved safety. The reversible insertion and extraction of Li^+ for a $Li/LiNi_{1/2}Mn_{1/2}O_2$ cell takes place on the basis of Ni^{2+}/Ni^{4+} redox couple. In 2006, Kang et al. [98] synthesized a $LiNi_{1/2}Mn_{1/2}O_2$ cathode with very little intralayer disordering by ion-exchange method. The as-obtained cathode demonstrated superior rate capability by delivering an ultrahigh specific capacity of 180 mA h g^{-1} under a high discharge rate of 6 C. Considering the similar ionic radii of Li^+ and Ni^{2+}, the divalent nickel ions might migrate from the TMs to the Li slabs resulting in the "cation disorder". Computing results demonstrate that more than 75% of Ni ions migrate into the Li vacancies while Li enters into the Mn layers [99]. On one hand, the cation mixing greatly impairs the Li ions transport and thus leads to poor rate capability. On the other hand, TM migrations would engender a surface phase transformation to defect-spinel as well as large irreversible capacity [100]. It's worth noting that migration energy is tightly linked to the Ni environments in the TM layers, and thus solely regulating the Ni^{2+} content of a material is supposed to be an efficient strategy to reduce Ni migration. Multi-doped materials particularly those with gradient core/shell/surface structures and variable TM concentrations show exceptional thermal stability [101]. Based on the DFT calculations and experimental results, it is promising to design a material with the features of Mn rich at the surface and Ni rich in the bulk to realize high energy density.

3.2.3.8 Mixed-Nickel-Cobalt-Manganese Oxides

$LiCoO_2$ as a mature cathode material has been widely used in portable electronic devices [102]. However, the high price of cobalt gives rise to a concern about the large-scale application of $LiCoO_2$. $LiNiO_2$ is considered as a promising alternative for large-scale storage applications due to its similar crystal structure and the low cost of Ni resources. Since it is very difficult to prepare pure phase $LiNiO_2$, the NCA or $LiNi_{1-x-y}Co_xMn_yO_2$ (NCM) family has been widely studied [103–105]. In general, the $Ni^{2+/3+}$ and/or $Ni^{3+/4+}$ redox couples contribute to a high capacity while Mn stabi-

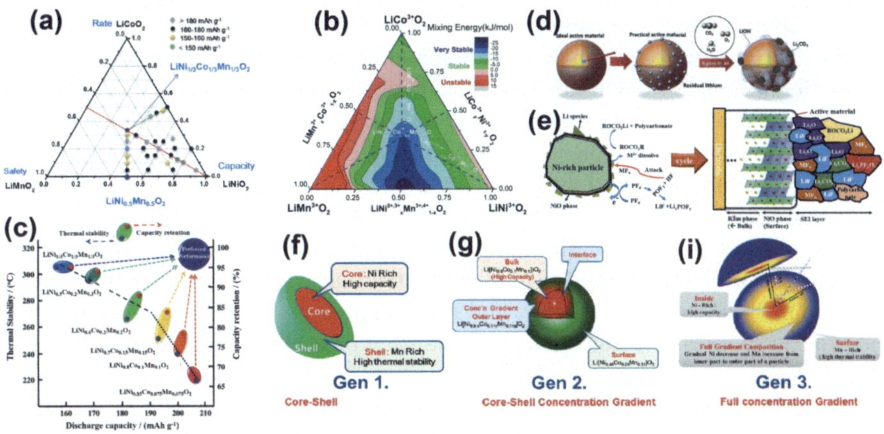

Fig. 3.5 **a** Compositional phase diagrams. Reproduced with the permission from Ref. [106], Copyright © 2015 Wiley-VCH, **b** corresponding mixing energy. Reproduced with the permission from Ref. [105], Copyright © 2016 American Chemical Society. **c** correlation between discharge capacity, thermal stability, and capacity retention of layered structure ternary NCM, **d** illustration of the surface changes of Ni-rich cathode materials after exposure in air, **e** illustration of the surface interphases of Ni-rich cathode materials. Reproduced with the permission from Ref. [106], Copyright © 2015 Wiley-VCH. **f** Core-shell structure, **g** core-shell concentration gradient, and **i** full concentration gradient structure ternary layered oxides. Reproduced with the permission from Ref. [51], Copyright © 2017 American Chemical Society

lizes the layered structure with prolonging lifespan and enhanced safety, as well as Co reinforces the layered ordering with enhanced rate capability (Fig. 3.5a, b) [105, 106].

We all know that the electrochemical performance and thermal stability of NCM is directly related to the Ni content. The higher Ni content of the layered structure results in higher utilization of lithium. $LiNi_{1/3}Co_{1/3}Mn_{1/3}O_2$, $LiNi_{0.5}Co_{0.2}Mn_{0.3}O_2$, $LiNi_{0.6}Co_{0.2}Mn_{0.2}O_2$, $LiNi_{0.75}Co_{0.15}Mn_{0.15}O_2$, $LiNi_{0.8}Co_{0.1}Mn_{0.1}O_2$ and $LiNi_{0.85}Co_{0.075}Mn_{0.075}O_2$ have been synthesized to investigate the role of each TM in NCM (Fig. 3.5c). As the Ni content increases, the specific capacity also increases. For NCM, when the Ni content is 1/3, 0.5, 0.6, 0.7, 0.8 and 0.85, the initial discharge capacity is 163, 175, 187, 194, 203 and 206 mA h g^{-1}, respectively [55, 104]. When the content of Ni is 1 (x = 0, y = 0), the capacity approaches 220 mA h g^{-1} with almost 80% reversible extraction of Li in the host structure. However, when the Ni content is higher than 0.6, Ni-rich cathode materials are sensitive to thermal, emerging the phase transition from layered to spinel and rock-salt structures, thereby resulting in severe capacity fading [107, 108]. Ni-rich NCM materials, especially the content of Ni greater than 0.6, inevitably react with H_2O and CO_2 in air to generate Li_2CO_3 and LiOH on the surface of the material (Fig. 3.5d). LiOH further reacts with $LiPF_6$ in the electrolyte to form hydrogen fluoride (HF) and Li_2CO_3, producing gas at high temperatures, and thus leading to increased impedance, decayed capacity and power (energy) (Fig. 3.5e) [109, 110]. In addition, because the ionic radius of Ni^{2+} (69 pm) is close to that of Li^+ (76 pm), NCM cathodes have to bear cation mixing obsession stemming from Li/Ni site-exchange upon cycling, and thus suffer from rapid capacity fading and severe structure deterioration [104, 111]. The presence of Co in the Ni-rich cathodes alleviates the capacity fading but at the expense of increased material cost and decreased capacity. Co ions migrate from the octahedron site to the tetrahedron site and stably occupy the position of the tetrahedron, inhibiting the transition from the spinel phase to the rock salt phase. Moreover, Co-doped cathodes exhibit improved electric conductivity, which fosters the high rate capability. The Mn^{4+} doping not only can lower the material cost but also can improve thermal stability in in-depth delithiated state. At high temperatures, the oxidation state of Mn is stable, and the occupied position remains unchanged, and Mn can suppress the transition from a layered phase to a spinel phase [106, 112, 113].

Due to high energy density, NCA become the important layered cathode materials. LIBs with $LiNi_{0.8}Co_{0.15}Al_{0.05}O_2$ cathode material have already been used in Tesla for electric vehicle application [114]. Al is the most commonly used doping element because of its stabilizing effect on Ni-rich cathode materials. Al^{3+} is electrochemically inactive in the TM layer, Co^{3+} and Al^{3+} can inhibit cation mixing in the Li layer and prevent multistep phase transition during Li^+ insertion and extraction. Due to Al^{3+} inactivity, a large amount of Al^{3+} will lower the reversible capacity. As a commercialized cathode material, the $LiNi_{0.8}Co_{0.15}Al_{0.05}O_2$ delivers a large discharge capacity of approximately 200 mA h g^{-1} and exhibits excellent cycling performance. However, the discharge capacity and cycling performance is highly dependent on the conditions of DOD (depth of discharge). In the ΔDOD of 0–100%, after a few cycles, clear microcracks were produced in the intersurface among primary grains. These cracks and separations make the loss of intergrains connectivity and possibly give rise to increased polarization. The rock salt structure ($Fm\bar{3}m$) NiO surface will form on the fresh grains and further result in the degradation of electrochemical properties. In contrast, there are almost no cracks generated in the test condition of ΔDOD of 0–60% even after 5000 cycles [115, 116]. Therefore, controlling the DOD is a good way to improve cycle performance.

Nickel-rich layered lithium TM oxides have been considered as promising cathode electrode materials for the next generation high energy density LIBs. In order to achieve the goals of a battery with an energy density higher than 250 Wh kg^{-1}, Ni-rich cathode materials pairing with high specific capacity Si/C anode materials is a good choice. Another proposed solution to address the issues regarding to poor structural stability and thermal stability for the Ni-rich cathodes is to develop core-shell structure, core-shell concentration gradient, and full concentration gradient structure ternary layered oxides (Fig. 3.5f–i) [51]. Although Ni-rich cathode materials still have challenges such as safety concerns, structural instability, and surface side reactions need to overcome, we believe that Ni-rich lithium transitional-metal layered oxides will be widely applied as next-generation high-energy cathode materials in the future.

3.2.3.9 Lithium-Rich Layered Oxides

Lithium-rich layered oxides (LLOs) electrode materials, with a formula $Li_{1+x}M_{1-x}O_2$ (M = Mn, Ni, and Co) and reversible capacity exceeding 250 mA h g^{-1}, have been emerged as promising alternative to substitute the current LIB cathodes [117]. Excess Li can be incorporated into the layered structure via a solid solution of monoclinic Li_2MnO_3 (C2/m space group) and rhombohedral $LiMO_2$ ($R\bar{3}m$ space group) (Fig. 3.6a, b) [118–120]. In addition, both of the interlayer space of Li_2MnO_3 and $LiMO_2$ are close to 4.7 Å that makes them be integrated at the atomic level. Li_2MnO_3 functions like a stabilizing agent within the structure of the electrode between 2 and 4.4 V due to manganese atom is tetravalent in Li_2MnO_3. The electrode usually needs to charge beyond 4.4 V even up to 4.6–4.9 V to remove Li_2O so that it can deliver a high capacity of >250 mA h g^{-1} (Fig. 3.6c) [121]. Although a high capacity (~400 mA h g^{-1}) was obtained during the initial charge, Li^+ from Li_2O cannot be reversibly reinserted into Li_2MnO_3. An electrochemically active MnO_2 is obtained after the extraction of Li_2O, which can be compatible with the $LiMO_2$.

Despite LLOs deliver a high initial capacity, these cathode materials bear the major challenges of low initial Coulombic efficiency (ICE), large voltage decay,

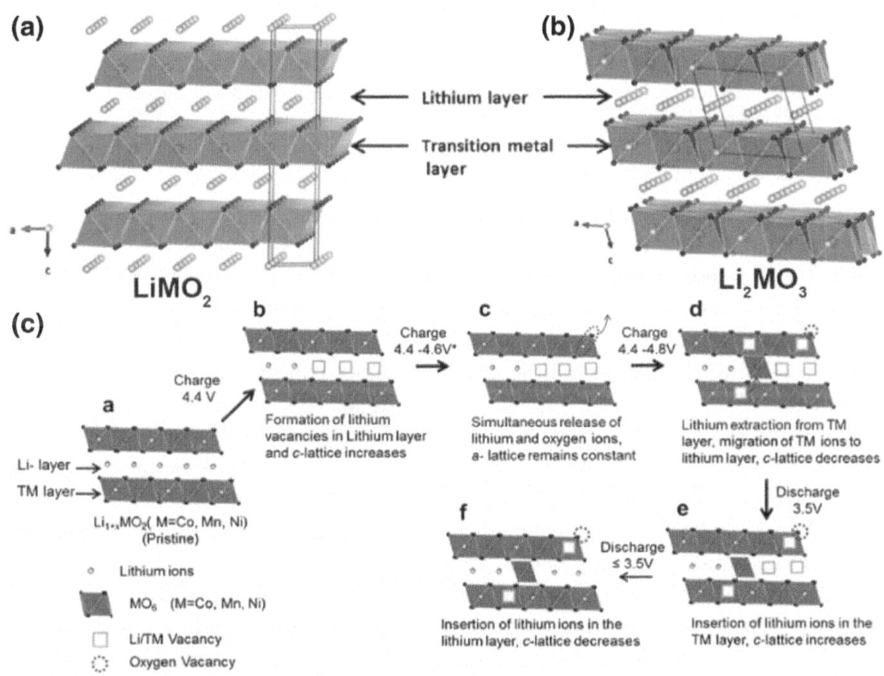

Fig. 3.6 **a** Crystal structure of $LiMO_2$ and **b** Li_2MnO_3. **c** Schematic illustration of the crystal structure variation during charge/discharge process. Reproduced with the permission from Ref. [120], Copyright © 2013 Elsevier

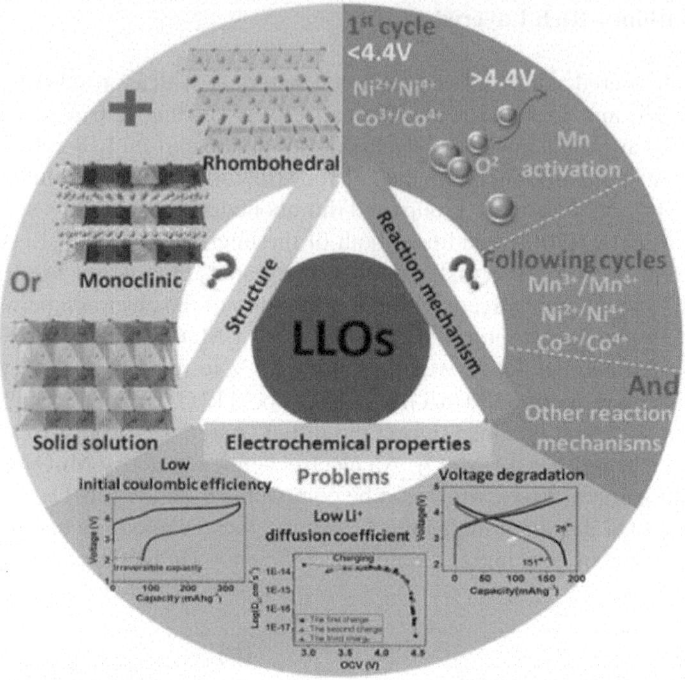

Fig. 3.7 Major challenges for developing high-energy LLOs. Reproduced with the permission from Ref. [123], Copyright © 2013 American Chemical Society

and poor electrode kinetics (Fig. 3.7) [122, 123]. Surface coating with metal oxides (Al_2O_3, CeO_2, ZrO_2), metal fluorides (AlF_3, SrF_2, ZrF_2), phosphates ($AlPO_4$, $CoPO_4$, Li–Ni–PO_4), carbon, metal element and polymer conductor is an effective strategy that has been extensively explored for improving the cycling stability of LLOs [124, 125]. Zhang et al. reported that the spinel $Li_4Mn_5O_{12}$ coating formed on LLOs via controlled $KMnO_4$ oxidation on the oxide precursor can prevent oxygen release at the deep delithiated state (Fig. 3.8a–c) [126]. With the spinel $Li_4Mn_5O_{12}$ coating, the undesired interfacial spinel-like phase was moderated and the stability of the inner LLO got improved. As a result, the heterostructure cathode demonstrates relatively competitive electrochemical performs, in terms of high initial discharge capacity (276 mA h g^{-1} at 0.05 C), and excellent capacity retention (83.1% after 300 cycles), good voltage stability, and favorable kinetics [126].

Although surface coating approaches work effectively on improving cycle life, surface coating could only improve the surficial electrode kinetics rather than in the bulk region. In addition, most surface coatings are electronically insulated and cannot prevent the crystal structure from collapsing at high charge/discharge voltage. To address these issues, constructing an interconnected conductive structure and chemical doping with cation (Al^{3+}, Mg^{2+}, Fe^{2+}, Ti^{3+}, Cr^{3+}) and anion (S-doping, F-doping) have been developed to improve the lithium diffusion coefficient and

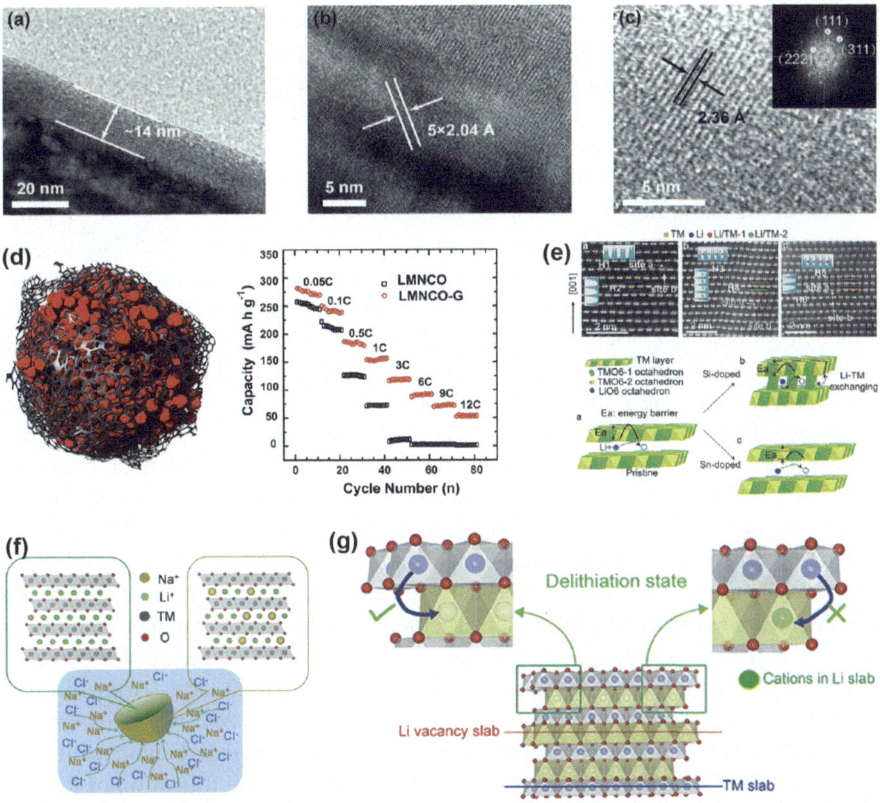

Fig. 3.8 Structural characterization of LLOs coated by spinel $Li_4Mn_5O_{12}$: **a** HTEM image of its surface, **b** HRTEM image of its inner region, and **c** the external layered and spinel region. Reproduced with the permission from Ref. [126], Copyright © 2018 Wiley-VCH. **d** Graphene-wrapped LLOs hybrid cathode and corresponding rate capability. Reproduced with the permission from Ref. [127], Copyright © 2012 American Chemical Society. **e** High-angel annular dark field images and histogram profiles of the pristine, Si-doped and Sn-doped samples, respectively. Reproduced with the permission from Ref. [128], Copyright © 2015 Wiley-VCH. **f** Schematic illustration of the structure design of gradient surface Na^+ doping Li-rich material. Reproduced with the permission from Ref. [129], Copyright © 2016 Wiley-VCH. **g** Schematic illustration of some cations in the Li slab that maintain the structural stability and inhibit TM-ion migration from the 3a site to the 3b site. Reproduced with the permission from Ref. [131], Copyright © Tsinghua University Press and Springer-Verlag Berlin Heidelberg 2017

electronic conductivity. For example, Jiang et al. engineered a graphene-wrapped LLOs hybrid cathode material with improved rate capability (Fig. 3.8d) [127]. Wang et al. demonstrated that Sn-doping approach can effectively elevate the rate capability and capacity of the Sn-doped LLOs cathode by expanding the interplanar spacing in layered structure (Fig. 3.8e) [128]. Zr-doping strategy has been applied not only to stabilize the crystal structure of $Li[Li_{0.2}Mn_{0.54}Ni_{0.13}Co_{0.13}]O_2$ but also enhance the lithium diffusion kinetic and electronic conductivity [124]. As for Na^+ doping, owing to thermodynamical Na^+ concentration diffusion, gradient surface Na^+ doping is obtained by the calcination of Li-rich materials in molten NaCl state to improve the kinetics of Li-rich cathode materials (Fig. 3.8f) [129]. Ding et al. reported gradient-surface Si/Sn-doping Li-rich materials [130]. Benefiting from the expanded interslab spacing, the doped cathode exhibits improved Li^+ diffusivity and rate performance. Shi et al. reported the introduction of extra Ni^{2+} into the LLOs cathode can slow down the voltage decay by suppressing the structural evolution (Fig. 3.8g) [131]. When the charging voltage window was limited remaining some Li^+ ions in the Li slab, the voltage of Li-rich materials decayed by only 0.04 V after 100 cycles. Gu et al. utilized aberration-corrected scanning transmission electron microscopy (STEM) to detect the atomic structure of LLOs cathodes which are composed of random mixture of $LiMO_2$ $R\bar{3}m$ and Li_2MO_3 *C2/m* phases [132]. Both phases experience the transition from the layered structure to the spinel structure in cycling. Rather than transforming from Li_2MO_3 *C2/m* to spinel by the removal of Li^+ and O^{2-}, which could generate large lattice strain and result in the decomposition of the parent lattice, transforming of $LiMO_2$ R3m to spinel through moving TM ions to the Li site would ensure the lattice, while simultaneously forming mosaic structured spinel grains within the parent particle. The AlF_3 coating could help slow down of the transition of layered structure to the spinel structure.

Aside from surface coating and doping modification, the composition of Li_2MnO_3 and $LiMO_2$ is greatly linked to the voltage decay and rate capability. Idemoto et al. investigated the relationship between the composition of $xLi(Li_{1/3}Mn_{2/3})O_2–(1–x)Li(Mn_{1/3}Ni_{1/3}Co_{1/3})O_2$ $(0 < x < 1)$ and electrochemical performance [133]. According to the Rietveld refinement of neutron diffraction patterns, when $x = 0.2$, Li was localized at 2b and 4 g sites, whereas for $x \geq 0.3$, Li existed at 2b sites and Mn occupied the 4 g sites. Thus, there was significant differences in the electrochemical performance between the composition of $x = 0.2$ and $x \geq 0.3$. The pair distribution function analysis and extended X-ray absorption fine structure further confirmed that the ordering of $LiMn_6$ and $LiMn_5Ni$ leads to different cathode performance.

3.2.4 Iron Compounds Including Oxides and Phosphates

Iron compounds have been developed as promising cathode materials for lithium cells due to the potentially low cost, plentiful natural abundance, and environmentally benignity of Fe. This section intends to discuss the categories of iron compounds including oxides and phosphates in cathode materials for LIBs.

3.2.4.1 Olivine Phase ($LiMPO_4$: M = Fe, Mn, Co, Ni)

Since the pioneering work by Padhi et al., olivine $LiMPO_4$ (M: Fe, Mn, Co and Ni) cathode materials received numerous attentions, attributing to their intrinsic stable structure and satisfactory electrochemical properties [134]. Olivine $LiFePO_4$ is made up of low cost and natural rich Fe and PO_4 moieties, which is beneficial for large-scale application [135]. The most outstanding feature of $LiFePO_4$ is that it can be used at very high rates, although both related phases demonstrate poor ionic and electronic conductivities [136]. Numerous efforts have been focused on establishing the relation between the structure and ionic-electronic conductivity. However, this material exhibits a low operating voltage at 3.45 V versus Li^+/Li, indicating insufficient energy storage performance, which largely limits its application. Compared to the extensively investigated $LiFePO_4$, $LiMnPO_4$ has a higher intercalation potential at 4.1 V versus Li^+/Li, contributing to nearly 1.2 times higher theoretical energy density than that of $LiFePO_4$ [137]. However, the intrinsic electronic conductivity of $LiMnPO_4$ is about five orders of magnitude lower than $LiFePO_4$, making the $LiMnPO_4$ exhibit a poor rate capability. $LiCoPO_4$ emerges into an alternative cathode material because of its relatively high operating voltage (~4.8 V vs. Li^+/Li), which brings about a superior energy density of about 800 Wh kg^{-1}, exceeding that of either $LiFePO_4$ or $LiMnO_4$ [138]. Besides, the volume change between the $CoPO_4$/$LiCoPO_4$ couple during the intercalation-deintercalation process is nearly 2%, which is much smaller than other olivine cathode materials. Even of this, $LiCoPO_4$ cathode get fewer attention than other cathode materials due to its intrinsic low electronic conductivity and ionic diffusivity stemming from the 1D ion transport channel. Among the lithium transition-metal phosphates, $LiNiPO_4$ cathode demonstrates the highest theoretical redox potential (~5.1 V) but is hindered by the sluggish kinetics of the electronic and Li^+ transport [139, 140]. Overall, slow lithium diffusion kinetics and poor electronic conductivity challenge the exploration of the olivine phase cathode materials, and therefore, high-conductive layer coatings and reducing NP size are required to improve the performance of lithium transition phosphates.

3.2.4.2 Other Iron Phosphate Phases ($LiMn_{1-x}Fe_xPO_4$, $Li_3Fe_2(PO_4)_3$)

In spite of the success, the practical application of $LiFePO_4$ cathode material is still limited by its low intercalation/deintercalation potential. Although the $LiMnPO_4$ owns a higher intercalation potential (4.1 V vs. Li^+/Li), the lower electronic conductivity severely limits the performance under high current rate. Therefore, doping $LiFePO_4$ with a portion of Mn, namely forming a $LiMn_{1-x}Fe_xPO_4$ ($0 < x < 1$), turns out to be an effective method to combine the advantages of both $LiFePO_4$ and $LiMnPO_4$ [141]. However, the resulting material still faces an obsession of poor electronic conductivity. Introducing conducting matrixes, such as carbon, graphene and carbon nanotubes (CNTs), have been proven to be the most direct and effective way to improve the electrochemical performance. Cao and his co-workers realized the improvement in conductivity and suppressing the dissolution of metal ions by successfully fabricating a core-shell structured $LiMn_{0.5}Fe_{0.5}PO_4$@C [142]. By anchoring the tunable $LiMn_{0.75}Fe_{0.25}PO_4$ particles on graphene sheets, the resulting composite cathode exhibits reduced resistance and enhanced rate capability due to the significantly improved electronic conductivity [143]. CNTs functions the same as graphene, which is confirmed by Zhou et al. [144]. Because the introduced multi-walled nanotubes (MWCNTs) can form an interconnected conducting network and facilitate rapid the electron transport, the as-obtained composite with 2% MWCNTs loading delivers the highest reversible capacity of 163.3 mA h g^{-1} at 0.1 C, which is almost 96% of the theoretical capacity (~170 mA h g^{-1}). When discharged at 50 C, the composite still exhibits obviously higher capacity (64.23 mA h g^{-1}) than the $LiFe_{0.6}Mn_{0.4}PO_4$/C without MWCNTs (12.8 mA h g^{-1}).

Except for $LiFePO_4$, $Li_3Fe_2(PO_4)_3$ is another kind of iron compound, which has been successfully used as cathode for LIBs [145]. Although the theoretical capacity of this Nasicon-type $Li_3Fe_2(PO_4)_3$ (128.2 mA h g^{-1}) is lower than the olivine $LiFePO_4$ (170 mA h g^{-1}), $Li_3Fe_2(PO_4)_3$ has the following competitive advantages: (i) superior ionic conductivity related to the disorder of Li^+ in the structure; (ii) stable structure; (iii) the synthetic process of $Li_3Fe_2(PO_4)_3$ is easier than $LiFePO_4$. (iv) 2 mol Li^+ can be reversibly intercalated into 1 mol $Li_3Fe_2(PO_4)_3$ due to Fe^{3+}/Fe^{2+} couple redox. According to the above-mentioned advantages, increasing attention will be paid to this Nasicon-type $Li_3Fe_2(PO_4)_3$ cathodic material.

3.2.5 *Summary*

In this part, we discussed the recent process on various positive electrode materials. Although they have achieved significant improvements in the past decades, their commercialization will be much tougher because there is no one cathode material that can simultaneously meet all requirements, in terms of the high energy density, high power density, long service life, low cost, and high safety. Olivine $LiFePO_4$ has emerged as one of the most promising cathode candidates for commercial LIBs owing to their stable structure, high safety, good resistance to overcharging, and low cost. It delivers a theoretical capacity of 170 mA h g^{-1} with an average voltage of 3.4 V based on the Fe^{2+}/Fe^{3+} redox couple, of which the energy density is estimated to be merely 578 W h kg^{-1}. Various strategies including elevate the capacities and widen operating voltages have been utilized to increase the energy density of the active cathode. The charge-discharge voltage for $LiNiPO_4$ is of ~5.2 V while is around 4.1 and 4.8 V for $LiMnPO_4$ and $LiCoPO_4$, respectively. Layered ternary metal oxides which integrates the merits of good cycling stability of $LiCoO_2$, high specific capacity of $LiNiO_2$, and high safety and low cost of $LiMnO_2$ would be no doubt to develop into a potential cathode material in LIBs. However, the reversible capacity, capacity retention and thermal stability of layered ternary metal oxides is directly linked to the ratio of Ni–Co–Mn in the cathode materials. The best-known layered ternary metal oxides refer to the NCM ratio of 333, 523, 622, and 811. In general, a high Ni content leads to high discharge capacity, but poor capacity retention and thermal stability. The Ni-rich layered oxides are found to be sensitive to CO_2/H_2O, as a result, byproduct of Li_2CO_3 and LiOH generate on the surface of cathode materials, which might result in the decomposition of electrolyte, the dissolution of metal ions, and thus the deterioration of cycling and storage lifespan. It is highlighted that doping or surface coating is indeed an efficient strategy to improve the structural and thermal stability. It is anticipated that LLOs may be an alternative cathode material in the near future due to their high capacity of 250–300 mA h g^{-1} and low cost once the low ICE, large voltage drop and poor rate capability are well addressed. In addition to exploring new materials, combining the synergistic effects of various cathode materials to form gradient core/shell/surface structures and variable TM concentrations can be regarded as a feasible avenue for future efforts to fabricate positive materials simultaneously possessing high specific energy, long-term endurability, and superior rate capability. Last but not the least, fabrication nanostructured electrode materials, optimization of the electrolyte composition and exploration of new binders should also be taken into consideration.

3.3 Anode Materials for Lithium-Ion Batteries

3.3.1 Carbonaceous Anode Materials

Carbonaceous materials have been widely studied as anode materials for LIBs due to their important virtues, such as abundant sources, low cost, eco-friendness, physical and chemical stability, good working potential, and good reversibility during lithium intercalation and de-intercalation and so on [146]. According to the degree of graphitization, carbonaceous materials can be divided into two categories, graphite (graphitized) carbon and non-graphitized carbon, like soft and hard carbons. Although conventional graphite carbon has been the most successfully commercialized anode material for LIBs, its relatively low theoretical capacity of 372 mA h g^{-1} and limited rate performance can hardly meet the growing demand of advanced devices nowadays [147]. To increase the energy and power densities of LIBs, great efforts have been paid on nanostructured carbonaceous anode materials.

3.3.1.1 1D Structures

Carbonaceous materials with 1D structure usually possess high specific surface area and directional electronic/ionic transportation, which lead to high electron collection efficiency [148]. Common 1D carbonaceous materials include CNT, CNF, carbon nanospring and so on. As 1D allotrope of graphite, CNT features cylindrical tubule of graphite sheet and has many advantages while used as anode material in LIBs, like high conductivity (10^6 S m^{-1} at 300 K for single walled carbon nanotubes (SWCNTs) and >10^5 S m^{-1} for MWCNTs), high rigidity (Young's modulus of the order of 1 TPa), low density and high tensile strength (up to 60 GPa). Compared to graphite, CNTs have higher capacity especially SWCNTs. Because Li^+ can not only diffuse into the active sites on the NT surface but also inside individual NTs and in the interstitial sites of close-packed bundles between SWCNTs, so SWCNTs can attain high reversible capacities (>1116 mA h g^{-1}) exceeding a LiC_2 stoichiometry. Recently, a measured capacity of 1050 mA h g^{-1} has been achieved by a purified SWCNTs electrodes with titanium contacts, which means a dramatic improvement of lithium intercalation over the LiC_6 limit of conventional graphite [149].

To further improve the performance of lithium rechargeable batteries based on CNT anode, great efforts have been focused on functionalization of CNT. Lee et al. reported a stable pseudo-capacitive groups functionalized CNT assembled by layer-by-layer (LBL) technique, which acted as an additive-free electrode and exhibited excellent energy/power capability (a reversible gravimetric capacity of ~200 mA h $g_{\text{electrode}}{}^{-1}$ while also delivering 100 kW $kg_{\text{electrode}}{}^{-1}$ of power) and stability (Fig. 3.9) [150]. Moreover, it's a universal strategy to use CNTs as a conductive and support matrix combining with high specific capacity electrode materials (e.g., silicon, germanium or other metal and metal compounds), which can make full use of the higher capacity of these materials and efficiently remit the problem of pulverization [149].

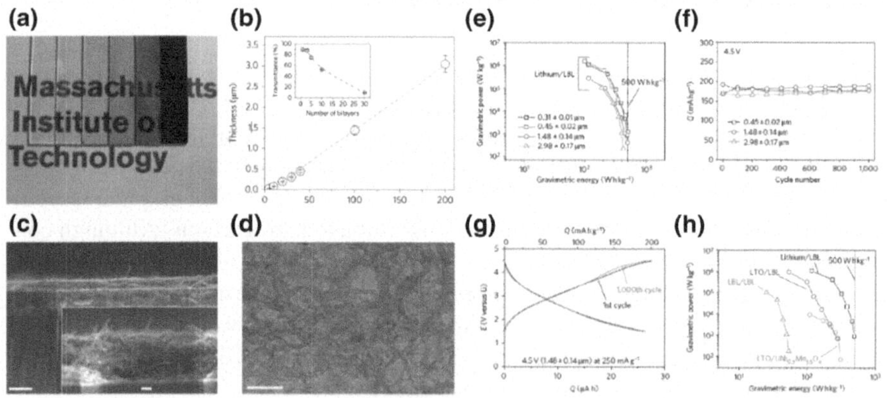

Fig. 3.9 **a** Digital image of representative MWCNT electrodes on $Li_4Ti_5O_{12}$-coated glass slides; **b** Thickness of the LBL-MWCNT electrodes as a function of the number of bilayers; **c** SEM cross-sectional image of an LBL-MWCNT electrode on an $Li_4Ti_5O_{12}$-coated glass slide after heat treatments; **d** TEM image of an LBL-MWCNT electrode slice, showing pore sizes of the order of ~20 nm; **e**–**h** Gravimetric energy and power densities, and cycle life of LBL-MWCNT electrodes obtained from measurements of two-electrode cells. Reproduced with the permission from Ref. [150], Copyright © 2010 Springer Nature

In addition to CNTs, CNF and carbon nanospring are also widely studied as anode materials for LIBs. CNFs can be synthesized by low-cost and facial techniques, like electrospinning, vapour deposition, arc discharge, and laser ablation [148]. For example, 1D hierarchical porous carbon fibers (HPCFs) were synthesized through a controlled carbonization of alginic acid fibers, the nanosized 1D hierarchical porous architecture contributed a lot to the rate capability and capacity retention of the as-synthesized HPCFs [151]. Wu et al. reported a catalytic chemical vapor deposition (CVD) technology preparing carbon nanospring with excellent elasticity that could efficiently buffer the volume variation during Li insertion/extraction process and it exhibited superior rate performance (160 mA h g^{-1} at 3 A g^{-1}) [152]. These works aimed at improving the power capability through morphological modification of carbonaceous materials (1D morphology and decreased particle size can efficiently decrease the diffusion distance of Li^+). Actually, there is another way to synthesize carbonaceous materials with special and favorable nanostructures, which derives from biomass products. Ye et al. reported a renewable nanofiber (NF) of cellulose with tunable diameters, which were extracted from bamboo pulps by high frequency ultrasound and subsequent selective oxidation, were assembled into a 3D aerogel framework with a controllable structure and a hierarchical porous structure. The 1D building blocks of NFs can provide multiple pathways for Li^+ and charge transport, making it possess outstanding cycling stability, high capacity, and good rate performance. The N-ACNF-50 aerogel with a 1D CNF diameter of 50 nm exhibited the highest capacity of 630.7 mA h g^{-1} at 1 A g^{-1}, the best rate capability (289 mA h g^{-1} at 20 A g^{-1}) and cycling performance (651 mA h g^{-1} at 1 A g^{-1} after 1000 cycles) [153]. To improve the energy capability, carbon nanospring and NF are also combined with high specific capacity materials in the place of CNT [147]. For instance, a novel Si/C hybrid was synthesized via an electrospinning-assisted process, where the Si NPs were well encapsulated in hollow graphitized CNFs derived from polydopamine (PDA), which formed a multi-layered graphene-like structure doped with heteroatoms and high electrical conductivity comparable to that of graphene after carbonization. The poor electrical conduction caused by the high content of insulating Si could be compensated by the high electrical conductivity of the C-PDA shell and the volume variation of Si may be buffered by the surrounding carbon to some extent [154].

3.3.1.2 2D Structures

Since the term graphene introduced in 1987, numerous efforts have been paid on the graphene NSs based anode materials especially in LIBs. As a two-dimensional (2D) single layer of carbon, graphene has many unusual physical, mechanical and chemical properties. It's found that even at ambient temperature electrons can move ballistically in graphene layer, which is attributed to the high quality of the sp^2 carbon lattice. This phenomenon endows graphene with highly conductive nature. Meanwhile, graphene has high surface area and large surface-to-volume ratio, and Li^+ could absorb not only on each side of graphene NSs, but also on the edges and covalent sites of the graphene nanoplatelets. These properties make graphene a suitable anode material for reversible lithium storage in LIBs [155]. However, a study of theoretical Li^+ storage has showed that the capacity of graphene is lower than graphite (372 mA h g^{-1}) when a single layer of graphene is taken into consideration, but when a number of graphene NSs are taken into consideration together, the capacity could be higher to 780 mA h g^{-1} (Li^+ absorb on both faces of graphene: Li_2C_6 stoichiometry) or to 1116 mA h g^{-1} (Li trapped at the benzene ring in a covalent bond: LiC_2 stoichiometry) [146]. Although theoretical capacity is good, unavoidable irreversible capacity and low Coulombic efficiency of graphene, which are caused by the undesirable reactions between Li^+ and oxygen containing functional groups and the formation of SEI layers on the defect sites, are still the barriers for its application. It has been reported that the existing of oxygen contained functional groups can significantly decreasing the reversible capacity, which can be confirmed by the much lower reversible capacity of graphene oxide (GO) (335 mA h g^{-1}) than that of reduced graphene oxide (rGO) (~1000 mA h g^{-1}) [156]. Graphene NSs are generally prepared through mechanical exfoliation of natural graphite, in which way oxygen contained functional groups can be introduced during the synthesis process, but the most commonly used solution chemical exfoliation methods cannot avoid the chemical oxidation of natural graphite, which is usually followed by a reduction treatment [148].

Reports about modification or decoration of graphene NSs are too numerous to enumerate. For example, a soft-chemistry approach was used to prepare naturally agglomerated and crumpled flower-petals like graphene NSs, which exhibited significantly enhanced reversible lithium storage capacity and good cycling performance as anode material in LIBs. Recently, a novel 3D holey-graphene framework was designed by systematically tailoring the porosity in the graphene backbone (Fig. 3.10a, d, e), which tremendously optimized the charge transport in the composite architecture and achieved high areal capacity and ultrahigh-rate capability at high mass loading (Fig. 3.10b, c, f) [157]. Moreover, composites of graphene and high capacity materials (such as Si and Sn [158, 159]) or metal compounds (TM oxides, sulfide, phosphides and so on) are reported to be very promising materials. Zhou et al. synthesized a silicon NPs/graphene sheets composite with Si NPs successfully inserted into graphene sheets via a novel method combining freeze-drying and thermal reduction [160]. In metal compounds/graphene composites, metal compounds anchored on the surface of graphene NSs prevent the coagulation of stacks while graphene NSs provide high surface area and improve conductivity of material, which synergistically facilitate lithium storage performance.

Another important 2D carbonaceous material is graphdiyne (GDY), which is a new carbon allotrope that has been synthesized recently. GDY is a one-atom thick 2D layer of strongly bonded carbon atoms that exhibits chemical stability and electrical conductivity [161]. GDY is composed of sp^2 and sp hybridized carbon atoms and is predicted to be the most stable one of the various diacetylenic non-natural carbon allotropes. Unlike wholly sp^2 hybridized carbon structures, the flat carbon (sp^2 and sp) network endows the GDY material with uniformly distributed pores, and tunable electronic properties [162]. GDY and its unusual structure promises to yield new

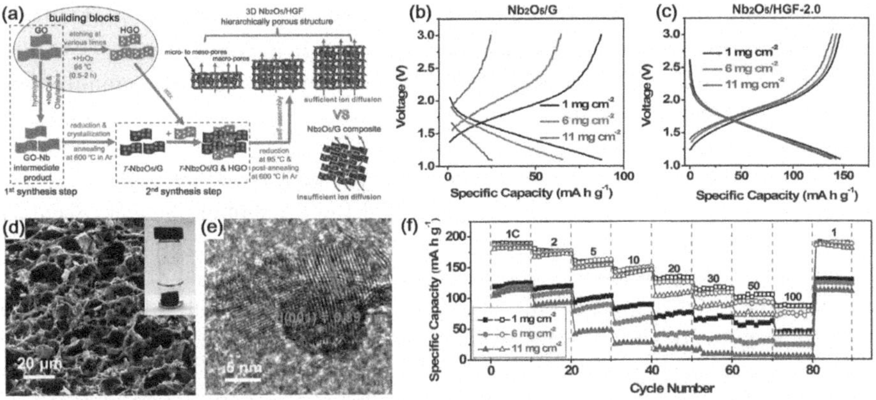

Fig. 3.10 **a** Illustration of the two-step process flow to prepare 3D hierarchically porous composite architecture; Galvanostatic charge-discharge curves for **b** the Nb_2O_5/graphene control electrode and **c** the Nb_2O_5/HGF-2.0 electrode at a rate of 10 C for the mass loadings of 1, 6, and 11 mg cm^{-2}. **d** Cross-sectional scanning electron microscopy image of Nb_2O_5/HGF composite showing 3D hierarchically porous structure. (Inset) A free-standing monolithic composite used to make the electrode. **e** HRTEM image of the orthorhombic Nb_2O_5 NPs. **f** Comparison of the rate performance between 1 and 100 C for Nb_2O_5/HGF-2.0 (open) and Nb_2O_5/graphene (solid) electrodes under different mass loadings (1, 6, and 11 mg cm^{-2}). Reproduced with the permission from Ref. [157], Copyright © 2016 The American Association for the Advancement of Science

functional materials [163]. For example, an carbon-rich frame hydrogen substituted transparent GDY film is fabricated through an in situ cross-coupling reaction of triethynylbenzene on copper foil, which can be directly used as free-standing flexible electrode for LIBs and can deliver a large reversible capacity of 1050 mA h g^{-1}. Besides, due to the extended π-conjugated system and the hierarchical pore bulk with large surface area, the rate and cycling performance of this electrode is also good [164].

3.3.1.3 3D Structures

Porous carbon is one of the most familiar 3D carbonaceous materials, which generally owns high specific surface area and interconnected 3D conductive networks as well as relatively good mechanical strength that can accommodate strain during the charge/discharge process. These properties make porous carbon an ideal anode material in LIBs. The porous nanostructure is the distinct feature of porous carbon that differs from 1D or 2D carbon nanostructures, which can significantly avoid the problem of irreversible capacity loss. Porous carbon can be segmented into microporous (<2 nm), mesoporous (2–50 nm) and macroporous (>50 nm) carbon depending on the pore sizes, which have a great influence on the performance of materials in anode electrode. To get porous carbon possessing appropriate pore size, template-assisted synthetic methods or chemical treatment followed by pyrolysis of precursor materials are usually adopted. A silicon NP/porous carbon hybrid NF was synthesized from a stable suspension of Si@SiO_x NPs in polyacrylonitrile (PAN)/N, N-dimethylformamide (DMF) solution via a facile single-nozzle electrospinning and subsequent calcination and HF etching process, where Si NPs was encapsulated in porous carbon nanostructure that could accommodate the volume variation and thus attain improved cycling performance and excellent rate capabilities [165].

Metal-organic frameworks (MOFs) generated from a supramolecular assembly of inorganic components (metal ions or metal clusters) with organic components (organic organometallic complexes) represent an emerging class of materials that have attracted extensive research interest because of their versatile functionalities and tunable porosities [166]. Due to their permanent nanoscale cavities and open channels, which are similar to mesoporous silicas and zeolites, MOFs exhibit a strong potential for use as templates and reactive precursors for synthesizing nanoporous carbon materials. Porous carbon derived from MOFs usually contains heteroatoms (such as phosphorous, boron, sulfur and nitrogen) and metals or their compounds due to the originally existed organic components and metals in MOFs, which could improve the electronic conductivity as well as the lithium electroactivity [167]. For example, Zheng et al. reported nitrogen-doped graphene particles analogues with a nitrogen content of up to 17.72 wt% that were prepared by the pyrolysis of a nitrogen-containing zeolitic imidazolate framework (ZIF) at 800 °C under a nitrogen atmosphere, the nitrogen content of which was much higher than that of most NDC materials (approximately 10%) (Fig. 3.11a). When used as anode material for LIBs, these particles retain a capacity of 2132 mA h g^{-1} after 50 cycles at a current density of 100 mA g^{-1}, and 785 mA h g^{-1} after 1000 cycles at 5 A g^{-1} (Fig. 3.11b–e) [168].

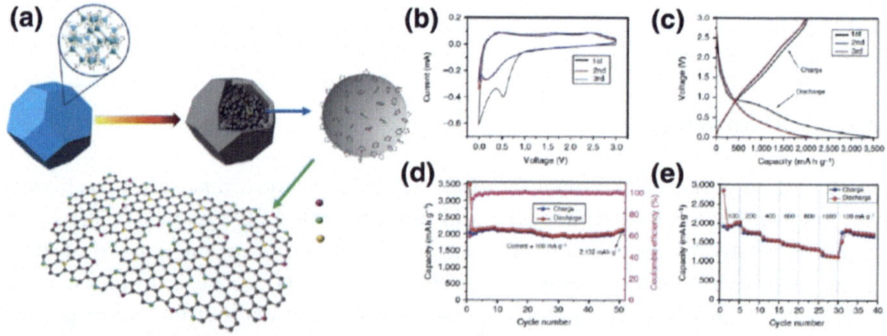

Fig. 3.11 **a** Schematic illustration of the synthesis procedure of N-doped graphene analogous particles and model of N-doping; **b** Cyclic voltammetry curve at 0.1 mV s^{-1} scan rate; **c** Galvanostatic charge–discharge profiles at a current density of 100 mA g^{-1}. The third curves exhibited significant overlap with the second charge and discharge curves; **d** Cycling performance at a current density of 100 mA g^{-1}; **e** Rate performance at different current densities from 100 to 1600 mA g^{-1}. Reproduced with the permission from Ref. [168], Copyright © 2014 Springer Nature

As the 3D allotrope of graphene, graphite also plays an important role in anode materials of LIBs. Graphite shows a lithiation potential of 0.2 V versus Li^+/Li, which is close to the lithium stripping voltage, which may cause a safety issue especially at high rate. Moreover, such a low lithiation potential range leads to the formation of electronically insulating SEI on the surface of graphite, which is likely to occur below 1.0 V versus Li^+/Li. It's worth mentioning that graphite can also be used as a conductive matrix. Si/graphene/ultrathin graphite framework electrode architecture was prepared by drop-casting a composite of graphene coated Si NPs on an ultrathin-graphite foam. The gravimetric capacity versus the mass of the whole electrode was 983 mA h g^{-1} on the first cycle and retained 370 mA h g^{-1} after 100 cycles for a Si loading density of 1.5 mg cm^{-2}, which was much higher than that of the graphite electrode used in a commercial 18650 cell with a gravimetric capacity of 244 mA h g^{-1}, and an order of magnitude higher than that of Si loaded on a metal foil [169].

3.3.1.4 Other Structures

Recently, the hard carbon has received much attention as an anode material for the large-scale LIB due to its superior cycling stability and high specific capacity, which represent a valid alternative to soft carbons. The term "hard carbon" normally represents those non-graphitizable carbons heat-treated at 1000–1100 °C and has random alignment of graphene sheets which provides many voids to insert Li^+. It is considered that lithium species are absorbed in the micropores as well as in the interlayer, which leads to the increase in capacity (~740 mA h g^{-1} can be delivered in case of single layers assuming the formation of Li_2C_6 compound) [170]. However, the formation of SEI on the surface and the hetero atoms or radicals in hard carbon lead to the high reversible capacity during the first cycle and ICE. Meanwhile, due to the disordered structure and relatively large particle size contributed to the slow lithium diffusion and lower conductivity than those ordered carbon lattice, namely very poor rate performance [171].

Many efforts have been paid to solve these problems, for example, by surface functionalization like oxidation, heteroatoms doping, or by using metal coating or a thin layer of soft carbon. Graphene is an ideal candidate to coat hard carbon, as it has been reported that hard carbon wrapped by the flexible and electrically conductive graphene results relatively low irreversible capacity, large reversible capacity and great rate performance [172]. Generally, the hard carbons are synthesized by pyrolysis of sucrose, glucose, polyvinyl chloride and so on. To find cheaper replacement for these industrial products, biomass wastes products are studied widely. Ou et al. used ox horn as a novel, low-cost and easily available precursor for the synthesis of a nitrogen containing carbon anode material, which attained high specific surface area and porous structure after a KOH activation during carbonization process. The ox horn derived carbonanode delivered excellent capacities of 1181, 750 and 304 mA h g^{-1} corresponding to current densities of 100 (0.26 C, 1 C = 372 mA g^{-1}), 500 (1.34 C) and 5000 mA g^{-1} (13.4 C), respectively [173].

3.3.2 Group IVA Elements and Compounds

Carbonaceous materials are commercially used in LIBs due to their low cost, high thermal and chemical stabilities, low electrochemical potential, and environmental benignity. However, the uptake of lithium in carbonaceous materials is based on an intercalation mechanism, in which only one Li atom can be stored per six carbon atoms, thus leading to a relatively low theoretical capacity of 372 mA h g^{-1}, which is too low to meet the rapidly growing demands for high-energy-density storage devices. Hence, there could be a breakthrough in capacity on anodes of LIBs by moving from a classical intercalation to an alloying reaction because the alloying reaction can store more Li compared with the intercalation reaction. Considerable attention has been devoted to alloy-type anode materials, owing to their high specific capacity and safety characteristics.

Among various alloy-type anode materials, Group IVA elements, including silicon (Si), germanium (Ge), and tin (Sn) hold great potential as anode materials for next-generation LIBs, which show striking capacity of 4200, 1625, 994 mA h g^{-1} for full lithiation to the $Li_{22}Si_5$, $Li_{22}Ge_5$, and $Li_{22}Sn_5$, respectively, based on alloy reactions between the elements and lithium. Besides, in the case of the M_{IVA} elements (M_{IVA} represents Si, Ge, and Sn), the Li-alloying voltages are about 0.4–0.6 V versus Li^+/Li, which could make the cut-off voltage much higher than the Li electrodeposition voltage during charging, and thus effectively avoiding the formation of lithium dendrites and improving the safety of LIBs. Furthermore, these M_{IVA} anode materials have other merits, such as abundant resources, environmental benignity for Si, high lithium diffusion coefficient for Ge, and high electronic conductivity for Sn. For these reason, M_{IVA} elements represent one of the development direction of high-capacity anode materials towards high energy density LIBs.

However, the volume expansion of the M_{IVA} elements are up to several hundred percent when lithiated to the $Li_{4.4}M_{IVA}$ phase. Along with the dramatic volume variation during cycling, there are also large stresses that would cause the cracking and even pulverization of M_{IVA} particles, which would impair the integrity of electrodes. The cracking and pulverization could lead to the electrode materials peel off from the current collectors, which damage the electrical contact of active materials to the current collector, thereby resulting in capacity decay of batteries. Another issue that gives rise to the fast capacity fade is the formation of an unstable SEI on the cracked M_{IVA} particles. The huge volume variation of electrodes during cycling will lead to breakdown of the SEI layer, and new fresh surface of these M_{IVA} anodes will be exposed to the organic electrolyte. Massive electrolytes will also be consumed in the formation of a thick SEI layer, which could be fatal for cycle performance of electrodes. In addition, SEI layers that are coated onto the surfaces and inside the cracked active particles will spread over the entire electrode and become thicker and thicker with charge/discharge cycling. Moreover, during the volume expansion/contraction of electrodes, the pristine big M_{IVA} particles will become smaller and smaller, and consequently deactivated, owing to the electronically insulating nature of the surrounded SEI layers, thus make the entire electrode non-capacitive.

According to the above discussion and analysis, it can be rationally considered that the key problems that lead to large irreversible capacity loss and to poor capacity retention of M_{IVA} elements include the huge volume variation upon cycling, the stability problem of SEI layers, and the invalidation of transport pathways for electrons. Another shortcoming of Si active material that cannot be ignored is its poor electrical conductivity, which leads to poor electrochemical reaction kinetics and resultantly a limited rate capability. In recent years, a large number of studies have been conducted for solving the mentioned problems. Significant breakthrough has been achieved by utilizing nanotechnologies with the aim of prolonging the cycle life and improving the rate performance. The volume change can be significantly buffered after downsizing the M_{IVA} particle to nanosize. In addition, nanostructured M_{IVA} materials could reduce transport length for Li^+ and electrons and facilitate a high electrode/electrolyte contact area, which is favorable for improving the lithium reaction rate. To improve the electrochemical performance of M_{IVA} anode materials, rational structure design is also essential, which influences the electron- and ion-transport pathways along the nanomaterials. In this section, we present various kinds of reasonable structure design on M_{IVA} nanomaterials, focusing on dimensional control in nanostructure design.

3.3.2.1 0D Structures (Nanoparticles)

Previous reports have shown that decreasing the dimension of M_{IVA} into nanometer range could reduce the mechanical stress generated during lithiation/delithiation and, thus, inhibit the tendency for fracture and pulverization. A great many of studies have shown that fabricating M_{IVA} NPs can enhance the electrochemical performance of M_{IVA} anode materials for LIBs. The key advance of NPs compared to their bulk counterparts is that it could effectively overcome the well-known pulverization problem, which is detrimental to the cycle life of anode materials. The significance of nanosize effects of M_{IVA} particles to LIBs performance has been investigated by several groups. Kim and coworkers synthesized 10 nm sized Si NPs for LIBs, and it showed capacity retention of 81% after 40 cycles, which was far superior to previous results [174]. Zhou and coworkers introduced to prepare crystalline embedded amorphous silicon NPs with an average size of 50 nm as anode materials for LIBs by a facile and large-scale fluidized bed reaction route [175]. The amorphous Si phase with isotropic expansion could facilitate the lithiation process and prevent the fracturing of Si particles, and the crystalline Si domains functioned as an efficient electronic conductor. The crystalline-amorphous-Si electrode could achieve excellent cycling stability with a high capacity retention of 94.7% after 200 cycles by increasing the operating potential.

The nanostructure could shorten the path lengths for Li^+ transport, thus providing high rate capability. However, the cyclability is still limited due to the aggregation of M_{IVA} NPs. To tackle this problem, another widely studied zero-dimensional (0D) structure design of NPs coated with protective layers, including carbon, metal layers and metal oxide layers, has been developed. Carbon could not only confine the aggregation of nanosized M_{IVA}, but also construct superior framework to buffer the huge volume change and enhance the electrical conductivity of the electrode. In addition, since extensions of the M_{IVA}-C interface and the satisfactory void space between the M_{IVA} and C phases are both critically important for good kinetics during lithiation/delithiation, which improve the cycling stability and rate capability of the electrodes. Jiang and coworkers designed PDA-coated SnO_2 nanocrystals (NCs), which were composed of hundreds of PDA-coated "corn-like" SnO_2 NPs (diameter ca. 5 nm) decorated along a "cob" [176]. The PDA-coated SnO_2 electrode showed the excellent rate capability, superior long-term stability over 300 cycles, and high $Sn \rightarrow SnO_2$ reversibility benefited from the corn-like nanostructure and the protective PDA coating. However, there always exist the porosity between Si NPs, which would reduce the volumetric energy density of the electrode. For this reason, one of the major challenges for designing high-capacity NP anode materials is to improve the tap density and maintain the cycling stability of those electrodes. A nano/micro-structured Si/C composite was designed to decrease the specific surface area and improve the tap density of Si/C materials [177]. The existing pores in micro carbon framework accommodated the huge volume expansion of nano-Si and allow for the rapid infiltration of electrolyte. With the optimized structure, the surface area of compact Si/C composites decreased to 10.5 $m^2\ g^{-1}$, corresponding to a high tap density, which made the electrode achieve an ultrahigh ICE of 91.2% and excellent cycling stability with 96.5% capacity retention after 100 cycles under a current density of 0.2 A g^{-1}. Smart metal or metal oxide coating of M_{IVA} NPs are also investigated to protect M_{IVA} NPs and shown to be good for improving the cyclability of LIBs. Stable surface coating metals could provide extra electrical conductivity for the composite electrode, and metal oxides usually possess a great mechanical property to bear the strain of M_{IVA} during its lithiation/delithiation. Yang et al. synthesized commercial Si NPs encapsulated by amorphous TiO_2 [178]. With this thin external layer, the composite materials showed greatly superior electrochemical performance and high-safety lithium storage. The amorphous TiO_2 shell (~3 nm) exhibited elastic behavior during lithium charging/discharging processes, maintaining high structural integrity. In addition, they found that the amorphous TiO_2 shell offered superior buffering properties compared to the crystalline TiO_2 layer for unprecedented cycling stability.

3.3.2.2 1D Structures

1D structured M_{IVA} anode materials have been explored to improve the electrochemical performance, such as nanowires (NWs) and NTs. Generally, 1D structures can buffer the mechanical strain during the volume change to a lower level in the radical direction to prevent the electrode materials pulverization. On the other hand, along the length direction, 1D structures are capable of providing efficient electron-transport pathways, allowing rapid charge transfer [179]. In addition, some NWs and NTs can be grown directly from current collectors, eliminating the need for binders and conductive additives.

Nanowires and Nanotubes

NWs and NTs can exhibit high discharge capacities and stable cycling life because there are adequate electrical contacts between the current collector and every NW or NT so that more active materials contribute to the capacity. Liu and coworkers synthesized Ge NWs using a CVD method via vapor-liquid-solid mechanism and reported a detailed study of the lithiation/delithiation behavior of Ge NWs using in situ TEM [180]. Fast reversible expansion and contraction of the NW during cycling were observed, indicating mechanical robustness and fast kinetics of a Ge anode in LIBs. Kennedy et al. reported the formation of high-density Ge NW arrays grown directly from the current collector as high-performance and high-capacity LIB anodes (Fig. 3.12) [181]. The average entangled Ge ligament diameter was ca. 5.6 nm. The anodes retained capacities of ~900 mA h g^{-1} after 1100 cycles with excellent rate performance characteristics, even at very high discharge rates of 20–100 C. This performance could be attributed to the complete restructuring of the NWs that occurred within the first 100 cycles to form a continuous porous network that was mechanically robust.

0D Nanoparticles Encapsulated in 1D Nanotubes

The strategy of compositing M_{IVA} NPs with 1D carbon materials, such as NTs, NWs and NFs, has been investigated as well. Those composites combine the advantages

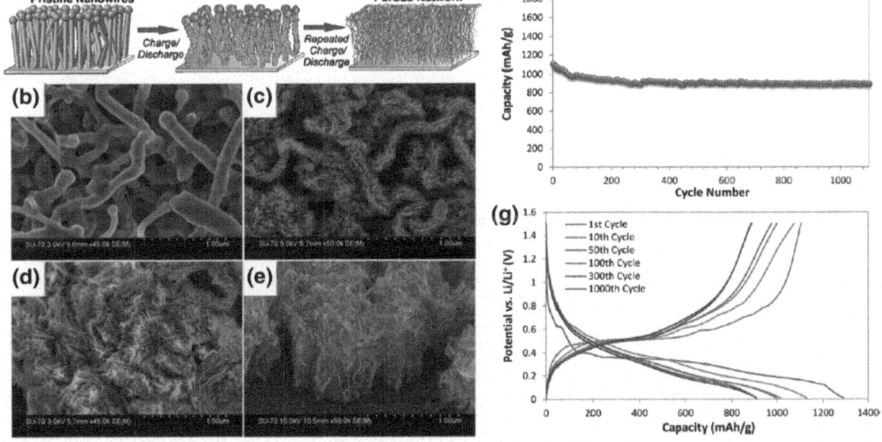

Fig. 3.12 **a** Schematic showing the cumulative effect of cycling on the NW architecture; **b**–**e** SEM images of Sn seeded Ge NWs after 1, 10, 20, and 100 cycles, respectively; **f** Discharge capacities of the Sn seeded Ge NW electrode over 1100 cycles at C/2 rate in the potential range of 0.01–1.5 V; **g** Voltage profiles of 1st, 10th, 50th, 100th, 300th, and 1000th cycle of the electrode cycled in (**f**). Reproduced with the permission from Ref. [181], Copyright © 2014 American Chemical Society

of the high capacity of M_{IVA} NPs and superior electrical conductivity of 1D carbon materials. In addition, those nanohybrid anodes can effectively suppress the agglomeration of M_{IVA} NPs uniformly dispersed in 1D carbon materials and tolerate the substantial volume change during the lithium alloying/dealloying reaction. Moreover, the conductive carbon matrix is beneficial to produce a stable SEI layer to prevent the electrolyte from further decomposition. An elegant structure was synthesized by confining Si NPs in porous CNFs (Si@PCNF) [165]. Besides, the coating of SiO_x was introduced on the surface of Si NPs to function as a solid source for HF etching to produce void space between Si NPs and carbon shell. Thus, a favorable integration of Si NPs between the high electro-conductive porous CNF sheaths was achieved to promote the electronic transportation, and sufficient void space could cushion the volume variation of Si during Li–Si alloying/dealloying. This novel Si-based hybrid material exhibited improved cycling performance (ca. 1104 mA g h^{-1} under 0.5 A g^{-1} after 100 cycles) as well as excellent rate capabilities (ca. 485 mA g h^{-1} under 10 A g^{-1}), promising great potential as an anode material for LIBs. A facile strategy was also reported for the fabrication of ultra-uniform SnO_x/carbon (U–SnO_x/C) nanohybrids by solvent replacement and subsequent electrospinning homogeneous dispersion of SnO_2 NPs in PAN/DMF solution [182]. The 1D nanostructure U–SnO_x/C held strong interaction between SnO_x and nitrogen-containing CNF matrix, which could effectively confine the uniformly embedded SnO_x. As a consequence, the U–SnO_x/C nanohybrids exhibited a high reversible capacity of 608 mA h g^{-1} after 200 cycles, with excellent rate capability. Nanostructured SnO_2/C microfibers were also prepared by thermal decomposition of tin alginate fibers produced via wet-spinning technique [183]. The as-obtained SnO_2/C microfibers consisted of nano-sized SnO_2 crystals with a mean diameter of 10–15 nm. The SnO_2/C microfibers possessed a high capacity and good rate capability compared with commercial SnO_2, since SnO_2 particles successfully dispersed in carbon matrix and the existing nanopores in the SnO_2/C microfibers.

Core-Shell Structures

To further promote the stability during volume variation, the core-shell design has been put forward. Structurally stable materials as supports can effectively protective M_{IVA} active materials. On the one hand, electrically conductive materials tend to be the select as the core to enhance the electronic transport; on the other hand, the stable materials acting as the shell is another powerful design, which not only provide efficient electron-transport pathways along the shell, but also control the formation of the SEI layers. Those core-shell structures have shown a large improvement in structural stability and electrochemical performance. Wang et al. developed a novel Si–C nanohybrid in which Si NWs dwelled in the hollow graphitic tubes (Si NW-d-GT) [184]. This hybrid structure held the robust line-to-line contact between Si and C, which created efficient channels for the fast transport of both electrons and Li^+ during cycling. The Si NW-d-GT exhibited good rate capability and remarkable cycling stability (~1100 mA h g^{-1} at 4200 mA g^{-1} over 1000 cycles). Cu–Si nanocable arrays were designed via directly anchored on a current collector to promote both the cycling stability and the high rate capability of Si as an anode of LIBs (Fig. 3.13) [185]. The conductive Cu core was grown from the current collector forming a 3D current collector substrate, which allowed for effective electron conduction to enhance the battery's high rate performance. Besides, the robust Cu core provided structural reinforcement to overcome the mechanical rupture during the volume changes of the Si shell. Furthermore, the space between the nanocables can accommodate large Si expansion and an additional coating with Al_2O_3 could stabilize the Si/electrolyte interface and trigger a stable SEI formation for long cycle lifetime. Additionally, this structure can possibly avoid any use of binders or conducting additives, reducing the extra weight to the battery. After coated with Al_2O_3, The Cu–Si–Al_2O_3 nanocables showed a remarkable high rate capability. Lin et al. developed novel twisted, aligned MWCNT/Si composite fiber anodes for flexible, wire-shaped LIBs (Fig. 3.14) [186]. A thin layer of silicon was coated on the outer surface of the aligned MWCNT to form a core-sheath structure that could effectively and simultaneously exploit the high specific capacity of the silicon section and high electrical conductivity of MWCNT, while the designed space with sizes of tens to hundreds of nanometers among the aligned composite NTs could effectively counterbalance the volume change of the silicon component. As a result, the aligned MWCNT/Si fiber electrode showed a combined high specific capacity and cyclic stability. In addition, this fiber electrode could be further made into a flexible wire-shaped battery, which offered unique advantages, such as being lightweight and weaveable compared with the conventional planar structure. The TiO_2 NT is electrochemically stable and has been reported as both durable anodes and effective surface coating material with volume expansion even less than that of graphite. Guan et al. coated the single-crystalline SnO_2 NWs with a hollow and conformal TiO_2 shell by using atomic layer deposition (ALD), in which optimization of the gap thickness and TiO_2 shell thickness toward superior performance was realized by ALD with great convenience [187]. The gap between the SnO_2 core and TiO_2 shell was designed to allow a nearly free expansion of the SnO_2 NW without severe pulverization and thus improved the cycling ability and rate capacity. The uniform hollow TiO_2 shell provided stable surface protection and the appropriate-sized gap effectively accommodated the expansion of the interior SnO_2 NW. Typically a relatively stable capacity of 393.3 mA h g^{-1} was achieved after 1000 charge-discharge cycles at a current density of 400 mA g^{-1}, and 241.2 mA h g^{-1} at 3200 mA g^{-1}.

3.3.2.3 2D Structures

Other promising structure design strategies for M_{IVA} anode materials have been also developed, such as 2D structures. Specific facets preferentially exposed in the 2D structure can enable fast ion and electron transfer for efficient lithium storage [188]. Furthermore, the ultrathin thickness of 2D nanomaterials offers good mechanical flexibility to endure the stress of M_{IVA}.

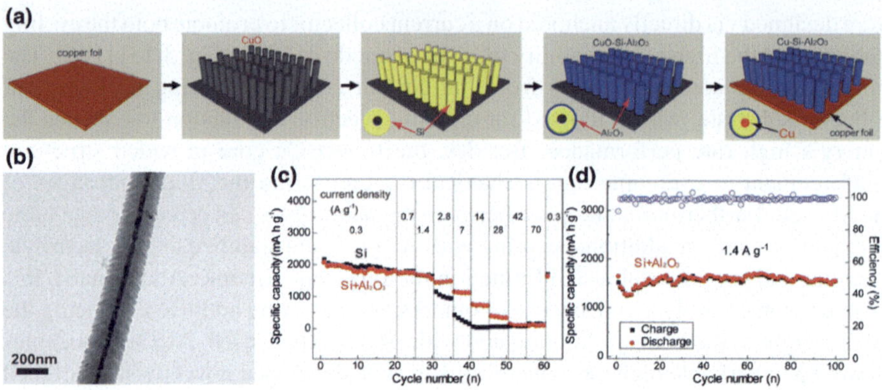

Fig. 3.13 **a** Schematic diagram showing the fabrication of Cu–Si and Cu–Si–Al_2O_3 nanocables; **b** Morphology characterization of Cu–Si–Al_2O_3 nanocable; **c** Comparison of the rate capabilities of Cu–Si–Al_2O_3 nanocables (red) and Cu–Si (black) cycled under different current densities; **d** Cycling behavior of Cu–Si–Al_2O_3 nanocables under a current density of 1.4 A g^{-1} for 100 cycles. Reproduced with the permission from Ref. [185], Copyright © 2011 WILEY-VCH

Thin Films

Thin films are able to accommodate stresses due to volume expansion and contraction during electrochemical cycling and maintain their mechanical integrity. Constructing thin film nanostructures to shorten the Li^+ diffusion pathway, facilitate interfacial charge transfer, and enlarge the electrode-electrolyte interface offers enhanced electrochemical performance LIBs.

Chen et al. first successfully synthesized mesoporous ultrathin 2D Si NSs through a soft-template and subsequent magnesiothermic reduction method [189]. The 2D Si NSs had a narrow mesoporous diameter distribution (2–6 nm), leading to fairly high specific surface areas (386.2 m^2 g^{-1}). As the LIB anode, the unique mesoporous and 2D structures significantly enhanced the cycling stability of Si. To further alleviate volume changes and promote kinetic behavior, mesoporous Si NSs were decorated with the homogeneous carbon layer. The Si/C nanocomposite anodes achieved remarkably enhanced lithium-storage properties (reversible capacity of 1575.5 mA h g^{-1} at 400 mA g^{-1} even after 100 cycles), cycle stability (capacity retention of 92% at 400 mA g^{-1} after 100 cycles), and rate capability (reversible of capacity of 1296.6 mA h g^{-1} at 4 A g^{-1} and retain the reversible capacity of 1072.2 mA h g^{-1} even after 500 cycles). The superior electrochemical performance of the Si/C nanocomposite electrode could be attributed to the synergistic effect between the unique 2D mesoporous nanostructure and carbon capsulation. Lin and coworkers fabricated "Welcome-mat"-like porous Si/Cu composite amorphous films by applying the predeposited Cu–NP-assembled film as the growth direction template for the subsequent deposition of a Si active layer with the cluster beam deposition technique [190]. When used as the binder-free anodes for LIBs, the acquired single-layer porous Si/Cu composite film exhibited a large reversible capacity of 3124 mA h g^{-1}

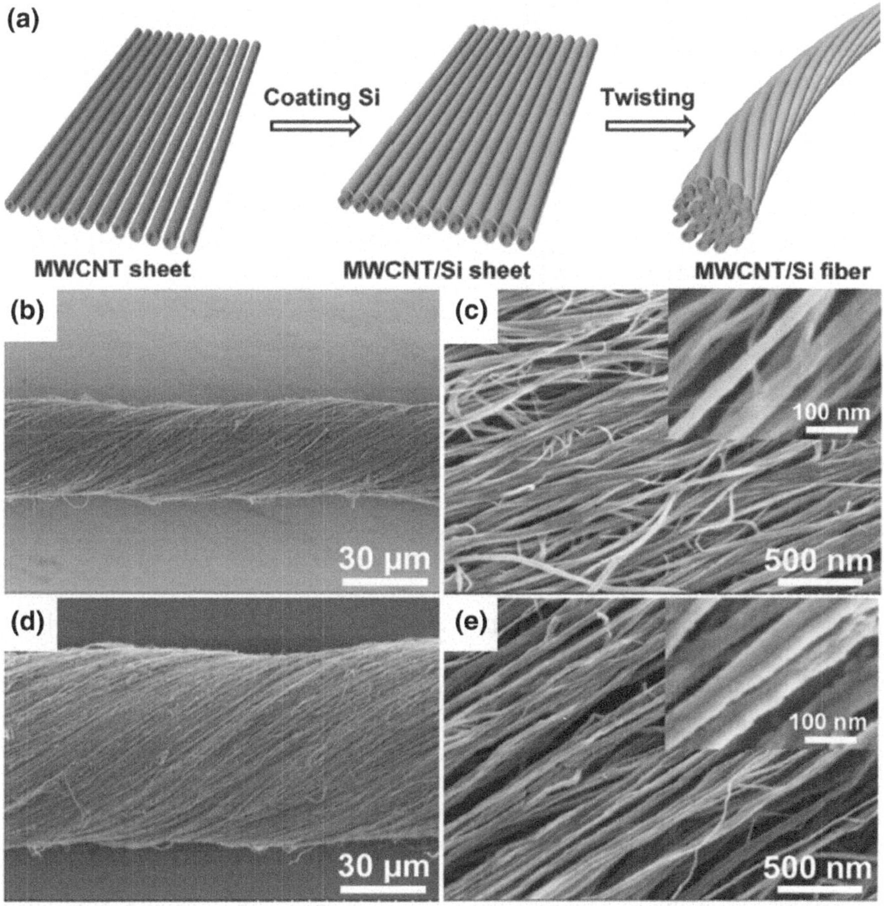

Fig. 3.14 **a** Schematic illustration to the preparation of the aligned MWCNT/Si composite fiber; **b–c** SEM images of a bare aligned MWCNT fiber at low and high magnifications, respectively; **d–e** SEM images of an aligned MWCNT/Si composite fiber with Si weight percentage of 38.1% at low and high magnifications, respectively. Reproduced with the permission from Ref. [186], Copyright © 2014 WILEY-VCH

after 1000 cycles at 1 A g^{-1} and excellent rate capability of 2086 mA h g^{-1} even at 20 A g^{-1} for 450 cycles. Furthermore, the multilayer construction of porous Si/Cu composite films was employed to improve the mass loading of Si active materials while keeping the specific capacity unchanged, which was promising for applications in LIBs. Suresh et al. used a carbon nanotube microfilm (CNM) instead of copper as the current collector substrate and by draping the Si film with an atomically thin sheet of graphene [191]. The CNM substrate was highly flexible and allowed the Si film to expand and contract during lithiation/delithiation cycles. This served to "unconstrain" the Si film and minimized stress buildup. The role of the monolayer graphene drape was to stabilize the SEI layer that was formed at the Si-electrolyte interface. When the Si film strained, the graphene layer would "slip" with respect to the Si, since there was a relatively weak van der Waals interaction between graphene and Si. Consequently, the underlying Si film was allowed to expand and contract without straining the mechanically fragile SEI layer, thereby leaving it intact. The graphene capping sheet also functioned as an interconnecting elastic web that helped to toughen the Si film and reduced its tendency to pulverize and delaminate. The graphene-draped Si films on CNM exhibited long cycle life (>1000 charge/discharge steps) with an average specific capacity of ~806 mA h g^{-1}.

Nanoparticles Anchored on 2D Layer Structures

One important strategy to fabricate 2D layer structure is compositing M_{IVA} with 2D materials, such as graphene, GO, rGO, and MXene, which provide a 2D electronic pathway for M_{IVA} NPs. Those materials always obtain enhanced electrochemical performance by means of the improved electronic conductivity, the reduced volume variation in the vertical direction of thin film, and high dispersion of M_{IVA} active materials.

The utilization of graphene in encapsulating M_{IVA} NPs as anode materials for LIBs is becoming more and more appealing due to its unique properties, such as excellent 2D electrical conductivity, superior mechanical flexibility, pronounced chemical and thermal stability, and large surface area. The design of anchoring M_{IVA} NPs on graphene can not only provide enough flexibility to accommodate huge volume changes, but also can enhance the conductivity of the whole electrode.

Our group have done numbers of interesting researches on composites of M_{IVA} with graphene including GO and rGO. A SnO_2 NC/nitrogen-doped rGO hybrid material (SnO_2 NC@N–rGO) was obtained by a in situ hydrazine monohydrate vapor reduction method for binding SnO_2 NCs in graphene sheets by Sn–N bonding [192]. The as-prepared SnO_2 NC@N–rGO held uniform SnO_2 NCs homogeneously confined in graphene sheets. The graphene coating promoted the electronic transport and restricted the growth of the SEI layer. The bond formed between graphene and SnO_2 NCs limited the agglomeration of tin NPs generated during lithiation. The void space and the nanometer-sized SnO_2 facilitated Li^+ diffusion. Also, the graphene sheets, along with the void space, worked as a cushion to buffer the huge volume change of SnO_2 NCs. The SnO_2 NC@N–rGO displayed the extremely high capacity of 1021 mA h g^{-1} (based on the total mass of the hybrid), excellent rate capability, and long cycle life. Besides, our group presented Sn NPs encapsulated in nitrogen-doped graphene sheets (Sn@N–rGO) through heat-treatment of the SnO_2 NCs/nitrogen-doped graphene hybrid [158]. The as-prepared Sn@N–rGO coated Sn surface through Sn–N and Sn–O bonding generated between Sn NPs and graphene. The graphene coating promoted the electron transport and helped to form a stable SEI layer as well as trapped the fractured Sn NPs formed during lithiation and delithiation, and the bonds generated between Sn NPs and graphene inhibited the aggregation of Sn NPs. The hybrid exhibited a reversible capacity of 481 mA h

g^{-1} after 100 cycles under 0.1 A g^{-1} and a charge capacity as high as 307 mA h g^{-1} under 2 A g^{-1}. In addition, our group had developed a poly(ethylene oxide)-assisted electrospinning method for accommodating Si NPs in hierarchical conducting networks consisting of graphene and carbon NPs to obtain a silicon-graphene composite (denoted as Si–G–C) [193]. The as-prepared Si–G–C held Si NPs uniformly dispersed in the hierarchical conducting networks. The nanodots formed by pyrolysis of the electrospun NFs could facilitate the Li^+ and electron transport and buffer the large volume change of Si NPs as well as restrict the Si NP agglomeration during cycling. The Si–G–C electrode exhibited excellent cycling performance (1344 mA h g^{-1} at 500 mA g^{-1} after 200 cycles) and great rate capability. Moreover, our group have developed an electrostatic attraction directed self-assembly approach for fabricating a nanocomposite of Si NPs encapsulated in graphene [194]. The complete fabrication of Si–NP@G nanocomposite included two self-assembly processes: (i) the Si NPs first adsorbed poly(diallydimethylammonium chloride) (PDDA) through electrostatic attraction to change the surface charge from negative to positive (Si-PDDA NPs); and, (ii) the obtained positively charged Si-PDDA NPs assembled with negatively charged GO by electrostatic attraction. This approach realized a uniform dispersion of Si NPs between two layers of graphene sheets. The as-obtained Si–NP@Graphene nanocomposite showed stable cycling performance (approximately 1205 mA h g^{-1} after 150 cycles) and excellent rate capability. Our group also used a facile spin-coating method to cast a spin-coated Si NP/graphene (SC–Si/G) film, in which graphene could act as both an efficient electronic conductor and effective binder with no need for other binders such as polyvinylidenefluoride or polytetrafluoroethylene [195]. The prepared SC–Si/G electrode could achieve a high-performance as an anode for LIBs benefiting from that the graphene enhanced the electronic conductivity of Si NPs and the void spaces between Si NPs could facilitate the Li^+ diffusion as well as cushion the volume expansion of Si NPs. As a result, the binder-free electrode showed a high capacity of 1611 mA h g^{-1} at 1 A g^{-1} after 200 cycles, a superior rate capability of 648 mA h g^{-1} at 10 A g^{-1}, and an excellent cycle life of 200 cycles with 74% capacity retention. Chang et al. reported on a novel Si-based anode architecture, namely, assembly of alternating Si/rGO layers on porous Ni foams [196]. Compared with the conventional Cu current collector, the 3D porous Ni foam current collector could improve mass loading of active materials, which enhanced the energy density of the whole electrode. RGO layers could effectively prevent aggregation of Si NPs and provided a conductive network for electronic transport, which improved the electrical conductivity, rate capability, and usage of Si. Moreover, the outmost continuous rGO layer prevented the top Si NP layer from detaching but allowed Li ions to pass through because the rate limiting step for Li^+ diffusion was through the Si NPs, which improved the stability of the anode. With the multilayered Si/rGO nanostructures, the electrode exhibited outstanding lithium storage performance with high reversible specific capacity (2300 mA h g^{-1} at 0.05 C, 700 mA h g^{-1} at 10 C; capacity based on Si/rGO total weight, respectively), fast rate capability (up to 10 C), and superior capacity retention during cycling (87% capacity retention with a rate of 10 C after 152 cycles up to 630, 780 mA h g^{-1} at 3 C after 300 cycles).

Exploiting graphene to fabricate sandwich structure could assure solid contact between the M_{IVA} particle and the graphene layer, it also facilitates a high electrode conductivity, and renders the elastomeric space needed to accommodate the volume changes of M_{IVA}, thus leading to an excellent electrochemical performance as the anode for LIBs. Our Group developed a two-step programmed method to load mono dispersed SnO_2 NPs on both sides of single layer graphene sheets (denoted as SnO_2–G) [159]. The SnO_2–G composite had near mono dispersion of the SnO_2 NCs as well as a high SnO_2 content of over 60 wt%. The SnO_2 NCs of about 3 nm were formed, and anchored by the oxygenate groups on the GO layer, forming uniform loading of the SnO_2 NCs on single layer GO sheets. These outstanding features were desirable and enabled the composite material to be an excellent anode material for LIBs. Wang et al. synthesized sandwiched structures consisting of carbon coated SnO_2 nanorod grafted on graphene based on a seed assisted hydrothermal growth to form graphene supported SnO_2 nanorods, followed by a nanocarbon coating [197]. Benefiting from the advanced sandwiched hierarchical structure and the intimate chemical bonding between each layers, the nano-hybrids showed an extremely high lithium storage capability of up to 1419 mA h g^{-1} benefiting from the advanced structural features. Wang et al. developed a facile strategy to fabricate N-doped graphene-SnO_2 (G–SnO_2) sandwich papers in which most of the SnO_2 NPs (32 wt%) were sandwiched between the graphene layers [198]. In this work, 7,7,8,8-tetracyanoquinodimethane ions played a key role in the formation of such N-doped sandwich structures, as they acted both as the nitrogen source and the complexing agent. All these notably enhanced electrochemical performances of this novel material with respect to commercial SnO_2 NPs could be attributed to the structural features that provided a large number of surface defects induced onto the graphene by N-doping, excellent electronic conductivity, short transportation length for both Li^+ and electrons, and enough elastomeric space to accommodate volume changes upon Li insertion/extraction.

Those results suggest that compositing M_{IVA} particles with graphene is effective in mitigating detrimental effects of volume expansion/contraction, improving electronic conductivity of charge transfer and charge diffusion during the lithiation/delithiation process, which represents a significant advancement for M_{IVA}-based materials in LIBs. Therefore, the utilization of graphene is promising to accommodate the volume variations of M_{IVA} NPs and even prevents the breakage of electrode integrity.

The MXene as an emerging 2D structure has been reported to exhibit metallic conductivity, negative surface charge in solutions, and hydrophilicity. Ahmed and coworkers fabricated a composite SnO_2/MXene anode for LIB applications [199]. The SnO_2/MXene anode exploited the high Li-ion capacity offered by SnO_2, while maintaining the structural and mechanical integrity of the conductive MXene platform. The SnO_2/MXene electrode composite design provided a high capacity thanks to conversion reaction of SnO_2, while MXene sheets provided a conductive network that accommodated the large volume changes that occurred in SnO_2 during the battery charge/discharge process. The thin HfO_2 passivation layer coated SnO_2/MXene anodes exhibited a stable specific capacity of 843 mA h g^{-1}.

Nanosheets Anchored on Nanotubes

For the composite of NSs anchored on NTs, the NT support can provide highly efficient electronic conductive networks for NSs active materials anchored on it. It is believed that these nanostructures with large exposed surface and short diffusion length could interact more efficiently with Li^+, which could enhance the electrochemical performance of M_{IVA} anode materials. Lian et al. synthesized a unique 3D hierarchical nanostructure C@SnS/SnO_2@CNFs (Fig. 3.15) [200]. The 3D hierarchical structure could provide a large surface, which was able to adequately contact with the electrolyte to facilitate the diffusion of Li^+ between the electrolyte and electrode materials. The carbon coating the SnS/SnO_2 heterostructures could improve the electrical conductivity. Meanwhile, SnS/SnO_2 heterostructures had the ability to encourage charge transfer. The 3D hierarchical structure, the coated carbon, and the SnS/SnO_2 heterostructures could alleviate the large volume effect without detachment from the current collector and prevent agglomeration. The C@SnS/SnO_2@CNFs electrode delivered high reversible capacity of 826.8 mA h g^{-1} at the current density of 1.0 A g^{-1} after 500 cycles.

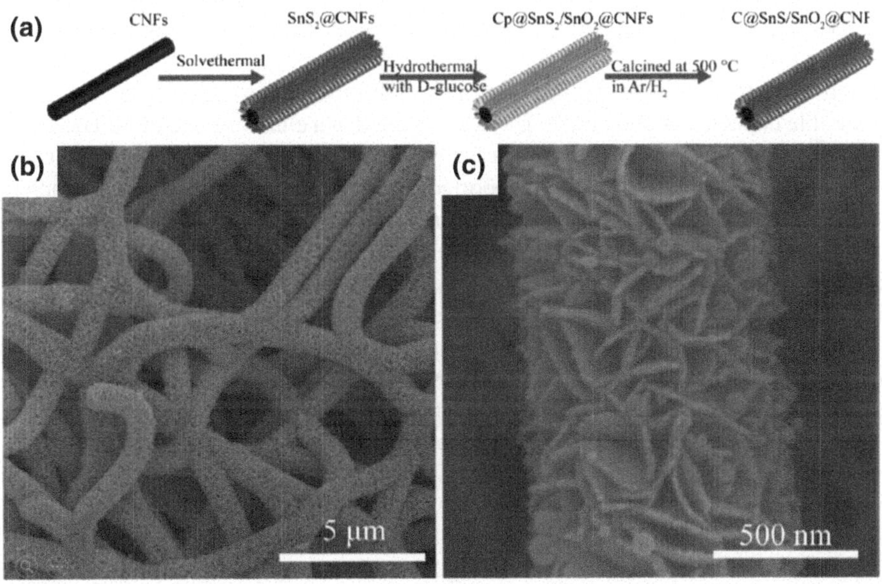

Fig. 3.15 **a** Schematic illustration of the formation of the C@SnS/SnO_2@CNFs; **b–c** SEM and HRSEM of C@SnS/SnO_2@CNFs. Reproduced with the permission from Ref. [200], Copyright © 2016 American Chemical Society

3.3.2.4 3D Structures

The 3D structure, owing to its high specific surface area, can remarkably mitigate the volume expansion and accelerate the electrolyte permeation, thus giving rise to a superior cycle performance. Benefiting from the 3D structures, the as-prepared 3D M_{IVA}-based anode materials show improved electrochemical performance.

3D Porous Structures

The porous structures can successfully provide sufficient void space for expansion/contraction of M_{IVA} and improve the contact between active materials and electrolyte. Li et al. combined in situ TEM and continuum media mechanical calculations to demonstrate that large (>20 μm) mesoporous silicon sponge prepared by the anodization method could limit the particle volume expansion at full lithiation to ~30% and prevent pulverization in bulk silicon particles [201]. The mesoporous silicon sponge delivered a capacity of up to ~750 mA h g^{-1} based on the total electrode weight with >80% capacity retention over 1000 cycles. A cost-efficient method to produce nanoporous Si particles from metallurgical Si was via ball-milling and inexpensive stain-etching [202]. An extensive study on the scanning transmission electron microscopy (STEM) tomography reconstruction revealed the highly porous feature of Si particles under proper etching conditions. Nanoporous Si anodes with a reversible capacity of 2900 mA h g^{-1} was attained at a charging rate of 400 mA g^{-1}, and a stable capacity above 1100 mA h g^{-1} was retained for extended 600 cycles tested at 2000 mA g^{-1}. Yi and coworkers reported a new, low-cost, large-scale approach to prepare a micro-sized Si–C composite with excellent performance as an anode materials for LIBs [203]. The composite showed a reversible capacity of 1459 mA h g^{-1} after 200 cycles at 1 A g^{-1} (97.8% capacity retention) and excellent high rate performance of 700 mA h g^{-1} at 12.8 A g^{-1}, and also had a high tap density of 0.78 g cm^{-3}. The excellent performance was attributed to the nanoscale size of primary particles and interconnected carbon and Si networks which could maintain internal electrical contact and sustain cycling stability. The synthesis method was low-cost and easy to scale up, and was thus believed to have great potential in practical production of high-performance Si materials for LIBs. Micro-sized materials are favorable for practical battery applications since they often enable higher tap density than nanosized materials and, as a result, are expected to offer higher volumetric capacity. Such micro-sized Si anode materials with nanoscale building blocks have been previously demonstrated with encouraging results. Because of these tradeoffs, it is desirable to develop new materials that combine the advantages of both micro-sized and nano-sized Si materials to improve the cycling performance, rate capability, and energy density of Si anodes. Huang and coworkers introduced a carbon-encapsulated porous SnO_2 submicrocubes (SMCs) yolk–shell structure (SnO_2 SMCs@C), in which a porous SnO_2 SMC formed a core with a carbon shell separated with a void space [204]. Porous SnO_2 cubes have been proven to possess better electrochemical performance than the sense NPs. The porous SnO_2 SMCs@C showed much increased capacity and cycling stability, demonstrating that the porous SnO_2 core was essential for better lithium-ion storage performance. Wang et al. demonstrated that a nanostructure of amorphous hierarchical porous GeO_x whose primary particles were ~3.7 nm diameter had a very stable capacity of ~1250 mA h g^{-1} for 600 cycles [205]. Furthermore, a full cell of the porous GeO_x coupled with a $Li(NiCoMn)_{1/3}O_2$ cathode exhibited high performance.

Nanostructures Deposited on 3D Porous Substrates

Another commonly used approach is to embed M_{IVA} nanomaterials into 3D conductive matrix, which serves as a physical scaffold to effectively counteract the pulverization of M_{IVA} and maintain the mechanical integrity of the composite electrode, thus improving the structure stability. Zhu and coworkers reported ultrasmall Sn NPs (~5 nm) embedded in nitrogen-doped porous carbon network (denoted as 5-Sn/C) with the Brunner-Emmet-Teller specific surface area of 286.3 m^2 g^{-1} [206]. The 5-Sn/C showed an initial discharge capacity of 1014 mA h g^{-1} and a capacity retention of 722 mA h g^{-1} after 200 cycles at the current density of 0.2 A g^{-1}. Furthermore, a reversible capacity of ~480 mA h g^{-1} was obtained at much higher current density of 5 A g^{-1}. Wang' group synthesized a nano-Sn/C composite with uniformly dispersed 10 nm nano-Sn within a spherical carbon matrix by an aerosol spray pyrolysis technique [207]. The nano-Sn/C composite sphere anodes maintained the initial capacity of 710 mA h g^{-1} after 130 cycles at 0.25 C. The nano-Sn/C composite sphere anodes could provide ~600 mA h g^{-1} even at a high rate of 20 C. The exceptional performance of those electrodes are attributed to that carbon matrix offers mechanical support, prevents M_{IVA} NP agglomeration and provides continuous path for Li ions and electrons inside the composite. Lee et al. developed a one-pot spray pyrolysis to synthesize the Si–SiO_x composite coated with outer carbon coating layer [208]. The control of reaction condition in the spray pyrolysis allowed the microstructure and composition of the Si–SiO_x composite to be adjusted, achieving good performance with respect to both ICE and cyclability. The optimal composite structure in terms of Si domain size and Si-to-O ratio exhibited an excellent reversible capacity of 1561.9 mA h g^{-1} at 0.06 C rate and ICE of 80.2% and 87.9% capacity retention after 100 cycles at 1 C rate. Zhang et al. prepared ordered mesoporous Si/C nanocomposites with ultrasmall (ca. 3 nm), uniform silicon NPs embedded inside a mesoporous carbon framework by a magnesiothermic reduction approach [209]. The large pore of the main channels and small pores inside the carbon frameworks can guarantee electrolyte transportation and facilitate Li^+ diffusion at the interface between the electrolyte and Si NPs. The internal void space generated inside the carbon framework during the magnesiothermic reduction process is just right to provide sufficient voids for the volume expansion of Si NPs during Li insertion. The mesoporous Si/C nanocomposites showed excellent performance with a high reversible capacity of 1790 mA h g^{-1}, excellent Coulombic efficiency (as high as ca. 99.5%) and rate capability, outstanding cycle stability (capacity remains as high as 1480 mA h g^{-1} after 1000 cycles at a high current density of 2 A g^{-1}). Our Group synthesized novel Si/C microspheres through our facile and low-cost production line, which was mainly composed of sand milling, ball milling and spray drying process [210]. Based on the porous Si/C structure, 3D carbon conducting networks are well-established in compacted Si/C microspheres to facilitate the fast access of electrons and Li^+ in porous Si/C granules. The optimal design for Si/C anodes contributes to improved electrochemical performance at ultrahigh mass loading due to the superior electrical conductivity and high tap density. Si/C anodes with various reversible capacities (450–750 mA h g^{-1}) could be fabricated by the flexible production process.

Some researchers have tried to embed M_{IVA} active materials into 3D hosts to fabricate hierarchical structures. The hierarchical structure can further buffer the volume expansion resulted by the cycling of M_{IVA}. Our Group designed the watermelon-inspired Si/C microspheres, which could alleviate the volume variation and particle fracture under a high pressing density [211]. The graphite framework functioned as the electrical highway and isolated layer to suppress the aggregation of nano-Si particles and the whole carbon shell also prevented the exposure of nano-Si under high pressing density. This Si/C anodes presented the appropriate reversible capacity of 620 mA h g^{-1}, excellent cycling stability over 500 cycles and superior ICE (89.2%) and average Coulombic efficiency (99.8%), due to the dual protection strategy including hierarchical buffer structure and optimized size distribution. Besides, the secondary structure for nano-Si particles could effectively enhance the tap density of nanomaterials, which was essential to practical applications of electrode materials in high energy density batteries. Our Group also reported the fabrication of silicon/NDC/CNT nano/micro-hierarchical structured spheres through a facile electrospray approach using rice husk as silicon source (Fig. 3.16) [212]. The unique hierarchical hybrid structure silicon/NDC/CNT spheres delivered a high reversible specific capacity of 1380 mA h g^{-1} at a current density of 0.5 A g^{-1}, and still maintained 1031 mA h g^{-1} after 100 cycles, in which the silicon NPs on the sizes of ~50 nm were homogenously encapsulated in the highly conductive, porous carbon matrix. The porous matrix built by CNTs and NDC, contributed to fast electronic transport and prevented silicon pulverization, which possessed good structure stability upon the synergistic lithiation/delithiation of the components. Besides, our group reported a novel Si NP-carbon NP/graphene composite [213]. In the composite, Si NPs were embedded in 3D conducting networks built by carbon NPs and graphene sheets. The addition of carbon NPs could effectively alleviate the aggregation of Si NPs by separating them from each other, and helped graphene sheets build efficient 3D conducting networks for Si NPs. Such Si–C/G composite showed much improved electrochemical properties in terms of specific capacity and cycling performance (ca. 1521 mA h g^{-1} at 0.2 C after 200 cycles), as well as a favorable high-rate capability. Cui's group prepared Si-nanolayer-embedded graphite/carbon hybrids using a CVD process with a scalable furnace [214]. This architecture increased compatibility between silicon and natural graphite, which inhibited severe side reactions caused by structural failure of crumbled graphite dust and uncombined residue of silicon particles by conventional mechanical milling. Therefore, the electrical interconnectivity and morphological integrity were preserved without severe cracks and contact losses. The structure showed a capacity retention of 96% after 100 cycles, with an

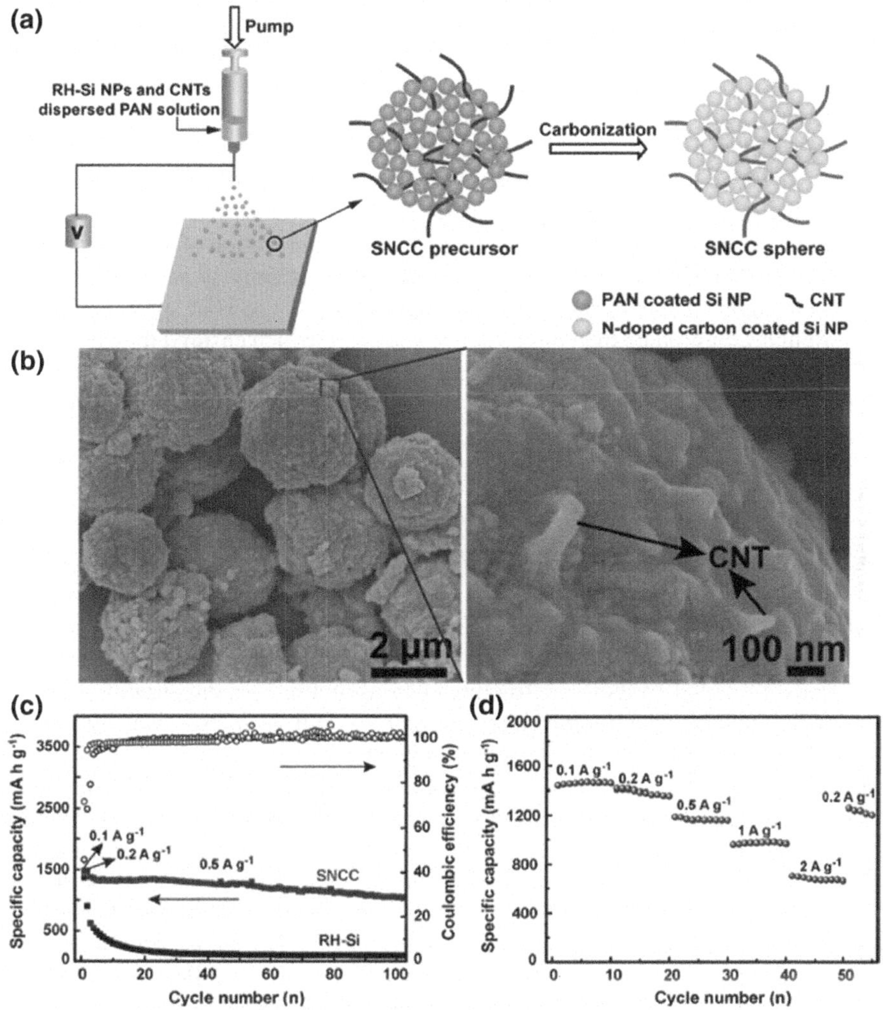

Fig. 3.16 **a** Schematic illustration of the preparation process for SNCC nano-micro-structured spheres; **b** SEM images of SNCC spheres; **c** Cycling performance and Coulombic efficiency of SNCC spheres electrode and RH–Si electrode at 0.2 A g^{-1} for the 1st and 2nd, then 0.5 A g^{-1} for latter cycles; **d** Rate performance of SNCC spheres electrode at various current densities. The voltage range is 0.01–1.5 V versus Li^+/Li. Reproduced with the permission from Ref. [212], Copyright © 2016 Elsevier

industrial electrode density of >1.6 g cm^{-3}, areal capacity loading of >3.3 mA h cm^{-2}, and <4 wt% binding materials in a slurry. In addition, a full cell using $LiCoO_2$ demonstrated a higher energy density (1043 Wh L^{-1}) than standard commercial graphite electrodes.

In addition, some researchers tried to design the 3D composite structure containing much void space to accept the increasing volume of M_{IVA} during cycling. Zhang and coworkers synthesized tin NPs encapsulated elastic hollow carbon spheres (TNHCs) [215]. In TNHCs, multiple tin NPs with a diameter of less than 100 nm were encapsulated in one thin hollow carbon sphere with a thickness of only about 20 nm, which led to both the content of Sn up to over 70% by weight and the void volume in carbon shell as high as about 70–80% by volume. The carbon shell had itself good electronic conductivity and prevented the aggregation of active materials, and especially thin carbon shell had good elasticity to effectively accommodate the strain of volume change during Li^+ insertion/extraction. The void space in TNHCs could right accommodate the volume after expansion, which ensured the good cycle characteristic as well as high volume energy density. This nanocomposites had very high specific capacity of >800 mA h g^{-1} in the initial 10 cycles, and >550 mA h g^{-1} after the 100th cycles, as well as excellent cycling performance, exhibiting a great potential as anode materials in LIBs.

3.3.2.5 Other Composite Structures

Hollow structures have been considered as advanced designs to effectively accommodate the volume expansion for anode materials in LIBs. The pre-constructed void space provided by hollow structures can accommodate the structural strain, thus leading to improved lithium storage properties. Hollow micro-/nano-structured materials have drawn growing interests owing to their unique structural features and unusual properties that endow them with great potential for LIBs. Available methods to prepare hollow structured materials can be generally categorized into templating and template-free approaches. The former one involves the use of various active or inactive particles to support the deposition/growth of shell materials, while the latter one utilizes several novel mechanisms, such as Kirkendall effect, galvanic replacement, chemical etching and inside-out Ostwald ripening, to form the hollow structure [216]. In our group, graphene enwrapped SnO_2 hollow NPs (SnO_2–HNS/G) was developed [217]. The hollow NPs acted as high-capacity lithium storage hosts and provided hollow interior spaces to buffer their large volume variations during lithium uptake/release processes, while the graphene coating facilitated the electronic transportation to reach the hollow NPs, and served as a flexible structure strengthening agent for the hollow NPs. The combination of the two components resulted in a superior SnO_2-based anode material (i.e. SnO_2–HNS/G) with much improved cycling performance (ca. 696 mA g h^{-1} under 0.5 A g^{-1} after 300 cycles) and high rate capability (ca. 307 mA g h^{-1} under 5 A g^{-1}). Lou's group have done plenty of researches on designing of hollow structure. In their reports, SnO_2/NDC (SnO_2/ND–C) submicroboxes were synthesized [218]. The as-synthesized SnO_2/N–C submicroboxes were highly porous with a high specific surface area of 125 m^2 g^{-1}, well-defined hollow structure (~400 nm in size) with a shell thickness of 40 nm, and ultrasmall SnO_2 NPs uniformly coated with NDC layer. As a result, the SnO_2/N–C submicroboxes showed outstanding electrochemical performance as an anode material for LIBs. A high reversible capacity of 491 mA h g^{-1} could be retained after 100 cycles at a current density of 0.5 A g^{-1}. Lou's group prepared porous SnO_2 microboxes with a high uniformity and well-defined non-spherical hollow structure via a selective leaching strategy using $ZnSn(OH)_6$ microboxes as the starting material [216]. The selective removal of Zn(II) species by an acid-washing process produced pure SnO_2 microboxes with a porous shell. After further wrapping with an elastic carbon matrix, the resultant SnO_2@C microboxes exhibited excellent electrochemical performance as an anode material for LIBs. Lou's group proposed a novel nanostructure for SnO_2-based anode materials, namely hierarchical tubular structures constructed by carbon-coated ultrathin SnO_2 nanoplates (SnO_2@C-HTs), which simultaneously integrated several desirable design rationales for high-energy anode materials based on low-dimensional ultrathin nanoplates, hollow tubular structure, and carbon nanocoating [219]. The as-synthesized SnO_2@C-HTs manifested high specific capacity and excellent cycling stability.

A common drawback of these simple hollow structures is the low packing density of the materials compared to the solid counterparts, which originated from the large empty space within the particles. As a result, electrodes fabricated with these hollow spheres typically suffered from relatively low volumetric energy and power densities, which are undesirable for practical applications. To make better utilization of the inner hollow cavity, construction of complex hollow structures could partly overcome the above-mentioned limitation of hollow particles. Lou's group designed bowl-like SnO_2@carbon hollow particles to reduce the excessive hollow interior space while retaining the general advantages of hollow structures [220]. As a result, the tap density of the bowl-like SnO_2@carbon hollow particles could increase about 30%. The as-prepared bowl-like SnO_2@carbon hollow particles with conformal carbon support exhibited excellent lithium storage properties in terms of high capacity, stable cyclability and excellent rate capability.

3.3.3 Transition-Metal Compounds

3.3.3.1 $Li_4Ti_5O_{12}$

Spinel lithium titanate ($Li_4Ti_5O_{12}$, LTO) has attracted considerable attentions as a potential anode material of LIBs for high power applications due to following outstanding features: (i) a flat charge/discharge plateaus (around 1.55 V vs. Li^+/Li) higher than that of commercial graphite anode materials, which can minimize the chance for the formation of SEI (usually occur below 1.0 V vs. Li^+/Li) and dendritic lithium, ensuring the safety; (ii) as a "zero strain" material, LTO has excellent structure and thermodynamic stability, there is almost no volume change during the insertion and extraction processes of lithium, because the lattice dimension won't change between the initial state of LTO and final state of $Li_7Ti_5O_{12}$, which enables a high cycling stability and long operational life of electrode material [221]. Despite these advantages, the electronic conductivity of LTO is relatively low (in the range of ~10^{-8}–~10^{-13} S cm^{-1}) which could result in severe polarization phenomenon (more seriously at higher cycling rates), and the Li^+ diffusion coefficient is also low (~10^{-8}–10^{-13} cm^2 s^{-1}) that leads to its poor rate performance. To improve its capacity and rate capability, considerable researches have been centered on the aliovalent ion doping, nanostructure design, and the conductive layer surface modification of LTO, or a combination of these strategies. The latter two are the most common approaches. Coating LTO with carbon layer (CNT, GO, carbon) is one of the most effective and facial way to improve the rate performance of LTO for the carbon coating layer, which can simultaneously enhance the electric conductivity and Li^+ diffusion efficiency. Li et al. reported a simple and scalable process to fabricate LTO NSs with NDC coating, and this material exhibited an excellent rate capability due to the thin carbon coating and porous NS structures, which produced a highly conductive pathway for electrons and fast transport channels for Li^+ [222]. However, the introduction of high-content carbon will increase the chance of SEI formation which may cause the safety issue, reduce the tap density and capacity of the whole material (the capacity that carbon contributes is negligible when the voltage is higher or equal than 1.0 V vs. Li^+/Li). Therefore, through a facile solution-based method, well-defined LTO NSs terminated with rutile-TiO_2 at the edges were synthesized and revealed directly at atomic resolution by an advanced spherical aberration imaging technique. The results showed that the carbon-free coating layer of rutile-TiO_2 efficiently improved the kinetics of LTO toward fast lithium insertion/extraction thus the composite attained high surface stability and much improved rate capability compared with the commonly used carbon coating in materials [223]. That's because of the presence of Ti^{3+} and the transformation from rutile-TiO_2 to Li_xTiO_2 inserted by Li^+ during the initial discharge process, which are able to enhance the electric conductivity of the entire material [224]. When it comes to varies nanostructures of LTO, some structures like NSs and porous frameworks usually possess large specific surface, low tap density that leads to high irreversible reaction and low volumetric energy density. Besides, these structures are relatively fragile which are susceptible to get damaged during the current battery assembly process pursuing high compact density induced by the high volumetric capacity. Nano-sized spheres of LTO solved these problems, Wang et al. reported a robust and precisely controllable strategy of synthesizing monodisperse LTO nanospheres with an average diameter of 120 nm by judiciously employing titanium nitride (TiN) as the titanium source. These uniform LTO nanospheres owned a tap density of 1.1 g cm^{-1} and exhibited outstanding rate capability, structure stability and remarkable capacity retention (92.6% after 500 cycles at 10 C) [225].

TM oxides, have been demonstrated to be promising anode materials for LIBs because of their high reversible capacity (500–1000 mA h g^{-1}) from the participation of a high number of electrons in the conversion reactions, as well as high safety, environmental benignity, low cost, but most TM oxides suffered from, low electrical conductivity, low Li^+ diffusion efficiency, high volume expansion and aggregation problem during charge/discharge process, making them usually exhibit poor cycling performance. There have reported numerous research to overcome these identified limits [146, 226].

Many papers have been focused on developing new methods to synthesize TM oxides with modified structures, improved electrochemical kinetics and power capability. Carbon coating or compositing with carbon are the most commonly used strategies to improve the electrical conductivity and rate capability, usually combined with novel nanostructures design.

Titanium dioxide (TiO_2, titania) has received much attention for application as a LIBs anode material due to its various advantages, such as cost-effective, high electrochemical activity, good mechanism and chemical stability and excellent safety. However, the poor electronic conductivity of TiO_2 has limited its application as a key problem. Therefore, many works focus on the design of synthetic approaches and engineering of structures, and the most commonly used strategies include exploiting 3D conductive networks, combination with conductive materials or designing porous structures [227]. For example, well-organized CNT@TiO_2 core/porous-shealth coaxial nanocables were synthesized by controlled hydrolysis of tetrabutyl titanate in the presence of CNTs. The CNT core provides sufficient electrons for the storage of Li in TiO_2 sheath while the CNT itself can also store Li, and this storage kinetics is improved by the presence of the nanoporous TiO_2 because the only very thin protection layer on TiO_2 (unlike free CNT) enables rapid access of Li ions from the liquid electrolyte [228].

Iron oxides, including haematite (α-Fe_2O_3) and magnetite (Fe_3O_4), can provide a theoretical capacity of 1007 and 926 mA h g^{-1}, respectively. They are synthesized to form various nanostructures. For example, a layered α-Fe_2O_3 nanodisk/rGO (α-Fe_2O_3/rGO) composite was synthesized by a silicate-anion-assisted in situ hydrothermal method. There are oxygen bridges in face-to-face manner between ultrathin α-Fe_2O_3 and rGO, which provides a larger effective interaction area between electrode materials and electrolytes, which shortens the Li^+ diffusion path and improves the kinetics of charge transportation. The composites exhibited excellent specific capacity of 931 mA h g^{-1} which was close to the theoretical capacity and much higher than the α-Fe_2O_3 nanodisks alone (10 mA h g^{-1}) or the α-Fe_2O_3/rGO-mix (11 mA h g^{-1}) [229]. The improvement can be ascribed to the synergistic effect of the layered structure of α-Fe_2O_3 and the face-to-face introduction of rGO. Other structures like nanospindles, NPs, nanospheres etc. [230–232] are also well studied.

Cobalt oxides materials including Co_3O_4 and CoO can provide theoretical capacities of 890 and 715 mA h g^{-1} in LIBs, respectively [233]. Similarly to other kinds of materials, a number of different forms and synthetic routes of cobalt oxides have been studied. Porous nanostructures, NSs, nanocubes, NWs, NTs, octahedral cages and NPs have been prepared by various synthetic routes, which can not only fasten the Li^+ diffusion within the free volume around the NPs, but also relax the volume-strain. A controllable nanocasting method was designed for the spatial and size-selective synthesis of well-dispersed Co_3O_4 NPs in mesoporous carbons. The obtained mesoporous peapod-like Co_3O_4@carbon NT arrays (Co_3O_4@CNT) have high surface areas of up to 750 m^2 g^{-1} and large pore sizes (Fig. 3.17a–c). The particle sizes of Co_3O_4 can be controlled between 3–7 nm by varying the thickness of the carbon layer, and their loading amount can be varied from 45 to 70 wt%. The resultant Co_3O_4@CNT material showed a high specific capacity of 780 mA h g^{-1}

Fig. 3.17 **a** SEM images of the composite Co_3O_4@CNT-2: **b** TEM, and **c** HRTEM images of thin microtome sections of the composite Co_3O_4@CNT-2; **d** Charge/discharge curves of the Co_3O_4@CNT-1 electrode at a current density of 0.1 A g^{-1}; **e** Cycling performance of the Co_3O_4@CNT-1, CMK-5, and mesoporous Co_3O_4 electrodes at constant current densities of 0.1 A g^{-1}; **f** Cycling performance of the Co_3O_4@CNT-1 electrode at different current densities from 0.1 to 5.0 A g^{-1}; **g** Schematic representation of the electrochemical reaction path on the Co_3O_4@CNT electrode materials. Reproduced with the permission from Ref. [234], Copyright © 2015 Wiley-VCH

(the volumetric capacity was calculated to be ~1370 mA h cm^{-3}) at a current density of 100 mA g^{-1}, an excellent rate capacity, and cycling performance (Fig. 3.17d–g) [234]. Similar with iron and cobalt oxides, other TM oxides like copper [226, 235] and chromium [236, 237] are also well studied as electrode materials for LIBs.

3.3.3.2 Transition Metal Sulfides

To date, metal sulfides have been extensively investigated for energy storage and conversion in view of their similar electrochemical mechanism to metal oxides as well as their abundance in nature and affordable cost. Among them, FeS_2 has been recognized as one of the most promising anode materials for LIBs due to its unique features, such as a high theoretical capacity of 890 mA h g^{-1}, low environmental impact (non-toxic elements with respect to Fe and S) and affordable cost (abundant and cheap). However, commercial FeS_2 suffers from poor cycling performance and low capacity retention due to lower conductivity and larger volume expansion during cycling processes, which prevents it from commercial applications of LIBs [238]. Various nano/micro structures of FeS_2 and its composites compositing with highly conductive materials (e.g., porous carbon, CNTs, and graphene) have been designed to overcome these problems. A novel composite of rGO and FeS_2 microparticles self-assembled from small size cubes as a high-performance anode material for LIBs has been prepared via a facile one-pot hydrothermal method. FeS_2 microparticles composed of small size cubic FeS_2 crystals well-dispersed in rGO sheets and formed interconnected networks, benefited from which the material exhibited a very large initial reversible capacity of 1147 mA h g^{-1} at a current rate of 100 mA g^{-1} and main-

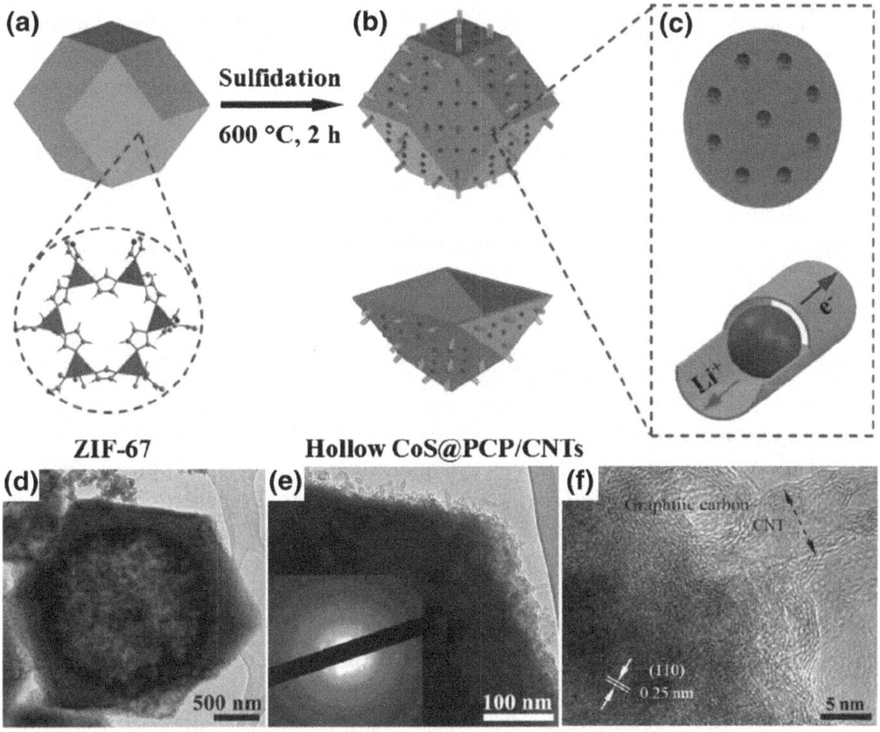

Fig. 3.18 Schematic illustration for the formation of 3D hollow CoS@PCP/CNTs: **a** ZIF-67 template; **b** hollow CoS@PCP/CNTs; and **c** magnified representation of an electron/lithium transportation channel of CoS@PCP/CNTs; **d** and **e** TEM images of CoS@PCP/CNTs polyhedron prepared at 600 °C (2 h), with the inset in (**e**) showing the selected area electron diffraction pattern; **f** HRTEM images of the CoS@PCP/CNTs polyhedron. Reproduced with the permission from Ref. [240], Copyright © 2015 WILEY-VCH

tains 1001.41 mA h g^{-1} over 60 cycles [239]. The available synthesis procedures for these TM sulfides/carbon hybrids usually involve multisteps and complicated synthetic conditions, and may have difficulty in constructing hybrids with strong coupling between metal sulfides and the carbonaceous matrix. Thus, many efficient approaches have been developed to synthesize these hybrids. For example, a novel one-step approach for the fabrication of 3D hollow cobalt sulfides@porous carbon polyhedra/CNTs hybrids (CoS@PCP/CNTs) through the simultaneous decomposition and sulfidation of a preformed Co-based ZIF-67 template was reported, which achieved nanostructure engineering and hybridization simultaneously (Fig. 3.18) [240].

Another kind of TM sulfides possess layered structure, in which, MoS_2 is the most representative one that S–Mo–S layers stacks together by van der Waals forces, analogous to that of graphite. Although such structure facilitates reversible Li^+ intercalation/extraction, the poor electronic/ionic conductivity between two adjacent

S–Mo–S sheets and their irreversible capacity still limits their further applications. Therefore, compositing with carbonaceous templates are studied a lot to solve these problems above. For instance, the fabrication of MoS_2–rGO NS–NS hybrids (MoS_2 NS–rGO) via a simple combination of a lithiation-assisted exfoliation process and a hydrazine monohydrate vapour reduction technique was reported. The as-prepared MoS_2 NS–rGO can not only prevent the agglomeration of MoS_2 NSs but also restrict the growth of Mo NPs during lithiation by the bond formed between MoS_2 NSs and graphene [241]. In addition, other layered TM sulfides like VS_2 and WS_2 are also well-studied [242, 243].

3.3.3.3 Transition Metal Phosphides

Transition metal phosphides attract much attention in LIBs due to their low polarization and good cycling stability. The electrochemical reaction in metal phosphides with lithium might be classified into two reaction mechanisms (intercalation and conversion reaction mechanisms), depending on electronegativity difference between metal and phosphorous atoms [244]. Copper, cobalt, iron, nickel and tin based phosphides are usually considered to belong to the second group, namely conversion mechanism. Metal phosphides can deliver high capacities between 500 and 1800 mA h g^{-1} and exhibit lower insertion potential than their oxide counterparts. However, MP_x usually have low electrical conductivity and high volume changes upon charge/discharge cycling [146]. To overcome these drawbacks, explorations are supposed to be conducted. Hierarchical, nanostructured nickel phosphide (h–Ni_2P) spheres are synthesized by a one-pot reaction from an organic-phase mixture of nickel acetylacetonate, trioctylphosphine, tri-n-octylamine, and oleylamine which is used as a surfactant to modify the surface morphology of Ni_2P spheres. Such hierarchical structure greatly increases the contact between electrode and electrolyte and can buffer volume expansion which increased the stability during cycling [245]. Similar lithium performance has been also found in the researches of cobalt and copper phosphides. Co_xP with various nanostructures (e.g., Co_2P rod and spheres, CoP hollow and solid particles) with amorphous carbon coating layer were fabricated through a facile thermal decomposing method, between which the CoP hollow NP depicted the best performance in terms of capacity retention and rate capability [246]. Size-tunable, hexagonal plate-like Cu_3P and janus-like Cu–Cu_3P NCs were synthesized using trioctylphosphine as phosphorus precursor copper and the phosphide plates were found to be capable of undergoing lithiation/delithiation through a displacement reaction while Cu–Cu_3P was not [247]. In addition to nanostructure engineering, fabricating into thin film is also a good choice. Air stable and carbon-free copper phosphide (CuP_3) of thicknesses (0.2, 0.4 μm) was synthesized over copper plates (of 10 mm diameter) by hybrid electrochemical deposition and low temperature solid-state reaction, and this binder-free electrode exhibited significantly improved capacity retention and rate capability characteristics.

3.3.4 Summary

In this part, we mainly discussed three typical anode materials with corresponding nanostructure designing strategies, including carbonaceous materials, alloy-type anode materials based on Group IVA elements and TM compounds. Conventional graphite carbon is considered as the most successfully commercialized anode material for LIBs because of its desirable electrical conductivity, structural stability and low cost. Many efforts have also been paid to develop other carbonaceous materials with higher energy and power density. Investigation about nanostructured carbonaceous materials have been intensively developed and achieved considerable developments. 1D CNT especially SWCNTs can attain high reversible capacities (>1116 mA h g^{-1}) exceeding a LiC_2 stoichiometry; 2D graphene can deliver higher capacity to 780 mA h g^{-1} (Li-ions absorb on both faces of graphene: Li_2C_6 stoichiometry) or to 1116 mA h g^{-1} (Li trapped at the benzene ring in a covalent bond: LiC_2 stoichiometry); 3D and other structure carbonaceous materials are also well studied. In general, carbonaceous materials is promising for practical application in LIBs due to its great conductivity, stability and low cost, but their capacity is limited. Therefore, alloy-type anode materials based on Group IVA elements has attracted much attention due to their high theoretical capacity. Unfortunately, their electrochemical performance is damaged by severe volume changes. Researchers have fabricated various nanostructures based on different design concepts, such as dimensional control, reserving hollow volume and enhancing conductivity and so on. Especially, Si-based anode materials have achieved delightful breakthroughs and get preliminary practical application researches. In addition, TM compounds have also been studied, including TM oxides, sulfides and phosphides. But they also face several drawbacks, such as low electrical conductivity, low Li^+ diffusion efficiency. Nanostructure design and combining with nanostructured carbonaceous materials can efficiently alleviate these problems. Remarkably, spinel LTO has attracted considerable attentions for its flat charge/discharge plateaus (~1.55 V vs. Li^+/Li) and zero strain feature but it also suffers the low conductivity and severe polarization.

References

1. Tarascon, J. M., & Armand, M. (2001). Issues and challenges facing rechargeable lithium batteries. *Nature, 414,* 359–367.
2. Whittingham, M. S. (1976). Electrical energy storage and intercalation chemistry. *Science, 192,* 1126–1127.
3. Rao, B. M. L., Francis, R. W., & Christopher, H. A. (1977). Lithium-aluminium electrodes. *Journal of the Electrochemical Society, 124,* 1490–1492.
4. Mizushima, K., Jones, P. C., Wiseman, P. J., & Goodenough, J. B. (1980). Li_xCoO_2 $(0<X<1)$: A new cathode material for batteries of high energy density. *Materials Research Bulletin, 15,* 783–789.
5. Thackeray, M. M., David, W. I. F., Bruce, P. G., & Goodenough, J. B. (1983). Lithium insertion into manganese spinels. *Materials Research Bulletin, 18,* 461–472.

6. Mohri, M., Yanagisawa, N., Tajima, Y., Tanaka, H., Mitate, T., Nakajima, S., et al. (1989). Rechargeable lithium battery based on pyrolytic carbon as a negative electrode. *Journal of Power Sources, 26,* 545–551.
7. Whittingham, M. S. (2004). Lithium batteries and cathode materials. *Chemical Reviews, 104,* 4271–4302.
8. Whittingham, M. S. (1976). The role of ternary phases in cathode reactions. *Journal of the Electrochemical Society, 123,* 315–320.
9. Jacobson, A. J., Chianelli, R. R., Rich, S. M., & Whittingham, M. S. (1979). Amorphous molybdenum trisulfide: A new lithium battery cathode. *Materials Research Bulletin, 14,* 1437–1448.
10. Sakuda, A., Taguchi, N., Takeuchi, T., Kobayashi, H., Sakaebe, H., Tatsumi, K., et al. (2013). Amorphous TiS4 positive electrode for lithium-sulfur secondary batteries. *Electrochemistry Communications, 31,* 71–75.
11. Trumbore, F. A. (1989). Niobium triselenide: A unique rechargeable positive electrode material. *Journal of Power Sources, 26,* 65–75.
12. Cao, A.-M., Hu, J.-S., Liang, H.-P., & Wan, L.-J. (2005). Self-assembled vanadium pentoxide (V_2O_5) hollow microspheres from nanorods and their application in lithium-ion batteries. *Angewandte Chemie International Edition, 44,* 4391–4395.
13. Yue, Y., & Liang, H. (2017). Micro- and nano-structured vanadium pentoxide (V_2O_5) for electrodes of lithium-ion batteries. *Advanced Energy Materials, 7,* 1602545.
14. Huang, X., Rui, X. H., Hng, H. H., & Yan, Q. Y. (2015). Vanadium pentoxide-based cathode materials for lithium-ion batteries: Morphology control, carbon hybridization, and cation doping. *Particle & Particle Systems Characterization, 32,* 276–294.
15. Tan, H. T., Rui, X., Sun, W., Yan, Q., & Lim, T. M. (2015). Vanadium-based nanostructure materials for secondary lithium battery applications. *Nanoscale, 7,* 14595–14607.
16. Liu, Y., Guan, D., Gao, G., Liang, X., Sun, W., Zhang, K., et al. (2017). Enhanced electrochemical performance of electrospun V_2O_5 nanotubes as cathodes for lithium ion batteries. *Journal of Alloys and Compounds, 726,* 922–929.
17. Wu, N., Du, W., Liu, G., Zhou, Z., Fu, H.-R., Tang, Q., et al. (2017). Synthesis of hierarchical sisal-like V2O5 with exposed stable 001 facets as long life cathode materials for advanced lithium-ion batteries. *ACS Applied Materials & Interfaces, 9,* 43681–43687.
18. Haris, M., Atiq, S., Abbas, S. M., Mahmood, A., Ramay, S. M., & Naseem, S. (2018). Acetylene black coated V_2O_5 nanocomposite with stable cyclability for lithium-ion batteries cathode. *Journal of Alloys and Compounds, 732,* 518–523.
19. Chen, M., Liang, X., Yin, J., Chen, Q., & Xia, X. (2018). Graphene foam supported V_2O_5/N-C core/shell arrays as advanced cathode for lithium ion storage. *Journal of Alloys and Compounds, 735,* 2022–2029.
20. Ding, Y.-L., Wen, Y., Wu, C., van Aken, P. A., Maier, J., & Yu, Y. (2015). 3D V_6O_{13} nanotextiles assembled from interconnected nanogrooves as cathode materials for high-energy lithium ion batteries. *Nano Letters, 15,* 1388–1394.
21. Zou, Z., Cheng, H., He, J., Long, F., Wu, Y., Yan, Z., et al. (2014). V_6O_{13} nanosheets synthesized from ethanol-aqueous solutions as high energy cathode material for lithium-ion batteries. *Electrochimica Acta, 135,* 175–180.
22. Ren, W., Zheng, Z., Luo, Y., Chen, W., Niu, C., Zhao, K., et al. (2015). An electrospun hierarchical LiV_3O_8 nanowire-in-network for high-rate and long-life lithium batteries. *Journal of Materials Chemistry A, 3,* 19850–19856.
23. Song, H., Liu, Y., Zhang, C., Liu, C., & Cao, G. (2015). Mo-doped LiV_3O_8 nanorod-assembled nanosheets as a high performance cathode material for lithium ion batteries. *Journal of Materials Chemistry A, 3,* 3547–3558.
24. Tan, H., Xu, L., Geng, H., Rui, X., Li, C., & Huang, S. (2018). Nanostructured $Li_3V_2(PO_4)_3$ cathodes. *Small (Weinheim an der Bergstrasse, Germany), 14,* 1800567.
25. Jian, Z. L., Hu, Y. S., Ji, X. L., & Chen, W. (2017). NASICON-structured materials for energy storage. *Advanced Materials, 29,* 1601925.

26. Rui, X. H., Li, C., Liu, J., Cheng, T., & Chen, C. H. (2010). The Li3V2(PO4)3/C composites with high-rate capability prepared by a maltose-based sol-gel route. *Electrochimica Acta, 55,* 6761–6767.
27. Liu, H., Gao, P., Fang, J., & Yang, G. (2011). $Li_3V_2(PO_4)_3$/graphene nanocomposites as cathode material for lithium ion batteries. *Chemical Communications, 47,* 9110–9112.
28. Rui, X., Yan, Q., Skyllas-Kazacos, M., & Lim, T. M. (2014). $Li_3V_2(PO_4)_3$ cathode materials for lithium-ion batteries: A review. *Journal of Power Sources, 258,* 19–38.
29. Kalluri, S., Yoon, M., Jo, M., Park, S., Myeong, S., Kim, J., et al. (2016). Surface engineering strategies of layered $LiCoO_2$ cathode material to realize high-energy and high-voltage li-ion cells. *Advanced Energy Materials, 7,* 1601507.
30. Jang, Y.-I., Huang, B., Wang, H., Sadoway, D. R., Ceder, G., Chiang, Y.-M., et al. (1999). $LiAl_yCo_{1-y}O_2$ ($R\bar{3}M$) intercalation cathode for rechargeable lithium batteries. *Journal of the Electrochemical Society, 146,* 862–868.
31. Tukamoto, H., & West, A. R. (1997). Electronic conductivity of $LiCoO_2$ and its enhancement by magnesium doping. *Journal of the Electrochemical Society, 144,* 3164–3168.
32. Lee, J.-G., Kim, B., Cho, J., Kim, Y.-W., & Park, B. (2004). Effect of $AlPO_4$-nanoparticle coating concentration on high-cutoff-voltage electrochemical performances in $LiCoO_2$. *Journal of the Electrochemical Society, 151,* A801–A805.
33. Cho, J., Kim, Y. J., & Park, B. (2001). $LiCoO_2$ cathode material that does not show a phase transition from hexagonal to monoclinic phase. *Journal of the Electrochemical Society, 148,* A1110–A1115.
34. Cho, J., Kim, Y. J., Kim, T.-J., & Park, B. (2001). Zero-strain intercalation cathode for rechargeable li-ion cell. *Angewandte Chemie International Edition, 40,* 3367–3369.
35. Cho, J., Kim, C.-S., & Yoo, S. I. (2000). Improvement of structural stability of $LiCoO_2$ cathode during electrochemical cycling by sol-gel coating of SnO_2. *Electrochemical and Solid-State Letters, 3,* 362–365.
36. Chen, Z., & Dahn, J. R. (2002). Effect of a ZrO_2 coating on the structure and electrochemistry of Li_xCoO_2 when cycled to 4.5 V. *Electrochemical and Solid-State Letters, 5,* A213–A216.
37. Hao, Q., Ma, H., Ju, Z., Li, G., Li, X., Xu, L., et al. (2011). Nano-cuo coated $LiCoO_2$: Synthesis, improved cycling stability and good performance at high rates. *Electrochimica Acta, 56,* 9027–9031.
38. Orikasa, Y., Takamatsu, D., Yamamoto, K., Koyama, Y., Mori, S., Masese, T., et al. (2014). Origin of surface coating effect for MgO on $LiCoO_2$ to improve the interfacial reaction between electrode and electrolyte. *Advanced Materials Interfaces, 1,* 1400195.
39. Lee, H. J., & Park, Y. J. (2013). Interface characterization of MgF_2-coated $LiCoO_2$ thin films. *Solid State Ionics, 230,* 86–91.
40. Sun, Y.-K., Yoon, C. S., Myung, S.-T., Belharouak, I., & Amine, K. (2009). Role of AlF_3 coating on $LiCoO_2$ particles during cycling to cutoff voltage above 4.5 V. *Journal of the Electrochemical Society, 156,* A1005–A1010.
41. Park, J. S., Mane, A. U., Elam, J. W., & Croy, J. R. (2015). Amorphous metal fluoride passivation coatings prepared by atomic layer deposition on $LiCoO_2$ for li-ion batteries. *Chemistry of Materials, 27,* 1917–1920.
42. Wang, Y., & Cao, G. (2008). Developments in nanostructured cathode materials for high-performance lithium-ion batteries. *Advanced Materials, 20,* 2251–2269.
43. Ceder, G., Chiang, Y. M., Sadoway, D. R., Aydinol, M. K., Jang, Y. I., & Huang, B. (1998). Identification of cathode materials for lithium batteries guided by first-principles calculations. *Nature, 392,* 694–696.
44. Lai, Y.-Q., Xu, M., Zhang, Z.-A., Gao, C.-H., Wang, P., & Yu, Z.-Y. (2016). Optimized structure stability and electrochemical performance of $LiNi_{0.8}Co_{0.15}Al_{0.05}O_2$ by sputtering nanoscale ZnO film. *Journal of Power Sources, 309,* 20–26.
45. Fang, T., Duh, J.-G., & Sheen, S.-R. (2005). Improving the electrochemical performance of $LiCoO_2$ cathode by nanocrystalline ZnO coating. *Journal of the Electrochemical Society, 152,* A1701–A1706.

46. Chung, K. Y., Yoon, W.-S., McBreen, J., Yang, X.-Q., Si Hyoung, O., Shin, H. C., et al. (2006). Structural studies on the effects of ZrO_2 coating on $LiCoO_2$ during cycling using in situ X-ray diffraction technique. *Journal of the Electrochemical Society, 153,* A2152–A2157.
47. Miyashiro, H., Yamanaka, A., Tabuchi, M., Seki, S., Nakayama, M., Ohno, Y., et al. (2006). Improvement of degradation at elevated temperature and at high state-of-charge storage by ZrO_2 coating on $LiCoO_2$. *Journal of the Electrochemical Society, 153,* A348–A353.
48. Miyashiro, M., Kobayashi, Y., Seki, S., Mita, Y., Usami, A., Nakayama, M., et al. (2005). Fabrication of all-solid-state lithium polymer secondary batteries using Al_2O_3-coated $LiCoO_2$. *Chemistry of Materials, 17,* 5603–5605.
49. Cho, J., Kim, Y. J., & Park, B. (2000). Novel $LiCoO_2$ cathode material with Al_2O_3 coating for a Li ion cell. *Chemistry of Materials, 12,* 3788–3791.
50. Goodenough, J. B., Wickham, D. G., & Croft, W. J. (1958). Some magnetic and crystallographic properties of the system $Li_x^+Ni_{1-2x}^{++}Ni_x^{+++}O$. *Journal of Physics and Chemistry of Solids, 5,* 107–116.
51. Myung, S.-T., Maglia, F., Park, K.-J., Yoon, C. S., Lamp, P., Kim, S.-J., et al. (2017). Nickel-rich layered cathode materials for automotive lithium-ion batteries: Achievements and perspectives. *ACS Energy Letters, 2,* 196–223.
52. Li, W., Song, B., & Manthiram, A. (2017). High-Voltage Positive Electrode Materials for Lithium-Ion Batteries. *Chemical Society Reviews, 46,* 3006–3059.
53. Manthiram, A., Vadivel, M. A., Sarkar, A., & Muraliganth, T. (2008). Nanostructured electrode materials for electrochemical energy storage and conversion. *Energy & Environmental Science, 1,* 621–638.
54. Dahn, J. R., von Sacken, U., Juzkow, M. W., & Al-Janaby, H. (1991). Rechargeable $LiNiO_2$/carbon cells. *Journal of the Electrochemical Society, 138,* 2207–2211.
55. Radin, M. D., Hy, S., Sina, M., Fang, C., Liu, H., Vinckeviciute, J., et al. (2017). Narrowing the gap between theoretical and practical capacities in Li-ion layered oxide cathode materials. *Advanced Energy Materials, 7,* 1602888.
56. Ohzuku, T., Ueda, A., & Nagayama, M. (1993). Electrochemistry and structural chemistry of $LiNiO_2$ (R3M) for 4 V secondary lithium cells. *Journal of the Electrochemical Society, 140,* 1862–1870.
57. Arai, H., Okada, S., Skurai, Y., & Yamaki, J.-I. (1997). Electrochemical and thermal behavior of $LiNi_{1-z}M_zO_2$. *Journal of the Electrochemical Society, 144,* 3117–3125.
58. Delmas, C., Saadoune, I., & Rougier, A. (1993). The cycling properties of the $Li_xNi_{1-y}Co_yO_2$ electrode. *Journal of Power Sources, 44,* 595–602.
59. Cho, J., & Park, B. (2001). Preparation and electrochemical/thermal properties of $LiNi_{0.74}Co_{0.26}O_2$ cathode material. *Journal of Power Sources, 92,* 35–39.
60. Arai, H., Okada, S., Sakurai, Y., & Yamaki, J.-I. (1998). Thermal behavior of $Li_{1-y}NiO_2$ and the decomposition mechanism. *Solid State Ionics, 109,* 295–302.
61. MacNeil, D. D., & Dahn, J. R. (2002). The reactions of $Li_{0.5}CoO_2$ with nonaqueous solvents at elevated temperatures. *Journal of the Electrochemical Society, 149,* A912–A919.
62. Guilmard, M., Croguennec, L., & Delmas, C. (2003). Thermal stability of lithium nickel oxide derivatives. Part II: $Li_xNi_{0.70}Co_{0.15}Al_{0.15}O_2$ and $Li_xNi_{0.90}Mn_{0.10}O_2$ ($X = 0.50$ and 0.30). Comparison with $Li_xNi_{1.02}O_2$ and $LixNi_{0.89}Al_{0.16}O_2$. *Chemistry of Materials, 15,* 4484–4493.
63. Park, S. H., Park, K. S., Sun, Y. K., Nahm, K. S., Lee, Y. S., & Yoshio, M. (2001). Structural and electrochemical characterization of lithium excess and Al-doped nickel oxides synthesized by the sol-gel method. *Electrochimica Acta, 46,* 1215–1222.
64. Chappel, E., Chouteau, G., Prado, G., & Delmas, C. (2003). Magnetic properties of $LiNi_{1-y}Fe_yO_2$. *Solid State Ionics, 159,* 273–278.
65. Kim, J., & Amine, K. (2001). The effect of tetravalent titanium substitution in $LiNi_{1-x}Ti_xO_2$ ($0.025 \leq X \leq 0.2$) system. *Electrochemistry Communications, 3,* 52–55.
66. Chang, C.-C., Kim, J. Y., & Kim, J. Y. (2000). Synthesis and electrochemical characterization of divalent cation-incorporated lithium nickel oxide. *Journal of the Electrochemical Society, 147,* 1722–1729.

67. Huang, Z.-F., Meng, X., Wang, C.-Z., Sun, Y., & Chen, G. (2006). First-principles calculations on the Jahn-Teller distortion in layered $LiMnO_2$. *Journal of Power Sources, 158,* 1394–1400.
68. Kim, T.-J., Son, D., Cho, J., & Park, B. (2006). Enhancement of the electrochemical properties of O-$LiMnO_2$ cathodes at elevated temperature by lithium and fluorine additions. *Journal of Power Sources, 154,* 268–272.
69. Huang, Z.-F., Wang, C.-Z., Meng, X., Wang, D.-P., & Chen, G. (2006). Effects of Al-doping on the stabilization of monoclinic $LiMnO_2$. *Journal of Solid State Chemistry, 179,* 1602–1609.
70. Cho, J., Kim, Y. J., Kim, T.-J., & Park, B. (2001). Enhanced structural stability of O-$LiMnO_2$ by sol-gel coating of Al_2O_3. *Chemistry of Materials, 13,* 18–20.
71. Spahr, M. E., Novák, P., Schnyder, B., Haas, O., & Nesper, R. (1998). Characterization of layered lithium nickel manganese oxides synthesized by a novel oxidative coprecipitation method and their electrochemical performance as lithium insertion electrode materials. *Journal of the Electrochemical Society, 145,* 1113–1121.
72. Yoshio, M., Noguchi, H., Itoh, J.-I., Okada, M., & Mouri, T. (2000). Preparation and properties of $LiCo_yMn_xNi_{1-x-y}O_2$ as a cathode for lithium ion batteries. *Journal of Power Sources, 90,* 176–181.
73. Armstrong, A. R., & Bruce, P. G. (1996). Synthesis of layered $LiMnO_2$ as an electrode for rechargeable lithium batteries. *Nature, 381,* 499–500.
74. Xiao, X., Wang, L., Wang, D., He, X., Peng, Q., & Li, Y. (2010). Hydrothermal synthesis of orthorhombic $LiMnO_2$ nano-particles and $LiMnO_2$ nanorods and comparison of their electrochemical performances. *Nano Research, 2,* 923–930.
75. Yi, X., Wang, X., Ju, B., Shu, H., Wen, W., Yu, R., et al. (2014). Effective enhancement of electrochemical performance for spherical spinel $LiMn_2O_4$ via Li ion conductive Li_2ZrO_3 coating. *Electrochimica Acta, 134,* 143–149.
76. Gao, X., Sha, Y., Lin, Q., Cai, R., Tade, M. O., & Shao, Z. (2015). Combustion-derived nanocrystalline $LiMn_2O_4$ as a promising cathode material for lithium-ion batteries. *Journal of Power Sources, 275,* 38–44.
77. Wang, Y., Shao, X., Xu, H., Xie, M., Deng, S., Wang, H., et al. (2013). Facile synthesis of porous $LiMn_2O_4$ spheres as cathode materials for high-power lithium ion batteries. *Journal of Power Sources, 226,* 140–148.
78. Xu, G., Liu, Z., Zhang, C., Cui, G., & Chen, L. (2015). Strategies for improving the cyclability and thermo-stability of $LiMn_2O_4$-based batteries at elevated temperatures. *Journal of Materials Chemistry A, 3,* 4092–4123.
79. Myung, S.-T., Komaba, S., & Kumagai, N. (2001). Enhanced structural stability and cyclability of Al-doped $LiMn_2O_4$ spinel synthesized by the emulsion drying method. *Journal of the Electrochemical Society, 148,* A482–A489.
80. Sun, Y. K., Yoon, C. S., Kim, C. K., Youn, S. G., Lee, Y. S., Yoshio, M., et al. (2001). Degradation mechanism of spinel $LiAl_{0.2}Mn_{1.8}O_4$ cathode materials on high temperature cycling. *Journal of Materials Chemistry, 11,* 2519–2522.
81. Guo, S., Zhang, S., He, X., Pu, W., Jiang, C., & Wan, C. (2008). Synthesis and characterization of Sn-doped $LiMn_2O_4$ cathode materials for rechargeable Li-ion batteries. *Journal of the Electrochemical Society, 155,* A760–A763.
82. Saitoh, M., Sano, M., Fujita, M., Sakata, M., Takata, M., & Nishibori, E. (2004). Studies of capacity losses in cycles and storages for a $Li_{1.1}Mn_{1.9}O_4$ positive electrode. *Journal of the Electrochemical Society, 151,* A17–A22.
83. Xia, Y., Wang, H., Zhang, Q., Nakamura, H., Noguchi, H., & Yoshio, M. (2007). Oxygen deficiency, a key factor in controlling the cycle performance of Mn-spinel cathode for lithium-ion batteries. *Journal of Power Sources, 166,* 485–491.
84. Takahashi, M., Yoshida, T., Ichikawa, A., Kitoh, K., Katsukawa, H., Zhang, Q., et al. (2006). Effect of oxygen deficiency reduction in Mg-doped Mn-spinel on its cell storage performance at high temperature. *Electrochimica Acta, 51,* 5508–5514.
85. Wei, Q., Wang, X., Yang, X., Ju, B., Hu, B., Shu, H., et al. (2013). Spherical concentration-gradient $LiMn_{1.87}Ni_{0.13}O_4$ spinel as a high performance cathode for lithium ion batteries. *Journal of Materials Chemistry A, 1,* 4010–4016.

86. Prabu, M., Reddy, M. V., Selvasekarapandian, S., Subba Rao, G. V., & Chowdari, B. V. R. (2013). (Li, Al)-Co-doped spinel, $Li(Li_{0.1}Al_{0.1}Mn_{1.8})O_4$ as high performance cathode for lithium ion batteries. *Electrochimica Acta, 88*, 745–755.
87. Tu, J., Zhao, X. B., Cao, G. S., Zhuang, D. G., Zhu, T. J., & Tu, J. P. (2006). Enhanced cycling stability of $LiMn_2O_4$ by surface modification with melting impregnation method. *Electrochimica Acta, 51,* 6456–6462.
88. Wang, G. X., Lindsay, M. J., Ionescu, M., Bradhurst, D. H., Dou, S. X., & Liu, H. K. (2001). Physical and electrochemical characterization of $LiNi_{0.8}Co_{0.2}O_2$ thin-film electrodes deposited by laser ablation. *Journal of Power Sources, 97–98,* 298–302.
89. Pouillerie, C., Perton, F., Biensan, P., Pérès, J. P., Broussely, M., & Delmas, C. (2001). Effect of magnesium substitution on the cycling behavior of lithium nickel cobalt oxide. *Journal of Power Sources, 96,* 293–302.
90. Pouillerie, C., Croguennec, L., & Delmas, C. (2000). The $Li_xNi_{1-y}Mg_yO_2$ (Y=0.05, 0.10) system: Structural modifications observed upon cycling. *Solid State Ionics, 132,* 15–29.
91. Madhavi, S., Subba Rao, G. V., Chowdari, B. V. R., & Li, S. F. Y. (2001). Effect of aluminium doping on cathodic behaviour of $LiNi_{0.7}Co_{0.3}O_2$. *Journal of Power Sources, 93,* 156–162.
92. Qi, R., Shi, J.-L., Zhang, X.-D., Zeng, X.-X., Yin, Y.-X., Xu, J., et al. (2017). Improving the stability of $LiNi_{0.80}Co_{0.15}Al_{0.05}O_2$ by $AlPO_4$ nanocoating for lithium-ion batteries. *Science China Chemistry, 60,* 1230–1235.
93. Robert Armstrong, A., & Gitzendanner, R. (1998). The intercalation compound $Li(Mn_{0.9}Co_{0.1})O_2$ as a positive electrode for rechargeable lithium batteries. *Chemical Communications*, 1833–1834.
94. Robertson, A. D., Armstrong, A. R., & Bruce, P. G. (2001). Layered $Li_xMn_{1-y}Co_yO_2$ intercalation electrodesinfluence of ion exchange on capacity and structure upon cycling. *Chemistry of Materials, 13,* 2380–2386.
95. Franger, S., Bach, S., Pereira-Ramos, J. P., & Baffier, N. (2000). Chemistry and electrochemistry of low-temperature manganese oxides as lithium intercalation compounds. *Journal of the Electrochemical Society, 147,* 3226–3230.
96. Liu, B., Xu, B., Wu, M. S., & Ouyang, C. Y. (2015). First-principles GGA+U study on structural and electronic properties in $LiMn_{0.5}Ni_{0.5}O_2$, $LiMn_{0.5}Co_{0.5}O_2$ and $LiCo_{0.5}Ni_{0.5}O_2$. *International Journal of Electrochemical Science, 11,* 432–445.
97. Ohzuku, T., & Makimura, Y. (2001). Layered lithium insertion material of $LiNi_{1/2}Mn_{1/2}O_2$: A possible alternative to $LiCoO_2$ for advanced lithium-ion batteries. *Chemistry Letters, 30,* 744–745.
98. Kang, K., Meng, Y. S., Bréger, J., Grey, C. P., & Ceder, G. (2006). Electrodes with high power and high capacity for rechargeable lithium batteries. *Science, 311,* 977–980.
99. Zhang, T., Li, D., Tao, Z., & Chen, J. (2013). Understanding electrode materials of rechargeable lithium batteries via DFT calculations. *Progress in Natural Science: Materials International, 23,* 256–272.
100. Xu, B., Fell, C. R., Chi, M., & Meng, Y. S. (2011). Identifying surface structural changes in layered Li-excess nickel manganese oxides in high voltage lithium ion batteries: A joint experimental and theoretical study. *Energy & Environmental Science, 4,* 2223–2233.
101. Sun, Y.-K., Myung, S.-T., Park, B.-C., Prakash, J., Belharouak, I., & Amine, K. (2009). High-energy cathode material for long-life and safe lithium batteries. *Nature Materials, 8,* 320–324.
102. Wang, P.-F., You, Y., Yin, Y.-X., & Guo, Y.-G. (2018). Layered oxide cathodes for sodium-ion batteries: Phase transition, air stability, and performance. *Advanced Energy Materials, 8,* 1701912.
103. Delmas, C., Menetrier, M., Croguennec, L., Saadoune, I., Rougier, A., Pouillerie, C., et al. (1999). An overview of the Li(Ni, M)O-2 systems: Syntheses, structures and properties. *Electrochimica Acta, 45,* 243–253.
104. Noh, H.-J., Youn, S., Yoon, C. S., & Sun, Y.-K. (2013). Comparison of the structural and electrochemical properties of layered $Li[Ni_xCo_yMn_z]O_2$ ($X = 1/3, 0.5, 0.6, 0.7, 0.8$ and 0.85) cathode material for lithium-ion batteries. *Journal of Power Sources, 233,* 121–130.

105. Liang, C., Kong, F., Longo, R. C., Kc, S., Kim, J.-S., Jeon, S., et al. (2016). Unraveling the origin of instability in Ni-Rich $LiNi_{1-2x}Co_xMn_xO_2$ (NCM) cathode materials. *The Journal of Physical Chemistry C, 120,* 6383–6393.
106. Liu, W., Oh, P., Liu, X., Lee, M.-J., Cho, W., Chae, S., et al. (2015). Nickel-rich layered lithium transition-metal oxide for high-energy lithium-ion batteries. *Angewandte Chemie International Edition, 54,* 4440–4457.
107. Myung, S. T., Lee, K. S., Yoon, C. S., Sun, Y. K., Amine, K., & Yashiro, H. (2010). Effect of AlF_3 coating on rhermal behavior of chemically delithiated $Li_{0.35}Ni_{1/3}Co_{1/3}Mn_{1/3}O_2$. *The Journal of Physical Chemistry C, 114,* 4710–4718.
108. Bak, S. M., Hu, E., Zhou, Y., Yu, X., Senanayake, S. D., Cho, S. J., et al. (2014). Structural changes and thermal stability of charged $LiNi_xMn_yCo_zO_2$ cathode materials studied by combined in situ time-resolved XRD and mass spectroscopy. *ACS Applied Materials & Interfaces, 6,* 22594–22601.
109. Oh, K.-Y., Siegel, J. B., Secondo, L., Kim, S. U., Samad, N. A., Qin, J., et al. (2014). Rate dependence of swelling in lithium-ion cells. *Journal of Power Sources, 267,* 197–202.
110. Liu, S., Xiong, L., & He, C. (2014). Long cycle life lithium ion battery with lithium nickel cobalt manganese oxide (NCM) cathode. *Journal of Power Sources, 261,* 285–291.
111. Venkatraman, S., Shin, Y., & Manthiram, A. (2003). Phase relationships and structural and chemical stabilities of charged $Li_{1-x}CoO_2$ delta and $Li_{1-x}Ni_{0.85}Co_{0.15}O_2$ delta cathodes. *Electrochemical and Solid-State Letters, 6,* A9–A12.
112. Lin, F., Markus, I. M., Nordlund, D., Weng, T. C., Asta, M. D., Xin, H. L., et al. (2014). Surface reconstruction and chemical evolution of stoichiometric layered cathode materials for lithium-ion batteries. *Nature Communications, 5,* 3529.
113. Myung, S.-T., Maglia, F., Park, K.-J., Yoon, C. S., Lamp, P., Kim, S.-J., et al. (2016). Nickel-rich layered cathode materials for automotive lithium-ion batteries: Achievements and perspectives. *ACS Energy Letters, 2,* 196–223.
114. Hou, P., Yin, J., Ding, M., Huang, J., & Xu, X. (2017). Surface/interfacial structure and chemistry of high-energy nickel-rich layered oxide cathodes: Advances and perspectives. *Small (Weinheim an der Bergstrasse, Germany), 13,* 1701802.
115. Bak, S.-M., Nam, K.-W., Chang, W., Yu, X., Hu, E., Hwang, S., et al. (2013). Correlating structural changes and gas evolution during the thermal decomposition of charged $Li_xNi_{0.8}Co_{0.15}Al_{0.05}O_2$ cathode materials. *Chemistry of Materials, 25,* 337–351.
116. Watanabe, S., Kinoshita, M., Hosokawa, T., Morigaki, K., & Nakura, K. (2014). Capacity fade of $LiAl_yNi_{1-x-y}Co_xO_2$ cathode for lithium-ion batteries during accelerated calendar and cycle life tests (Surface analysis of $LiAl_yNi_{1-x-y}Co_xO_2$ cathode after cycle tests in restricted depth of discharge ranges). *Journal of Power Sources, 258,* 210–217.
117. Manthiram, A., Knight, J. C., Myung, S.-T., Oh, S.-M., & Sun, Y.-K. (2015). Nickel-rich and lithium-rich layered oxide cathodes: Progress and perspectives. *Advanced Energy Materials, 6,* 1501010.
118. Robert, R., Villevieille, C., & Novák, P. (2014). Enhancement of the high potential specific charge in layered electrode materials for lithium-ion batteries. *Journal of Materials Chemistry A, 2,* 8589–8598.
119. Zheng, J., Myeong, S., Cho, W., Yan, P., Xiao, J., Wang, C., et al. (2016). Li- and Mn-rich cathode materials: Challenges to commercialization. *Advanced Energy Materials, 7,* 1601284.
120. Mohanty, D., Kalnaus, S., Meisner, R. A., Rhodes, K. J., Li, J., Payzant, E. A., et al. (2013). Structural transformation of a lithium-rich $Li_{1.2}Co_{0.1}Mn_{0.55}Ni_{0.15}O_2$ cathode during high voltage cycling resolved by in situ X-ray diffraction. *Journal of Power Sources, 229,* 239–248.
121. Liu, H., Harris, K. J., Jiang, M., Wu, Y., Goward, G. R., & Botton, G. A. (2018). Unraveling the rapid performance decay of layered high-energy cathodes: From nanoscale degradation to drastic bulk evolution. *ACS Nano, 12,* 2708–2718.
122. Cheng, X. B., Hou, T. Z., Zhang, R., Peng, H. J., Zhao, C. Z., Huang, J. Q., et al. (2016). Dendrite-free lithium deposition induced by uniformly distributed lithium ions for efficient lithium metal batteries. *Advanced Materials, 28,* 2888–2895.

123. Yu, H., & Zhou, H. (2013). High-energy cathode materials (Li_2MnO_3–$LiMO_2$) for lithium-ion batteries. *Journal of Physical Chemistry Letters, 4,* 1268–1280.
124. He, Z., Wang, Z., Chen, H., Huang, Z., Li, X., Guo, H., et al. (2015). Electrochemical performance of zirconium doped lithium rich layered $Li_{1.2}Mn_{0.54}Ni_{0.13}Co_{0.13}O_2$ oxide with porous hollow structure. *Journal of Power Sources, 299,* 334–341.
125. Liu, S., Wang, Z., Huang, Y., Ni, Z., Bai, J., Kang, S., et al. (2018). Fluorine doping and Al_2O_3 coating co-modified $Li[Li_{0.20}Ni_{0.133}Co_{0.133}Mn_{0.534}]O_2$ as high performance cathode material for lithium-ion batteries. *Journal of Alloys and Compounds, 731,* 636–645.
126. Zhang, X.-D., Shi, J.-L., Liang, J.-Y., Yin, Y.-X., Zhang, J.-N., Yu, X.-Q., et al. (2018). Suppressing surface lattice oxygen release of Li-rich cathode materials via heterostructured spinel $Li_4Mn_5O_{12}$ coating. *Advanced Materials, 30,* 1801751.
127. Jiang, K.-C., Wu, X.-L., Yin, Y.-X., Lee, J.-S., Kim, J., & Guo, Y.-G. (2012). Superior hybrid cathode material containing lithium-excess layered material and graphene for lithium-ion batteries. *ACS Applied Materials & Interfaces, 4,* 4858–4863.
128. Wang, Y., Yang, Z., Qian, Y., Gu, L., & Zhou, H. (2015). New insights into improving rate performance of lithium-rich cathode material. *Advanced Materials, 27,* 3915–3920.
129. Qing, R.-P., Shi, J.-L., Xiao, D.-D., Zhang, X.-D., Yin, Y.-X., Zhai, Y.-B., et al. (2016). Enhancing the kinetics of Li-rich cathode materials through the pinning effects of gradient surface Na^+ doping. *Advanced Energy Materials, 6,* 1501914.
130. Ding, Z., Xu, M., Liu, J., Huang, Q., Chen, L., Wang, P., et al. (2017). Understanding the enhanced kinetics of gradient-chemical-doped lithium-rich cathode material. *ACS Applied Materials & Interfaces, 9,* 20519–20526.
131. Shi, J.-L., Xiao, D.-D., Zhang, X.-D., Yin, Y.-X., Guo, Y.-G., Gu, L., et al. (2017). Improving the structural stability of Li-rich cathode materials via reservation of cations in the Li-slab for Li-ion batteries. *Nano Research, 10,* 4201–4209.
132. Gu, M., Belharouak, I., Zheng, J., Wu, H., Xiao, J., Genc, A., et al. (2013). Formation of the spinel phase in the layered composite cathode used in Li-ion batteries. *ACS Nano, 7,* 760–767.
133. Idemoto, Y., Inoue, M., & Kitamura, N. (2014). Composition dependence of average and local structure of $xLi(Li_{1/3}Mn_{2/3})O_2$–$(1-x)Li(Mn_{1/3}Ni_{1/3}Co_{1/3})O_2$ active cathode material for Li ion batteries. *Journal of Power Sources, 259,* 195–202.
134. Padhi, A. K., Nanjundaswamy, K. S., & Goodenough, J. B. (1997). Phospho-olivines as positive-electrode materials for rechargeable lithium batteries. *Journal of the Electrochemical Society, 144,* 1188–1194.
135. Wu, X.-L., Jiang, L.-Y., Cao, F.-F., Guo, Y.-G., & Wan, L.-J. (2009). Lifepo4 nanoparticles embedded in a nanoporous carbon matrix: Superior cathode material for electrochemical energy-storage devices. *Advanced Materials, 21,* 2710–2714.
136. Su, J., Wu, X.-L., Yang, C.-P., Lee, J.-S., Kim, J., & Guo, Y.-G. (2012). Self-assembled $LiFePO_4$/C nano/microspheres by using phytic acid as phosphorus source. *The Journal of Physical Chemistry C, 116,* 5019–5024.
137. Damen, L., De Giorgio, F., Monaco, S., Veronesi, F., & Mastragostino, M. (2012). Synthesis and characterization of carbon-coated Limnpo4 and $LiMnPO_4$ and $LiMn_{1-x}Fe_xPO_4$ (X = 0.2, 0.3) materials for lithium-ion batteries. *Journal of Power Sources, 218,* 250–253.
138. Liu, L., Zhang, H., Chen, X., Fang, L., Bai, Y., Liu, R., et al. (2015). Unique synthesis of sandwiched graphene@$(Li_{0.893}Fe_{0.036})Co(PO_4)$ nanoparticles as high-performance cathode materials for lithium-ion batteries. *Journal of Materials Chemistry A, 3,* 12320–12327.
139. Feng, Y., Zhang, H., Fang, L., Ouyang, Y., & Wang, Y. (2015). Designed synthesis of a unique single-crystal Fe-doped $LiNiPO_4$ nanomesh as an enhanced cathode for lithium ion batteries. *Journal of Materials Chemistry A, 3,* 15969–15976.
140. Örnek, A., & Kazancioglu, M. Z. (2016). A novel and effective strategy for producing core-shell $LiNiPO_4$/C cathode material for excellent electrochemical stability using a long-time and low-level microwave approach. *Scripta Materialia, 122,* 45–49.
141. Hu, J., Jiang, Y., Cui, S., Duan, Y., Liu, T., Guo, H., et al. (2016). 3D-printed cathodes of $LiMn_{1-x}Fe_xPO_4$ nanocrystals achieve both ultrahigh rate and high capacity for advanced lithium-ion battery. *Advanced Energy Materials, 6,* 1600856.

142. Chi, Z.-X., Zhang, W., Wang, X.-S., Cheng, F.-Q., Chen, J.-T., Cao, A.-M., et al. (2014). Accurate surface control of core-shell structured $LiMn_{0.5}Fe_{0.5}PO_4$@C for improved battery performance. *Journal of Materials Chemistry A, 2,* 17359–17365.
143. Wang, H., Yang, Y., Liang, Y., Cui, L.-F., Sanchez Casalongue, H., Li, Y., Hong, G., Cui, Y., & Dai, H. (2011). $LiMn_{1-x}Fe_xPO_4$ nanorods grown on graphene sheets for ultrahigh-rate-performance lithium ion batteries. *Angewandte Chemie International Edition, 50,* 7364–7368.
144. Mi, Y., Gao, P., Liu, W., Zhang, W., & Zhou, H. (2014). Carbon nanotube-loaded mesoporous $LiFe_{0.6}Mn_{0.4}PO_4$/C microspheres as high performance cathodes for lithium-ion batteries. *Journal of Power Sources, 267,* 459–468.
145. Karami, H., & Taala, F. (2011). Synthesis, characterization and application of $Li_3Fe_2(PO_4)_3$ nanoparticles as cathode of lithium-ion rechargeable batteries. *Journal of Power Sources, 196,* 6400–6411.
146. Goriparti, S., Miele, E., De Angelis, F., Di Fabrizio, E., Zaccaria, R. P., & Capiglia, C. (2014). Review on recent progress of nanostructured anode materials for Li-Ion batteries. *Journal of Power Sources, 257,* 421–443.
147. Qie, L., Chen, W.-M., Wang, Z.-H., Shao, Q.-G., Li, X., Yuan, L.-X., et al. (2012). Nitrogen-doped porous carbon nanofiber webs as anodes for lithium ion batteries with a superhigh capacity and rate capability. *Advanced Materials, 24,* 2047–2050.
148. Roy, P., & Srivastava, S. K. (2015). Nanostructured anode materials for lithium ion batteries. *Journal of Materials Chemistry A, 3,* 2454–2484.
149. de las Casas, C., & Li, W. (2012). A review of application of carbon nanotubes for lithium ion battery anode material. *Journal of Power Sources, 208,* 74–85.
150. Lee, S. W., Yabuuchi, N., Gallant, B. M., Chen, S., Kim, B.-S., Hammond, P. T., et al. (2010). High-power lithium batteries from functionalized carbon-nanotube electrodes. *Nature Nanotechnology, 5,* 531–537.
151. Wu, X.-L., Chen, L.-L., Xin, S., Yin, Y.-X., Guo, Y.-G., Kong, Q.-S., et al. (2010). Preparation and Li storage properties of hierarchical porous carbon fibers derived from alginic acid. *ChemSusChem, 3,* 703–707.
152. Wu, X.-L., Liu, Q., Guo, Y.-G., & Song, W.-G. (2009). Superior storage performance of carbon nanosprings as anode materials for lithium-ion batteries. *Electrochemistry Communications, 11,* 1468–1471.
153. Ye, G., Zhu, X., Chen, S., Li, D., Yin, Y., Lu, Y., et al. (2017). Nanoscale engineering of nitrogen-doped carbon nanofiber aerogels for enhanced lithium ion storage. *Journal of Materials Chemistry A, 5,* 8247–8254.
154. Kong, J., Yee, W. A., Wei, Y., Yang, L., Ang, J. M., Phua, S. L., et al. (2013). Silicon nanoparticles encapsulated in hollow graphitized carbon nanofibers for lithium ion battery anodes. *Nanoscale, 5,* 2967–2973.
155. Wang, G., Shen, X., Yao, J., & Park, J. (2009). Graphene nanosheets for enhanced lithium storage in lithium ion batteries. *Carbon, 47,* 2049–2053.
156. Pan, D. Y., Wang, S., Zhao, B., Wu, M. H., Zhang, H. J., Wang, Y., et al. (2009). Li storage properties of disordered graphene nanosheets. *Chemistry of Materials, 21,* 3136–3142.
157. Sun, H., Mei, L., Liang, J., Zhao, Z., Lee, C., Fei, H., et al. (2017). Three-dimensional holey-graphene/niobia composite architectures for ultrahigh-rate energy storage. *Science, 356,* 599–604.
158. Zhou, X., Bao, J., Dai, Z., & Guo, Y.-G. (2013). Tin nanoparticles impregnated in nitrogen-doped graphene for lithium-ion battery anodes. *The Journal of Physical Chemistry C, 117,* 25367–25373.
159. Zhang, L.-S., Jiang, L.-Y., Yan, H.-J., Wang, W. D., Wang, W., Song, W.-G., et al. (2010). Mono dispersed SnO_2 nanoparticles on both sides of single layer graphene sheets as anode materials in Li-Ion batteries. *Journal of Materials Chemistry, 20,* 5462–5467.
160. Zhou, X., Yin, Y. X., Wan, L. J., & Guo, Y. G. (2012). Facile synthesis of silicon nanoparticles inserted into graphene sheets as improved anode materials for lithium-ion batteries. *Chemical Communications, 48,* 2198–2200.

161. Wang, K., Wang, N., He, J., Yang, Z., Shen, X., & Huang, C. (2017). Graphdiyne nanowalls as anode for lithium—ion batteries and capacitors exhibit superior cyclic stability. *Electrochimica Acta, 253,* 506–516.
162. Huang, C., Zhang, S., Liu, H., Li, Y., Cui, G., & Li, Y. (2015). Graphdiyne for high capacity and long-life lithium storage. *Nano Energy, 11,* 481–489.
163. Zhang, S., Du, H., He, J., Huang, C., Liu, H., Cui, G., et al. (2016). Nitrogen-doped graphdiyne applied for lithium-ion storage. *ACS Applied Materials & Interfaces, 8,* 8467–8473.
164. He, J., Wang, N., Cui, Z., Du, H., Fu, L., Huang, C., et al. (2017). Hydrogen substituted graphdiyne as carbon-rich flexible electrode for lithium and sodium ion batteries. *Nature Communications, 8,* 1172.
165. Zhou, X., Wan, L. J., & Guo, Y. G. (2013). Electrospun silicon nanoparticle/porous carbon hybrid nanofibers for lithium-ion batteries. *Small (Weinheim an der Bergstrasse, Germany), 9,* 2684–2688.
166. Yang, S. J., Nam, S., Kim, T., Im, J. H., Jung, H., Kang, J. H., et al. (2013). Preparation and exceptional lithium anodic performance of porous carbon-coated Zno quantum dots derived from a metal-organic framework. *Journal of the American Chemical Society, 135,* 7394–7397.
167. Yan, Y., Yin, Y. X., Xin, S., Guo, Y. G., & Wan, L. J. (2012). Ionothermal synthesis of sulfur-doped porous carbons hybridized with graphene as superior anode materials for lithium-ion batteries. *Chemical Communications, 48,* 10663–10665.
168. Zheng, F., Yang, Y., & Chen, Q. (2014). High lithium anodic performance of highly nitrogen-doped porous carbon prepared from a metal-organic framework. *Nature Communications, 5,* 5261.
169. Wu, H., Yu, G., Pan, L., Liu, N., McDowell, M. T., Bao, Z., et al. (2013). Stable Li-ion battery anodes by in-situ polymerization of conducting hydrogel to conformally coat silicon nanoparticles. *Nature Communications, 4,* 1943.
170. Fujimoto, H., Tokumitsu, K., Mabuchi, A., Chinnasamy, N., & Kasuh, T. (2010). The anode performance of the hard carbon for the lithium ion battery derived from the oxygen-containing aromatic precursors. *Journal of Power Sources, 195,* 7452–7456.
171. Guo, B., Shu, J., Wang, Z., Yang, H., Shi, L., Liu, Y., et al. (2008). Electrochemical reduction of nano-SiO_2 in hard carbon as anode material for lithium ion batteries. *Electrochemistry Communications, 10,* 1876–1878.
172. Hou, G., Cheng, B., Cao, Y., Yao, M., Li, B., Zhang, C., et al. (2016). Scalable production of 3D plum-pudding-like Si/C spheres: Towards practical application in Li-ion batteries. *Nano Energy, 24,* 111–120.
173. Yang, J., Wang, Y.-X., Chou, S.-L., Zhang, R., Xu, Y., Fan, J., et al. (2015). Yolk-shell silicon-mesoporous carbon anode with compact solid electrolyte interphase film for superior lithium-ion batteries. *Nano Energy, 18,* 133–142.
174. Kim, H., Seo, M., Park, M. H., & Cho, J. (2010). A critical size of silicon nano-anodes for lithium rechargeable batteries. *Angewandte Chemie International Edition, 49,* 2146–2149.
175. Zhou, Y., Guo, H., Yan, G., Wang, Z., Li, X., Yang, Z., et al. (2018). Fluidized bed reaction towards crystalline embedded amorphous Si anode with much enhanced cycling stability. *Chemical Communications (Cambridge, England), 54,* 3755–3758.
176. Jiang, B., He, Y., Li, B., Zhao, S., Wang, S., He, Y. B., et al. (2017). Polymer-templated formation of polydopamine-coated SnO_2 nanocrystals: Anodes for cyclable lithium-ion batteries. *Angewandte Chemie International Edition, 56,* 1869–1872.
177. Xu, Q., Li, J. Y., Yin, Y. X., Kong, Y. M., Guo, Y. G., & Wan, L. J. (2016). Nano/micro-structured Si/C anodes with high initial coulombic efficiency in Li-ion batteries. *Chemistry An Asian Journal, 11,* 1205–1209.
178. Yang, J., Wang, Y., Li, W., Wang, L., Fan, Y., Jiang, W., et al. (2017). Amorphous TiO_2 shells: A vital elastic buffering layer on silicon nanoparticles for high-performance and safe lithium storage. *Advanced Materials, 29,* 1700523.
179. Li, W., Sun, X., & Yu, Y. (2017). Si-, Ge-, Sn-based anode materials for lithium-ion batteries: From structure design to electrochemical performance. *Small Methods, 1,* 1600037.

180. Liu, X. H., Huang, S., Picraux, S. T., Li, J., Zhu, T., & Huang, J. Y. (2011). Reversible nanopore formation in Ge nanowires during lithiation-delithiation cycling: An in situ transmission electron microscopy study. *Nano Letters, 11,* 3991–3997.
181. Kennedy, T., Mullane, E., Geaney, H., Osiak, M., O'Dwyer, C., & Ryan, K. M. (2014). High-performance germanium nanowire-based lithium-ion battery anodes extending over 1000 cycles through in situ formation of a continuous porous network. *Nano Letters, 14,* 716–723.
182. Zhou, X., Dai, Z., Liu, S., Bao, J., & Guo, Y. G. (2014). Ultra-uniform SnO_x/carbon nanohybrids toward advanced lithium-ion battery anodes. *Advanced Materials, 26,* 3943–3949.
183. Yin, Y.-X., Xin, S., Wan, L.-J., Li, C.-J., & Guo, Y.-G. (2012). Synthesis of nanostructured SnO_2/C microfibers with improved performances as anode material for Li-ion batteries. *Journal of Nanoscience and Nanotechnology, 12,* 2581–2585.
184. Wang, B., Li, X., Zhang, X., Luo, B., Zhang, Y., & Zhi, L. (2013). Contact-engineered and void-involved silicon/carbon nanohybrids as lithium-ion-battery anodes. *Advanced Materials, 25,* 3560–3565.
185. Cao, F. F., Deng, J. W., Xin, S., Ji, H. X., Schmidt, O. G., Wan, L. J., et al. (2011). Cu-Si nanocable arrays as high-rate anode materials for lithium-ion batteries. *Advanced Materials, 23,* 4415–4420.
186. Lin, H., Weng, W., Ren, J., Qiu, L., Zhang, Z., Chen, P., et al. (2014). Twisted aligned carbon nanotube/silicon composite fiber anode for flexible wire-shaped lithium-ion battery. *Advanced Materials, 26,* 1217–1222.
187. Guan, C., Wang, X., Zhang, Q., Fan, Z., Zhang, H., & Fan, H. J. (2014). Highly stable and reversible lithium storage in SnO_2 nanowires surface coated with a uniform hollow shell by atomic layer deposition. *Nano Letters, 14,* 4852–4858.
188. Lu, Y., Yu, L., & Lou, X. W. (2018). Nanostructured conversion-type anode materials for advanced lithium-ion batteries. *Chem, 4,* 972–996.
189. Chen, S., Chen, Z., Xu, X., Cao, C., Xia, M., & Luo, Y. (2018). Scalable 2D mesoporous silicon nanosheets for high-performance lithium-ion battery anode. *Small (Weinheim an der Bergstrasse, Germany), 14,* e1703361.
190. Lin, L., Ma, Y., Xie, Q., Wang, L., Zhang, Q., & Peng, D. L. (2017). Copper-nanoparticle-induced porous Si/Cu composite films as an anode for lithium ion batteries. *ACS Nano, 11,* 6893–6903.
191. Suresh, S., Wu, Z. P., Bartolucci, S. F., Basu, S., Mukherjee, R., Gupta, T., et al. (2017). Protecting silicon film anodes in lithium-ion batteries using an atomically thin graphene drape. *ACS Nano, 11,* 5051–5061.
192. Zhou, X., Wan, L. J., & Guo, Y. G. (2013). Binding SnO_2 nanocrystals in nitrogen-doped graphene sheets as anode materials for lithium-ion batteries. *Advanced Materials, 25,* 2152–2157.
193. Zhou, X., & Guo, Y.-G. (2013). A PEO-assisted electrospun silicon-graphene composite as an anode material for lithium-ion batteries. *Journal of Materials Chemistry A, 1,* 9019–9023.
194. Zhou, X., Yin, Y.-X., Wan, L.-J., & Guo, Y.-G. (2012). Self-assembled nanocomposite of silicon nanoparticles encapsulated in graphene through electrostatic attraction for lithium-ion batteries. *Advanced Energy Materials, 2,* 1086–1090.
195. Zhou, X., Cao, A.-M., Wan, L.-J., & Guo, Y.-G. (2012). Spin-coated silicon nanoparticle/graphene electrode as a binder-free anode for high-performance lithium-ion batteries. *Nano Research, 5,* 845–853.
196. Chang, J., Huang, X., Zhou, G., Cui, S., Hallac, P. B., Jiang, J., et al. (2014). Multilayered Si nanoparticle/reduced graphene oxide hybrid as a high-performance lithium-ion battery anode. *Advanced Materials, 26,* 758–764.
197. Wang, D., Yang, J., Li, X., Geng, D., Li, R., Cai, M., et al. (2013). Layer by layer assembly of sandwiched graphene/SnO_2 nanorod/carbon nanostructures with ultrahigh lithium ion storage properties. *Energy & Environmental Science, 6,* 2900–2906.
198. Wang, X., Cao, X., Bourgeois, L., Guan, H., Chen, S., Zhong, Y., et al. (2012). N-doped graphene-SnO_2 sandwich paper for high-performance lithium-ion batteries. *Advanced Functional Materials, 22,* 2682–2690.

199. Ahmed, B., Anjum, D. H., Gogotsi, Y., & Alshareef, H. N. (2017). Atomic layer deposition of SnO_2 on mxene for Li-ion battery anodes. *Nano Energy, 34,* 249–256.
200. Lian, Q., Zhou, G., Zeng, X., Wu, C., Wei, Y., Cui, C., et al. (2016). Carbon coated SnS/SnO_2 heterostructures wrapping on Cnfs as an improved-performance anode for Li-ion batteries: Lithiation-induced structural optimization upon cycling. *ACS Applied Materials & Interfaces, 8,* 30256–30263.
201. Li, X., Gu, M., Hu, S., Kennard, R., Yan, P., Chen, X., et al. (2014). Mesoporous silicon sponge as an anti-pulverization structure for high-performance lithium-ion battery anodes. *Nature Communications, 5,* 4105.
202. Ge, M., Lu, Y., Ercius, P., Rong, J., Fang, X., Mecklenburg, M., et al. (2014). Large-scale fabrication, 3D tomography, and lithium-ion battery application of porous silicon. *Nano Letters, 14,* 261–268.
203. Yi, R., Dai, F., Gordin, M. L., Chen, S., & Wang, D. (2013). Micro-sized Si-C composite with interconnected nanoscale building blocks as high-performance anodes for practical application in lithium-ion batteries. *Advanced Energy Materials, 3,* 295–300.
204. Huang, B., Li, X., Pei, Y., Li, S., Cao, X., Masse, R. C., et al. (2016). Novel carbon-encapsulated porous SnO_2 anode for lithium-ion batteries with much improved cyclic stability. *Small (Weinheim an der Bergstrasse, Germany), 12,* 1945–1955.
205. Wang, X. L., Han, W. Q., Chen, H., Bai, J., Tyson, T. A., Yu, X. Q., et al. (2011). Amorphous hierarchical porous Geo(X) as high-capacity anodes for Li ion batteries with very long cycling life. *Journal of the American Chemical Society, 133,* 20692–20695.
206. Zhu, Z., Wang, S., Du, J., Jin, Q., Zhang, T., Cheng, F., et al. (2014). Ultrasmall Sn nanoparticles embedded in nitrogen-doped porous carbon as high-performance anode for lithium-ion batteries. *Nano Letters, 14,* 153–157.
207. Xu, Y., Liu, Q., Zhu, Y., Liu, Y., Langrock, A., Zachariah, M. R., et al. (2013). Uniform nano-Sn/C composite anodes for lithium ion batteries. *Nano Letters, 13,* 470–474.
208. Lee, S. J., Kim, H. J., Hwang, T. H., Choi, S., Park, S. H., Deniz, E., et al. (2017). Delicate structural control of Si-SiO_x-C composite via high-speed spray pyrolysis for Li-ion battery anodes. *Nano Letters, 17,* 1870–1876.
209. Zhang, R., Du, Y., Li, D., Shen, D., Yang, J., Guo, Z., et al. (2014). Highly reversible and large lithium storage in mesoporous Si/C nanocomposite anodes with silicon nanoparticles embedded in a carbon framework. *Advanced Materials, 26,* 6749–6755.
210. Xu, Q., Sun, J.-K., Li, J.-Y., Yin, Y.-X., & Guo, Y.-G. (2018). Scalable synthesis of spherical Si/C granules with 3D conducting networks as ultrahigh loading anodes in lithium-ion batteries. *Energy Storage Materials, 12,* 54–60.
211. Xu, Q., Li, J.-Y., Sun, J.-K., Yin, Y.-X., Wan, L.-J., & Guo, Y.-G. (2017). Watermelon-inspired Si/C microspheres with hierarchical buffer structures for densely compacted lithium-ion battery anodes. *Advanced Energy Materials, 7,* 1601481.
212. Zhang, Y.-C., You, Y., Xin, S., Yin, Y.-X., Zhang, J., Wang, P., et al. (2016). Rice husk-derived hierarchical silicon/nitrogen-doped carbon/carbon nanotube spheres as low-cost and high-capacity anodes for lithium-ion batteries. *Nano Energy, 25,* 120–127.
213. Zhou, X., Yin, Y. X., Cao, A. M., Wan, L. J., & Guo, Y. G. (2012). Efficient 3D conducting networks built by graphene sheets and carbon nanoparticles for high-performance silicon anode. *ACS Applied Materials & Interfaces, 4,* 2824–2828.
214. Ko, M., Chae, S., Ma, J., Kim, N., Lee, H.-W., Cui, Y., et al. (2016). Scalable synthesis of silicon-nanolayer-embedded graphite for high-energy lithium-ion batteries. *Nature Energy, 1,* 16113.
215. Zhang, W.-M., Hu, J.-S., Guo, Y.-G., Zheng, S.-F., Zhong, L.-S., Song, W.-G., et al. (2008). Tin-nanoparticles encapsulated in elastic hollow carbon spheres for high-performance anode material in lithium-ion batteries. *Advanced Materials, 20,* 1160–1165.
216. Zhang, L., Wu, H. B., Liu, B., & Lou, X. W. (2014). Formation of porous SnO_2 microboxes via selective leaching for highly reversible lithium storage. *Energy & Environmental Science, 7,* 1013–1017.

217. Zhou, X., Yin, Y.-X., Wan, L.-J., & Guo, Y.-G. (2012). A robust composite of SnO_2 hollow nanospheres enwrapped by graphene as a high-capacity anode material for lithium-ion batteries. *Journal of Materials Chemistry, 22,* 17456–17459.
218. Zhou, X., Yu, L., & Lou, X. W. D. (2016). Formation of uniform N-doped carbon-coated SnO_2 submicroboxes with enhanced lithium storage properties. *Advanced Energy Materials, 6,* 1600451.
219. Zhang, L., Zhang, G., Wu, H. B., Yu, L., & Lou, X. W. (2013). Hierarchical tubular structures constructed by carbon-coated SnO_2 nanoplates for highly reversible lithium storage. *Advanced Materials, 25,* 2589–2593.
220. Liang, J., Yu, X. Y., Zhou, H., Wu, H. B., Ding, S., & Lou, X. W. (2014). Bowl-like SnO_2@Carbon hollow particles as an advanced anode material for lithium-ion batteries. *Angewandte Chemie International Edition, 53,* 12803–12807.
221. Zhao, B., Ran, R., Liu, M., & Shao, Z. (2015). A comprehensive review of $Li_4Ti_5O_{12}$-based electrodes for lithium-ion batteries: The latest advancements and future perspectives. *Materials Science and Engineering: R: Reports, 98,* 1–71.
222. Li, N., Zhou, G., Li, F., Wen, L., & Cheng, H.-M. (2013). A self-standing and flexible electrode of $Li_4Ti_5O_{12}$ nanosheets with a N-doped carbon coating for high rate lithium ion batteries. *Advanced Functional Materials, 23,* 5429–5435.
223. Wang, Y. Q., Gu, L., Guo, Y. G., Li, H., He, X. Q., Tsukimoto, S., et al. (2012). Rutile-TiO_2 nanocoating for a high-rate $Li_4Ti_5O_{12}$ anode of a lithium-ion battery. *Journal of the American Chemical Society, 134,* 7874–7879.
224. Wang, P., Zhang, G., Cheng, J., You, Y., Li, Y. K., Ding, C., et al. (2017). Facile synthesis of carbon-coated spinel $Li_4Ti_5O_{12}$/rutile-TiO_2 composites as an improved anode material in full lithium-ion batteries with $LiFePO_4$@N-doped carbon cathode. *ACS Applied Materials & Interfacess, 9,* 6138–6143.
225. Wang, C., Wang, S., Tang, L., He, Y.-B., Gan, L., Li, J., et al. (2016). A robust strategy for crafting monodisperse $Li_4Ti_5O_{12}$ nanospheres as superior rate anode for lithium ion batteries. *Nano Energy, 21,* 133–144.
226. Zheng, S.-F., Hu, J.-S., Zhong, L.-S., Song, W.-G., Wan, L.-J., & Guo, Y.-G. (2008). Introducing dual functional Cnt networks into Cuo nanomicrospheres toward superior electrode materials for lithium-ion batteries. *Chemistry of Materials, 20,* 3617–3622.
227. Cao, F.-F., Wu, X.-L., Xin, S., Guo, Y.-G., & Wan, L.-J. (2010). Facile synthesis of mesoporous TiO_2-C nanosphere as an improved anode material for superior high rate 1.5 V rechargeable Li ion batteries containing $LiFePO_4$-C cathode. *The Journal of Physical Chemistry C, 114,* 10308–10313.
228. Cao, F.-F., Guo, Y.-G., Zheng, S.-F., Wu, X.-L., Jiang, L.-Y., Bi, R.-R., et al. (2010). Symbiotic coaxial nanocables: Facile synthesis and an efficient and elegant morphological solution to the lithium storage problem. *Chemistry of Materials, 22,* 1908–1914.
229. Qu, J., Yin, Y. X., Wang, Y. Q., Yan, Y., Guo, Y. G., & Song, W. G. (2013). Layer structured alpha-Fe_2O_3 nanodisk/reduced graphene oxide composites as high-performance anode materials for lithium-ion batteries. *ACS Applied Materials & Interfacess, 5,* 3932–3936.
230. Zhang, W. M., Wu, X. L., Hu, J. S., Guo, Y. G., & Wan, L. J. (2008). Carbon coated Fe_3O_4 nanospindles as a superior anode material for lithium-ion batteries. *Advanced Functional Materials, 18,* 3941–3946.
231. Cui, Z.-M., Hang, L.-Y., Song, W.-G., & Guo, Y.-G. (2009). High-yield gas-liquid interfacial synthesis of highly dispersed Fe_3O_4 nanocrystals and their application in lithium-ion batteries. *Chemistry of Materials, 21,* 1162–1166.
232. Wu, X.-L., Guo, Y.-G., Wan, L.-J., & Hu, C.-W. (2008). Alpha-Fe_2O_3 nanostructures: Inorganic salt-controlled synthesis and their electrochemical performance toward lithium storage. *The Journal of Physical Chemistry C, 112,* 16824–16829.
233. Wang, X., Wu, X.-L., Guo, Y.-G., Zhong, Y., Cao, X., Ma, Y., et al. (2010). Synthesis and lithium storage properties of Co_3O_4 nanosheet-assembled multishelled hollow spheres. *Advanced Functional Materials, 20,* 1680–1686.

234. Gu, D., Li, W., Wang, F., Bongard, H., Spliethoff, B., Schmidt, W., et al. (2015). Controllable synthesis of mesoporous peapod-like Co_3O_4@Carbon nanotube arrays for high-performance lithium-ion batteries. *Angewandte Chemie International Edition, 54,* 7060–7064.
235. Wang, B., Wu, X.-L., Shu, C.-Y., Guo, Y.-G., & Wang, C.-R. (2010). Synthesis of Cuo/graphene nanocomposite as a high-performance anode material for lithium-ion batteries. *Journal of Materials Chemistry, 20,* 10661–10664.
236. Jiang, L.-Y., Xin, S., Wu, X.-L., Li, H., Guo, Y.-G., & Wan, L.-J. (2010). Non-sacrificial template synthesis of Cr_2O_3-C hierarchical core/shell nanospheres and their application as anode materials in lithium-ion batteries. *Journal of Materials Chemistry, 20,* 7565–7569.
237. Zhao, G., Wen, T., Zhang, J., Li, J., Dong, H., Wang, X., et al. (2014). Two-dimensional Cr_2O_3 and interconnected graphene-Cr_2O_3 nanosheets: Synthesis and their application in lithium storage. *Journal of Materials Chemistry A, 2,* 944–948.
238. Xu, L., Hu, Y., Zhang, H., Jiang, H., & Li, C. (2016). Confined synthesis of FeS_2 nanoparticles encapsulated in carbon nanotube hybrids for ultrastable lithium-ion batteries. *ACS Sustainable Chemistry Engineering, 4,* 4251–4255.
239. Wen, X., Wei, X., Yang, L., & Shen, P. K. (2015). Self-assembled FeS_2 cubes anchored on reduced graphene oxide as an anode material for lithium ion batteries. *Journal of Materials Chemistry A, 3,* 2090–2096.
240. Wu, R., Wang, D. P., Rui, X., Liu, B., Zhou, K., Law, A. W., et al. (2015). In-situ formation of hollow hybrids composed of cobalt sulfides embedded within porous carbon polyhedra/carbon nanotubes for high-performance lithium-ion batteries. *Advanced Materials, 27,* 3038–3044.
241. Zhou, X., Wan, L. J., & Guo, Y. G. (2013). Synthesis of MoS_2 nanosheet-graphene nanosheet hybrid materials for stable lithium storage. *Chemical Communications, 49,* 1838–1840.
242. Fang, W., Zhao, H., Xie, Y., Fang, J., Xu, J., & Chen, Z. (2015). Facile hydrothermal synthesis of VS_2/graphene nanocomposites with superior high-rate capability as lithium-ion battery cathodes. *ACS Applied Materials & Interfaces, 7,* 13044–13052.
243. Huang, J., Wang, X., Li, J., Cao, L., Xu, Z., & Wei, H. (2016). WS_2-super P nanocomposites anode material with enhanced cycling stability for lithium ion batteries. *Journal of Alloys and Compounds, 673,* 60–66.
244. Park, G., Sim, S., Lee, J., & Lee, S.-M. (2015). Effect of silicon doping on the electrochemical properties of MoP_2 nano-cluster anode for lithium ion batteries. *Journal of Alloys and Compounds, 639,* 296–300.
245. Lu, Y., Tu, J. P., Xiang, J. Y., Wang, X. L., Zhang, J., Mai, Y. J., et al. (2011). Improved electrochemical performance of self-assembled hierarchical nanostructured nickel phosphide as a negative electrode for lithium ion batteries. *The Journal of Physical Chemistry C, 115,* 23760–23767.
246. Yang, D., Zhu, J., Rui, X., Tan, H., Cai, R., Hoster, H. E., et al. (2013). Synthesis of cobalt phosphides and their application as anodes for lithium ion batteries. *ACS Applied Materials & Interfacess, 5,* 1093–1099.
247. De Trizio, L., Figuerola, A., Manna, L., Genovese, A., George, C., Brescia, R., et al. (2012). Size-tunable, hexagonal plate-like Cu_3P and janus-like Cu-Cu_3P nanocrystals. *ACS Nano, 6,* 32–41.

Chapter 4
Nanostructures and Nanomaterials for Lithium Metal Batteries

Chun-Peng Yang and Yu-Guo Guo

Abstract Lithium metal batteries consist of a high-capacity cathode (such as oxygen and sulfur) and a Li metal anode and can deliver extremely high theoretical energy densities. The Li metal batteries were proposed earlier than the Li-ion batteries, but the Li metal anode was considered unsafe and the cathodes were hardly reversible. To meet the ever-increasing demand for the high energy density in batteries, Li metal batteries have been recently revisited and gained great interest. The nanotechnology plays a critical role in improving the performance and safety of the Li metal batteries. This chapter introduces the nano engineering in Li metal batteries, including nanostructures and nanomaterials utilized in oxygen cathodes, sulfur cathodes, and Li metal anodes. Rationally designed nanostructures with micropores, mesopores, or macropores are essential to host, constrain, protect, and improve the electrodes per electrode material properties. The nanomaterials, such as porous carbon, carbon nanotubes, graphene, polymer nanofilms, and inorganic nanomaterials, all show different functions such as constructing 3D conductive networks, interface protection, catalysis, and so on. This chapter reviews the nanostructures and nanomaterials in promoting the Li metal batteries, as well as their drawbacks, to provide insights to the nanotechnology in boosting the development of high-energy-density Li metal batteries.

Abbreviations

Li–S battery	Lithium-sulfur battery
Li-air battery	Lithium-air battery
Li–O_2 battery	Lithium–oxygen battery
LISICON	Lithium Super Ionic Conductor
IUPAC	International Union of Pure and Applied Chemistry

C.-P. Yang
University of Maryland, College Park, MD 20742, USA

Y.-G. Guo (✉)
Institute of Chemistry, Chinese Academy of Sciences, Beijing 100190, People's Republic of China
e-mail: ygguo@iccas.ac.cn

Y.-G. Guo, *Nanostructures and Nanomaterials for Batteries*,
https://doi.org/10.1007/978-981-13-6233-0_4

ORR	Oxygen reduction reaction
OER	Oxygen evolution reaction
LiTFSI	Bis(trifluoromethylsulfonyl)amine lithium salt
DOL	1,3-dioxolane
DME	Dimethoxymethane
TEGDME	Tetraethylene glycol dimethyl ether
FEC	Fluoroethylene carbonate
PAN	Polyacrylonitrile
PMMA	Poly(methyl methacrylate)
CNT	Carbon nanotube
CNF	Carbon nanofiber
CVD	Chemical vapor deposition
AAO	Anodic aluminum oxide
MWCNT	Multiwall carbon nanotube
GO	Graphene oxide
rGO	Reduced graphene oxide
ALD	Atomic layer deposition
MOF	Metal-organic framework
SEI	Solid electrolyte interphase
MAG	Massive artificial graphite
SEM	Scanning electron microscopy
PPA	Polyphosphoric acid
PEDOT	Poly(3,4-ethylenedioxythiophene)
PEG	Polyethylene glycol
PDMS	Poly(dimethylsiloxane)
PVdF-co-HFP	Poly(vinylidene fluoride-co-hexafluoropropylene)
CPL	Composite protective layer

4.1 Introduction to Lithium Metal Batteries

The huge demand for energy consumption has brought great threat to the sustainable development of the earth. The development of novel energy storage and conversion systems is required for effective utilization of renewable energy sources. Rechargeable battery systems provide a feasible option to achieve this objective due to their efficient and reliable energy storage. Various battery systems have been developed and commercialized over the past few decades, such as lead-acid batteries, Ni-based batteries, and Li-ion battery. As elaborated in previous chapters, Li-ion batteries have recently gain great success in the market of portable devices. Nevertheless, with the rapid development of the electric vehicle and other energy-storage demands, Li-ion batteries cannot meet the increasing requirements for energy density. The ever-increasing demand for efficient energy storage has stimulated the exploration of new high-energy-density battery systems. Li metal batteries have been proposed because of their extremely high energy densities. Different from Li-ion batteries,

Table 4.1 The cell reactions and specific energy of various batteries

System	Cell reaction	Theoretical specific energy ($Wh\ kg^{-1}$)	Practical specific energy ($Wh\ kg^{-1}$)	Cycle durability (cycles)
Lead-Acid	$Pd + PdO_2 + 2HSO_4^- + 2H^+ \leftrightarrow 2PdSO_4 + 2H_2O$	170	30–50	500–800
Ni–Cd	$2NiO(OH) + Cd + 2H_2O \leftrightarrow 2Ni(OH)_2 + Cd(OH)_2$	245	45–80	2000
Ni–MH	$xNi(OH)_2 + M \leftrightarrow xNiOOH + MH_x$	280	60–120	500–2000
Li-ion	$LiCoO_2 + 3C \leftrightarrow ½LiC_6 + Li_{½}CoO_2$	387	150–200	2000
Zn-air	$2Zn + O_2 \rightarrow 2ZnO$	1084	~470	
Li–S	$2Li + S \leftrightarrow Li_2S$	2600	~500	
Li-air	$2Li + O_2 \rightarrow Li_2O_2$	3500	~500	

the Li metal batteries use Li metal anodes, which have an extremely high specific capacity (3860 mA h g^{-1}). Li-air and Li–S batteries are the most well-known Li metal batteries and are considered as the next-generation battery systems. As shown in Table 4.1, Li-air and Li–S batteries have the most attractive theoretical specific energy, 3500 and 2600 Wh kg^{-1}, respectively, and are expected to fulfil the requirements of many high-energy applications, such as electric vehicles. Moreover, the cathode resources of Li-air and Li–S batteries are abundant, with air being free and S being cheap, promising to reduce the prices of the high-energy batteries.

However, the high theoretical energy density of the Li–S and Li-air batteries are too far to be achieved yet. The sulfur is electrically insulating, leading to a poor electrochemical activity. The utilization of S cathodes and thus the practical energy density of the Li–S battery are very limited. The air electrode is even worse since the air electrode is in a gas state. In addition, the electrochemical reactions of Li-air and Li–S batteries are conversion reaction, which is different with the intercalation reaction in Li-ion cathode/anode materials. The conversion reactions in the Li metal batteries cause significant structural changes in the electrodes, making the long-term cycles extremely difficult in the Li metal batteries.

To best exploit the emerging Li metal batteries, researchers have devoted tremendous efforts to understand the fundamental chemistries of the electrochemical systems. Based on the theoretical studies, battery material design and structure engineering have remarkably promoted the performance of the high-energy Li metal batteries. In particular, nanotechnology has boosted the development of the Li-air and Li–S batteries. Use of nanomaterials with suitable nanostructures in the electrodes and other battery components are demonstrated with great success in enhancing the electrochemical performance of the batteries. This chapter elaborates the nanostructures and nanomaterials in air cathodes, S cathodes, and Li metal anodes of the Li metal batteries.

4.2 Nanostructures in Air Cathodes of Li-Air Batteries

The Li-air battery was first reported by Abraham et al. in an organic electrolyte [1], and has since received enormous attentions due to its extremely high theoretical energy density (~3500 Wh kg^{-1}). The Li-air battery uses a Li metal anode, an air (or oxygen) cathode that is infused in open porous conductive electrode, and an organic or aqueous electrolyte. Currently pure O_2 is often used instead of air and is hence also called Li–O_2 battery. The electrochemical reactions of the Li–O_2 batteries in organic and aqueous electrolytes are different and the Li-air batteries can be divided into two categories depending on the electrolytes used: aqueous and non-aqueous Li–O_2 batteries (although there are also hybrid and solid-state Li–O_2 batteries). Figure 4.1 shows the electrochemical reaction mechanisms, cell configurations, and the discharge/charge profiles of the two different Li–O_2 batteries. The output voltage and energy density of the aqueous Li–O_2 battery is higher than that of the non-aqueous system. The discharge product LiOH is dissolvable in water, thus mitigating the electrode clogging issue and overpotential. However, as illustrated in Fig. 4.1a, a Li-ion conductive membrane (such as Lithium Super Ionic Conductor, LISICON) is indispensable in the aqueous Li–O_2 battery to separate Li from the aqueous electrolyte [2]. The LISICON membrane might not function as an effective and hermetic passivation layer after extended cycles, finally leading to the failure of the aqueous Li–O_2 battery. The most research efforts on the Li–O_2 batteries have been devoted to the non-aqueous (aprotic) Li–O_2 batteries [3].

A typical non-aqueous Li–O_2 battery consists of a Li metal anode, a separator saturated an organic electrolyte, and a porous air cathode. The cathode electrochemical pathway is based on the reversible redox reaction $2Li^+ + 2e^- + O_2 \leftrightarrow Li_2O_2$ ($E^0 = 2.96$ V vs. Li^+/Li), as illustrated in Fig. 4.1b. The energy density and cycle capability of Li-air batteries are greatly dependent on the air electrode. During the discharge process, O_2 is reduced (oxygen reduction reaction, ORR) in the porous cathode and consequently combined with the Li ions to form Li_2O_2 product, which is insoluble and insulated. The Li_2O_2 sluggishly decomposes to release O_2 in the subsequent charge process (oxygen evolution reaction, OER).

A three-phase boundary is involved in the oxygen reduction and evolution reactions between the air, the liquid electrolyte, and the solid electrode that contains the catalyst and the conducting substrate. The insoluble and insulated Li_2O_2 is likely to precipitate on the surface of the electrode and block the triple-phase reaction interface, finally ceasing the electrochemical reactions. It is prerequisite to design catalysts with a high specific surface area and substrate with a nanoporous structure. Therefore, rational design of the nanostructure in the cathode is expected to increase the interface area and enhance air diffusion and transportation. Below a

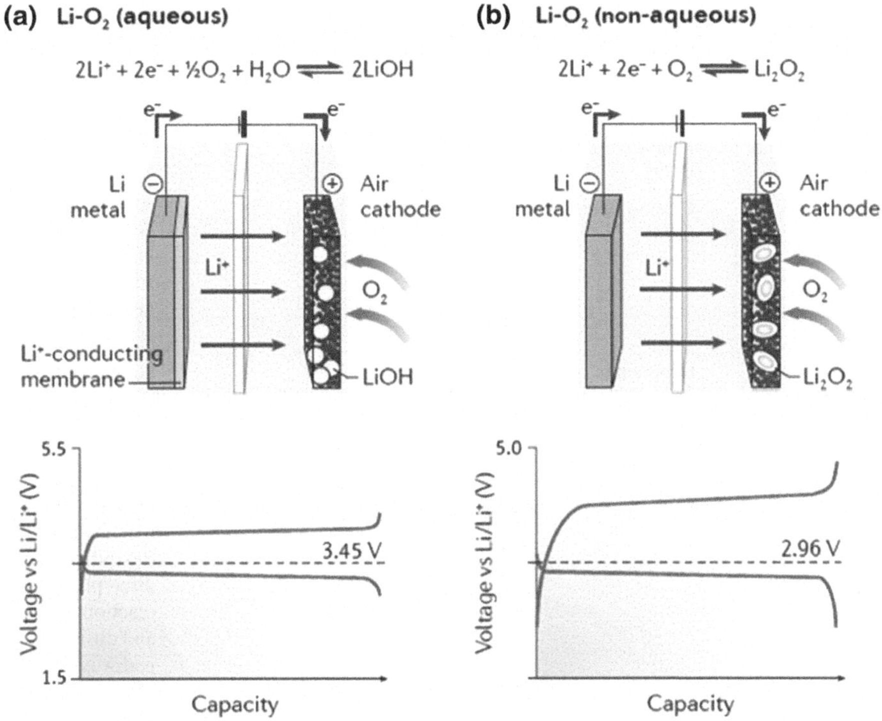

Fig. 4.1 The electrochemical reaction mechanisms, schematic cell configurations, and the discharge/charge profiles of the Li–O_2 batteries in **a** aqueous and **b** non-aqueous systems. Reproduced with permission from Ref. [169], Copyright © 2016, Nature Publishing Group

brief overview of the nanomaterials and nanostructures in air electrodes is given mainly including the supporting materials and catalyst materials of the air electrode.

4.2.1 Nanostructured Substrate Materials in Air Cathodes

An ideal cathode substrate material for the Li–O_2 battery are expected to have a high specific surface area, good conductivity, chemical and electrochemical stability, and low reactivity with electrolytes in Li-air battery systems [4]. In this respect, nanostructured carbon materials are the electrode materials of choice for current Li–O_2 batteries, because of their large surface areas, well-defined pore structures, excellent conductivity, and low cost. Considerable efforts have been made on porous carbon materials for Li–O_2 batteries to control the carbon structure, morphology, porosity, defects, and so on. Various nanostructured carbon-based electrodes have been synthesized, some of which are summarized in Table 4.2. Relationships between the electrochemical performance and carbon nanostructures have been investigated.

Table 4.2 Nano-structured carbon materials for cathodes of Li–O_2 batteries

Category	Material	Capacity (mA h g^{-1})	Current density (mA g^{-1})	Features and advantages
Carbon nanotubes and carbon nanofibers	Aligned carbon nanotube fibrils [172]	1000	1000	Carbon nanotubes have high chemical stability, high conductivity, high tensile strength and large surface area
	Carbon fibers [173]	4720	43	
Graphene nanosheets	Graphene nanosheets [174]	2332	50	Graphene has superior electrical conductivity, large surface area, remarkable mechanical flexibility, ideal three-phase reaction area and diffusion pathways for the electrolyte and oxygen
	Hierarchically Porous Graphene [175]	15,000	0.1	
Hierarchical carbon architectures	3D ordered meso-porous/macroporous carbon sphere arrays [176]	1000	50	Mesopores improve the electrolyte immersion and facilitate ion diffusion and electron-transfer process. Macropores provide sufficient space for Li_2O_2 accommodation
	Graphene foam [177]	5500	200	
	carbon nanotube sponge [178]	9092		
Doped carbon	Vertically aligned N-doped coral-like carbon fiber [179]	1000	100	Defects, functional groups, and dopants act as the active sites for oxygen reduction reaction, inducing the catalytic activity of carbon materials
	N-doped graphene [180]	10,400	200	
	3D porous N-doped graphene aerogels [181]	1000	300	

Due to different electrochemical measurement standards, the performances in different publications differ tremendously. However, the advantages of the carbon nanomaterials in air electrodes are consistent. The nanoporous structure of carbon accommodates discharge products and provides channels for ion transfer and oxygen diffusion. The carbon nanomaterials can also act as electrocatalysts towards oxygen reduction and evolution in non-aqueous Li–O_2 batteries.

Although carbon exhibits great potential as a substrate material for the Li-air batteries, research on carbon electrodes in Li–O_2 batteries shows instability of the carbon material despite the impression of its stability. Carbon material in air cathodes could undergo oxidation reaction to form Li_2CO_3 when charged up to 3.5 V (vs. Li/Li^+) [5]. The carbon decomposition will further promote the decomposition of organic electrolytes. The application of carbon materials in stable Li–O_2 batteries is hence limited. A stable coating on the carbon materials is required to prevent the reaction of Li_2O_2 with the electrolyte and carbon. Non-carbon electrode materials have also been explored for Li–O_2 batteries, such as TiC material and nanoporous gold [6, 7].

4.2.2 Nanocatalysts in Air Cathodes

In Li-air batteries, large overpotential or polarization is frequently observed at the cathode with charging potential at ~4.5 V and discharge potential ~2.5 V, leading to poor energy efficiency. Electrocatalysts that can promote both ORR during discharge and OER during charge have been employed to reduce the reaction overpotential and to improve the efficiency. The catalyst nanomaterials developed in Li-air batteries include carbon nanomaterial, transition metal oxides, and noble metals. The typical catalyst materials, their nanostructure and features are summarized in Table 4.3. The ideal catalyst should have excellent catalytic activity, high electrical conductivity, high porosity, large specific surface area and low cost. Hybrid catalyst materials have also been designed noble metal or transition metal oxide nanomaterials with various carbon substrates. The hybrid catalyst nanomaterials are promised with superior performance due to their synergistic catalytic effect. The new concepts of catalysis materials have greatly advanced Li-air batteries.

The Li-air battery is one of the most promising future energy-storage systems with greatly opportunities as well as challenges. Despite the preceding achievements made in the O_2 electrodes by nanotechnology in catalysts and supports, Li–O_2 battery technology is still at its infant stage. Some fundamental mechanisms in Li-air batteries remain unclear. Numerous practical problems in the electrodes, catalysts, electrolytes, and the device are yet to be addressed to realize reversible Li-air batteries. Although many obstacles lie in the way toward commercially feasible Li-air batteries, the enormous advantages of Li-air batteries that far outperform other conventional rechargeable battery systems. More efforts into the research on Li-air batteries are worthwhile. Special attentions should be paid to the novel nanomaterial and nanostructure design in the air electrodes with more fundamental understanding of the reaction in Li-air batteries.

Table 4.3 Nanocatalysts for the cathodes of Li-air batteries

Types	Typical catalyst materials	Morphology	References	Comment
Carbon nanomaterials	N-doped carbon materials	0D carbon black	[182, 183]	Carbon materials have certain catalytical activity. N-doping improve the ORR kinetics and discharge capacity but no OER activity or charging capacity
		1D nanofibers and nanotubes		
		2D graphene		
		3D porous carbon materials		
Noble metals	Au	Nanocrystals	[184]	Noble metals have high selectivity, oxidation resistance, and corrosion resistance. Noble metals oxides have excellent electronic conductivity and catalytic performance. The drawback of noble metals is their high cost
	Pt–Au	Nano-alloy	[185]	
	AgPd–Pd	1D porous nanotubes	[186]	
	RuO_2	Nanosheets	[187]	
Transition metal oxides	NiO	Mesoporous nanosheets	[188]	Transition metal oxides have low cost, abundance and good catalytic activity. The high surface area and tailored structure of the nano catalyst materials further increase catalytic activity and the efficiency of the batteries
	δ-MnO_2	Porous submicrotubes	[189]	
	Co_3O_4	Ultrathin nanosheets	[190]	
	Fe_2O_3	Nanoflakes	[191]	
	$FeCo_2O_4$	Porous nanorods	[192]	
	$LaNiO_3$	Nanocubes	[193]	
	$NiCo_2O_4$	3D foam	[194]	

4.3 Nanostructures in Sulfur Cathodes of Li–S Batteries

Sulfur is an abundant element is earth. Elemental sulfur is cheap, easy-producing, widely accessible, inexpensive, non-toxic. Considering the advantages of sulfur and its high specific capacity, sulfur is an extraordinary cathode material for rechargeable batteries. However, due to the insulting nature of sulfur, sulfur cathode has gain little success until the utilization of nanoporous carbon host material in 2009. Recent research shows that nanomaterials are indispensable in sulfur cathodes. They can improve the electronic conductivity of the electrode, suppress the polysulfide dissolution, and alleviate the volume change of sulfur cathode.

4.3.1 Electrochemistry, Advantages, and Issues of Li–S Batteries

Different from the conventional Li-ion battery cathode materials, which are based on Li-ion intercalation reactions, the electrochemistry of sulfur cathode is a conversion reaction. During lithiation (i.e., Li^+ reacts with the S cathode), elemental S is converted to Li_2S as a final product, whose molecular and crystalline structure is completely altered after the reaction. In different electrochemical environments, the electrochemical reaction mechanisms of Li–S batteries are slightly different. The most commonly known and most sufficiently studied system is the elemental S_8, which is the thermodynamically stable allotrope of element S, in ether-based organic electrolytes, such as 1 M LiTFSI dissolved in DOL/DME (1/1, v/v) or TEGDME. The electrochemical reaction of a typical Li–S battery is

Anode	$Li \leftrightarrow Li^+ + e^-$
Cathode	$S_8 + 16Li^+ + 16e^- \leftrightarrow 8Li_2S$
Overall	$S_8 + 16Li \leftrightarrow 8Li_2S$

The lithiation process of the S cathode is stepwise, evolving as $S_8 \rightarrow Li_2S_8 \rightarrow Li_2S_6 \rightarrow Li_2S_4 \rightarrow Li_2S_2 \rightarrow Li_2S$ with two discharge plateaus, 2.3 and 2.1 V, as shown in Fig. 4.2.

The stepwise reaction mechanism of the Li–S battery has been well demonstrated by numerous experiments. Due to the nature of S and the intermediate products, the Li–S battery is faced with the following challenges.

(1) Poor conductivity. The element S and final product Li_2S are both electronic and ionic insulators. Thus, the S cathode has a very low electrochemical activity.
(2) Dissolution of intermediate polysulfides and "shuttle effect". Among the intermediary products during the discharge process, the high-order lithium polysulfides (Li_2S_n, $4 \leq n \leq 8$) are soluble in the ether-based electrolytes. The dissolution of the polysulfides is good in terms of enhancing the activity of the

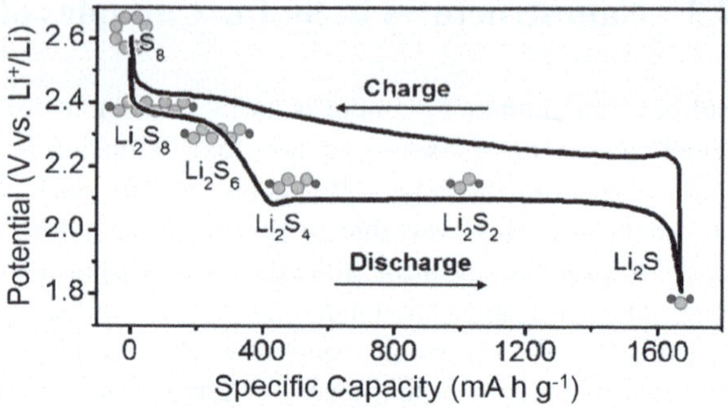

Fig. 4.2 Typical discharge/charge voltage profile of Li–S battery and the evolution of the reaction products. During lithiation (discharge) the cyclo-S_8 converts to Li_2S_8, Li_2S_6, Li_2S_4, Li_2S_2, and Li_2S, successively

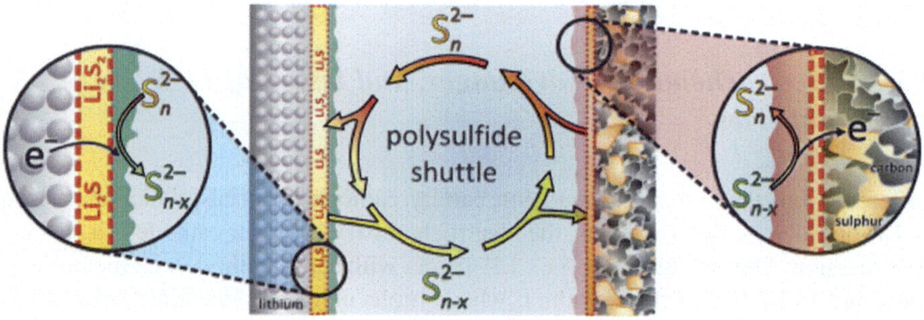

Fig. 4.3 Schematic illustration of the "shuttle effect" of Li–S battery with polysulfide dissolution. Reproduced with permission from Ref. [170], Copyright © 2014, Elsevier

S cathode, but incurs many problems to the Li–S battery. The dissolution of Li_2S_n from the cathode to the electrolyte causes loss of the active materials in the cathode (and thus the loss of capacity). The dissolved Li_2S_n can diffuse to the anode and corrode the Li anode and hamper the cycle stability of the cell. What is worse, when the dissolved Li_2S_n generated in the cathode diffuses to the anode, it will be reduced to Li_2S_{n-x}. The Li_2S_{n-x} formed in the anode side diffuses back to the cathode because of the concentration gradient and be oxidized to Li_2S_n during the charge process, associated with a charge plateau at 2.4 V. The Li_2S_n will diffuse to the anode again. The repeated migration of the polysulfide between the cathode and anode is so called "shuttle effect", as schematically shown in Fig. 4.3. The shuttle effect contributes to a long or even endless charge plateau at 2.4 V without effective electrochemical reaction. Thus, the shuttle effect notably lengthens the charging time, lowers the Coulombic efficiency and wastes the charge energy. The Li_2S_n could even not be charged to element S, leading to cell failure.

(3) Volume change. The element S expands by ~79% of its volume after being reduced to Li_2S, according the density of α-S and crystalline Li_2S [8]. The volume change during repeated discharge/charge cycles could cause the detachment of the S cathode.

With the knowledge of the reaction mechanism and the associated problems above, researchers have reported various approaches to improve the electrochemical performance of the S cathode. Among various methods, the most extensive and effective one is to confine the S molecules in a porous host material, especially porous carbon host materials [8, 9]. Hence, various nanoporous materials have been designed and employed for high-performance S cathodes.

4.3.2 *Design of Nanostructures for S Cathodes*

The poor electron/ion conductivity of S cathode, dissolving problem of polysulfides, and volume expansion problem of the S cathode are the vital roadblocks to the practical application of Li–S battery. As a host material, the nanoporous substrate plays several key roles in improve the cathode performance. First, the conductive host materials, usually carbon host materials, can provide a conductive network to enhance the electron conductivity of the cathode. The nanopores reduce ion transfer path and increase surface area in the nanoporous host improves the cathode-electrolyte contact, improving the ionic diffusivity as well. Compared with bulk S materials, the S molecules stored in the porous channels possesses enhanced electrochemical activity owing to the diminished particle size. Second, the space confinement of the nanopores retard or even prevent the dissolution of polysulfides and suppress the shuttle effect, thereby contributing to favorable cycle life of the cathode. Third, the nanopores in the host materials constrain the volume expansion of the S, allowing accommodation of the strain induced by volume change, free from fracture or pulverization of sulfur. Below we discuss the effect of the nanoporous host materials for the S cathodes, overview nanomaterial design in S cathodes, and show how indispensable the nanostructure hosts are in the S cathodes.

4.3.2.1 Effect of Pore Size

According to IUPAC, a material having pores smaller than 2 nm in diameter is defined as microporous material, a material containing pores with diameters between 2 and 50 nm is mesoporous material, and a material having pores larger than 50 nm in diameter is macroporous material. As illustrated in Fig. 4.4, all three kinds of nanomaterials have been utilized for confining the S cathode materials with their own advantages and disadvantages. In general, the microporous host materials have a stronger confinement for the S cathode materials and thus better electrochemical performance. However, the S content in microporous host material is relatively low because the pore volume of the micropores is limited. On the contrary, the host materials with larger pores (mesopores and even macropores) can encapsulate higher content of S while weaker space confinement for the S molecules, resulting in inferior electrochemical performance. Here, we take some typical nanoporous carbon

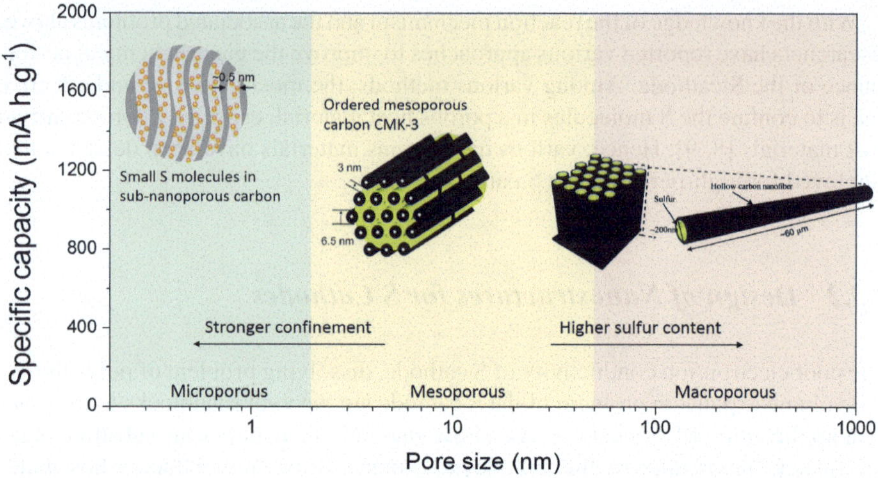

Fig. 4.4 Representative S cathodes hosted in nanoporous materials with different pore sizes. Left: small S molecules (S_2–S_4) confined in the sub-nanoporous channels (pore size ~0.5 nm) of the substrate (Reproduced with permission from Ref. [11], Copyright © 2012, American Chemical Society); middle: S/C composite material, in which the S (yellow) in confined in the ordered mesoporous channel of CMK-3 (black) (Reproduced with permission from Ref. [10], Copyright © 2009, Nature Publishing Group); right: high aspect ratio hollow carbon nanofiber–sulfur composite (Reproduced with permission from Ref. [16], Copyright © 2011, American Chemical Society)

materials as examples to elaborate the effect of the pore size on the electrochemistry and performance of the S cathodes.

Mesoporous Carbon

The porous host material is so important for rechargeable S cathode. It is not until Nazar et al. proposed to use a porous carbon for S cathode that Li–S batteries gain a breakthrough in cycling performance [10]. The porous carbon is a typical mesoporous carbon materials, ordered porous carbon CMK-3. The middle schematic of Fig. 4.4 illustrates the structure of the S-impregnated CMK-3 host material. CMK-3 is synthesized by a nanocasting method using SBA-15 as template. The CMK-3 consists of hexagonally ordered carbon nanorods that is 6–7 nm thick and separated by empty nanochannels of ~3 nm. The mesoporous structure strongly constrains the S cathode, prevents the dissolution of polysulfides, and provides pathways for electron and Li ion transfer.

The method for infiltrating S into the porous host materials has also been widely adopted by other researchers. The S/C composite material was synthesized by a melt-diffusion strategy. The mixture of S and C (mass ratio 7:3) was heated at 155 °C, where the viscosity of S is the lowest. The molten S was imbibed into the nanochannels by capillary forces and solidified to form S/C cathode material. The melt-diffusion method provides a facile way to synthesize nanopore-encapsulated S cathodes. The S cathode confined in the mesoporous carbon can delivers a high reversible specific capacity of 1005 mA h g^{-1}, which was never achieved with bulk S cathode for decades without the mesoporous carbon confinement.

The demonstration of the rechargeable S cathodes using the mesoporous carbon-sulfur composite has boosted the research on Li–S batteries. Many mesoporous carbon materials have been developed and examined as the host materials for high-performance S cathodes. Through investigating the electrochemistry of S cathode materials with mesoporous carbon matrices, the electrochemistry of S during lithiation and delithiation is well demonstrated, as discussed in Fig. 4.2. It is now generally acknowledged that the best suitable electrolytes for the mesoporous confined S cathode are ether-based electrolytes (e.g., 1 M LiTFSI dissolved in DOL/DME). The mesoporous host materials establish good contact for the carbon substrate (electron conductive network), the liquid electrolyte (Li ion diffuse path), and the active material (S or the discharged product Li_2S_x). The contact of the S and liquid electrolyte allows a slight dissolution of the polysulfides and leads to a liquid reaction mechanism involving a solid-liquid-solid process, which is in favor of the electrochemical activity of the S cathode. The dissolved polysulfides, however, are confined in the mesoporous host with pore size <50 nm and do not cause significant loss of active materials or shuttle effect. Moreover, the mesoporous substrate has a sufficient volume to host sulfur with content of up to 70 wt % in the composite while maintaining strong constraint on polysulfides and interaction between sulfur and the porous channels. Therefore, the mesoporous carbon has been widely studied for Li–S batteries with high energy density and good cycling performance.

Microporous Carbon

As the host materials for the S cathode, the smaller pores have the stronger space confinement effect for the S molecules. The strong confinement of micropores, or more accurately, pores smaller than 1 nm (sub-nanopores), can even alter the electrochemical reaction mechanism of the S cathode. As discussed above, the dissolution problems of the S cathode stems from the step-wise reaction of the cyclo-S_8 molecules in the cathode. The S_8 allotrope is the thermodynamic stable and most widely existing variation. However, when the S is confined by sub-nanoporous host materials, the molecule of S can be changed. Because the molecule size of the S_5–S_8 are greater than 0.5 nm (e.g., S_8 is 0.6876 nm), the large S molecules (S_5–S_8) cannot be accommodated in the micropores of 0.5 nm. As schematically illustrated in Fig. 4.4 (left), the S molecules stored in the sub-nanopores (~0.5 nm) are only small S molecules, S_2–S_4, whose dimension is smaller than 0.5 nm [11]. The small S_2–S_4 molecules are metastable in are confined in the pores smaller than 0.5 nm due to the nanosize space confinement. The small S molecules in the sub-nanopores displays a single-plateau discharge reaction at ~1.9 V. The upper-level discharge plateau (2.3 V, as shown in Fig. 4.2), which is attributed to the conversion from S_8 to Li_2S_4, is absent, because the large S molecules (S_5–S_8) cannot be store in the sub-nanoporous channels. Therefore, the shuttle effect of the lithium polysufides is eliminated. The small S molecules in the sub-nanopores exhibit high electrochemical activity and cycling stability.

It is noteworthy that the smaller S molecules can work not only ether-based electrolytes, which is commonly used in Li–S batteries, but also in carbonate-based electrolytes. This is an extraordinary change realized by the confinement of the sub-nanoporous host. In fact, the polysulfide anions are nucleophilic and can react with the carbonyl groups in the carbonate solvent [12]. Since there are not high-ordered S molecules in the sub-nanopores, no high-ordered polysulfides are formed to react with the carbonate solvents during the electrochemical reactions. Another explanation is that the nanopores <0.5 nm is effective to spatially block the carbonate solvent molecules because the molecular size of the carbonate solvent is larger than 0.5 nm. Thus, the nanopores block the carbonate molecules from the S cathode and avoid the detrimental nucleophilic reaction. It is believed that, due to the microporous confinement, the small S molecules do not contact with the solvents. Li ions diffuse and react with S as solid-state Li ions [13]. In other words, the electrochemical reaction of the small S molecules is a solid-state reaction.

The solid-state reaction of the S cathode in microporous host materials is also observed when the S is confined in the nanochannels of carbon nanotubes. When the S is confined in single-walled or double-walled carbon nanotubes, it is no longer cyclo-S_8, but chain-like 1D S molecules. The electrochemical test attests to the high electrochemical activity and extinctive electrochemistry of S chains compared with conventional S_8. The lithiation of the long S chains leads to degrade to short chains during discharge with a solid-phase. An advantage of the solid-phase reaction is structure cycling stability while the disadvantage is a sluggish kinetics. This study of the electrochemistry of S chains exhibits how the microporous host materials changes the electrochemistry of the Li–S batteries, improve the electrochemical performance, and creates various possibilities.

The concept of the chain-like smaller S cathode in sub-nanoporous host materials has attracted broad interest due to its unique electrochemical property and excellent electrochemical performance. Researchers have thereafter developed numerous microporous carbon materials for hosting small S molecule cathode. The host materials were modified by optimizing the host structure, synthesis process, materials cost, etc. [14, 15]. The electrochemical performance of the S cathode in microporous host materials is improved remarkably. Studies have shown that the S cathode accommodated in the micropores have great advantages in terms of specific capacity (based on the mass of S), cycle stability, and rate capability. However, due to the small pore size and low pore volume of the microporous substrate, the S content in the microporous substrates is low, not exceeding 55%. Thus, the overall capacity of the cathode (i.e., specific capacity based on the mass of S/C composite material) is small and the energy density using micropore-hosted S is limited.

Macroporous Carbon

Compared with the microporous and mesoporous carbon materials, the macroporous carbon has less constrain effect on the S cathode. The advantage of the macroporous texture is that it provides a high pore volume for hosting a high content of S while maintaining relatively good electrochemical performance. The S in macropores shares the same electrochemistry and problems as the bulk S or the S in mesoporous host materials. Therefore, the efforts made on macroporous carbon-S composite cathodes are (i) maximizing the S content in the composite and S mass loading in the electrode, (ii) minimizing the polysulfide dissolution and shuttling, (iii) preserving fast transport of Li ions, and (iv) enhancing the electronic conducing network. The outstanding advantage of S in macroporous host materials is the S content in the composite materials and the overall capacity (i.e., the specific capacity based on the mass of the S/C composite, not just the mass of S).

As an example, Fig. 4.4 (right image) shows a sulfur cathode materials confined in macroporous carbon, which is a hollow carbon nanofiber [16]. The hollow carbon nanofibers were derived from the anodic aluminum oxide (AAO) templates through carbonization of polystyrene. The diameter of the hollow nanofibers is ~200 nm and the length is ~60 μm. The macropores enable a high S content of 75% in the electrode with space for volume expansion during cycling. The nanofibers with the high aspect ratio constrain the S and polysulfides within the hollow nanofibers minimizing polysulfide dissolution. These features attributes to a high specific capacity of the S/C cathode material ~1000 mA h g^{-1}. Due to the limited confinement effect of the macropores, the shuttle effect is not eliminated in the hollow carbon nanofiber–sulfur composite cathode, which is indicated by an overcharge at 2.4 V and a low Coulombic efficiency (~85%, discharge/charge). The shuttle effect was suppressed by using $LiNO_3$ additive in the electrolyte, improving the Coulombic efficiency to 98–99%.

Summary

Based on the analysis and the examples of the different nanoporous host materials, one can summarize the features of the host nanomaterials with different pore sizes in Table 4.4.

Understanding the effect of the various pore sizes on the nano host materials, it is expected that the development of hierarchically porous carbon materials with micro/meso/macropores with desired physical/chemical properties for optimal electrochemical performances. To take advantage of the ameliorating features of microporous and mesoporous carbon, a highly ordered meso-microporous core-shell carbon host has been designed using a SBA-15 template and sucrose as the carbon source [13]. The ordered mesoporous core possesses a large pore volume for high sulfur loading and abundant ionic/electronic conductive channels for enhanced reaction kinetics. The microporous carbon shell serves as strong physical barrier and absorber to prevent dissolution of polysulfides and shuttle effect. In addition, the microporous carbon shell stores small sulfur molecules (S_{2-4}), contributing to additional capacities. Owing to its highly optimized structure, it displayed long cyclability over 200 cycles and presented a high capacity of 837 mA h g^{-1}, improved than compared to the other similar porous carbon nanomaterials. Other micro-mesoporous hierarchical nanostructures have also been synthesized, such as template assisted porous carbon [17, 18], micro-mesoporous carbon spheres [19], and biomass-derived porous carbon [20]. Macroporous carbons with large open pores, in contrast, are not widely employed as sulfur host materials, because of their weak confinement to the polysulfides. However, hollow nanostructures with a macroporous core and a micro- or

Table 4.4 Features of nanoporous carbon materials with different pore sizes as the host materials of S cathodes

	Micropores		Mesopores	Macropores
Pore size (nm)	<1	1–2	2–50	>50
Electrochemistry	Single plateau, solid reaction mechanism	Stepwise reaction, two discharge plateaus, solid-liquid-solid reactions involving polysulfide dissolution		
Compatible electrolyte	Carbonate-based or ether-based electrolytes	Mainly ether-based electrolytes		
Function	Encapsulating sulfur, suppressing polysulfide dissolution, improving electrochemical activity	Accommodating sulfur, trapping dissolved polysulfides, suppressing capacity fade	Accommodating more sulfur, trapping dissolved polysulfides, facilitating charge transport	Electrolyte pathway for good electrolyte immersion. Large pore volume
Advantages	Good cycling stability	Good cycling stability	High sulfur content, high output voltages	
Disadvantages	Low sulfur contents, low output voltages	Low sulfur content, possible polysulfide dissolution	Inferior cycling stability, possible polysulfide dissolution	Inferior cycling stability, severe shuttle effect

mesoporous shell can be an effective sulfur host, providing high sulfur loading, high utilization of sulfur, and good cycling stability [21].

The effect of the pores with different sizes applies not only to the carbon nanoparticles, but also in various nanomaterials as the host in S cathodes. Below we discuss the sulfur host nanomaterials in the perspective of materials, including synthesis, morphology and performance.

4.3.2.2 Nanomaterials for Sulfur Cathodes

To date, numerous nanomaterials have been employed as the host materials for S cathodes. The frequently host materials for S cathode are carbon materials, including porous carbon spheres, carbon nanotubes, carbon nanofibers, and graphene-based materials. The carbon materials are favorable in electrode materials for obvious advantages. The carbon nanomaterials are chemically and electrochemically stable, superior to most other substrate materials. The carbon materials have excellent electron conductivity (among the highest) while low density (much lower than most metal). Moreover, the synthesis of porous carbon materials, carbon nanotubes,

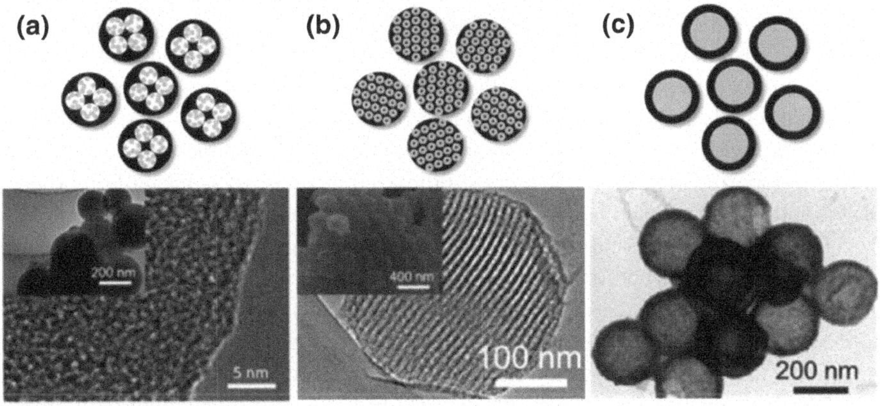

Fig. 4.5 Design of nanoporous spherical carbon for S/C composite cathode materials: **a** microporous carbon spheres, **b** spherical ordered mesoporous carbon nanoparticles, and **c** porous hollow carbon. **a** Reproduced with permission from Ref. [22], Copyright © 2010, Royal Society of Chemistry. **b** Reproduced with permission from Ref. [24], Copyright © 2012, Wiley. **c** Reproduced with permission from Ref. [25], Copyright © 2011, Wiley

graphene has been well-established. The nanofabrication method for carbon host materials greatly boosted the investigation on sulfur/carbon composite materials with advanced nanostructures and improved electrochemical performance. Here, we summarize a few typical carbon host materials, their synthesis methods, and the resulting S/C composite cathodes. One can see how the electrochemical performance of the S/C cathode is strongly dependent on the nanostructure design.

Porous Carbon Nanoparticles
Carbon nanoparticles with a porous structure have been widely studied for decades. For the use in S cathodes, the porous structure of the carbon nanoparticles is prominent. As discussed above, nanoporous carbon materials are classified to microporous, mesoporous, and macroporous carbon depending the pore size. Figure 4.5 shows the representative carbon nanospheres with micropores, mesopores, and macropores.

Microporous carbon can be obtained by pyrolyzation of carbonaceous precursors and/or etching by KOH, CO_2, etc. The microporous carbon in Fig. 4.5a was derived from sucrose [22]. It has a spherical morphology with size of 200–300 nm and has micropores of 0.7 nm for accommodating S cathode. The S cathode was loaded by a two-step heat treatment, melt-diffusion and vaporizing. As discussed above, the microporous carbon has a constrain effect on the S material, but a limited pore volume for S (42% in this work). The microporous carbon spheres with 51% S showed much smaller capacity (~300 mA h g^{-1}) than the ones with 42% loading (~900 mA h g^{-1}). The micropores have strong adsorption to the S and polysulfides in the pores but the excess S out of the micropores is counterproductive. Therefore, microporous carbon is limited to low content S/C composite cathode materials.

Mesoporous carbon can be synthesized via various methods, frequently employing templates, such as colloidal silica (e.g., SBA-15) [10], sodium silicate [23], tetraethylorthosilicate and block copolymer [19, 24]. Plentiful materials have been used as the carbon resources, such as carbohydrates (e.g., glucose, sucrose, starch), polymers (e.g., polypyrrole, polyacrylonitrile (PAN), polystyrene), resin (e.g., phenol formaldehyde), and many other organic or biomass precursors. The CMK-3 is a typical and wide-used mesoporous carbon material [10]. The ordered mesoporous carbon nanoparticles in Fig. 4.5b were synthesized using resol as the carbon precursor, tetraethylorthosilicate and a block copolymer Pluronic F127 as a soft template [24]. The obtained spherical ordered mesoporous carbon has a diameter of 300 nm, 2D-hexagonal mesostructured (*P6mm*) pore of 3.1–6.0 nm, pore volume of 2.32 cm^3 g^{-1}, and surface area of 2445 m^2 g^{-1}. The mesoporous carbon can be loaded with more S, compared with microporous carbon. A S/C composite with 70% S was made by melt and infiltrate S in the spherical ordered mesoporous carbon in Fig. 4.5b, which showed a high reversible capacity up to 1200 mA h g^{-1} and excellent Coulombic efficiency.

Macroporous carbon materials can be synthesized via many methods, such as carbonization of natural biomass, freeze drying, pyrolysis of aerogel, and template methods. For example, the porous hollow carbon in Fig. 4.5c was fabricated by a frequently used template synthesis method [25]. The petroleum pitch (carbon precursor) was carbonized on porous silica template (hard template), followed by HF etching the silica template. The hollow carbon sphere has an internal void space of ~200 nm and a mesoporous shell with pore size of 3 nm. A vapor-phase infusion technique was used to encapsulate the S in the interior void and the porous shell, with 70% total S content. The interior macropores can host a large content of S but has a weak interaction with the S and polysulfides; the mesoporous shells accommodate less amount of S but sequester the S and minimize the polysulfide dissolution and shuttling. S cathode encapsulated in the porous hollow carbon can be cycled at 850 mA g^{-1} over 100 cycles with a high capacity of ~1000 mA h g^{-1} [25].

Carbon Nanotubes and Nanofibers

Carbon nanotubes (CNTs) and carbon nanofibers (CNFs) are 1D host materials for S cathodes. As host materials, the 1D conductive carbon with high aspect ratios can provide an excellent electron conductivity along the nanotubes/nanofibers and reduce the ion diffuse path along the diameter for rapid Li ion transport. The 1D structure is also advantageous in terms of mechanical strength compared with dispersed nanoparticles. The CNTs are generally synthesized via chemical vapor deposition (CVD) and are commercially available. They can be prepared by templates (mainly anodic aluminum oxide, AAO) as well for particular uses. CNFs are mainly fabricated by electrospinning and carbonizing polymer nanofibers.

Multiwall carbon nanotubes (MWCNTs) have an excellent electron conductivity due to their high graphitization level (sp^2 hybridized). The highly conductive MWCNT can form a 3D conductive network to enhance the activity of the S cathode. The large specific surface area of MWCNT is beneficial for absorbing polysulfides. In Fig. 4.6a, Manthiram and co-workers report a self-weaving S-MWCNT composite cathode, which has an intertwined MWCNT network for electron and ion conducting pathways [26]. The S-MWCNT cathode exhibits high capacities of 1352 mA h g^{-1}

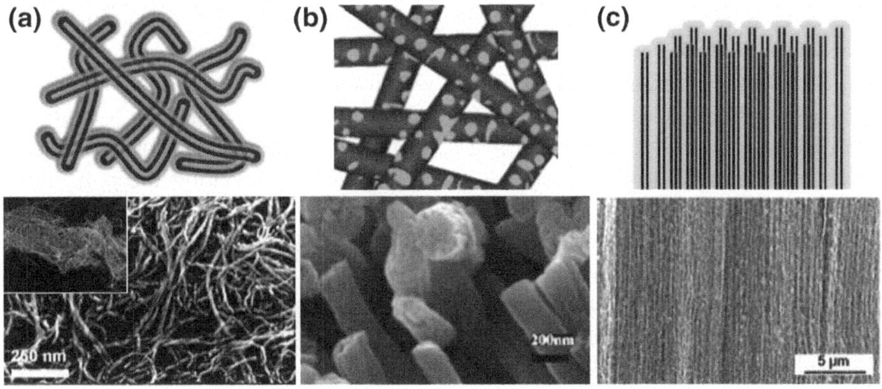

Fig. 4.6 Designs of carbon nanotubes and carbon nanofibers for sulfur cathode: **a** Multiwall carbon nanotubes as an interwoven conductive network, **b** Porous carbon nanofiber-encapsulated sulfur cathode, **c** Vertical aligned CNTs with an open, ordered and straight pore structure for high sulfur loading. **a** Reproduced with permission from Ref. [26], Copyright © 2012, Royal Society of Chemistry. **b** Reproduced with permission from Ref. [28], Copyright © 2011, Royal Society of Chemistry. **c** Reproduced with permission from Ref. [30], Copyright © 2012, Royal Society of Chemistry

at 1 C and 1012 mA h g^{-1} at 4 C. Due to the self-weaving behavior of MWCNTs, the cell is free from binders and current collectors, delivering high energy density.

Despite the outstanding advantages of CNTs as the host material for S cathode, utilization of CNTs as the host material for S is not so successful as the porous carbon materials discussed above. The main drawback of the CNT containing sulfur cathodes is their poor bonding relation with sulfur. In addition, the CNTs have few close pores for hosting S, except for the interior channels that can hardly encapsulate S. Thus, the S is not constrained in the nanopores but coated outside of the MWCNT, exposed to the liquid electrolyte with weak space confinement. Simply using the CNTs as the host materials in S cathode could lead to severe shuttle effect and poor cycling stability. Limited sulfur loading also hinders the use of CNT as S host materials. Instead of using as the host materials, one can find CNTs very beneficial when CNTs are used as conductive additives because of the high conductivity of the CNTs. Due to the 1D structure, one can easily fabricated free-standing CNT membranes as a current collector for the S cathode, which will be discussed below.

Different from the common-sense CNTs (i.e., CVD-synthesized sp^2 hybridized CNTs with high conductivity and fewer defects), disordered CNTs synthesized from PAN filled AAO template have also been used as sulfur host [27]. The disordered CNTs has less graphitic tube walls and bamboo-like hollow voids, which are accessible for S vapor at 160, 300, and 500 °C. The S cathode in the disordered CNTs shows reasonably good cycle stability with suppressed shuttle effect.

CNFs are morphologically similar to CNTs but have no interior hollow channels. The CNFs have good electrical conductivity and high mechanical strength. Different from CNTs, CNFs are sp^2 and sp^3 hybridized amorphous carbon with certain pores on the fibers, which are akin to the nanoporous carbon in properties. Thus, CNFs have been studied as a desirable host for S cathodes. In Fig. 4.6b, the porous CNFs were synthesized via carbonizing electrospun PAN/PMMA bicomponent nanofibers [28]. By electrospinning and partial decomposition of the PMMA blend, porous CNFs were obtained for loading S. The S/CNF composite with 42% S exhibited high discharge capacity, 1400 mA h g^{-1} at a 0.05 C. The CNFs homogeneously disperse and immobilize S on their porous structures, alleviating the polysulfide shuttle phenomenon.

By the above approach of electrospinning and partial decomposition, the obtained porous CNFs consist of micropores and mesopores with broad pore size distributions and relatively small BET surface areas. Mesoporous CNFs with high porosity could be a more suitable sulfur host material. Using silica templates, mesoporous CNFs with a bimodal porosity, an extremely high surface area (1928 m^2 g^{-1}) and a high inner pore volume (2.41 cm^3 g^{-1}) were synthesized for high S loading and better electrochemical performance. Porosity optimization of the nanofibers significant for sulfur electrodes [29].

The 1D carbon host materials, especially CNTs, generally lacks sufficient pore volume or space confinement for high sulfur loading. Increasing the S content in the electrode, while maintaining its high specific capacity, is one of the most challenging tasks to exploit the high energy density of Li–S batteries. An interconnected 3D architecture built by the 1D CNTs is expected to load a large amount of sulfur with good distribution and high conductivity for improved energy density. Figure 4.6c displays a typical substrate of vertical aligned CNTs prepared by CVD, enabling high sulfur loadings up to 70 wt% [30]. The highly aligned CNTs with an open, ordered and straight pore structure can load more sulfur up to 90 wt% [31]. The increasing sulfur content enhanced the energy density of the Li–S batteries and volumetric discharge capacity from 200.1 to 1116.0 mA h cm^{-3}. The CNT/sulfur composites with 90 wt% S content delivered an energy density of 1249.2 Wh kg^{-1} at a power density of 134.3 Wh kg^{-1} [31].

Graphene and Its Derivatives

Graphene is a 2D carbon with excellent electrical conductivity nature (106 S cm^{-1}), mechanical strength and flexibility, high surface area (2600 m^2 g^{-1}), and chemical stability [32]. It has been widely used in Li-ion batteries as described in previous chapters. These merits also render graphene and its derivatives one of the most promising candidates for Li–S batteries. The graphene (and its derivatives) are expected to solve the insulating problem of sulfur, physically trap or chemically absorb the polysulfides, and mitigate the volume change of S cathodes during the charging/discharging process. So far, various sulfur/graphene composite materials have been designed to boost the electrochemical performance of Li–S batteries [33–50]. There are different categories of graphene, such as graphene oxide (GO), reduced graphene oxide (rGO), thermally exfoliated graphene, functionalized graphene, and CVD-grown graphene. Despite the different synthesis method and properties, these graphene materials have been explored for sulfur loading for Li–S batteries. A few representative graphene/sulfur composite materials are shown in Fig. 4.7.

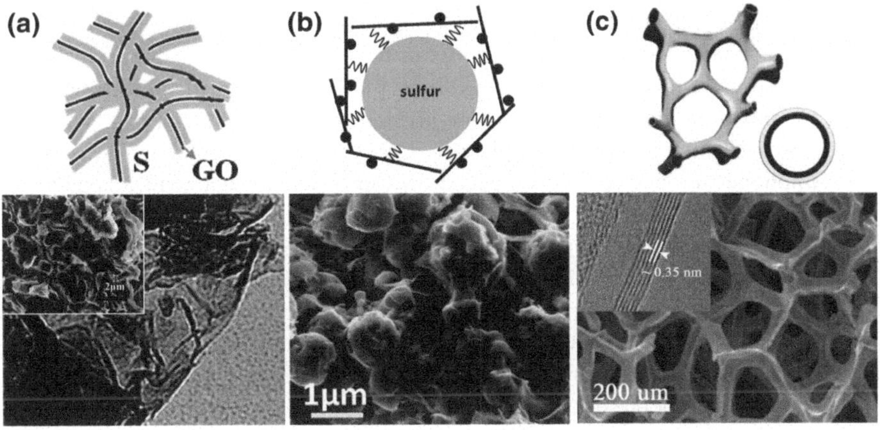

Fig. 4.7 Designs of graphene and its derivatives for hosting S cathodes: **a** Sulfur cathode immobilized on GO nanosheets. **b** Graphene–sulfur composite material by wrapping PEG-coated submicrometer sulfur particles with graphene sheets decorated by carbon black nanoparticles. **c** Sulfur infiltrated in the interconnected 3D graphene foam. **a** Reproduced with permission from Ref. [34], Copyright © 2011, American Chemical Society. **b** Reproduced with permission from Ref. [33], Copyright © 2011, American Chemical Society. **c** Reproduced with permission from Ref. [45], Copyright © 2014, Royal Society of Chemistry

Figure 4.7a shows a uniform and thin sulfur cathode coated on GO sheets [34]. GO is oxidized and exfoliated from graphite and is rich of functional groups such as epoxide, carbonyl, hydroxyl, phenol, and organosulfate. The GO-S nanocomposite materials in Fig. 4.7a was synthesized via a chemical reaction-deposition and thermal treatment process. Approximately 66 wt% S was immobilized on the GO after heat treatment. The GO-S nanocomposite cathode displays a high capacity of ~1000 mA h g^{-1}. The GO has a good conductivity and high surface area, providing good electron/ion accessibility for sulfur. The functional groups on the GO surface are crucial to immobilize S species and effectively confine any polysulfides from dissolving. The GO nanosheets also alleviate the volume change of the S cathode during the electrochemical reactions, leading to enhanced discharge capacity and cycle performance. Nevertheless, the functionalities of the GO substrate could only interact with the sulfur that is intimately on the GO surface. Dissolution of outer layer sulfur species could still occur and cause the capacity decay.

In addition to being used as the substrate for immobilizing the sulfur, graphene can be applied to coat or wrap the sulfur and suppress the dissolution of active sulfur species. In Fig. 4.7b, carbon black nanoparticle-decorated graphene is wrapped on submicrometer sulfur particles, which were first coated by polyethylene glycol (PEG) [33]. The PEG/graphene coating material accommodates the volume expansion of the S cathode during discharge, trap the soluble polysulfides, and enhances the cathode conductivity. The graphene coating remarkably improved the specific capacity and cycling stability of the S particles.

Furthermore, 3D graphene foam or sponge nanostructures have been synthesized as interconnected 3D graphene network for high-loading S cathodes. The 3D graphene framework is often synthesized by CVD graphene on the porous host followed by substrate etching. The 3D few-layered graphene foam in Fig. 4.7c was formed on a Ni metallic foam template by CVD followed by etching away of Ni [45]. Graphene of ~5 layers was stacked on top of each other with a spacing of 0.34–0.35 nm. The graphene foam shows a hierarchically porous structure, micropores, mesopores, and macropores, for the effect host of S cathode. The excellent electrical conductivity and 3D interconnected graphene structure promotes rapid electron and ion transport of the S cathode. The S cathode accommodated in the graphene foam shows high-rate discharge capacity retention for 400 cycles at a high current density of 3200 mA g^{-1} and a Coulombic efficiency of ~96.2%. Moreover, owing to the 3D graphene structure, this cathode system requires no additional binder, conductive additives, or current collector, greatly increasing the energy density of the Li–S battery.

The 3D graphene foams can provide a highly conductive network with robust mechanical support and sufficient space for a high sulfur loading. The graphene foam is expected to be used in flexible and high-energy-density Li–S batteries. Using the graphene foam host, the sulfur loading can be improved from 3.3 to 10.1 mg cm^{-2}. The S cathode with 10.1 mg cm^{-2} sulfur loading in the graphene foam delivered an extremely high areal capacity of 13.4 mA h cm^{-2}, much higher than the most Li–S batteries with low S loadings. The S-graphene foam can deliver 9.3 mA h cm^{-2} areal capacity at a high current density of 1500 mA g^{-1}, which is promising for flexible Li–S batteries with high energy density [46].

Hybrid Carbon Nanomaterials

Porous carbon nanoparticles ("0D"), CNTs and CNFs (1D), and graphene (2D) are the typical varieties of carbon. As discussed above, each material has its own advantages and disadvantages. Carbon nanomaterials combining these simple varieties, such as micro/meso-porous carbon with CNTs or graphene [51] and CNT-graphene hybrid materials [52] have been fabricated and show synergistic effect on the S cathode, with balanced performance of confinement, conductivity, S content, and so on. Three kinds of carbon nanomaterials with high graphitic degree and high surface area, that is, Ketjen carbon black ("0D"), CNTs (1D), and graphene (2D), were integrated into a high-efficient 3D conducting network [53]. The typically different dimension of nanocarbon entities can provide sufficient electrical conductivity and electrical contact, as sulfur was homogeneously distributed in the multidimensional carbon network. The mixture of the carbon nanomaterials affords a high sulfur loading of as high as 90 wt% with high specific capacity, 1115 mA h g^{-1} based on the S/C composite material.

It is worth noting that high cost, low yields and complex synthetic procedures are often required to synthesize such hybrid or hierarchical carbon nanomaterials, unfavorite for large-scale applications. Further research should be conducted to find the optimal pore size and volume, considering the tradeoff between confinement effect and sulfur loading. Attention should also be paid to exploring simple and scalable approaches to optimal nanoporous carbon host materials at a low cost.

Non-carbon Nanomaterials

As discussed above, to confine the sulfur molecules in nanoporous host materials, more effort has been devoted to designing nanoporous carbon materials. Generally, carbon is considered non-polar in nature and cannot form chemical bonding to the polar and ionic sulfide species. The carbon host materials mainly rely on physical confinement on the sulfur and lithium sulfides. Over long-term cycling, the hydrophilic lithium polysulfides are likely to diffuse out from the hydrophobic carbon pores. Some research shows that heteroatom doping into the carbon matrix, such as doping N, B, S, etc., could enhance the interfacial bonding of the carbon substrate and the sulfides to some extent [54–56]. A more effective approach for chemical absorbing polysulfides is to utilize non-carbon and polar nanomaterials as additives, interface modification materials, and the host materials. Based on this hypothesis, non-carbon nanomaterials, including metal oxides, metal sulfides, metal organic framework, and so on, which are discussed in this section.

Titania is a popular oxide material in Li-ion batteries. A mixture of mesoporous TiO_2 and mesoporous carbon was reported as a polysulfide reservoir in Li–S batteries [57]. The soluble polysulfides were trapped via chemical interactions and physical confinement of TiO_2. The mesoporous TiO_2 also provides short transport paths for electrons and Li ions. The S cathode containing TiO_2 with a pore diameter of 5 nm exhibits a discharge capacity retention that is 37% greater than a cell without the TiO_2 additive after 100 cycles. Furthermore, a yolk-shell S–TiO_2 nanostructure was designed as a cathode material [58]. The internal space effectively accommodates the volume expansion of sulfur, resulting in an intact TiO_2 shell to minimize polysulfide dissolution. The S–TiO_2 yolk-shell nanostructures delivered an initial capacity of 1030 mA h g^{-1} at 0.5 C and showed a very low capacity decay rate of 0.033% per cycle after 1000 cycles at 0.5 C.

The well-known drawback of the metal oxides as the host materials for S is their poor electronic conductivity. In order to address the insulating property of TiO_2, Magnéli-phase titanium oxide (Ti_nO_{2n-1}, e.g., Ti_4O_7) with much higher electronic conductivity has been synthesized to host sulfur, which can form strong chemical absorption with lithium polysulfides and good electrical conductivity for the cathode [59, 60]. The Magnéli-phase Ti_4O_7 is reported to have a bulk metallic conductivity as high as 2×10^3 S cm^{-1} at 298 K. The inherently polar, high surface area Ti_4O_7 cathode host mitigates polysulfide dissolution by forming an excellent interface with Li_2S, which has been demonstrated by X-ray photoelectron spectroscopy, *Operando* X-ray absorption near-edge structure, lithium polysulfide adsorption experiments, and density functional theory calculations. The Ti_4O_7–S cathode delivers high specific capacities at various C-rates (1342, 1044, and 623 mA h g^{-1} at 0.02, 0.1, and 0.5 C, respectively) and remarkable capacity retention of 99% (100 cycles at 0.1 C) [60]. The performance of the Ti_4O_7–S can be further improved by nanostructure design and surface carbon coating. For example, Ti_4O_7 nanoparticles with interconnected pore structures were synthesized porous poly(styrene-*b*-2-vinylpyridine) particles as template. The well-defined Ti_4O_7 nanospheres contains mesopores of 2–5 and 15–25 nm with a specific surface area of 592 $m^2\ g^{-1}$ providing a polysulfide reservoir and sufficient polar surface for binding polysulfides. Thus, a high capacity of 1219 mA h g^{-1} with a low capacity-decay rate of 0.094% per cycle was obtained in this S–Ti_4O_7 composite.

In addition to titania, many other metal oxide nanomaterials have been explored in S cathodes for trapping polysulfides. Representative metal oxides that have been studied are SiO_2 [61], Al_2O_3 [62], ZnO [63], Fe_2O_3 [64], MnO_2 [65], and so on. In particular, atomic layer deposition (ALD) provides an easy approach to metal oxide coating layers with nanoscale thickness, high uniformity, and good crystallinity for S cathodes. The metal oxide nanolayer can be coated on the carbon substrate by ALD functioning as a reactivation component [62], or on the surface of a S/C-composite electrode as a thin conformal Li-ion-permeable Al_2O_3 layer [66]. The Al_2O_3 coating layer could prevent the dissolution of polysulfides, while preserving electrical connectivity due to its atomic thickness.

Compared with metal oxides, metal sulfides possess even stronger affinity for lithium polysulfides, which is favorable as host materials to trap polysulfides and mitigate the shuttle effect. There are many metallic or half-metallic phases of metal chalcogenides with benign electrical conductivities, such as pyrite, spinel, and NiAs structures [67]. Various metal sulfides have been employed as the hosting or coating materials in S cathodes, including hollow TiS_2 foam or coating [68, 69], Co_3S_4 polyhedra [70], Co_9S_8 [71], pyrite CoS_2 [72], pyrite FeS_2 [73], and MoS_2 [74].

In addition to the metal oxide and metal sulfide materials, some other nanomaterials, such as carbides (MXene nanosheets) [75] and nitrides (mesoporous TiN) [76], have been employed as the host materials for S cathodes. Metal-organic frameworks (MOFs), which are a category of nanoporous materials with abundant and well-defined porosity, high surface area, and strong affinity for (poly)sulfides, have also been developed to accommodate S cathode and constrain polysulfides [77].

In general, the non-carbon host materials exhibit promising capability to absorb and trap polysulfides, superior to carbon host materials in this regard. The intrinsic drawbacks of the non-carbon materials, including oxides, sulfide and MOF, are their low electronic conductivities and high density compared with carbon. Therefore, the S cathodes encapsulated in the non-carbon nanomaterials are limited to low rate cycling and low sulfur content.

Summary

Several critical factors are listed below for the rational design of nano host materials in S cathodes:

(1) Pore size: suitable micropores or small mesopores should be used in the host materials (or at least in the shell of the host materials) for strong space confinement for S.
(2) Pore volume: large pore volume is desired to improve the S content in the composite cathode.
(3) Conductivity: a preferential content of graphitic carbon to facilitate electron transport in the insulating S/Li_2S_x.

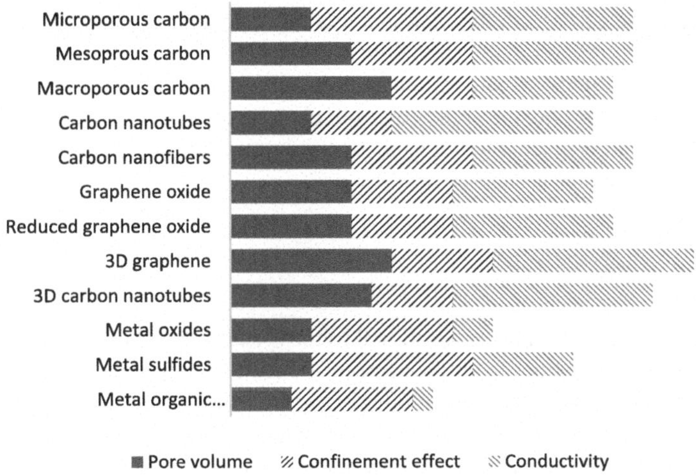

Fig. 4.8 Summary of host nanomaterials for S cathodes and a general comparison in three aspects: pore volume (available S content), conductivity, and confinement effect on the sulfur

(4) Li ion access: short ion diffusion pathways should be preserved for Li ions from the electrolyte to the sulfur.
(5) Facile fabrication and inexpensive scale-up: simple and scalable approaches to optimal nanoporous carbon host materials at a low cost are important for low-cost S cathodes.

Figure 4.8 lists the properties of various host nanomaterials for the S cathode. Each kind of nano host material has its own advantages and disadvantages in terms of pore volume, confinement effect, and conductivity. An optimal sulfur composite cathode material will benefit from both fundamental understanding and rational material synthesis. Rational design of the hosting nanomaterials, in combination of optimal pore size and volume, nanostructures, and suitable carbon and polar non-carbon host materials, is expected to synergistically improve the electrical conductivity and polysulfide affinity of the cathode, enhancing the rate capability and cycling stability. For example, hollow carbon nanofibers (1D carbon material) filled with MnO_2 nanosheets (2D non-carbon material) have been designed to host sulfur with the physical and chemical entrapment strategy [78]. The S cathode can benefit from both kinds of host materials for a high sulfur content (71%), mass loading (3.5 mg cm^{-2}), specific capacity (1161 mA h g^{-1}), and cycle life (300 cycles).

4.3.3 Nanostructures for Other Components in Li–S Batteries

The Li–S batteries resemble Li-ion batteries in battery assembly to a large extent. However, due to their high capacity, low conductivity and the conversion electrochemical reaction of the S cathode, the cell components and configurations of Li–S batteries can be improved from the conventional Li-ion batteries. The Li–S battery could benefit from material and structure innovation in all cell components in addition to the S cathode. Several attractive novel structures for the cathode/cell have explored. Here the use of nanomaterials in conductive additives, current collectors, interlayers, and separators is presented below.

4.3.3.1 Electrode

In the positive electrode side, in additional to the S cathode and its host material, key components are conductive additive, current collector, and binder. The conductive additive is usually a commercially available carbon black with high graphitic degree and excellent conductivity, such as Super P®, Ketjen Black®, and Picactif®. CNTs and graphene materials have also been studied as the additive materials in S cathodes.

In most cells, a flat Al foil is used as the current collector for the S cathodes. Advanced 3D current collectors have been fabricated using nanomaterials like CNTs or graphene, with high conductivity, high surface contact area, and other features such as binder-free and/or flexibility. In Li–S batteries, 3D current collectors could be especially interesting because the 3D nanostructure serves as a container for sulfur and the high surface area enhances the sulfur-substrate interaction. So far, carbon materials, metal foam, and porous biomaterials have been utilized as 3D current collectors [79–83]. Improved electrochemical performances with higher sulfur loading were reported in the S cathodes with the 3D current collectors.

Among possible 3D current collectors, the CNTs and graphene are outstanding for their robust mechanical strength, structural integrity, and excellent conductivity, which are required for free-standing and binder-free current collectors in Li–S batteries. One of the earliest investigated free-standing electrodes was realized by suing mesoporous graphene paper, which is very flexible with a highly porous structure and large pore distribution [84]. Graphene-CNT composite materials have since been frequently synthesized as a free-standing current collector and host for S cathode [85–87]. The free-standing current collectors are advantageous for sulfur cathodes are promising because they ensure the excellent mechanical properties of electrodes and buffer the volume change, as well as reducing the unnecessary mass of metal current collectors and binders.

4.3.3.2 Interlayer

The porous nanomaterials discussed above have been mainly used to confine lithium polysulfides in the local nanostructure around the cathode. However, even with good host materials that confines sulfur, the dissolution of polysulfides in organic electrolytes appears to be inevitable. If the lithium polysulfides do diffuse into the electrolyte, an interception layer inserted between the cathode and the separator, which is called interlayer, would be beneficial to inhibit further migration of polysulfides to the anode. Manthiram's group first proposed the use of nanoporous membranes as interlayers in Li–S batteries in 2012 [88]. The interlayer has two main functions. One is to capture and localize the migrating polysulfides, preventing the loss of sulfur species and suppress the shuttle effect; the other function is to serve as a "top current collector" that offers additional electron pathways on the top surface of the cathode. Carbon is kind of suitable materials for interlayers because they possess high electrical conductivity, low density, and tunable porosity, which suit well for the bifunctional interlayers. Manthiram and co-workers reported a series of interlayers made of various carbon nanomaterials for Li–S batteries, such as microporous carbon paper [88, 89], mesoporous carbon paper, MWCNT paper [90], interwoven CNFs [91], and porous biomaterials [83, 92]. The carbon interlayers successfully enhance the cycling stability of the Li–S batteries and are applicable to S cathodes with high S content and high areal mass loading. To design an interlayer nanomaterial, the porosity, electrical conductivity, surface chemistry and thickness of the interlayer material are considered as the critical parameters of the interlayer. Moreover, as the addition of the interlayer increase the mass and volume of inactive materials in the battery, the interlayer could lower the energy density. Hence, special attention should be paid to the mass and thickness of the interlayers and the trade-off between the interception effect and energy density.

A further development of the Li–S battery configuration is a sandwiched structure that constrains the S cathode between two layers of carbon, which ensures the electrical contact and polysulfide capture. For example, a graphene-sulfur-graphene sandwiched structure has been reported with pure S filled in between a graphene current collector and a graphene interlayer [93]. The Li–S cell with sandwiched configuration outperforms the cell using conventional foil current collectors and the cell without the graphene interlayer. Since the extra carbon layers in the sandwich configuration reduce the S content in the electrode, high-content S cathode or even pure S cathode has to be employed in the sandwich cathode to guarantee a high energy density.

4.3.3.3 Separator

The modification on the separators in Li–S batteries is usually for the same purpose as the introduction of interlayers, that is, to block the shuttle of polysulfides. Nanoporous materials with proper pore sizes are cast on the separator as a sieve to separate polysulfide ions from solvents and Li ions. Different from interlayers, the separators are not electron conductive. Hence, in addition to carbon nanomaterials, many other insulting nanomaterials are utilized to cast on the separator as an ionic sieve. For example, MOF-based materials have a large surface area and highly ordered pores with tunable porosity but is naturally insulating. MOF is not an excellent host material for S cathode as discussed above, nor suitable interlayer materials, but can be used to modify the separator. A MOF@GO-modified separator acts as an ionic sieve in Li–S batteries, which selectively sieves Li ions while efficiently suppressing undesired polysulfides migrating to the anode side [94]. The MOF has a pore size of ~9 Å, which is smaller than the polysulfides (Li_2S_n, 4 < n ≤ 8) and can preclude the polysulfides from migrating to the anode. The porous framework on the separator contributes to a stable cycling performance of the Li–S battery.

4.3.4 Summary

The contribution of the nanomaterials for high-performance Li–S batteries is indisputable. The nanomaterials are now well acknowledged indispensable to enhance the activity of the S cathodes, trap Li polysulfides, and alleviate the volume changes of S cathodes. With the development of well-designed nanostructured S cathode, the problems of the nanomaterials in S cathodes surface. The utilization of the host materials (primarily carbon host materials) significantly lowers the energy density of the Li–S batteries. If a Li–S battery is ever to outperform a commercial Li-ion battery with an areal capacity of ~4 mA h cm^{-2} and a discharge voltage of ~3.6 V, the Li–S battery must possess an areal capacity of over 6 mA h cm^{-2}, which require a sulfur loading over 5 mg cm^{-2} and a sulfur content of over 70 wt% [95]. The trade-off between the nanomaterials and energy density in S cathodes in turn upgrades the criteria for nanomaterials in Li–S batteries. Recently, more efforts have been made to design of nanomaterials for high-loading and high-content S cathodes toward high-energy-density Li–S batteries.

4.4 Nanostructures in Li Metal Anodes

Li metal batteries use Li metal anodes because the Li metal anode has a theoretically highest capacity (3860 mA h g^{-1}) and lowest reduction potential (−3.04 V vs. standard hydrogen electrode). These features of Li metal endow Li metal battery with a high discharging voltage and thus a high energy density. The Li metal anode is indispensable in high-energy Li metal batteries. However, the development of the Li metal anode has long been hindered by several issues.

(1) The intrinsic formation and growth of Li dendrites during Li plating. The dendritic Li could penetrate the separator and cause a short circuit of the cell. The short circuit leads to cell thermal runaway, even electrolyte combustion and cell explosion, imposing the Li metal anode on wide-ranging safety concerns. This is the most mortal obstruction for practical applications of Li metal anode.
(2) The interfacial instability of Li metal in the electrolytes. The intrinsic high reactivity of Li metal results in corrosive reactions occurring on the Li metal surface in contact electrolytes. The side reactions on the Li metal surface will irreversibly consume the Li metal and the electrolyte, increase the interfacial resistance, and reduce the Coulombic efficiency. The active surface of Li can be passivated by

a solid electrolyte interphase (SEI) film, which is formed spontaneously but can be damage because of the Li dendrites.

(3) The dendritic Li could also lead to an electrical detachment of Li after repeated plating/stripping and become "dead Li". The inactive dead Li further decreases the Coulombic efficiency and shortens the cycle life of the Li metal battery.

(4) Huge volume change of the electrode during Li plating/stripping. The volume change issue of Li metal anode is even more severe than that of the silicon anode because Li plating can lead to virtually infinite volume expansion. The volume effect further give rise to the interfacial instability of the Li metal anodes.

The Li metal anode has been proposed for long, but the research on Li metal anodes has not been as prosperous until recently owing to the boost of nanotechnology. Nanotechnology is extensively used to resolve the intrinsic problems of Li metal anodes: dendrite growth, interface reactivity, and volume change. Below we elaborate the effect of the nanomaterials in Li metal anodes from two perspectives: the materials hosting the Li metal and the coating layers passivating the surface.

4.4.1 Nano/Micro Host Materials for Li Metal Anodes

Although the Li foil can be directly used as an anode in the Li metal battery, the corrosion by the electrolyte and formation of dendrites on the Li foil surface could finally damage the Li anode and fail the cell. Nano/micro-structured host materials, which has great importance in the S cathode, air cathode, and Li-ion battery, gradually gain new interest in the Li metal anode. The basis of nanostructured Li metal anode is that the nano/micro-structured host materials provide increased effective electroactive area for Li anode, which lowers the local current density during Li plating. Another significant advantage of the nano/micro host structure for is the spacing confinement for Li anode, which can to some extent buffer the huge volume change of Li anode during plating (expansion) and stripping (shrinkage). Therefore, the rapid developing nanotechnology has reinvigorated the Li metal anode. Because the Li metal has a different nucleation and growth nature from S cathodes and other electrode materials, the pore size of the host structure is significantly larger, not limited to nanopores. Various nano/micro structured materials have been proposed as the hosts for Li metal anode.

4.4.1.1 Copper Host

Cu has long been used as the current collector for anodes in Li-ion batteries. Thus, porous Cu, in lieu of Cu foil, has been explored as a 3D current collector to accommodate the Li metal anode [96]. As schematically shown in Fig. 4.9a, nucleation of Li metal on a planar substrate (Cu foil or Li foil) forms a tip that induces more Li deposition on it and causes formation of Li dendrites. Compared with the planar

(2D) current collector, the 3D Cu host possesses a large electroactive area to reduce the local current density, preserved porous volume to accommodate Li anode and alleviate the volume change. The 3D Cu current collector has been first reported by Guo's group, with a submicron fibrous Cu skeleton, to homogenize the deposition of Li metal anode (Fig. 4.9b). The 3D current collector has a median pore diameter of 2.1 μm and an areal pore volume (pores < 5 μm) of 1.5×10^{-3} cm^3 cm^{-2} and a high areal pore area. Experiment shows that the Li deposition on the 3D Cu forms a flat surface without observable dendrites after depositing 2 mA h cm^{-2} of Li meal at 0.5 mA cm^{-2} (Fig. 4.9b), in contrast to the dendrite formation of Li on planar Cu foils. The pore structure plays a critical role in constructing effective 3D current collector. Owing to the submicron-structure of the 3D Cu host, the possibility of dendrite formation is mitigated, resulting in extended short-circuit time, which is tested in by unidirectionally plating Li until short circuit (Fig. 4.9d). The Li metal anodes with the submicron-structured Cu current collector apparently exhibits longer cycles life. Also, due the augmented surface area in the 3D current collector, the 3D structured Li anode shows smaller voltage hysteresis and improved Coulombic efficiency than the planar anode.

As a stable, inert, and mechanically robust substrate for Li metal anode, 3D Cu substrates with nano/micro structures have since received numerous attentions. Various 3D Cu substrates with different morphologies were fabricated via different approaches, among which four typical 3D Cu hosts are presented in Fig. 4.10. Figure 4.10a shows a 3D porous Cu current collector fabricated by dealloying a Cu–Zn alloy and Li metal anode deposited in the porous Cu current collector [97]. The 3D Cu has an interlinked porous framework to accommodate Li, suppress the Li dendrite growth and alleviate the volume changes. By adjusting the dealloying time, the porous Cu with different pore sizes has been studied. It is found that suitable pore size is critical to achieve desirable a 3D current collector for Li metal anode. Figure 4.10b shows Li deposition in a free-standing Cu nanowire network forming a Li–Cu composite without forming significant Li dendrites [98]. The Cu nanowires have an open porous structure and facilitate the large-capacity Li deposition into the entire network. With the Cu nanowire current collector, a Li anode of 7.5 mA h cm^{-2} was demonstrated plating in the host without the growth of dendritic Li. Only until 10 mA h cm^{-2} of Li was plated, part of Li was deposited on top of the 3D host forming Li dendrites, indicating the accommodation capacity limit of the Cu nanowire current collector.

More complex and rational 3D Cu structures have been developed as the current collectors of Li metal anodes. As shown in Fig. 4.10c, through a colloidal crystal template technique, a 3D highly porous Cu framework was grown on commercial Cu foam, forming a hierarchical porous Cu architecture [99]. The hierarchical structure could provide with proper porous structures for efficient Li constrain. The Li anode with the hierarchically porous Cu substrate shows an improved Coulombic efficiency, 98% over 200 cycles at 1 mA cm^{-2}. Further studies on the porous Cu substrate implies that vertically aligned microchannels, compared with the random pores (random pore radii, depth, and distributions), could better regulate the current density distributions and thereby suppressing the growth of Li dendrites. Vertically aligned microchannels

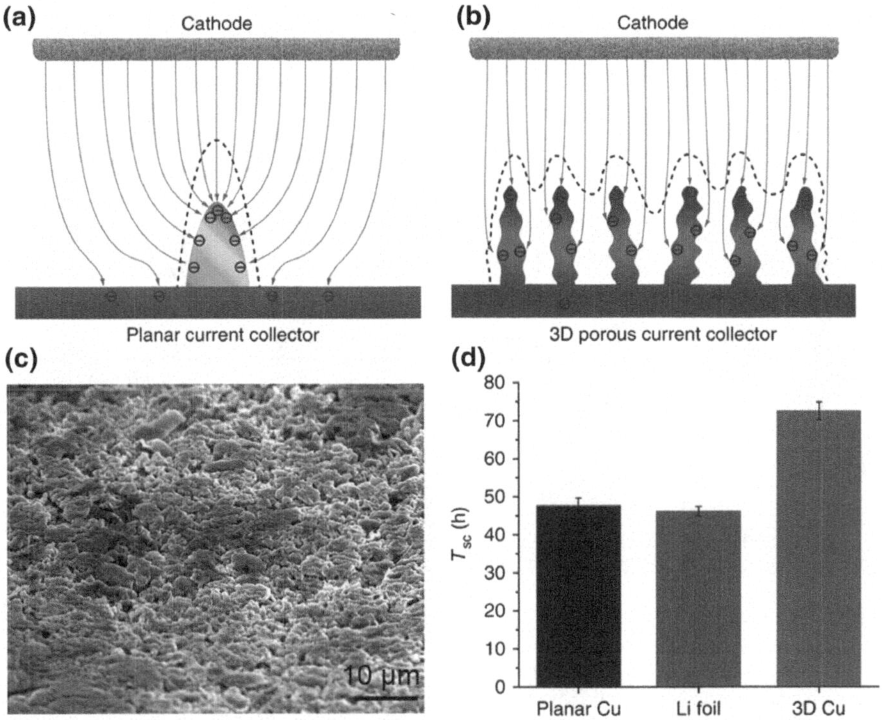

Fig. 4.9 **a** Scheme of electrochemical Li deposition behaviour on (**a**) planar current collector forming Li dendrites and **b** 3D current collector forming uniform Li deposition. **c** Side view scanning electron microscopy (SEM) image of Li deposited on a 3D Cu foil with a submicron skeleton. **d** Average short-circuit time T_{sc} by unidirectional galvanostatic Li plating onto planar Cu foils, Li foils and 3D Cu foils at 0.5 mA cm^{-2}. Reproduced with permission from Ref. [96], Copyright © 2009, Nature Publishing Group

were fabricated in Cu current collectors and shown in Fig. 4.10d [100]. Vertically aligned pores were fabricated with pore radii of 5, 7.5, 10, and 15 μm with different spacings and pore depths of 25, 30, and 50 μm, to reveal the effect of the porous parameters on the Li anode properties. The porous Cu with 5 μm pore radius, 12 μm pore spacing, and 50 μm pore depth exhibit the most benefit to the depositing behavior and cycling performance of the anode. The aligned pores decrease the local current density, homogenize the current distribution, and constrain the Li anode. In addition to the porous Cu substrates discussed above, various nano/micro structured Cu with different morphologies and pore structures have been developed as the 3D hosts to improve the Li metal anodes. The improvement of Li anode by the ingenious porous Cu current collectors attest to the importance of nano/micro structure for the Li metal anode.

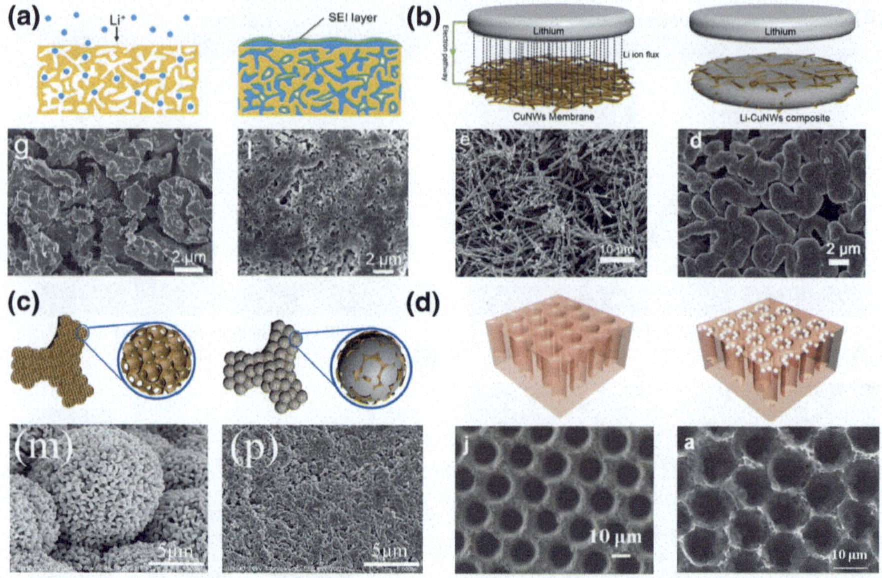

Fig. 4.10 Li metal anode deposition in 3D Cu substrates with nano/micro structures: **a** 3D porous Cu current collector derived by dealloying Cu–Zn alloy, **b** free-standing Cu nanowire network current collector, **c** hierarchically bicontinuous porous skeleton with highly porous Cu architecture grown on a Cu foam substrate, **d** porous Cu with vertically aligned microchannels. In each panel, the left column shows the scheme and SEM image of the pristine 3D Cu substrate and the right column shows corresponding scheme and SEM image of the 3D Cu substrate after Li deposition. **a** Reproduced with permission from Ref. [97], Copyright © 2015, Wiley. **b** Reproduced with permission from Ref. [98], Copyright © 2016, American Chemical Society. **c** Reproduced with permission from Ref. [99], Copyright © 2018, American Chemical Society. **d** Reproduced with permission from Ref. [100], Copyright © 2017, Wiley

4.4.1.2 Carbon Host

Although Cu has been widely applied as the current collectors for Li-ion battery anodes, Cu is high-density and expensive, adversely affecting its use as the host material for Li metal anode. Therefore, porous carbon materials, which are light, inexpensive, conductive, and stable, have been extensively studied as the host materials for Li metal anode. As have been proved in cathodes, carbon materials can be made with numerous structures, various porosities, and different chemical properties. The superb properties of nano/micro porous carbons make them perfectly hosting Li metal anodes. So far, researchers have explored various porous carbon materials to host the Li metal anode. In Fig. 4.11, four typical carbon frameworks are shown with schematics and SEM images as examples of Li anode hosts.

Graphite is a well-known material in battery community but is often considered not suitable for hosting Li metal. In Fig. 4.11a, small 2D graphite flakes pack into a secondary particle with a 3D structure, massive artificial graphite (MAG) [101]. The

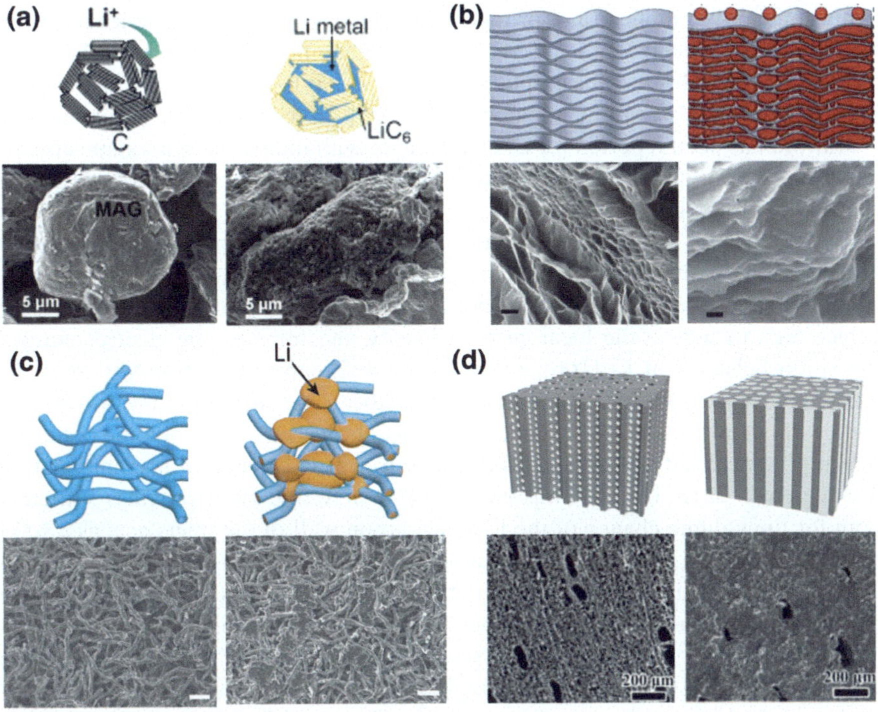

Fig. 4.11 Li metal anode deposition carbon substrates with nano/micro structures: **a** Massive artificial graphite (MAG), which consists of 3D packing of small 2D graphite flakes into a secondary particle, **b** porous graphene network, **c** flexible and free-standing hollow carbon fibers, **d** wood-derived carbon modified by MgO nanoparticles. In each panel, the left column shows the scheme and SEM image of the pristine carbon substrate and the right column shows corresponding scheme and SEM image of the carbon substrate after Li deposition. **a** Reproduced with permission from Ref. [101], Copyright © 2016, Cell Press. **b** Reproduced with permission from Ref. [102], Copyright © 2014, Nature Publishing Group. **c** Reproduced with permission from Ref. [106], Copyright © 2017, Cell Press. **d** Reproduced with permission from Ref. [108], Copyright © 2018, Elsevier

interior structure of MAG has many edge planes and some empty space. Li-metal deposition would preferably take place inside individual MAG particles, reducing the chance of Li-metal dendrite formation as well as the side chemical reaction with electrolyte. With a controlled capacity of 744 mA h g^{-1}, the 3D graphite electrode exhibited excellent reversibility and high Coulombic efficiency of ~98.4% with stable cycling for 50 cycles. Compared with graphite, graphene network is more widely used because it has more surface area and porosity. Figure 4.11b shows Li plating with the interior of a porous graphene network induced by defects in the graphene [102]. The free-standing graphene network was obtained by a thermal shock to GO paper with defect sites in the graphene lattice. The plated Li metal is trapped with the sub-surface of the graphene network, free from dendrite problems. The graphene network with defects acts as a caged entrapment for lithium metal and prevents dendrite growth. The highly conductive graphene also provides a conductive network for efficient electron transfer in the Li metal anode. Due to the outstanding properties (high conductivity, mechanical strength, etc.), the graphene framework has been intensively investigated as the host material for Li metal anode [103–105]. In particular, 3D graphene on a Ni scaffold has been reported for Li metal anode with interfacial synergistic effects, that is, the 3D porous stricture provides high surface area while the graphene acts as an artificial protection layer [105].

Compared with graphite or graphene, interconnected carbon nanofibers could provide a large volume for high-capacity Li anode. Figure 4.11c presents a light-weight, flexible, and free-sanding 3D hollow carbon fiber network, which has a large surface area to reduce the local current density and improve the electrochemical deposition behavior of Li [106]. Even for an areal capacity of 6 mA h cm^{-2}, Li is confined within the interspace among the fibers and inside the hollow tubular fibers without uncontrollable Li dendrites. Since the effective pore volume of the 3D carbon fiber scaffold is ~1.83×10^{-2} cm^3 cm^{-2}, it is capable of accommodating ~9.8 mg cm^{-2} of Li, i.e., 37.7 mA h cm^{-2}. The large free space provides sufficient room for the volume change of the Li. The carbon wall of the fibers acts as a solid barrier to confine Li inside the hollow space, thus inhibiting the formation of Li dendrites. Carbon fiber material, without the inner hollow structure, has also been explored as the host of Li metal anode [107]. The carbon fiber substrate has an average pore size of ~72 μm, pore volume of 0.04 cm^3 cm^{-2}, and large surface area available for Li deposition. The high pore volume facilitates high-capacity Li anode and the large surface area lowers the local current density toward uniform Li deposition. A high areal capacity of 8 mA h cm^{-2} has been realized with a good Coulombic efficiency of (98%) after 50 cycles in the carbon fiber hosted Li metal anode. Another type of 1D carbon nanomaterial, carbon nanotube, is not so suitable as the host material for Li metal anode, due to the relatively small pore volume in CNTs in comparison with CNFs.

To realize high-capacity Li metal anodes, large pore volume in the carbon host materials are desirable. Various innovative carbon nanomaterials have been synthesized for hosting Li metal anode and suppressing dendrite growth. As shown in Fig. 4.11d, balsa wood-derived porous carbon matrix was used as Li host [108]. The natural channels of the wood-derived carbon homogenize the Li^+ flux distribution to some extent and the pores suppress the serious volumetric change of the Li metal anode during cell cycling. In addition, MgO nanoparticles were introduced into the wood-derived porous carbon to favor Li nucleation and accommodation with the wood channels.

4.4.1.3 Other Host Materials

Inorganic interlayers with nano-/micro-pores have also been designed developed as protecting hosts for Li metal anodes. A 3D glass fiber cloth was reported as a modification interlayer host between the electrode and the separator to stabilize the Li metal anode [109]. The non-conductive glass fiber framework allows Li growth only from the electron conductor (former Li metal). The glass fiber matrix assists in homogenizing Li-ion flux. Li metal is deposited uniformly underneath glass fiber layer and grows within the matrix without forming any dendrites. The dendrite-free Li deposition stabilizes the SEI film of the Li metal anode and improves the Coulombic efficiency of the anode from 80 to 97% at a current density of 1.0 mA cm^{-2}. These nanoporous layers serve as surface protection layers for uniform Li-ion flux as well as Li metal hosts to inhibit the growth of Li dendrites. It is believed that ceramic membranes (such as the anodic aluminum oxide membrane) with pores smaller than the diameter of mossy Li whiskers can lead to a transition of the Li formation mechanism and block the mossy Li [110].

More complicated nanomaterials, which combine carbon, metal, and/or oxide nanomaterials, have also been designed for host Li metal anodes. For example, a composite microcage with a carbon nanotube core and a porous silica sheath was designed to accommodate Li metal [111]. The conductive CNT core provides electron access and the insulating SiO_2 porous sheath transports Li-ion flux. Thus, the Li metal is trapped in the hollow SiO_2 microspheres.

In most Li host structures, the host materials can be classified into non-conducting for both electrons and Li ions (polymer, glass fiber, etc.), electron-conducting (Cu, carbon materials), and ion-conducting only matrix. It is speculated that mixed ion and electron conducting network would better promote Li ion and electron transport in the host and lower interface resistance. Based on this theory, Li-ion conductor $Li_{6.4}La_3Zr_2Al_{0.2}O_{12}$ nanoparticles were incorporated into 3D carbon nanofiber scaffold as a mixed ion/electron conducting network for Li metal anode. In such a mixed conducting scaffold, the Li metal anodes can deliver an areal capacity of 16 mA h cm^{-2}, volumetric capacity of 1600 mA h cm^{-3}, and remain stable for more than 1000 h cycling at 5 mA cm^{-2}.

Similar to the composite nanomaterials used in other electrode systems, the sophisticated nanostructures could combine advantages of the components but also increase the risk and cost of synthesis.

4.4.1.4 Li Nucleation Nanoseeds

To date the most frequently studied host materials for Li metal anode are carbon and copper, as discussed above. However, the Li metal has a poor affinity to Cu and C. Significant nucleation overpotentials have been observed when Li nucleates on Cu or C substrates, indicating that Cu and C are unfavorable substrates for Li nucleation. Cui and co-workers discovered that certain materials (Pt, Au, Ag, Zn, Mg, Al, etc.) with some solubility in Li at room temperature leads to a zero nucleation overpotential, while other materials (c, Cu, Ni, Sn, Si, etc.) has a nucleation barrier [112]. The materials without overpotential can serve as heterogeneous nucleation seeds for spatial control of Li metal deposition. As illustrated in Fig. 4.12a, because of the Au nanoparticles as nucleation seeds in the hollow carbon shells, Li is selectively nucleated and grown in the carbon spheres, thus encapsulating the Li anode in the host material. For hollow carbon spheres without the nanoseeds, the deposition of

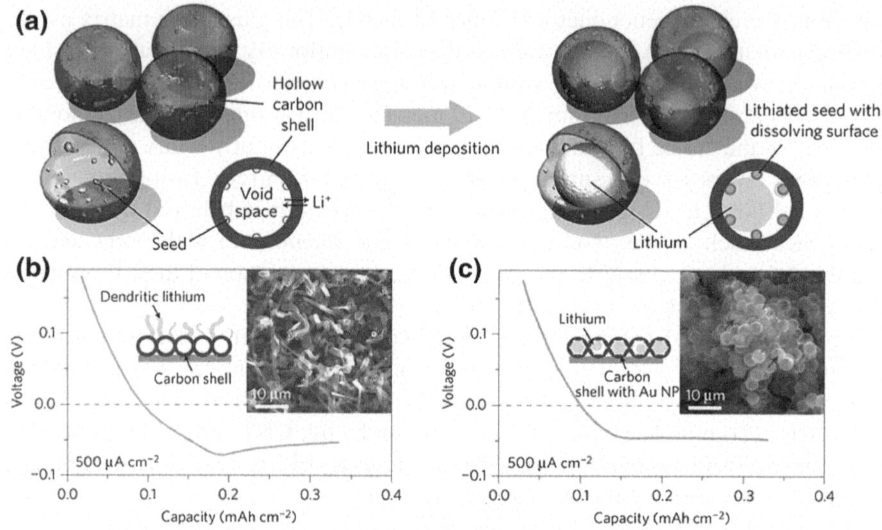

Fig. 4.12 **a** Schematic of Li metal nanocapsules design. Au nanoparticles are loaded inside hollow carbon spheres, where a large void space is reserved for Li metal. Li is expected to nucleate from the Au seed and regulated in the hollow carbon shells. **b** Voltage profile during Li deposition on hollow carbon shells without (**b**) and with **c** Au nanoparticles at 0.5 mA cm^{-2}. The inset shows the SEM image after deposition as well as the corresponding schematic. Fully filled and partially filled carbon spheres are marked by blue and red arrows, respectively. Reproduced with permission from Ref. [112], Copyright © 2009, Nature Publishing Group

Li exhibits a nucleation overpotential of 20 mV at 0.5 mA cm^{-2}, (Fig. 4.12b). The Li is deposited outside the hollow carbon sphere, exhibiting a dendritic morphology (inset SEM image in Fig. 4.12b). In contrast, the with the Au nanoseeds inside the carbon spheres, a much flatter potential profile without a nucleation overpotential is observed during Li deposition at 0.5 mA cm^{-2} (Fig. 4.12c). The morphology of deposited Li with the nanoseeds is a smooth sphere surface free of Li metal dendrites (Fig. 4.12c, inset), suggesting that Li metal deposition is guided by the Au nanoseeds inside the hollow spherical nanocapsules.

The strategy to direct Li deposition via seeds has inspired many studies on fabricating nanoseeds on the Li host materials for directed deposition and regulated growth of Li metal anode. Similar to Au nanoparticles, ultrafine Ag nanoparticles has been synthesized on carbon nanofibers as the nanoseeds to direct the deposition of Li. The Ag nanoparticles with a particles size of ~40 nm anchored on carbon nanofibers effectively reduce the nucleation overpotential of Li and guide the Li deposition in the 3D carbon matrix uniformly, free from Li dendrites. A stable and reversible Li metal anode is achieved in virtue of the Ag nanoseeds in the 3D substrate, showing a low overpotential (~0.025 V) for a long cycle life.

Besides the metal nanoparticles, ZnO is utilized as the lithiophilic functional seed for Li deposition. ZnO has been synthesized as nanoparticles decorated on carbon fibers [113], ZnO quantum dots on hierarchical porous carbon [114], and a thin coating layer by ALD [115]. During the initial Li deposition, Li ions react with ZnO via $Li^+ + ZnO \rightarrow Li_2O + Zn$, and subsequently alloy with zinc to form LiZn, which is highly conductive and lithiophilic. Therefore, the ZnO serves as nucleation sites, guiding the uniform deposition of lithium within the host structure. Other oxide nanoparticles used as the lithiophilic seeds include MgO, SnO_2, and so on [108, 116]. Similarly, N-doping sites in graphene show a strong interaction with Li atoms. Li prefers to be adsorbed and nucleated on the lithiophilic N-doping sites [117]. The concept to use nano size nucleation sites as the seeds to direct Li deposition into the 3D hosts outstands the merits of the nanostructure design.

4.4.1.5 Li Preloaded Host

The efforts discussed above have improved the stability and safety of the Li metal anodes significantly by adopting the 3D nanostructures. However, with these Li free hosts, the Li source in the anode depends completely on Li from the cathode side. The Li-free substrates could risk of Li depletion after cycling when paired with cathode materials such as $LiCoO_2$ and $LiFePO_4$ because of poor Coulombic efficiency. What is worse, the 3D substrates without pre-loaded Li cannot couple with the high-capacity S and O_2 cathodes, which contain no Li in the cathodes. Thus, pre-loaded Li metal anodes have been proposed by thermally infusing Li into the 3D nano/micro-structured host materials [115, 118–120], which are feasible for use in batteries, particularly those without Li in cathode materials (S and O_2).

To infiltrate molten Li into the lithiophobic porous substrates (porous carbon), a lithiophilic nano-coating layer is usually necessary for effective Li infusion. ZnO and Si are applicable nano-coating materials. Coating ZnO and Si has been well-established by nano-processing methods, such as ALD and CVD. For example, after applying a lithiophilic coating of Si on carbon nanofibers, molten Li metal can be easily coated on the substrate surface or encapsulated inside the 3D structure [118]. The electrode reduces effective current density and effective accommodates of Li volume variation, exhibiting a stable cycling performance with smaller overpotential (<90 mV) at 3 mA cm^{-2}.

Aside from lithiophilic coating, Li can be infused into the nanoscale gaps of layered reduced graphene oxide (rGO) due to the carbonyl and alkoxy groups on the surface of rGO [120]. The formed Li/rGO anode can deliver a high capacity up to 3390 mA h g^{-1} with an excellent flexibility and small dimensional change (~20%) during cycling.

The molten Li can also be infused into surface-modified polymer substrate. The ALD technique enables nanometer-precise ZnO coating layer on the polyimide substrate and greatly reinforce the surface lithiophilicity for Li infiltration. Because of the insulating feature of polyimide membrane, the infused Li metal itself, instead of the matrix, serves as the electrically conductive media to transport electrons. Thus, the exposed insulating surface after Li stripping prevents the Li depositing on the top of the matrix directly, achieving a surface dendrite-free electrode [119].

Superior to the above randomly porous structures, well-aligned channels with low tortuosity can guide Li-ion flux move along the vertical direction of the electrode other than the parallel or wound direction, thereby mitigating uncontrolled Li deposition [121]. Recently, a low-tortuosity and channel-guided Li metal anode was designed by adopting carbonized wood as a matrix with pre-loaded Li metal. The molten Li can be easily infused into and encapsulated within the void space of the ZnO-coated carbonized wood channels, forming a Li/carbonized wood electrode. The Li anode accommodated in the carbonized wood channels displays an exceptional cycle stability with a low overpotential (90 mV at 3 mA cm^{-2}) over 225 cycles, indicating that the well-aligned channels can serve as an excellent guide to regulate the Li stripping/plating process and provide good confinement to constrain Li volume change [115].

4.4.1.6 Summary

The pore volume, surface area, and pore size are three key parameters to design a rational 3D hosts for Li metal anodes in terms of nano/micro architecture construction. The pore volume of the 3D host should be large as it determines the capacity (amount) of the Li metal anode, over which the 3D substrate loses its effectiveness. The surface area is beneficial to reduce local current density but could adversely escalate the SEI formation. The pore size significantly affects the deposition behavior and reversibility of the Li metal anode. Too large pores are not effective in space confinement and reducing local current density while too small pores are unable to accommodate the submicron Li grains and possible to escalate SEI formation. So far, the effect of the porosity has not been sufficiently investigated, though various porous substrates have been reported as the host materials for Li anode. The pore size of the substrates in literature varies from nanometers (such as graphene) to hundreds of micrometers (such as metal foam and carbonized wood), but should be rationally selected and designed. Some studies suggest that the suitable pore size for Li anode host could be 0.1–10 μm. However, further studies are yet to be done to optimize the nano/micro structures for the Li metal anode hosts. Other features, such as homogeneous Li-ion flux, capability to resist Li dendrite growth, etc., are perceived as the advantages of nano/micro hosted Li anodes.

However, the nano/micro Li hosts have their own disadvantages. Wide concern is that the increased interface area could instead promote interfacial side reactions and boost SEI formation. In addition, the energy density of the nano/micro hosted Li anode can be encumbered by the extra weight and volume of the host materials.

4.4.2 Protective Nano Surface Layers for Li Metal Anodes

Same as the anodes in Li-ion batteries, a SEI film is generated spontaneously on the Li metal anode surface in the electrolyte. The SEI film is electronically insulative but ionically conductive, thus preventing further electrolyte corrosion to the Li metal. To obtain stable Li metal anode, it is critical to form a stable SEI film on the Li metal in the electrolyte. However, the SEI film on Li metal has a complicated composition and is usually susceptible to fracture. The growth of Li dendrites and huge volume change of Li metal anode would easily crack the SEI film; Li dissolution forms a pitted morphology, which also exposes fresh Li to the organic electrolyte [122, 123]. The SEI film repairs instantaneously at the dendrites or pits. The ever-growing of SEI film improve the interface resistance and consumes the electrolyte. Hence an ideal SEI film on Li metal anode should be thin yet hermetic, chemical and electrochemical stable, highly Li-ion conductive, and mechanically robust [124]. Since the spontaneous SEI film is not stable enough, several surface pretreatments have been applied to form protective nano layers (sometimes called "artificial SEI film") on the Li metal anode surface.

4.4.2.1 Electrochemical Reinforcement of SEI Film

As the in situ formed SEI film on the Li metal anode in the conventional electrolyte has certain disadvantages, it is reasonable to modify the electrolyte to form enhanced SEI film during the electrochemical cycling, especially during the initial cycles. By adjusting the Li salt or solvent of the electrolyte [10, 125–132], changing the concentration of the salt in the electrolyte [133–135], and/or adding functional additives [136–141], the SEI film on the Li anode can be modified with more stable composition, reinforced mechanical strength, easy Li-ion transfer, and other possible merits. The reinforced SEI film during electrochemical cycling in the modified electrolyte is associated with delicate electrochemical and chemical reactions.

Here, a commonly used additive, $LiNO_3$, for protection on the Li anodes in Li–S batteries is taken as an example to show the mechanism of the electrochemical reinforcement of the SEI film. The $LiNO_3$ additive can passivate the Li metal anode to inhibit the reaction between Li metal and polysulfides and thus eliminate the "shuttle effect" (dissolved lithium polysulfides shuttle and redox repeatedly between the S cathode and Li metal anode) [142]. Recently, extensive studies have been reported using $LiNO_3$ for Li anode passivation [136–138]. Because of the strong oxidation of $LiNO_3$, the additive creates a surface film composed of LiN_xO_y in addition to the native film (Li_2CO_3, LiOH, Li_2O, etc.). The LiN_xO_y film keeps growing during cycling and can cause increase of the resistance. When the $LiNO_3$ additive cooperates with Li_2S_n ($n = 4$–8) additives, or dissolved Li_2S_n in Li–S batteries, a synergetic effect is found as reduced products of $LiNO_3$ and Li_2S_n co-precipitate and further convert to a smooth and stable SEI film on the Li surface. By tuning the concentrations of the $LiNO_3$ and Li_2S_n, the Li metal anode can benefit most from the synergetic passivation effect of the additives. Using an ether-based electrolyte, 1 M LiTFSI in DOL/DME, with the addition of 0.18 M Li_2S_8 and 5 wt% $LiNO_3$, the Li metal anode is free from dendrite problems with a capacity up to 6 mA h cm^{-3} at a current density of 2 mA cm^{-2}. The Li metal anode also exhibits excellent cycle stability with a high Coulombic efficiency (>99%) for 300 cycles at 2 mA cm^{-2} [137]. These appealing results demonstrate the beneficial synergetic effect of the $LiNO_3$ and Li_2S_n additives for Li metal batteries. A ternary-salt (LiTFSI–$LiNO_3$–Li_2S_5) electrolyte has been proposed based on the synergetic effect for dendrite-free Li metal anode [143].

In addition to the in situ SEI enhancement during electrochemical cycling, protective SEI films can be ex situ created by electrochemical pretreatment [144, 145]. For example, the Li metal anode can form an artificial protective film by electrochemically pretreating in a symmetric cell with a TEGDME-FEC electrolyte before reassembled in Li-air batteries [144].

The electrochemical reinforcement of the SEI film is an upgrade to the in situ formed native SEI film. The electrochemically reinforced SEI film with modified electrolytes contributes to significant improvements of the performance, but generally does not affect the primary properties of the SEI film, such as thickness and mechanical properties.

4.4.2.2 Chemical Passivation Layer

Because of the extremely high activity of the Li metal, Li anode reacts with the organic electrolyte instantly when the Li is exposed to the organic electrolyte, forming the SEI film. Due to the severe volume change while cycling and growth of Li dendrites during Li plating, the in situ formed SEI film could easily break and expose the fresh Li to the organic electrolyte, generating new SEI film (SEI repair). The break and repair of the SEI film deteriorate the cycling performance and increase the cell impedance. Therefore, a chemically and electrochemically stable, mechanically robust and flexible protective layer (or "artificial SEI film") is desired to provide additional enhancement for the Li metal anode. Chemical pretreatment is a facile way to create a nanoscale passivation layer on the Li metal as an artificial SEI film with similar properties yet stronger strength and better dendrite-suppressing capability.

Solution processing is a convenient yet effective method to modify the surface chemistry and fabricate a nanoscale thin film on the Li metal. To enhance the stability of the SEI film, an artificial Li_3PO_4 layer was fabricated on the Li metal by pretreating Li metal in polyphosphoric acid (PPA) [146]. Through controlling the pretreating conditions, a uniform Li_3PO_4 SEI layer with an effective thickness of ~50 nm was generated on the Li metal replacing the pristine native film (mainly composed of Li_2CO_3, LiOH, Li_2O, and etc.). The Li_3PO_4 layer is chemically stable in the organic electrolyte and possesses a high Young's modulus (10–11 GPa), significantly strengthening of the SEI film and suppressing the growth of dendritic Li. A model study indicates that the electrolyte/separator with a shear modulus greater than 6 GPa is considered sufficient to inhibit Li dendrite growth [147, 148]. The Li metal anode with the Li_3PO_4 SEI film shows extended cycling life. Similar chemical pretreating approaches were conducted in RTIL [149], tetraethoxysilane [150], and so on to form durable artificial SEI films.

Reaction between Li metal and certain atmosphere can also create a passivation layer on the Li surface. A layer of Li-ion conductor Li_3N was formed on Li metal as a protective film by in situ reaction between Li metal and N_2 gas [151, 152]. The tight and dense Li_3N film improves the Li cycling efficiency, homogenous Li deposition, and the electrochemical performance of the Li–S battery. A more complicated chemical pretreating approach has been reported. Li foil was exposed to tetrahydrofuran, O_2 atmosphere, and trimethylsilyl chloride in sequence to produce a thin and dense protective film [153].

Superior to the simple pretreatment in solutions (or gas atmosphere), an advanced surface modification method, atomic layer deposition (ALD), has been utilized for Li surface passivation. ALD uses alternating and self-limiting chemical reactions between precursor vapors and the substrate surface to grow a thin film in an atomic scale. The ALD technique can enable atomically precise surface control and form conformal coating films. Moreover, ALD is usually conducted in an inert environment at relatively low temperatures (lower than the melting point of Li), allowing surface coating on Li metal by the ALD method. ALD coating is believed an ideal surface pretreating method suited to Li metal passivation [154–156]. Al_2O_3 can be directly deposited on Li metal foil via ALD and form stable and ion-conductive $LiAlO_x$ solid electrolyte after lithiation. The Al_2O_3 coating layer via ALD on Li metal shows improved stability against the corrosion by moisture, CO_2, and organic solvents. The Li protection effect is found proportional to the thickness of the ALD coating layer [154]. In Li–S batteries, the Li metal with the Al_2O_3 layer exhibits improved stability in the corrosive sulfur-rich electrolyte (containing dissolved sulfur and polysulfides). The Li metal anode with 14 nm Al_2O_3 is found to reduce the self-discharge effect and initial capacity loss significantly in the Li–S battery compared with the bare Li metal anode. The ALD-Al_2O_3 protection layer also improves the capacity retention of the Li–S battery due to the reduced Li metal corrosion. Even an ultrathin (~2 nm) ALD-Al_2O_3 layer can suppress the Li metal dendrite growth during the initial cycles [155], indicating the favorable protection for Li metal anode via the ALD surface engineering. Besides Al_2O_3, an ultrathin and ion-conducting Li_xAl_yS film coated via ALD has been found effective for stabilizing the Li-electrolyte interface and suppressing Li dendrite growth [156]. LiF is electrochemically stable and electronically insulating, and has a relatively high shear modulus of 55 GPa. The ALD technique has also enable conformal LiF coating layer for passivating Li metal anodes [157].

Magnetron sputtering has also been employed to deposit a conformal Li_3PO_4 thin film on Li metal, which has reduced grain boundary, controllable thickness, and isotropic amorphous nature [158]. The nanoscale amorphous protective films render Li deposition homogeneous and improve the anode lifespan.

4.4.2.3 Physical Coating Interlayer

Above we discussed the artificial SEI film generated on the Li metal via various methods: solution treating, gas reactions, ALD coating and so on. Aside from the chemically formed surface film, a protective interlayer that is physically coated, cast, or deposited on the Li metal with rational properties and nanostructures can also improve the Li plating behavior and suppress/alleviate the growth of dendrites. Compared with the electrochemical or chemical artificial SEI film, the "physical interlayers" are generally thicker layers on the Li metal anode with little chemical bonding but a weak interaction with the Li metal. The physically coated interlayers have relatively stronger shear moduli for suppressing Li dendrites. The protective physical interlayer is less complex in coating methods but has more choices in material and structure design. The coating layer can be carbon materials, polymer, inorganic, and combination of them. These materials come with various nanostructures, such as nanospheres, nanoporous films, nanofibers, and so on. Here we elucidate the design of the physical coating interlayers from the perspective of nanomaterials and nanostructure.

One of the most widely-utilized interlayer materials is the carbon with various allotropes and structures. An innovative example of the ex situ protective layer is a monolayer consisting of interconnected hollow carbon nanospheres designed by Zheng et al. [159]. As illustrated in Fig. 4.13a, the monolayer is loosely attached to the current collector surface and moves up and down together with the SEI film during Li plating and stripping. The hollow carbon nanopheres are synthesized via a template synthesis, using vertical deposition of polystyrene nanoparticles. The carbon wall has a thickness of ~20 nm and the hemispherical carbon nanospheres are interconnected to form a thin film (Fig. 4.13b). The robust carbon nanosphere monolayer can suppress Li dendrite growth and has mechanical flexibility to accommodate the volume change during Li plating. With the monolayer coating, the deposited Li metal forms columns rather than dendrites, with a diameter of 3.0 ± 0.3 μm (Fig. 4.13c). Electrochemical cycling at 0.25, 0.5, or 1 mA h cm^{-2} of the nanosphere-coated Li anode shows high Coulombic efficiency (~99%) for over 150 cycles, demonstrating the interfacial nanoengineering on Li metal anode a promising strategy.

The interfacial protection effect was also investigated using hexagonal boron nitride, graphene, and other 2D layered materials [160, 161]. As a 2D layered materials, graphene film has extremely high Young's modulus to suppress Li dendrites and excellent chemical stability to passivate the Li-electrolyte interface. The boron nitride thin films have a high mechanical strength to suppress Li dendrites and excellent chemical stability to passivate the Li-electrolyte interface.

Robust and flexible polymer films are promising as protective coating layers for Li metal anodes. Polymer coating layers with high viscoelastic and self-healing properties are extremely desired for dendrite suppression [162]. However, to facilitate better Li ion transfer through the polymer protective layer, the coating polymer should be ion conductive. Poly(3,4-ethylenedioxythiophene)-*co*-poly(ethylene glycol) (PEDOT-*co*-PEG) is a promising candidate in this regard. PEDOT-*co*-PEG is strongly adhesive to the Li metal to mechanically suppress the growth of Li dendrites, and more importantly, PEG is highly ion conductive for Li ion transport. The PEDOT-*co*-PEG protection layer coated on the Li anode forms a less resistive SEI film, stabilizes the interface of the Li anode, and suppresses the undesired corrosion reaction.

Nevertheless, the polymers have poor Li-ion conductivities as the coating layers on Li anodes for high-current cycling. Zhu et al. proposed a modified poly(dimethylsiloxane) (PDMS) thin film with nanopores serving as a protective layer for improving Li metal anode (Fig. 4.13d). The PDMS is cast on the Li metal and nanopores with the size of 40–100 nm are created in the PDMS film by hydrofluoric acid etching to provide pathways for Li ion transport (Fig. 4.13e). Thus, the PDMS film can enable uniform Li deposition and suppress Li dendrite formation by

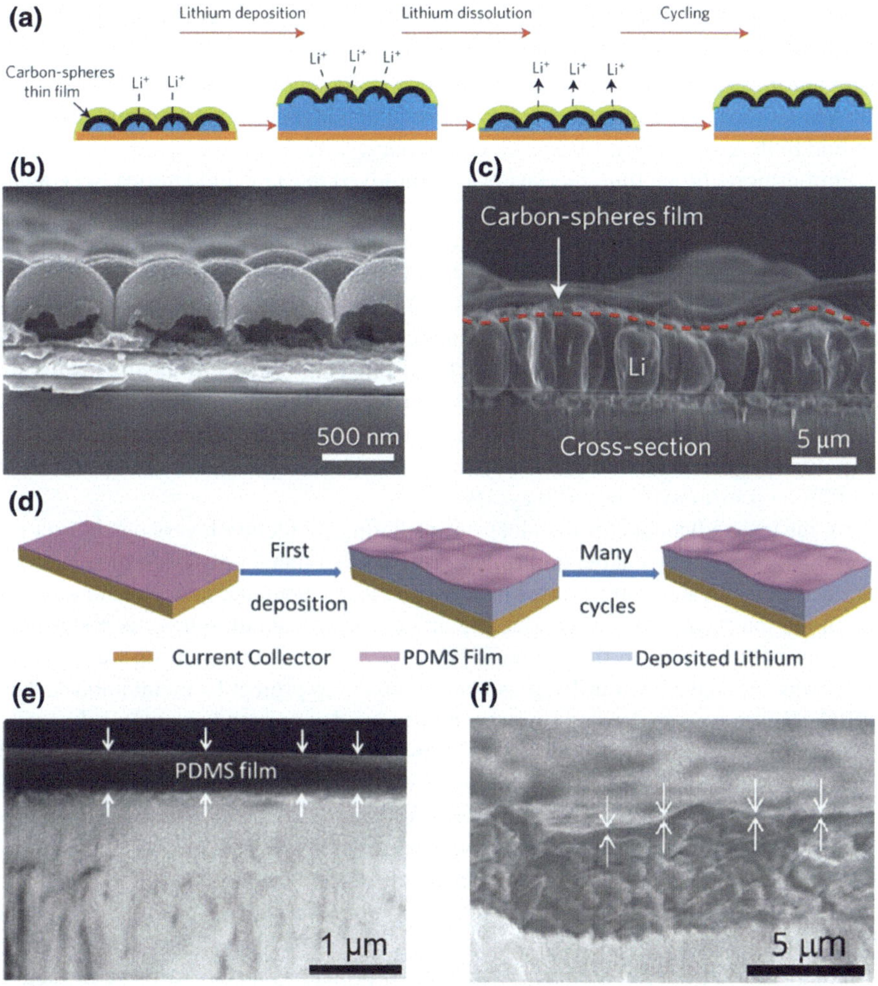

Fig. 4.13 Physical coating protective interlayers for Li metal anodes. **a** Schematics of Li deposition and dissolution on a Cu substrate modified by a hollow carbon nanosphere layer. **b** Cross-sectional SEM image of the hollow carbon nanospheres. **c** Cross-sectional SEM image showing the Li deposition elevating the hollow carbon nanosphere layer. **a–c** Reproduced with permission from Ref. [159], Copyright © 2014, Nature Publishing Group. **d** Schematic representation of Li deposition on the current collector coated with the PDMS thin film, which suppresses the dendrite formation and enables uniform deposition for many cycles. **e** Cross-sectional SEM image of the PDMS film. **f** Cross-sectional SEM image of initial Li deposition on PDMS (indicated by the arrows) coated Cu foil. **d–f** Reproduced with permission from Ref. [171], Copyright © 2016, Wiley

the high shear modulus and flexibility. The Li metal anode modified by the PDMS film exhibits smooth and flat surface after 1 mA h cm^{-2} Li deposition (Fig. 4.13f). The modified Li metal anode has a Coulombic efficiency of ~95% for 200 cycles. The nanoporous non-conductive polymer protecting layer provides a complementary method to develop dendrite-free Li metal anodes.

Furthermore, protecting interlayers are employed in Li metal batteries, serving as surface protection layers for uniform Li-ion flux as well as Li metal hosts to inhibit the growth of Li dendrites. A polyimide film with aligned nano- and micro-sized channels has been reported to coat on the Li metal as a protective interlayer for a stable Li deposition [121]. Homogenous Li^+ flux through the aligned nanochannels of the polyimide film leads to a uniform Li nucleation and thus improved cycle life and Coulombic efficiency (97.6% for 240 cycles at 1 mA cm^{-2}). Electronically insulating 3D oxidized polyacrylonitrile nanofibers on Cu current collector was fabricated by electrospinning to guide uniform Li deposition [163]. Li metal is deposited within the 3D polymer layer and shows stable cycling performance with a Coulombic efficiency of ~97% at 3 mA cm^{-2} over 120 cycles.

An intrinsic advantage of the electron-insulating protecting layers (such as polymer films) is that they enable controlled growth of Li metal only from the electron-conducting substrates and Li metal is deposited underneath the interlayer or embedded in the interlayer. An ion-conducting and electron-insulating layer is better than non-conducting layers in this scenario because it can facilitate Li-ion transfer. Li-ion conductors have been utilized as coating layers to protect Li metal anodes. The LISICON film was applied as Li-ion conductor and separator for aqueous Li metal batteries (mostly aqueous Li-air battery) [164]. The LISICON film was also studied as coating layer combined with a sandwiched polymer membrane saturated with Li salt [165]. The ceramic Li-ion conductor is utilized to isolate Li from aqueous electrolyte for a stable and safe aqueous Li metal battery. However, it remains unclear how the ceramic Li-ion conductor affects the Li deposition behavior and the Li interface stability, and how the stiff ceramic would accommodate the Li volume expansion during Li plating.

Compared with the ceramic ion-conductive protecting layers, a more attractive approach is to fabricate a polymer-inorganic composite protective layer (CPL) that conducts Li ions on the Li metal anode. As schematically depicted in Fig. 4.14, the CPL is composed of inorganic nanoparticles in a flexible polymer matrix and is expected to have favorable ionic conductivity for Li-ion migration and desired flexibility to accommodate Li volume change. The CPL not only mechanically suppresses Li dendrite but also high ionic conductivity for even distribution of Li-ion flux across the electrode surface, resulting in more uniform Li plating/stripping behavior. A facile strategy is to dispersed inorganic nanoparticles in ion-conductive polymer membrane as the CPL. For example, the Al_2O_3 nanoparticles (~500 nm) dispersed in PVdF-co-HFP polymer was cast on the Li metal as the CPL, exhibiting good shear modulus and ionic conductivity, except that Al_2O_3 nanoparticles are not ionic conductive [166]. A more rational design of the CPL could be an artificial SEI film that contains Li-ion conductors dispersed in a flexible polymer matrix, greatly enhancing the mechanical strength of the coating layer [167]. The inorganic nanofillers

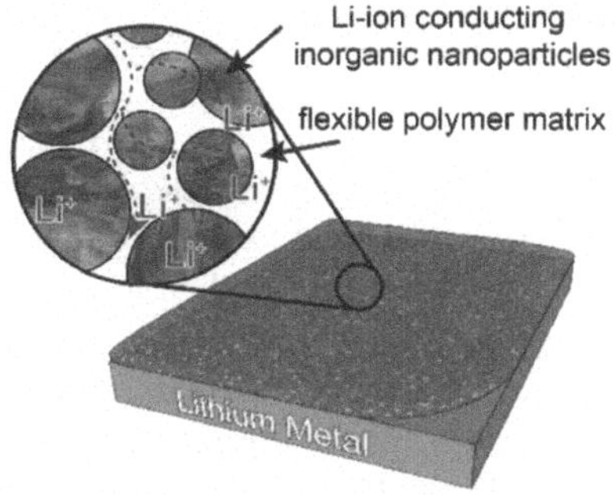

Fig. 4.14 Schematic illustration of a rationally designed artificial SEI protective layer for Li metal anode. Reproduced with permission from Ref. [167], Copyright © 2017, Wiley

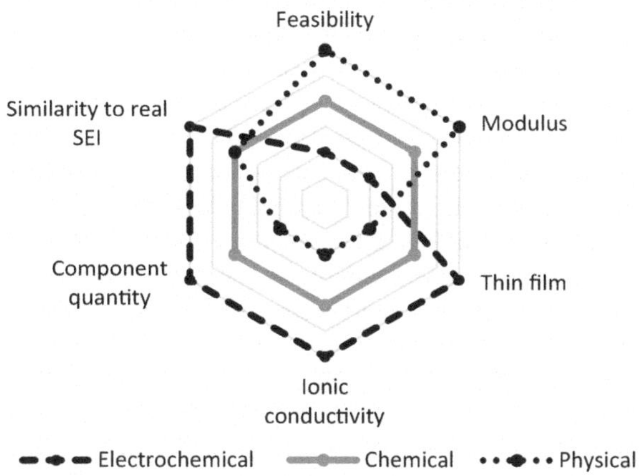

Fig. 4.15 Summary for the properties of the protective interlayers for Li metal anodes via electrochemical, chemical, and physical treating methods. The higher value in the Radar chart represents the outstanding advantage

are either Li-ion conductors (e.g., garnet-type Li-ion conductor [168]), or converted to Li-ion conductors (e.g., Li_3N converted from Cu_3N [167]). Thus, the high ionic conductivity of the coating layer is ensured for Li-ion transfer and deposition. The electronically insulative layers block electron passage and constrain the Li deposition between the interlayer and the current collector.

4.4.2.4 Summary

Figure 4.15 summarizes and compares the properties of the protective interlayers for Li metal anodes via different approaches. The three surface pretreatments on Li metal have advantages from different perspectives and each has its application scenario. More studies should be done to determine the optimal modification interlayers for Li metal anode, which should provide homogeneous Li-ion flux for a uniform Li deposition, mechanical strength to suppress the dendrite growth, flexibility to accommodate the volume change, and chemical stability for Li-electrolyte interface stability.

4.5 Summary

The elements in Group 16 have been extensively studied as cathode materials and are still attracting great interest. The new electrode materials involving novel electrochemistries and Li storage mechanisms can offer great prospective in energy density. The wide accessibility, low cost, abundant sources, and most importantly, high energy density of the O_2 and S cathodes have endowed them the most promising next-generation cathode materials. Apparently, utilization of these high-capacity cathode materials is faced with crucial challenges, which have been elucidated above. To couple the O_2 and S cathodes with Li metal anodes in rechargeable full cells, dendritic problems and associated issues of the Li metal anode are yet to be resolved. Fortunately, the flourishing nanotechnology has offered a great opportunity to develop fantastic nanomaterials and nanostructures that can markedly improve the high-capacity electrodes. This chapter reveals the principle of nano effect in the O_2 and S cathodes and the Li metal anode, and introduces the design of nanomaterials in the electrodes of Li metal batteries, which are recapitulated in Table 4.5. The nanomaterials play a vital role in enabling and enhancing the high-capacity Li metal batteries.

Although fundamental principles for the nanomaterials in the Li metal battery electrodes have been widely acknowledged, some concerns do exist as to commercially utilize the nano-structured high-capacity electrodes. As discussed above, it is concerned that the substrate materials in Li-air batteries could be instable, the S host nanomaterials in S cathodes could reduce the energy density of the Li–S batteries, the nanomaterials could lead to complicated interfacial reactions, and so on. The commercial applications of nanomaterials are also limited by the high cost, complex synthesis procedures, and capability of scaling-up of the nanomaterials, which have long been the problems of nanomaterials.

Among Li metal batteries, Li–O_2 battery is still some way from practical application (let alone Li-air battery). Li–S batteries are closer to commercialization, but are still limited to low energy density and short lifespan. Although the commercialization of Li metal batteries remains challenging, the nanotechnology approach is the most promising approach to make this happen. Future attention should be paid to further understand the effect of the nanostructure in the electrochemistry and electrochemical performance. High-performance nano-structured electrodes should also realize mass loading of the electrodes, tap density, and low cost while preserving high activity of the electrode materials. More innovation in the electrochemistry and nanomaterials will undoubtedly promote the electrochemical energy storage.

Table 4.5 Nanomaterials and their roles in the electrodes of Li metal batteries

Electrode	Nanomaterials	Key roles	Comments
O_2 cathode	Nanoporous substrate materials	Nanomaterials as electrode substrates increase the interface area, enhance air diffusion and transportation, and provide 3D electrical conductive network	Some fundamental issues in Li-air batteries remain unsolved. The stability of some substrate materials and electrolytes and the actual efficacy of some catalysts are debatable. There is a long way to go before applying these nanomaterials in commercial Li-air batteries
	Nano catalyst materials	Catalytic nanomaterials reduce the reaction overpotential and improve the energy efficiency	
Sulfur cathode	Sulfur hosting nanomaterials, including nanoporous carbon, CNTs, graphene, and non-carbon materials	The nano-structure hosting materials can (i) provide electrical conductive network for the insulting sulfur, (ii) suppress the dissolution of polysulfides and associated shuttle effect, and (iii) alleviate the volume change of the sulfur cathode	The sulfur hosting nanomaterials have greatly promoted the development of rechargeable Li–S batteries. More research effort should be made to increase the sulfur loading in the composite and sulfur mass loading in the electrode for high energy density
Lithium metal anode	Nano/micro-porous host materials for Li metal	The host materials suppress the growth of Li dendrites, lower the local current density, and alleviate the volume change of the Li anode during cycling	The nanotechnology including nano hosting and nano coating has revived the Li metal anode. However, the use of nano hosting materials in Li anodes is still controversial, as the nanomaterials could increase the instable surface area. More attentions should also be paid to the safety of the Li meta anode
	Nano/micro coating layers	Protective coating layers stabilize the Li surface and prohibit Li dendrite penetration	

References

1. Abraham, K. M., & Jiang, Z. (1996). A polymer electrolyte-based rechargeable lithium/oxygen battery. *Journal of the Electrochemical Society, 143,* 1–5.
2. Hasegawa, S., Imanishi, N., Zhang, T., et al. (2009). Study on lithium/air secondary batteries—Stability of NASICON-type lithium ion conducting glass-ceramics with water. *Journal of Power Sources, 189,* 371–377.
3. Shao, Y., Ding, F., Xiao, J., et al. (2013). Making Li-air batteries rechargeable: Material challenges. *Advanced Functional Materials, 23,* 987–1004.
4. Wang, J., Li, Y., & Sun, X. (2013). Challenges and opportunities of nanostructured materials for aprotic rechargeable lithium–air batteries. *Nano Energy, 2,* 443–467.
5. Ottakam Thotiyl, M. M., Freunberger, S. A., Peng, Z., et al. (2013). The carbon electrode in nonaqueous Li–O_2 cells. *Journal of the American Chemical Society, 135,* 494–500.
6. Ottakam Thotiyl, M. M., Freunberger, S. A., Peng, Z., et al. (2013). A stable cathode for the aprotic Li–O_2 battery. *Nature Materials, 12,* 1050–1056.
7. Peng, Z., Freunberger, S. A., Chen, Y., et al. (2012). A reversible and higher-rate Li-O_2 battery. *Science, 337,* 563–566.
8. Yin, Y.-X., Xin, S., Guo, Y.-G., et al. (2013). Lithium-sulfur batteries: Electrochemistry, materials, and prospects. *Angewandte Chemie International Edition, 52,* 13186–13200.
9. Evers, S., & Nazar, L. F. (2012). New approaches for high energy density lithium-sulfur battery cathodes. *Accounts of Chemical Research, 46,* 1135–1143.
10. Ji, X., Lee, K., & Nazar, L. F. (2009). A highly ordered nanostructured carbon-sulphur cathode for lithium-sulphur batteries. *Nature Materials, 8,* 500–506.
11. Xin, S., Gu, L., Zhao, N.-H., et al. (2012). Smaller sulfur molecules promise better lithium-sulfur batteries. *Journal of the American Chemical Society, 134,* 18510–18513.
12. Gao, J., Lowe, M. A., Kiya, Y., et al. (2011). Effects of liquid electrolytes on the charge-discharge performance of rechargeable lithium/sulfur batteries: Electrochemical and in-situ X-ray absorption spectroscopic studies. *The Journal of Physical Chemistry C, 115,* 25132–25137.
13. Li, Z., Jiang, Y., Yuan, L., et al. (2014). A highly ordered meso@ microporous carbon-supported sulfur@ smaller sulfur core–shell structured cathode for Li–S batteries. *ACS Nano, 8,* 9295–9303.
14. Xin, S., Yin, Y. X., Wan, L. J., et al. (2013). Encapsulation of sulfur in a hollow porous carbon substrate for superior Li-S batteries with long lifespan. *Particle & Particle Systems Characterization, 30,* 321–325.
15. Ye, H., Yin, Y.-X., Xin, S., et al. (2013). Tuning the porous structure of carbon hosts for loading sulfur toward long lifespan cathode materials for Li–S batteries. *Journal of Materials Chemistry A, 1,* 6602–6608.
16. Zheng, G., Yang, Y., Cha, J. J., et al. (2011). Hollow carbon nanofiber-encapsulated sulfur cathodes for high specific capacity rechargeable lithium batteries. *Nano Letters, 11,* 4462–4467.
17. Lee, J. T., Zhao, Y., Thieme, S., et al. (2013). Sulfur-infiltrated micro- and mesoporous silicon carbide-derived carbon cathode for high-performance lithium sulfur batteries. *Advanced Materials, 25,* 4573–4579.
18. Liang, C., Dudney, N. J., & Howe, J. Y. (2009). Hierarchically structured sulfur/carbon nanocomposite material for high-energy lithium battery. *Chemistry of Materials, 21,* 4724–4730.
19. He, G., Ji, X., & Nazar, L. (2011). High "C" rate Li-S cathodes: Sulfur imbibed bimodal porous carbons. *Energy & Environmental Science, 4,* 2878–2883.
20. Rybarczyk, M. K., Peng, H.-J., Tang, C., et al. (2016). Porous carbon derived from rice husks as sustainable bioresources: Insights into the role of micro-/mesoporous hierarchy in hosting active species for lithium–sulphur batteries. *Green Chemistry, 18,* 5169–5179.
21. Sun, Q., He, B., Zhang, X.-Q., et al. (2015). Engineering of hollow core-shell interlinked carbon spheres for highly stable lithium-sulfur batteries. *ACS Nano, 9,* 8504–8513.

22. Zhang, B., Qin, X., Li, G. R., et al. (2010). Enhancement of long stability of sulfur cathode by encapsulating sulfur into micropores of carbon spheres. *Energy & Environmental Science, 3,* 1531–1537.
23. Wang, J., Chew, S. Y., Zhao, Z. W., et al. (2008). Sulfur–mesoporous carbon composites in conjunction with a novel ionic liquid electrolyte for lithium rechargeable batteries. *Carbon, 46,* 229–235.
24. Schuster, J., He, G., Mandlmeier, B., et al. (2012). Spherical ordered mesoporous carbon nanoparticles with high porosity for lithium-sulfur batteries. *Angewandte Chemie International Edition, 51,* 3591–3595.
25. Jayaprakash, N., Shen, J., Moganty, S. S., et al. (2011). Porous hollow Carbon@Sulfur composites for high-power lithium-sulfur batteries. *Angewandte Chemie International Edition, 50,* 5904–5908.
26. Su, Y.-S., Fu, Y., & Manthiram, A. (2012). Self-weaving sulfur–carbon composite cathodes for high rate lithium–sulfur batteries. *Physical Chemistry Chemical Physics, 14,* 14495–14499.
27. Guo, J., Xu, Y., & Wang, C. (2011). Sulfur-impregnated disordered carbon nanotubes cathode for lithium-sulfur batteries. *Nano Letters, 11,* 4288–4294.
28. Ji, L., Rao, M., Aloni, S., et al. (2011). Porous carbon nanofiber–sulfur composite electrodes for lithium/sulfur cells. *Energy & Environmental Science, 4,* 5053–5059.
29. He, G., Mandlmeier, B., Schuster, J., et al. (2014). Bimodal mesoporous carbon nanofibers with high porosity: Freestanding and embedded in membranes for lithium-sulfur batteries. *Chemistry of Materials, 26,* 3879–3886.
30. Dörfler, S., Hagen, M., Althues, H., et al. (2012). High capacity vertical aligned carbon nanotube/sulfur composite cathodes for lithium–sulfur batteries. *Chemical Communications, 48,* 4097–4099.
31. Cheng, X.-B., Huang, J.-Q., Zhang, Q., et al. (2014). Aligned carbon nanotube/sulfur composite cathodes with high sulfur content for lithium–sulfur batteries. *Nano Energy, 4,* 65–72.
32. Mkhoyan, K. A., Contryman, A. W., Silcox, J., et al. (2009). Atomic and electronic structure of graphene-oxide. *Nano Letters, 9,* 1058–1063.
33. Wang, H., Yang, Y., Liang, Y., et al. (2011). Graphene-wrapped sulfur particles as a rechargeable lithium-sulfur battery cathode material with high capacity and cycling stability. *Nano Letters, 11,* 2644–2647.
34. Ji, L., Rao, M., Zheng, H., et al. (2011). Graphene oxide as a sulfur immobilizer in high performance lithium/sulfur cells. *Journal of the American Chemical Society, 133,* 18522–18525.
35. Xu, H., Deng, Y., Shi, Z., et al. (2013). Graphene-encapsulated sulfur (GES) composites with a core–shell structure as superior cathode materials for lithium–sulfur batteries. *Journal of Materials Chemistry A, 1,* 15142–15149.
36. Wang, J.-Z., Lu, L., Choucair, M., et al. (2011). Sulfur-graphene composite for rechargeable lithium batteries. *Journal of Power Sources, 196,* 7030–7034.
37. Wei, Z.-K., Chen, J.-J., Qin, L.-L., et al. (2012). Two-step hydrothermal method for synthesis of sulfur-graphene hybrid and its application in lithium sulfur batteries. *Journal of the Electrochemical Society, 159,* A1236–A1239.
38. Xu, J., Shui, J., Wang, J., et al. (2014). Sulfur-graphene nanostructured cathodes via ball-milling for high-performance lithium-sulfur batteries. *ACS Nano, 8,* 10920–10930.
39. Zhou, G., Yin, L.-C., Wang, D.-W., et al. (2013). Fibrous hybrid of graphene and sulfur nanocrystals for high-performance lithium-sulfur batteries. *ACS Nano, 7,* 5367–5375.
40. Zu, C., & Manthiram, A. (2013). Hydroxylated graphene-sulfur nanocomposites for high-rate lithium-sulfur batteries. *Advanced Energy Materials, 3,* 1008–1012.
41. Hou, Y., Li, J., Gao, X., et al. (2016). 3D dual-confined sulfur encapsulated in porous carbon nanosheets and wrapped with graphene aerogels as a cathode for advanced lithium sulfur batteries. *Nanoscale, 8,* 8228–8235.
42. Li, H., Yang, X., Wang, X., et al. (2015). Dense integration of graphene and sulfur through the soft approach for compact lithium/sulfur battery cathode. *Nano Energy, 12,* 468–475.
43. Zhang, C., Liu, D.-H., Lv, W., et al. (2015). A high-density graphene–sulfur assembly: A promising cathode for compact Li–S batteries. *Nanoscale, 7,* 5592–5597.

44. Lu, S., Cheng, Y., Wu, X., et al. (2013). Significantly improved long-cycle stability in high-rate Li–S batteries enabled by coaxial graphene wrapping over sulfur-coated carbon nanofibers. *Nano Letters, 13,* 2485–2489.
45. Xi, K., Kidambi, P. R., Chen, R., et al. (2014). Binder free three-dimensional sulphur/few-layer graphene foam cathode with enhanced high-rate capability for rechargeable lithium sulphur batteries. *Nanoscale, 6,* 5746–5753.
46. Zhou, G., Li, L., Ma, C., et al. (2015). A graphene foam electrode with high sulfur loading for flexible and high energy Li-S batteries. *Nano Energy, 11,* 356–365.
47. Deng, W., Zhou, X., Fang, Q., et al. (2016). Graphene/sulfur composites with a foam-like porous architecture and controllable pore size for high performance lithium-sulfur batteries. *ChemNanoMat, 2,* 952–958.
48. Zhang, K., Xie, K., Yuan, K., et al. (2017). Enabling effective polysulfide trapping and high sulfur loading via a pyrrole modified graphene foam host for advanced lithium–sulfur batteries. *Journal of Materials Chemistry A, 5,* 7309–7315.
49. Li, N., Zheng, M., Lu, H., et al. (2012). High-rate lithium–sulfur batteries promoted by reduced graphene oxide coating. *Chemical Communications, 48,* 4106–4108.
50. Moon, J., Park, J., Jeon, C., et al. (2015). An electrochemical approach to graphene oxide coated sulfur for long cycle life. *Nanoscale, 7,* 13249–13255.
51. Ding, B., Yuan, C., Shen, L., et al. (2013). Chemically tailoring the nanostructure of graphene nanosheets to confine sulfur for high-performance lithium-sulfur batteries. *Journal of Materials Chemistry A, 1,* 1096–1101.
52. Zhao, M.-Q., Liu, X.-F., Zhang, Q., et al. (2012). Graphene/single-walled carbon nanotube hybrids: One-step catalytic growth and applications for high-rate Li–S batteries. *ACS Nano, 6,* 10759–10769.
53. Du, W.-C., Yin, Y.-X., Zeng, X.-X., et al. (2016). Wet chemistry synthesis of multidimensional nanocarbon-sulfur hybrid materials with ultrahigh sulfur loading for lithium-sulfur batteries. *ACS Applied Materials & Interfaces, 8,* 3584–3590.
54. Song, J., Gordin, M. L., Xu, T., et al. (2015). Strong lithium polysulfide chemisorption on electroactive sites of nitrogen-doped carbon composites for high-performance lithium-sulfur battery cathodes. *Angewandte Chemie International Edition, 54,* 4325–4329.
55. Yang, C.-P., Yin, Y.-X., Ye, H., et al. (2014). Insight into the effect of boron doping on sulfur/carbon cathode in lithium–sulfur batteries. *ACS Applied Materials & Interfaces, 6,* 8789–8795.
56. Balach, J., Singh, H. K., Gomoll, S., et al. (2016). Synergistically enhanced polysulfide chemisorption using a flexible hybrid separator with N and S dual-doped mesoporous carbon coating for advanced lithium-sulfur batteries. *ACS Applied Materials & Interfaces, 8,* 14586–14595.
57. Evers, S., Yim, T., & Nazar, L. F. (2012). Understanding the nature of absorption/adsorption in nanoporous polysulfide sorbents for the Li–S battery. *The Journal of Physical Chemistry C, 116,* 19653–19658.
58. Wei Seh, Z., Li, W., Cha, J. J., et al. (2013). Sulphur–TiO_2 yolk–shell nanoarchitecture with internal void space for long-cycle lithium–sulphur batteries. *Nature Communications, 4,* 1331.
59. Pang, Q., Kundu, D., Cuisinier, M., et al. (2014). Surface-enhanced redox chemistry of polysulphides on a metallic and polar host for lithium-sulphur batteries. *Nature Communications, 5,* 4759.
60. Tao, X., Wang, J., Ying, Z., et al. (2014). Strong sulfur binding with conducting magnéli-phase Ti_nO_{2n-1} nanomaterials for improving lithium-sulfur batteries. *Nano Letters, 14,* 5288–5294.
61. Ji, X., Evers, S., Black, R., et al. (2011). Stabilizing lithium–sulphur cathodes using polysulphide reservoirs. *Nature Communications, 2,* 325.
62. Han, X., Xu, Y., Chen, X., et al. (2013). Reactivation of dissolved polysulfides in Li–S batteries based on atomic layer deposition of Al_2O_3 in nanoporous carbon cloth. *Nano Energy, 2,* 1197–1206.
63. Zhao, T., Ye, Y., Peng, X., et al. (2016). Advanced lithium-sulfur batteries enabled by a bio-inspired polysulfide adsorptive brush. *Advanced Functional Materials, 26,* 8418–8426.

64. He, J., Luo, L., Chen, Y., et al. (2017). Yolk-shelled C@Fe3O4 nanoboxes as efficient sulfur hosts for high-performance lithium-sulfur batteries. *Advanced Materials, 29,* 1702707.
65. Liang, X., Kwok, C. Y., Lodi-Marzano, F., et al. (2016). Tuning transition metal oxide-sulfur interactions for long life lithium sulfur batteries: The "Goldilocks" principle. *Advanced Energy Materials, 6,* 1501636.
66. Kim, H., Lee, J. T., Lee, D.-C., et al. (2013). Plasma-enhanced atomic layer deposition of ultrathin oxide coatings for stabilized lithium-sulfur batteries. *Advanced Energy Materials, 3,* 1308–1315.
67. Liu, X., Huang, J.-Q., Zhang, Q., et al. (2017). Nanostructured metal oxides and sulfides for lithium-sulfur batteries. *Advanced Materials, 29,* 1601759.
68. Seh, Z. W., Yu, J. H., Li, W., et al. (2014). Two-dimensional layered transition metal disulphides for effective encapsulation of high-capacity lithium sulphide cathodes. *Nature Communications, 5,* 5017.
69. Ma, L., Wei, S., Zhuang, H. L., et al. (2015). Hybrid cathode architectures for lithium batteries based on TiS_2 and sulfur. *Journal of Materials Chemistry A, 3,* 19857–19866.
70. Xu, H., & Manthiram, A. (2017). Hollow cobalt sulfide polyhedra-enabled long-life, high areal-capacity lithium-sulfur batteries. *Nano Energy, 33,* 124–129.
71. Pang, Q., Kundu, D., & Nazar, L. F. (2016). A graphene-like metallic cathode host for long-life and high-loading lithium–sulfur batteries. *Materials Horizons, 3,* 130–136.
72. Yuan, Z., Peng, H.-J., Hou, T.-Z., et al. (2016). Powering lithium-sulfur battery performance by propelling polysulfide redox at sulfiphilic hosts. *Nano Letters, 16,* 519–527.
73. Zhang, S. S., & Tran, D. T. (2016). Pyrite FeS_2 as an efficient adsorbent of lithium polysulphide for improved lithium–sulphur batteries. *Journal of Materials Chemistry A, 4,* 4371–4374.
74. Tang, W., Chen, Z., Tian, B., et al. (2017). In situ observation and electrochemical study of encapsulated sulfur nanoparticles by MoS2 flakes. *Journal of the American Chemical Society, 139,* 10133–10141.
75. Liang, X., Garsuch, A., & Nazar, L. F. (2015). Sulfur cathodes based on conductive MXene nanosheets for high-performance lithium-sulfur batteries. *Angewandte Chemie International Edition, 54,* 3907–3911.
76. Cui, Z., Zu, C., Zhou, W., et al. (2016). Mesoporous titanium nitride-enabled highly stable lithium-sulfur batteries. *Advanced Materials, 28,* 6926–6931.
77. Zhou, J., Li, R., Fan, X., et al. (2014). Rational design of a metal–organic framework host for sulfur storage in fast, long-cycle Li–S batteries. *Energy & Environmental Science, 7,* 2715–2724.
78. Li, Z., Zhang, J., & Lou, X. W. (2015). Hollow carbon nanofibers filled with MnO2 nanosheets as efficient sulfur hosts for lithium-sulfur batteries. *Angewandte Chemie International Edition, 54,* 12886–12890.
79. Chung, S.-H., & Manthiram, A. (2013). Nano-cellular carbon current collectors with stable cyclability for Li–S batteries. *Journal of Materials Chemistry A, 1,* 9590–9596.
80. Zhang, S. S., & Tran, D. T. (2012). A proof-of-concept lithium/sulfur liquid battery with exceptionally high capacity density. *Journal of Power Sources, 211,* 169–172.
81. Fu, Y., Su, Y.-S., & Manthiram, A. (2013). Highly reversible lithium/dissolved polysulfide batteries with carbon nanotube electrodes. *Angewandte Chemie International Edition, 52,* 6930–6935.
82. Barchasz, C., Mesguich, F., Dijon, J., et al. (2012). Novel positive electrode architecture for rechargeable lithium/sulfur batteries. *Journal of Power Sources, 211,* 19–26.
83. Chung, S.-H., & Manthiram, A. (2014). Carbonized eggshell membrane as a natural polysulfide reservoir for highly reversible Li-S batteries. *Advanced Materials, 26,* 1360–1365.
84. Huang, X., Sun, B., Li, K., et al. (2013). Mesoporous graphene paper immobilised sulfur as a flexible electrode for lithium–sulfur batteries. *Journal of Materials Chemistry A, 1,* 13484–13489.
85. Sun, L., Kong, W., Jiang, Y., et al. (2015). Super-aligned carbon nanotube/graphene hybrid materials as a framework for sulfur cathodes in high performance lithium sulfur batteries. *Journal of Materials Chemistry A, 3,* 5305–5312.

86. Chen, Y., Lu, S., Wu, X., et al. (2015). Flexible carbon nanotube–graphene/sulfur composite film: Free-standing cathode for high-performance lithium/sulfur batteries. *The Journal of Physical Chemistry C, 119,* 10288–10294.
87. Zhu, L., Peng, H.-J., Liang, J., et al. (2015). Interconnected carbon nanotube/graphene nanosphere scaffolds as free-standing paper electrode for high-rate and ultra-stable lithium–sulfur batteries. *Nano Energy, 11,* 746–755.
88. Su, Y.-S., & Manthiram, A. (2012). Lithium-sulphur batteries with a microporous carbon paper as a bifunctional interlayer. *Nature Communications, 3,* 1166.
89. Su, Y.-S., Fu, Y., Cochell, T., et al. (2013). A strategic approach to recharging lithium-sulphur batteries for long cycle life. *Nature Communications, 4,* 2985.
90. Su, J., Wu, X.-L., Yang, C.-P., et al. (2012). Self-assembled $LiFePO_4$/C nano/microspheres by using phytic acid as phosphorus source. *The Journal of Physical Chemistry C, 116,* 5019–5024.
91. Zu, C., Su, Y.-S., Fu, Y., et al. (2013). Improved lithium-sulfur cells with a treated carbon paper interlayer. *Physical Chemistry Chemical Physics, 15,* 2291–2297.
92. Manthiram, A., Fu, Y., Chung, S.-H., et al. (2014). Rechargeable lithium-sulfur batteries. *Chemical Reviews, 114,* 11751–11787.
93. Zhou, G., Pei, S., Li, L., et al. (2014). A graphene-pure-sulfur sandwich structure for ultrafast, long-life lithium-sulfur batteries. *Advanced Materials, 26,* 625–631.
94. Bai, S., Liu, X., Zhu, K., et al. (2016). Metal–organic framework-based separator for lithium–sulfur batteries. *Nature Energy, 1,* 16094.
95. Xu, Z.-L., Kim, J.-K., & Kang, K. (2018). Carbon nanomaterials for advanced lithium sulfur batteries. *Nano Today, 19,* 84–107.
96. Yang, C. P., Yin, Y. X., Zhang, S. F., et al. (2015). Accommodating lithium into 3D current collectors with a submicron skeleton towards long-life lithium metal anodes. *Nature Communications, 6,* 8058.
97. Yun, Q., He, Y.-B., Lv, W., et al. (2016). Chemical dealloying derived 3D porous current collector for Li metal anodes. *Advanced Materials, 28,* 6932–6939.
98. Lu, L.-L., Ge, J., Yang, J.-N., et al. (2016). Free-standing copper nanowire network current collector for improving lithium anode performance. *Nano Letters, 16,* 4431–4437.
99. Ke, X., Cheng, Y., Liu, J., et al. (2018). Hierarchically bicontinuous porous copper as advanced 3D skeleton for stable lithium storage. *ACS Applied Materials & Interfaces, 10,* 13552–13561.
100. Wang, S.-H., Yin, Y.-X., Zuo, T.-T., et al. (2017). Stable Li metal anodes via regulating lithium plating/stripping in vertically aligned microchannels. *Advanced Materials, 29,* 1703729.
101. Sun, Y., Zheng, G., Seh, Zhi W., et al. (2016). Graphite-encapsulated Li-metal hybrid anodes for high-capacity Li batteries. *Chem, 1,* 287–297.
102. Mukherjee, R., Thomas, A. V., Datta, D., et al. (2014). Defect-induced plating of lithium metal within porous graphene networks. *Nature Communications, 5,* 3710.
103. Cheng, X.-B., Peng, H.-J., Huang, J.-Q., et al. (2015). Dual-phase lithium metal anode containing a polysulfide-induced solid electrolyte interphase and nanostructured graphene framework for lithium-sulfur batteries. *ACS Nano, 9,* 6373–6382.
104. Zhang, R., Cheng, X.-B., Zhao, C.-Z., et al. (2016). Conductive nanostructured scaffolds render low local current density to inhibit lithium dendrite growth. *Advanced Materials, 28,* 2155–2162.
105. Xie, K., Wei, W., Yuan, K., et al. (2016). Toward dendrite-free lithium deposition via structural and interfacial synergistic effects of 3D Graphene@Ni scaffold. *ACS Applied Materials & Interfaces, 8,* 26091–26097.
106. Liu, L., Yin, Y.-X., Li, J.-Y., et al. (2017). Free-standing hollow carbon fibers as high-capacity containers for stable lithium metal anodes. *Joule, 1,* 563–575.
107. Zuo, T. T., Wu, X. W., Yang, C. P., et al. (2017). Graphitized carbon fibers as multifunctional 3D current collectors for high areal capacity Li anodes. *Advanced Materials, 29,* 1700389.
108. Jin, C., Sheng, O., Lu, Y., et al. (2018). Metal oxide nanoparticles induced step-edge nucleation of stable Li metal anode working under an ultrahigh current density of 15 mA cm^{-2}. *Nano Energy, 45,* 203–209.

109. Cheng, X.-B., Hou, T.-Z., Zhang, R., et al. (2016). Dendrite-free lithium deposition induced by uniformly distributed lithium ions for efficient lithium metal batteries. *Advanced Materials, 28,* 2888–2895.
110. Bai, P., Li, J., Brushett, F. R., et al. (2016). Transition of lithium growth mechanisms in liquid electrolytes. *Energy & Environmental Science, 9,* 3221–3229.
111. Zuo, T.-T., Yin, Y.-X., Wang, S.-H., et al. (2018). Trapping lithium into hollow silica microspheres with a carbon nanotube core for dendrite-free lithium metal anodes. *Nano Letters, 18,* 297–301.
112. Yan, K., Lu, Z., Lee, H.-W., et al. (2016). Selective deposition and stable encapsulation of lithium through heterogeneous seeded growth. *Nature Energy, 1,* 16010.
113. Deng, W., Zhu, W., Zhou, X., et al. (2018). Highly reversible Li plating confined in three-dimensional interconnected microchannels toward high-rate and stable metallic lithium anodes. *ACS Applied Materials & Interfaces, 10,* 20387–20395.
114. Jin, C., Sheng, O., Luo, J., et al. (2017). 3D lithium metal embedded within lithiophilic porous matrix for stable lithium metal batteries. *Nano Energy, 37,* 177–186.
115. Zhang, Y., Luo, W., Wang, C., et al. (2017). High capacity, low tortuosity and channel-guided lithium metal anode. *Proceedings of the National Academy of Sciences, 114,* 3584–3589.
116. Zhang, Y., Wang, C., Pastel, G., et al. (2018). 3D wettable framework for dendrite-free alkali metal anodes. *Advanced Energy Materials, 8,* 1800635.
117. Zhang, R., Chen, X.-R., Chen, X., et al. (2017). Lithiophilic sites in doped graphene guide uniform lithium nucleation for dendrite-free lithium metal anodes. *Angewandte Chemie International Edition, 56,* 7764–7768.
118. Liang, Z., Lin, D., Zhao, J., et al. (2016). Composite lithium metal anode by melt infusion of lithium into a 3D conducting scaffold with lithiophilic coating. *Proceedings of the National Academy of Sciences, 113,* 2862–2867.
119. Liu, Y., Lin, D., Liang, Z., et al. (2016). Lithium-coated polymeric matrix as a minimum volume-change and dendrite-free lithium metal anode. *Nature Communications, 7,* 10992.
120. Lin, D., Liu, Y., Liang, Z., et al. (2016). Layered reduced graphene oxide with nanoscale interlayer gaps as a stable host for lithium metal anodes. *Nature Nanotechnology, 11,* 626–632.
121. Liu, W., Lin, D., Pei, A., et al. (2016). Stabilizing lithium metal anodes by uniform Li-ion flux distribution in nanochannel confinement. *Journal of the American Chemical Society, 138,* 15443–15450.
122. Lu, D., Shao, Y., Lozano, T., et al. (2015). Failure mechanism for fast-charged lithium metal batteries with liquid electrolytes. *Advanced Energy Materials, 5,* 1400993.
123. Wood, K. N., Kazyak, E., Chadwick, A. F., et al. (2016). Dendrites and pits: Untangling the complex behavior of lithium metal anodes through operando video microscopy. *ACS Central Science, 2,* 790–801.
124. Kim, H., Jeong, G., Kim, Y.-U., et al. (2013). Metallic anodes for next generation secondary batteries. *Chemical Society Reviews, 42,* 9011–9034.
125. Xu, K. (2004). Nonaqueous liquid electrolytes for lithium-based rechargeable batteries. *Chemical Reviews, 104,* 4303–4418.
126. Xu, D., Wang, Z. L., Xu, J. J., et al. (2012). Novel DMSO-based electrolyte for high performance rechargeable Li-O_2 batteries. *Chemical Communications, 48,* 6948–6950.
127. Johnson, L., Li, C., Liu, Z., et al. (2014). The role of LiO_2 solubility in O_2 reduction in aprotic solvents and its consequences for Li–O_2 batteries. *Nature Chemistry, 6,* 1091–1099.
128. MacFarlane, D. R., Forsyth, M., Howlett, P. C., et al. (2007). Ionic liquids in electrochemical devices and processes: Managing interfacial electrochemistry. *Accounts of Chemical Research, 40,* 1165–1173.
129. Armand, M., Endres, F., MacFarlane, D. R., et al. (2009). Ionic-liquid materials for the electrochemical challenges of the future. *Nature Materials, 8,* 621–629.
130. Lu, Y., Das, S. K., Moganty, S. S., et al. (2012). Ionic liquid-nanoparticle hybrid electrolytes and their application in secondary lithium-metal batteries. *Advanced Materials, 24,* 4430–4435.

131. Hu, J. J., Long, G. K., Liu, S., et al. (2014). A LiFSI-LiTFSI binary-salt electrolyte to achieve high capacity and cycle stability for a Li-S battery. *Chemical Communications, 50,* 14647–14650.
132. Miao, R., Yang, J., Feng, X., et al. (2014). Novel dual-salts electrolyte solution for dendrite-free lithium-metal based rechargeable batteries with high cycle reversibility. *Journal of Power Sources, 271,* 291–297.
133. Aurbach, D., & Zaban, A. (1993). Impedance spectroscopy of lithium electrodes: Part 1. General behavior in propylene carbonate solutions and the correlation to surface chemistry and cycling efficiency. *Journal of Electroanalytical Chemistry, 348,* 155–179.
134. Qian, J., Henderson, W. A., Xu, W., et al. (2015). High rate and stable cycling of lithium metal anode. *Nature Communications, 6,* 6362.
135. Suo, L., Hu, Y.-S. S., Li, H., et al. (2013). A new class of solvent-in-salt electrolyte for high-energy rechargeable metallic lithium batteries. *Nature Communications, 4,* 1481.
136. Xiong, S., Xie, K., Diao, Y., et al. (2014). Characterization of the solid electrolyte interphase on lithium anode for preventing the shuttle mechanism in lithium–sulfur batteries. *Journal of Power Sources, 246,* 840–845.
137. Li, W., Yao, H., Yan, K., et al. (2015). The synergetic effect of lithium polysulfide and lithium nitrate to prevent lithium dendrite growth. *Nature Communications, 6,* 7436.
138. Yan, C., Cheng, X.-B., Zhao, C.-Z., et al. (2016). Lithium metal protection through in-situ formed solid electrolyte interphase in lithium-sulfur batteries: The role of polysulfides on lithium anode. *Journal of Power Sources, 327,* 212–220.
139. Ding, F., Xu, W., Graff, G. L., et al. (2013). Dendrite-free lithium deposition via self-healing electrostatic shield mechanism. *Journal of the American Chemical Society, 135,* 4450–4456.
140. Wu, F., Qian, J., Chen, R., et al. (2014). An effective approach to protect lithium anode and improve cycle performance for Li-S batteries. *ACS Applied Materials & Interfaces, 6,* 15542–15549.
141. Jia, W., Fan, C., Wang, L., et al. (2016). Extremely accessible potassium nitrate (KNO_3) as the highly efficient electrolyte additive in lithium battery. *ACS Applied Materials & Interfaces, 8,* 15399–15405.
142. Zhang, S. S. (2012). Role of $LiNO_3$ in rechargeable lithium/sulfur battery. *Electrochimica Acta, 70,* 344–348.
143. Zhao, C.-Z., Cheng, X.-B., Zhang, R., et al. (2016). Li_2S_5-based ternary-salt electrolyte for robust lithium metal anode. *Energy Storage Materials, 3,* 77–84.
144. Liu, Q. C., Xu, J. J., Yuan, S., et al. (2015). Artificial protection film on lithium metal anode toward long-cycle-life lithium-oxygen batteries. *Advanced Materials, 27,* 5241–5247.
145. Cheng, X.-B., Yan, C., Chen, X., et al. (2017). Implantable solid electrolyte interphase in lithium-metal batteries. *Chem, 2,* 258–270.
146. Li, N. W., Yin, Y. X., Yang, C. P., et al. (2016). An artificial solid electrolyte interphase layer for stable lithium metal anodes. *Advanced Materials, 28,* 1853–1858.
147. Monroe, C., & Newman, J. (2005). The impact of elastic deformation on deposition kinetics at lithium/polymer interfaces. *Journal of the Electrochemical Society, 152,* A396–A404.
148. Stone, G. M., Mullin, S. A., Teran, A. A., et al. (2012). Resolution of the modulus versus adhesion dilemma in solid polymer electrolytes for rechargeable lithium metal batteries. *Journal of the Electrochemical Society, 159,* A222–A227.
149. Basile, A., Bhatt, A. I., & O'Mullane, A. P. (2016). Stabilizing lithium metal using ionic liquids for long-lived batteries. *Nature Communications, 7,* 11794.
150. Umeda, G. A., Menke, E., Richard, M., et al. (2011). Protection of lithium metal surfaces using tetraethoxysilane. *Journal of Materials Chemistry, 21,* 1593–1599.
151. Wu, M., Wen, Z., Liu, Y., et al. (2011). Electrochemical behaviors of a Li_3N modified Li metal electrode in secondary lithium batteries. *Journal of Power Sources, 196,* 8091–8097.
152. Ma, G., Wen, Z., Wu, M., et al. (2014). A lithium anode protection guided highly-stable lithium-sulfur battery. *Chemical Communications, 50,* 14209–14212.
153. Wu, M., Wen, Z., Jin, J., et al. (2016). Trimethylsilyl chloride-modified Li anode for enhanced performance of Li–S cells. *ACS Applied Materials & Interfaces, 8,* 16386–16395.

154. Kozen, A. C., Lin, C. F., Pearse, A. J., et al. (2015). Next-generation lithium metal anode engineering via atomic layer deposition. *ACS Nano, 9,* 5884–5892.
155. Kazyak, E., Wood, K. N., & Dasgupta, N. P. (2015). Improved cycle life and stability of lithium metal anodes through ultrathin atomic layer deposition surface treatments. *Chemistry of Materials, 27,* 6457–6462.
156. Cao, Y., Meng, X., & Elam, J. W. (2016). Atomic layer deposition of LixAlyS solid-state electrolytes for stabilizing lithium-metal anodes. *ChemElectroChem, 3,* 858–863.
157. Chen, L., Chen, K.-S., Chen, X., et al. (2018). Novel ALD chemistry enabled low-temperature synthesis of lithium fluoride coatings for durable lithium anodes. *ACS Applied Materials & Interfaces, 10,* 26972–26981.
158. Wang, L., Wang, Q., Jia, W., et al. (2017). Li metal coated with amorphous Li_3PO_4 via magnetron sputtering for stable and long-cycle life lithium metal batteries. *Journal of Power Sources, 342,* 175–182.
159. Zheng, G., Lee, S., Liang, Z., et al. (2014). Interconnected hollow carbon nanospheres for stable lithium metal anodes. *Nature Nanotechnology, 9,* 618–623.
160. Yan, K., Lee, H.-W., Gao, T., et al. (2014). Ultrathin two-dimensional atomic crystals as stable interfacial layer for improvement of lithium metal anode. *Nano Letters, 14,* 6016–6022.
161. Tian, H., Seh, Z. W., Yan, K., et al. (2017). Theoretical investigation of 2D layered materials as protective films for lithium and sodium metal anodes. *Advanced Energy Materials, 7,* 1602528.
162. Zheng, G., Wang, C., Pei, A., et al. (2016). High-performance lithium metal negative electrode with a soft and flowable polymer coating. *ACS Energy Letters, 1,* 1247–1255.
163. Liang, Z., Zheng, G., Liu, C., et al. (2015). Polymer nanofiber-guided uniform lithium deposition for battery electrodes. *Nano Letters, 15,* 2910–2916.
164. Wang, Y., He, P., & Zhou, H. (2011). A lithium–air capacitor–battery based on a hybrid electrolyte. *Energy & Environmental Science, 4,* 4994–4999.
165. Wang, X., Hou, Y., Zhu, Y., et al. (2013). An aqueous rechargeable lithium battery using coated Li metal as anode. *Scientific Reports, 3,* 1401.
166. Lee, H., Lee, D. J., Kim, Y.-J., et al. (2015). A simple composite protective layer coating that enhances the cycling stability of lithium metal batteries. *Journal of Power Sources, 284,* 103–108.
167. Liu, Y., Lin, D., Yuen, P. Y., et al. (2017). An artificial solid electrolyte interphase with high Li-Ion conductivity, mechanical strength, and flexibility for stable lithium metal anodes. *Advanced Materials, 29,* 1605531.
168. Yang, C., Liu, B., Jiang, F., et al. (2017). Garnet/polymer hybrid ion-conducting protective layer for stable lithium metal anode. *Nano Research, 10,* 4256–4265.
169. Choi, J., & Aurbach, D. (2016). Promise and reality of post-lithium-ion batteries with high energy densities. *Nature Reviews Materials, 1,* 16013.
170. Busche, M., Adelhelm, P., Sommer, H., et al. (2014). Systematical electrochemical study on the parasitic shuttle-effect in lithium-sulfur-cells at different temperatures and different rates. *Journal of Power Sources, 259,* 289–299.
171. Zhu, B., Jin, Y., Hu, X., et al. (2016). Poly(dimethylsiloxane) thin film as a stable interfacial layer for high-performance lithium-metal battery anodes. *Advanced Materials, 29,* 1603755.
172. Lim, H.-D., Park, K.-Y., Song, H., et al. (2013). Enhanced power and rechargeability of a $Li-O_2$ battery based on a hierarchical-fibril CNT electrode. *Advanced Materials, 25,* 1348–1352.
173. Mitchell, R. R., Gallant, B. M., Thompson, C. V., et al. (2011). All-carbon-nanofiber electrodes for high-energy rechargeable $Li–O_2$ batteries. *Energy & Environmental Science, 4,* 2952–2958.
174. Sun, B., Wang, B., Su, D., et al. (2012). Graphene nanosheets as cathode catalysts for lithium-air batteries with an enhanced electrochemical performance. *Carbon, 50,* 727–733.
175. Xiao, J., Mei, D., Li, X., et al. (2011). Hierarchically porous graphene as a lithium-air battery electrode. *Nano Letters, 11,* 5071–5078.

176. Guo, Z., Zhou, D., Dong, X., et al. (2013). Ordered hierarchical mesoporous/macroporous carbon: A high-performance catalyst for rechargeable Li–O_2 batteries. *Advanced Materials, 25,* 5668–5672.
177. Zhao, G., Zhang, L., Lv, J., et al. (2016). A graphitic foam framework with hierarchical pore structure as self-supported electrodes of Li–O_2 batteries and Li ion batteries. *Journal of Materials Chemistry A, 4,* 1399–1407.
178. Shen, Y., Sun, D., Yu, L., et al. (2013). A high-capacity lithium–air battery with Pd modified carbon nanotube sponge cathode working in regular air. *Carbon, 62,* 288–295.
179. Shui, J., Du, F., Xue, C., et al. (2014). Vertically aligned N-doped coral-like carbon fiber arrays as efficient air electrodes for high-performance nonaqueous Li–O_2 batteries. *ACS Nano, 8,* 3015–3022.
180. Han, J., Guo, X., Ito, Y., et al. (2016). Effect of chemical doping on cathodic performance of bicontinuous nanoporous graphene for Li-O_2 batteries. *Advanced Energy Materials, 6,* 1501870.
181. Zhao, C., Yu, C., Liu, S., et al. (2015). 3D porous N-doped graphene frameworks made of interconnected nanocages for ultrahigh-rate and long-life Li–O_2 batteries. *Advanced Functional Materials, 25,* 6913–6920.
182. Li, Y., Wang, J., Li, X., et al. (2011). Nitrogen-doped carbon nanotubes as cathode for lithium–air batteries. *Electrochemistry Communications, 13,* 668–672.
183. Ding, N., Chien, S. W., Hor, T. S. A., et al. (2014). Influence of carbon pore size on the discharge capacity of Li–O_2 batteries. *Journal of Materials Chemistry A, 2,* 12433–12441.
184. Su, D., Dou, S., & Wang, G. (2015). Gold nanocrystals with variable index facets as highly effective cathode catalysts for lithium–oxygen batteries. *NPG Asia Materials, 7,* e155.
185. Lu, Y.-C., Xu, Z., Gasteiger, H. A., et al. (2010). Platinum−gold nanoparticles: A highly active bifunctional electrocatalyst for rechargeable lithium−air batteries. *Journal of the American Chemical Society, 132,* 12170–12171.
186. Luo, W.-B., Gao, X.-W., Chou, S.-L., et al. (2015). Porous AgPd–Pd composite nanotubes as highly efficient electrocatalysts for lithium-oxygen batteries. *Advanced Materials, 27,* 6862–6869.
187. Liao, K., Wang, X., Sun, Y., et al. (2015). An oxygen cathode with stable full discharge–charge capability based on 2D conducting oxide. *Energy & Environmental Science, 8,* 1992–1997.
188. Tong, S., Zheng, M., Lu, Y., et al. (2015). Mesoporous NiO with a single-crystalline structure utilized as a noble metal-free catalyst for non-aqueous Li–O_2 batteries. *Journal of Materials Chemistry A, 3,* 16177–16182.
189. Hu, X., Cheng, F., Han, X., et al. (2015). Oxygen bubble-templated hierarchical porous ε-MnO2 as a superior catalyst for rechargeable Li–O_2 batteries. *Small (Weinheim an der Bergstrasse, Germany), 11,* 809–813.
190. Wu, F., Zhang, X., Zhao, T., et al. (2015). Hierarchical mesoporous/macroporous Co_3O_4 ultrathin nanosheets as free-standing catalysts for rechargeable lithium–oxygen batteries. *Journal of Materials Chemistry A, 3,* 17620–17626.
191. Zhang, Z., Zhou, G., Chen, W., et al. (2014). Facile synthesis of Fe_2O_3 nanoflakes and their electrochemical properties for Li-air batteries. *ECS Electrochemistry Letters, 3,* A8–A10.
192. Mohamed, S. G., Tsai, Y.-Q., Chen, C.-J., et al. (2015). Ternary spinel MCo_2O_4 (M = Mn, Fe, Ni, and Zn) porous nanorods as bifunctional cathode materials for lithium–O2 batteries. *ACS Applied Materials & Interfaces, 7,* 12038–12046.
193. Zhang, J., Zhao, Y., Zhao, X., et al. (2014). Porous perovskite $LaNiO_3$ nanocubes as cathode catalysts for Li-O_2 batteries with low charge potential. *Scientific reports, 4,* 6005.
194. Liu, L., Wang, J., Hou, Y., et al. (2016). Self-assembled 3D foam-like $NiCo_2O_4$ as efficient catalyst for lithium oxygen batteries. *Small (Weinheim an der Bergstrasse, Germany), 12,* 602–611.

Chapter 5
Nanostructures and Nanomaterials for Solid-State Batteries

Junpei Yue, Shu-Hua Wang and Yu-Guo Guo

Abstract (All-)solid-state batteries promise higher energy densities, longer shelf life, and lower packaging cost than batteries with conventional liquid electrolytes. The fast development of superionic conductors flourishes the studies on all-solid-state batteries, and therefore the history and general knowledge about solid-state electrolytes (SSEs) are firstly summarized in this chapter. And then, more attentions are paid to the application of nanostructures and nanomaterials to deal with the incompatibility between electrode and SSEs and the limited kinetic process in composite electrodes. Last but not the least, an overall consideration on constructing all-solid-state batteries including temperature and stress is illustrated.

Abbreviation

SSEs	Solid-state electrolytes
LGPS	$Li_{10}Ge_2PS_{12}$
PEO	Poly(ethylene oxide)
PAN	Poly(acrylonitrile)
PMMA	Poly(methyl methacrylate)
PVDF	Poly(vinylidene fluoride)
PVDF-HFP	Poly(vinylidene fluoride-hexafluoropropylene)
LiTFSI	Lithium bis(trifluoromethanesulfonyl)imide
LiPON	Lithium phosphorus oxynitride
LISICON	Li^+ super ion conductor
NASICON	Na^+ super ion conductor
LLTO	$Li_{0.33}La_{0.56}TiO_3$
LATP	$Li_{1.3}Al_{0.3}Ti_{1.7}(PO_4)_3$

J. Yue · Y.-G. Guo (✉)
Institute of Chemistry, Chinese Academy of Sciences, Beijing 100190, People's Republic of China
e-mail: ygguo@iccas.ac.cn

S.-H. Wang
Shandong University, Jinan 250100, Shandong, People's Republic of China

Y.-G. Guo, *Nanostructures and Nanomaterials for Batteries*,
https://doi.org/10.1007/978-981-13-6233-0_5

LAGP	$Li_{1.3}Al_{0.3}Ge_{1.7}(PO_4)_3$
LLZO	$Li_7La_3Zr_2O_{12}$
LSV	Linear sweep voltammetry
CV	Cyclic voltammetry
EIS	Electrochemical impedance spectroscopy
NCM	Lithium nickel cobalt manganese oxide
NCA	Lithium nickel cobalt aluminum oxide
LCO	$LiCoO_2$
rGO	Reduced graphene oxide
CNT	Carbon nanotube
LIBs	Lithium ion batteries
AFM	Atomic force microscopy

The volatile and highly flammable conventional electrolytes mainly contribute to catching fire and explosion of current lithium ion batteries, which promotes the studies on safe solid-state electrolytes. Furthermore, solid-state electrolytes with high transference number, high strength and safety are widely regarded as the enabler for lithium metal anode, which presents a high specific capacity and the lowest redox potential. However, the ion and electron transport in electrode, dendrites and volume changes during lithium plating and striping, and compatibility between electrode and electrolytes are major challenges lying on the practical application of solid-state lithium batteries. Liquid electrolytes or gel electrolytes are normally used to deal with the compatibility and kinetic limitation in solid-state batteries and these batteries are named by semi-solid-state batteries or quasi-solid-state batteries, whereas the ones containing no liquid phases are named by all-solid-state batteries. Semi-solid-state or quasi-solid-state batteries can be regarded as a compromise to the reality at the current stage due to the limitation of technology and knowledge to achieve all-solid-state batteries. Here, we use solid-state batteries to cover all these solid systems.

Nanostructures and nanomaterials represent the most promising approaches to improve the interface issues and enhance ion/electron transport in solid-state batteries and further to realize a satisfactory performance of solid-state batteries. For instance, a combination of a 3D current collector and gradient polymerization of polymer electrolytes enables a good interface contact in the cathode and a rigid surface in anode to inhibit lithium dendrites. Hence, in this chapter, apart from a special attention on the ionic conduction (mechanism, design principle, and characterization), more attentions were paid to the nanostructure and nanomaterials to deal with the kinetic and interfacial issues in solid-state batteries.

5.1 Solid-State Electrolytes

Solid-state electrolytes can offer great advantages in safety and stability of batteries but meanwhile most troubles or issues faced in solid-state batteries are induced by them. The development of solid-state electrolytes flourished the studies of solid-state batteries and the current state of solid-state electrolytes determines the future of solid-state batteries. The most surprising achievement recently in solid-state electrolytes is the discovery of a Li^+ superionic conductor—$Li_{10}Ge_2PS_{12}$ (LGPS), which exhibits a comparable ionic conductivity with the one of commercial aprotic electrolytes [1]. This study encouraged more and more efforts on understanding and design of the solid-state electrolytes. In this section, several aspects involved in solid-state electrolytes, including the development and current state, the ionic conduction mechanism and design principles, the thermal and electrochemical stability, and characterization methods will be illustrated.

5.1.1 History and Current State of Solid-State Electrolytes

With respect to safety issues caused by flammable liquid electrolytes, the solid-state electrolyte has been one of the hottest topics in the field of energy storage and the achievement of superionic materials in past decades provides a great convenience for realization of solid-state electrochemical devices. In this section, a brief history and current state of superionic materials will be introduced.

The beginning study about the ionic conductivity can be traced back to 1883, at which time Michael Faraday stated that the conducting power of silver sulfide (Ag_2S) and lead fluoride (PbF_2) can be enhanced with increased temperature. After nearly 100 years, the diffusion theory based on defects in crystalline materials was proposed, which provides macroscopic insights into the ionic conduction. The energy crisis at the middle of last century highly encouraged the development of batteries, such as high temperature Na–S batteries, fuel cells, which provoked the development of O^{2-}, Ag^+, Na^+, and Li^+ ionic conductors. Two well-known examples for superionic conductors are AgI and PbF_2, as stated by Michael Faraday. For AgI, the ionic conductivity can reach up to 2 S cm^{-1} after the phase transition from β-phase (hexagonal wurtzite: $P6_3mc$) or γ-phases (cubic zinc blende: F43m) to α-phase at around 147 °C [2]. In order to stabilize the α-AgI at room temperature, numerous efforts have been made and one of the most successful approaches is chemical doping. As a consequence, $RbAg_4I_5$ was discovered with a Ag^+ conductivity of 0.26 S cm^{-1} at room temperature in 1960s [3, 4]. Based on this superionic conductor, an all-solid-state battery or capacitor was constructed and presented an excellent power density (25 mC cm^{-3} at 1240 C) [5]. Later on, more and more studies about H^+, O^{2-}, Na^+ and Li^+ conductors appeared due to the demands of electrochemical energy storage and conversion based on Na^+ and Li^+ ions. For instance, the solid-state electrolytes based on β-alumina for high temperature Na–S batteries was explored in the mid of 1960s [6]. The β-alumina presents a crystalline structure with special anisotropic ionic diffusion pathway and can be applied in both Li^+ and Na^+ conductors. Although Li^+ ions possess smaller ionic radius, the conductivity for Li^+ conductor based on β-alumina is two orders of magnitude lower than that in Na^+ or Ag^+ ones [6–8]. With the increasing demands on the energy density and power density, lithium ion batteries have become the prevalent electrochemical energy storage and conversion system. With respect to this, the focus of this part will be devoted to Li^+ conductors.

Generally speaking, solid-state electrolytes can be basically divided into two categories: inorganic and polymer-based materials. Inorganic solid-state electrolytes can possess higher ionic conductivity for the rigid ionic transport pathway at room temperature, while polymer-based solid-state batteries can be more easily fabricated for the flexible feature and low cost.

The first study about polymer-based solid-state electrolytes was reported in 1973 by Fenton et al., where alkali-metal ions were dissolved into poly(ethylene oxide) (PEO) [9]. And further, the intensive study about the ionic conductivity was carried out by Armand et al., who illustrated the possible application of this polymer-based electrolytes in battery field [10]. Latter, the polymer matrix was extended to poly(acrylonitrile) (PAN), poly(methyl methacrylate) (PMMA), poly(vinylidene fluoride) (PVDF), poly(vinylidene fluoride-hexafluoropropylene) (PVDF-HFP). The hottest spot in this field is mainly focused on improving the ionic conductivity at ambient temperature, enhancing the mechanical strength and Li^+ ions transference number. It has been shown that the ion migration in polymer matrix is mainly dependent on the polymer chain assisted hopping mechanism, which normally occurs in the amorphous part of polymers. Hence the conductivity of solid polymer electrolytes at room temperature is rather low [11]. The first approach to deal with this issue is adding liquid plasticizers to lower the glass-transition temperature, such as ethylene carbonate, propylene carbonate etc. On the other side, adding liquid plasticizers would decrease the mechanical properties of electrolyte, which is not favorable for dendrite-free lithium deposition. Later on, nanosized inorganic fillers such as TiO_2, SiO_2, Al_2O_3 etc. were developed, and the conductivity at room temperature can reach up to 10^{-5} S cm^{-1} [12]. Recently, the development of inorganic Li^+-superionic conductors provided more and advanced choices for composite electrolytes (polymer and inorganic electrolytes). Several critical parameters for inorganic fillers, such as particle size, weight ratio between fillers and polymers, morphology of fillers, should be optimized for the best ionic conductivity. The other possible and effective approach to decrease the crystallinity and increase the ionic conductivity is cross-linking, for the intrinsic properties of polymers can be improved via cross-linking with other polymers. Anyway, how to balance the mechanical properties and ionic conductivity is eternal topic in solid polymer electrolytes. Except from the host matrix, the choice of lithium salts is essential for the ionic conductivity. With respect of the thermal and electrochemical stability, dissociation constant, and solubility, $LiClO_4$, $LiPF_6$, and LiTFSI (Lithium bis(trifluoromethanesulfonyl)imide) are the common choices for polymer electrolytes.

The other issue for polymer electrolytes is the low transference number. The lithium ions associate with polar functional group in matrix and transfer with the motion of the chains, while the anions are free to move as well. On the account of this, the transference number of polymer electrolytes is generally lower than 0.5. The low transference number would result in a strong concentration polarization and facilitate lithium dendrite growth according to the model from Chazalviel [13]. With this respect, more and more efforts are paid to increase the transference number. Generally speaking, the approach to enhance the transference number is decreasing the mobility of anion by inhibiting the anions on polymer matrix or additives/fillers. For instance, the anions can be covalently linking to the polymeric matrix to obtain a single-ion conductor, which can exhibit quite high transference number (e.g., 0.85). Meanwhile the mechanical properties can be improved by introducing polystyrene group to form co-block polymer with a stress of 10 MPa [14].

Similar with polymer electrolytes, the study about inorganic Li^+ ion conductors started at 1970s as well and the initial study focus is lithium compounds such as Li_3N, Li_2S, LiI, and Li_3PO_4 etc. In order to improve the conductivity of these inorganic compounds, some inorganic fillers were applied as well, such as Al_2O_3, P_2O_5, P_2S_5, SiS_2, GeS_2, and B_2S_3. The most successful system is sulfide-based material, for instance, a Li^+ ionic conductor composed of 70 mol% Li_2S and 30 mol% P_2S_5 can present as high Li^+-ion conductivity as 17 mS cm^{-1} after heat treatment [15]. A modified Li_3PO_4 electrolyte with nitrogen (lithium phosphorus oxynitride, LiPON) can be obtained via radio-frequency magnetron sputtering, which can stand for the usage of $LiNi_{0.5}Mn_{1.5}O_4$ as cathode and metallic lithium as anode although the ionic conductivity only reaches 10^{-6} S cm^{-1} [16, 17]. The sulfide-based material with special components after proper heat treatment can present the crystalline structure and this structure is similar with the one containing oxygen (Li_3PS_4 vs Li_3PO_4), hence the sulfide electrolytes can be regarded as the substitution of oxygen with sulfur [18]. This substitution possesses bifunctional impact on the ionic conductivity: enhanced diffusion bottleneck size because of the bigger ionic radius of S^{2-} than that of O^{2-} and decreased dissociating energy for the lower electronegativity of sulfur than oxygen. One of famous Li^+ ion conductors is γ-Li_3PO_4 based materials, such as $Li_{14}Zn(GeO_4)_4$, known as Li^+ super ion conductor (LISICON) [19]. After the total substitution of oxygen with sulfur, the materials were called by thio-LISICON. The breakthrough in Li^+ ion conductor is the discovery of one of thio-LISICON materials—$Li_{10}Ge_2PS_{12}$ (LGPS), which exhibits a Li^+ ion conductivity of 12 mS cm^{-1}, comparable with the commonly-used aprotic electrolytes. To decrease the cost, Sn and Si can be used to substitute Ge in LGPS and the material containing Sn or Si can present quite high conductivity as well. Recently a newly-developed Li^+ ionic conductor based on LGPS is $Li_{9.54}Si_{1.74}P_{1.44}S_{11.7}Cl_{0.3}$ in 2016, which presents the ionic conductivity of 25 mS cm^{-1} [20].

Although sulfide-based solid-state electrolytes present the highest ionic conductivity, the narrow electrochemical window and poor thermal and chemical stability are the main drawbacks. The exploration about oxygen-based solid electrolytes has never ceased, for instance, NASICON-structured (Na^+ super ion conductor) solid electrolytes were developed in 1970s, perovskite typed electrolytes were found in

1980s, and garnet-typed solid electrolytes were proposed in 2000s [21–23]. The highest ionic conductivity among perovskite-structured oxide electrolytes occurs in $Li_{0.33}La_{0.56}TiO_3$ (LLTO), reaching up to 1.5×10^{-3} and 2×10^{-5} S cm^{-1} for bulk conductivity and overall conductivity, respectively. NASICON-structured electrolytes, such as $Li_{1.3}Al_{0.3}Ti_{1.7}(PO_4)_3$ (LATP), and $Li_{1.3}Al_{0.3}Ge_{1.7}(PO_4)_3$ (LAGP), can present a bulk conductivity as high as 3×10^{-3} S cm^{-1} [24]. But these superionic conductors possess a detrimental drawback, which is unstable with metallic lithium resulting in a mixed (electronic and ionic) conducting layer. This mixed conducting layer would eventually lead to a death of a battery due to the short circuit. Among these, Li-rich garnet-structured electrolytes ($Li_7La_3Zr_2O_{12}$) exhibit promising properties, for instance: a high ionic conductivity (more than 10^{-3} S cm^{-1}) and a wide electrochemical window (stable with metallic lithium). Furthermore, some other new types, such as hydrides and halides, were studied as well. For instance, $LiBH_4$ can possess an ionic conductivity of as high as 1 mS cm^{-1} at the temperature of higher than 110 °C, which induces the phase transition from the orthorhombic phase to hexagonal phase [25].

As mentioned above, the different-structure solid-state electrolytes possess varied physicochemical properties, which determine their practical application. The physicochemical parameters including ionic conductivity, chemical and electrochemical stability, and mechanical properties, are summarized in some special reviews as well as in Table 5.1.

Table 5.1 Various properties of solid-state electrolytes

Categories	Materials	$\sigma_{Li+,max}$ (S cm^{-1})	Young's modulus (GPa)	E_{stable} (V vs Li^+/Li)
Oxides	LLTO	10^{-3}	–	2–3.8
	LATP	10^{-3}	143	2.5–4.8
	$Li_7La_3Zr_2O_{12}$	10^{-3}	150	0.05–3.2
Sulfides	Li_3PS_4; LGPS; Li_2S:P_2S_5;	10^{-2}	20–40	2–2.5
Hydrides	$LiBH_4$; $LiNH_2$	10^{-4}	–	0.5–2
Halides	Antiperovskite Li_3OCl	10^{-3}	–	0–3
Thin film	LiPON	10^{-6}	77	0.8–1.2
Polymers	PEO	10^{-4}	<10	to 4.0
	PAN			to 4.3
	PVDF			to 4.5

5.1.2 Ion Transport Mechanism in Solid-State Electrolytes

After polymer electrolytes were proposed in 1973 and later applied in the battery field in 1978, the study about the mechanism of ionic diffusion was intensively conducted and a common acknowledgement was achieved that the segmental motion in amorphous phase of polymeric matrix drives the ions which formed complex with the moieties (such as C–O–C, C–NH–C, and C–S–C) to migrate, see Fig. 5.1. It is well known that one of cyclic oligomers (crown ethers) possess excellent coordination with metal ions and linear oligomers can form complexes with metal ions as well. Hence, the interaction between cations and functional moieties in polymeric chains is crucial for the dissolution and dissociation of salts. Low lattice energy of the salt and a high dielectric constant of the polymer host are prerequisite for fast ion diffusion. Hence, lithium salts, such as $LiClO_4$ and LiTFSI, and host polymers such as PEO, PAN, and PVDF are normally used in polymer electrolytes. Apart from the kinds of salts and host polymers, the ratio between salts and polymers determines the dissociation of salts and further the ionic conductivity. The optimal salt concentration needs a patient seeking because the dissociation of salts decreases with the increase of salt concentration. In some PEO systems, the optimal molar ratio Li/O (the molar ratio between lithium salts and ethylene oxide units) is around 0.04. Since the migration of cations is dependent on the segmental motion of polymer chains, an increase of segmental motion would induce a high ionic conductivity.

The polymer architectures and temperature play important roles for the segmental motions. An empirical relationship between temperature and ionic conductivity (σ) for polymer electrolytes can be described by Vogel–Tammann–Fulcher (VTF) behaviors:

$$\sigma = \sigma_0 T^{-1/2} \exp(-B/(T - T_0)) \tag{5.1}$$

where B has the dimensions of energy and can be regarded as pseudo-activation energy. T_0 is a reference temperature, which is normally 10–50 K below glass

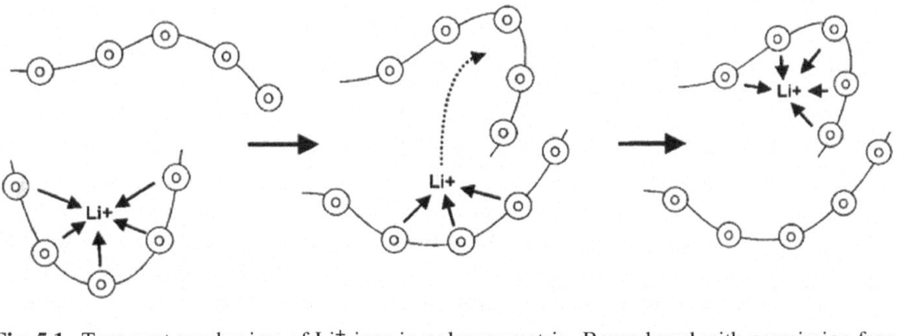

Fig. 5.1 Transport mechanism of Li^+ ions in polymer matrix. Reproduced with permission from Ref. [11], Copyright ©1998 Wiley-VCH

transition temperature (T_g) of the host polymer. The segmental motion would be fast given the low T_g. Hence, the design principle of the polymer architecture is to obtain the polymer with a low T_g and the detailed approaches include cross-linking with special polymer, such as polystyrene or siloxane, adding liquid plasticizers or inorganic fillers.

The ionic diffusion in crystalline materials possesses rigid diffusion pathway, which can be one-dimension, two-dimension or three-dimension dependent on the crystalline structure. The famous ionic conductors with two-dimensional diffusion pathway are Li_3N and β-alumina. The ionic conductivity in Li_3N presents strong anisotropic properties, e.g., 1.2×10^{-3} S cm^{-1} in the direction perpendicular to c-axis while two orders of magnitude lower in parallel direction at room temperature [26]. From the macroscopic view, ions can hop from the site they reside to an adjacent occupancy overcoming certain activation energy and these motions can be separate or cooperative. The defects in perfect crystals can be created by thermal generation or charged impurities, which can be divided into two categories: Schottky defect (cation and anion pairs simultaneously leave their lattice sites and vacancy pairs are left) and Frenkel defect (cations or anions leave their lattice sites into interstitial sites and vacancies are left). When Schottky defects predominate in given structure, ions can migrate directly into adjacent vacancy sites (Fig. 5.2a), while the ions diffuse through the interstitial sites in the case of Frenkel defects. It is notable that the mechanisms can be mixed or cooperative, for instance, Frenkel defect mechanism could be only one stage in a long range conduction process. Furthermore, cooperative ion migration can be found in many fast ionic conductors. Recent report according to the first principle calculation has shown that the fast ionic conductivity is originated from the multi-ion concerted migration, which would induce a low activation energy [27]. Cooperative ion migrations such as knock-on mechanism can be found in Na^+-β-alumina as well, where Na^+ located at the interstitial site in conduction plane can migrate to nearby site only given the Na^+ ions lying at the nearby site move first. Ions occasionally leaving their sites to partially occupied crystallographically equivalent sites must pass through a common face or bottleneck points. Whether the shortest length from face center to corner is shorter or larger than the sum radii of mobile ions and peripheral ions, the thermal energy must be supplied to open the face or derive the ion to leave their present site and through the saddle position of the potential-energy curve (Fig. 5.2b). The theoretical description about the ionic conductivity and activation energy can be semi-quantitatively achieved by a random walk theory for the occasional hopping of ions is apparently at random.

Generally, the ionic conductivity σ is proportional with the ion concentration c, charged numbers q, and mobility u of mobile ions:

$$\sigma = cqu. \tag{5.2}$$

According to the Einstein relationship

$$u = qD/kT, \tag{5.3}$$

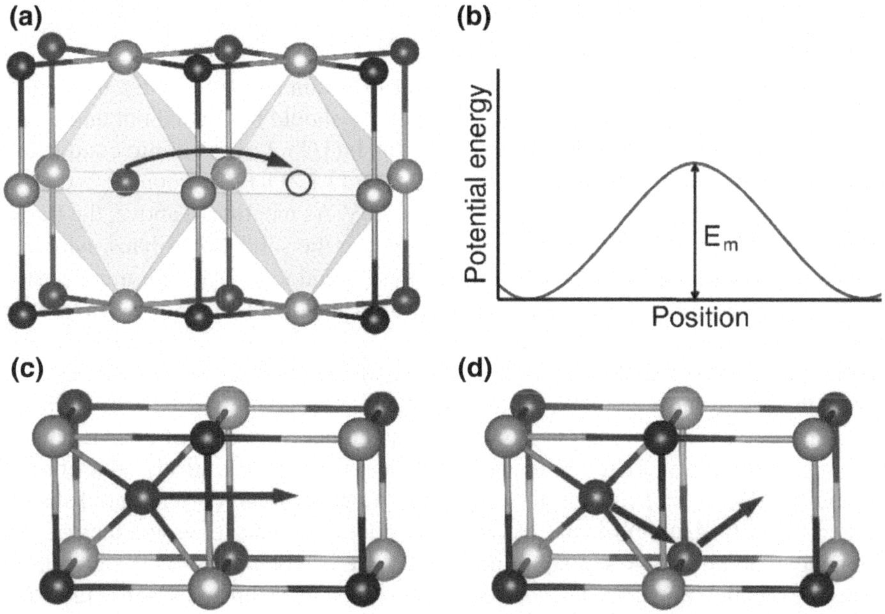

Fig. 5.2 Several hoping mechanisms in the ionic motions. **a** Vacancy hopping mechanism and corresponding potential energy profile for the hopping (**b**); **c** Interstitial hopping mechanism; **d** Knock-on hopping mechanism. Reproduced with permission from Ref. [8], Copyright © 2016 American Chemical Society

where D is the diffusion coefficient, k is Boltzmann constant and T is temperature. Hence, the conductivity σ can be related with the diffusion coefficient D, giving

$$\sigma = cq^2D/kT. \quad (5.4)$$

The diffusion coefficient D can be described according to the kinetic express:

$$D = D_0 \exp(-\Delta G_m/RT), \quad (5.5)$$

ΔG_m is the Gibbs free energy of ion motion. Combining Eqs. (5.4) and (5.5) gives normally seen equation

$$\sigma = (A/T) \exp(-E_a/RT), \quad (5.6)$$

where the prefactor A and E_a can be described by:

$$A = \left(cq^2D_0/k\right) \exp(\Delta S_m/k) \quad (5.7)$$

$$E_a = \Delta H_m. \quad (5.8)$$

ΔS_m and ΔH_m are the entropy and the enthalpy of this motion, respectively. E_a is called the activation energy, which indicates the energy barrier for this motion and determines the ionic conductivity. In order to obtain the value of E_a, the experiment of temperature-dependent ionic conductivity should be carried out and it can be deduced from the plot slope of ln(σT) versus ln(1/T). For fast ionic conductors (in the range of mS cm^{-1}) such as LGPS, LATP, and LLZO, E_a is around 0.2–0.3 eV and for $RbAg_4I_5$(ca. 0.25 S cm^{-1}) it is only 0.07 eV. As mentioned above, the transference number is an important factor for assessing the solid electrolytes, which is defined by the current fraction carried by the designated ions. The high transference number can alleviate the concentration polarization and facilitate the uniform lithium deposition. In inorganic solid electrolytes, the transference number is nearly unity, but the charge carriers include electrons which could result in strong current leakage. Hence, the electronic conductivity must be taken into account when designing inorganic electrolytes. The measurements of the transference number can be carried out in four ways, complex alternative current impedance spectroscopy, steady-state current method developed by Bruce et al., isothermal transient ionic current and Tubandt method [28]. To distinguish the electronic conductivity and ionic conductivity in inorganic solid electrolytes, a small potential pulse can be placed on the measured electrolytes with ionic blocking electrodes, developed by Hebb and Wagner [29, 30]. The detailed discussion about these methods would be elaborated below.

Aside from the bulk conductivity, the surface or interface conductivity is crucial for nanomaterials. For two materials or two grains of a material, the charge carrier concentration at the interface or surface can be totally different with the one in the bulk. Due to different chemical potentials of different materials, the mobile charges at the surface could be accumulated or depleted at the surface to alleviate the potential difference. The accumulation or depletion regions are called space charge layers. In most cases, space charge layer would create a barrier for charge transfer or diffusion, and therefore lower the ionic conductivity or the reaction. Some strategies can be applied to alleviate this effect in solid-state batteries and detailed discussion will be given in Sect. 5.1.5. Nanostructure can affect the space charge layer and further the conductivity can be tuned through the structure tuning. A famous example is epitaxial BaF_2/CaF_2 multilayers with ultrathin layer thickness, which present an enhanced ionic conductivity due to the significant enlarging of the space charge layer, which was reported by Sata et al. [31]. Nanomaterials present different electronic structure and more pronounced interfacial effects on ionic conductivity. Nanostructured AgI as a good example to study ionic conductivity can present much higher ionic conductivity than bulk AgI and α-AgI can be stable even at annealed temperature of 100 °C in AgI nanoplates [32]. Furthermore, nanoconfined systems such as $LiBH_4$ in ordered mesoporous silica, LiI or Li_2ZnI_4 in nanoporous Al_2O_3 can present several orders of magnitude higher ionic conductivity than the one without nanoporous matrix [33, 34]. Except from the nanosizing effect on space charge layer, doping can significantly shorten or narrow the space charge layer. For example, 8 mol% yttrium doped zirconium oxide possesses the space charge layer thickness of only 2–3 nm at 500 °C, which has little effect on the ionic conductivity [35]. Therefore, the effect from doping with optimized concentration on the space charge layer is a big topic in nanoionics. Overall speaking, size effects and grain boundary should be taken into account when designing or optimizing the ionic conductivity.

5.1.3 Optimization of Ionic Conductivity

5.1.3.1 From LISICON to Thio-LISICON

LISICON structured solid-state electrolytes were proposed by Hong in 1977, which are based on γ-Li_3PO_4 (orthorhombic structure, Pmnb) [19]. According to the Eqs. 5.6 and 5.7, the concentration of mobile ions or vacancies plays a key role in enhancing ionic conductivity, and a normal approach to optimize this is doping with aliovalent elements. For instance, the P^{5+} ions were partially substituted by Si^{4+} to form a $Li_{3+x}P_{1-x}Si_xO_4$ system in γ-Li_3PO_4 and therefore the excess Li^+ ions were regarded to locate at migratory sites [36]. The conductivity can be significantly improved after partially doping with Li_4SiO_4 and one half of Li_3PO_4 with Li_4SiO_4 present the highest ionic conductivity and lowest activation energy. Moreover, the skeleton-typed structure composed of PO_4 and SiO_4 tetrahedrons possesses cavities for ion diffusion and these cavities can be optimized by doping as well. Therefore, a series of lithium salts (Li_3AO_4) composed of penta-valence ions A^{5+}, such as P^{5+}, V^{5+}, As^{5+} were partially substituted with the one (Li_4BO_4) containing tetra-valence ions B^{4+}, such as Ti^{4+}, Ge^{4+}, Si^{4+} etc. to hunt for a big bottleneck size and a fast ionic conductivity. The experimental data showed that the $Li_{3.5}V_{0.5}Ti_{0.5}O_4$ presents an ionic conductivity of 4×10^{-5} S cm^{-1} at room temperature [37].

In spite of numerous attempts and optimization for this family of materials, the ionic conductivity still cannot meet the practical requirement. In 2000s, Kanno et al. first reported the substitutions of oxygen with sulfur in LISICON structured materials and called this by thio-LISICON, but the first attempt is not that successful. Although a series of materials were prepared, the highest ionic conductivity is only 6.5×10^{-5} S cm^{-1} for the system of Li_2S-GeS_2-Ga_2S_3 [18]. Surprisingly, a breakthrough was achieved, where Li_2S-GeS_2-P_2S_5 solid-solution system ($Li_{4-x}Ge_{1-x}P_xS_4$) was developed and the highest conductivity can reach 2.2×10^{-3} S cm^{-1} for $x = 0.75$ with a low activation energy of 0.21 eV [38]. Compared with oxides, sulfide solid electrolytes present high ionic conductivity due to the bigger ionic radius and polarizability of S^{2-} ion. And then a solid-state lithium ion battery with graphite as anode and $LiCoO_2$ as cathode was developed and exhibited a comparable electrochemical performance with commercial lithium ion batteries [39].

Here a special attention was focused on LGPS which is regarded as the milestone study in Li^+ superionic conductors for its conductivity (12 mS cm^{-1}) at room temperature comparable with commercial aprotic electrolytes [1]. LGPS is one of derivates in the Li_2S-GeS_2-P_2S_5 and crystallizes in tetragonal structure with the space group $P4_2/nmc$ (Fig. 5.3a), where Ge atoms and half of P atoms occupy tetrahedral 4d sites surrounded by S atoms and the other half of P atoms locate at tetrahedral 2b sites. The lithium ions located at tetrahedral 8f and 16 h sites migrate in the 1D channel constructed by the abovementioned tetrahedrons with low activation energy of 0.17 eV (Fig. 5.3b). Aside from the fast 1D channel, it is reported that the lithium ions can migrate in ab plane as well in spite of the activation energy of 0.28 eV [40]. Although Ge^{4+} in LGPS can be substituted by a bigger tetra-valent ion Sn^{4+}, it does not exhibit better ionic conductivity (4 mS cm^{-1}) than the one in LGPS [41]. A further optimized sulfide electrolyte ($Li_{9.54}Si_{1.74}P_{1.44}S_{11.7}Cl_{0.3}$) was developed and the ionic conductivity (25 mS cm^{-1}) and electrochemical stability get improved [20].

Another case for solid electrolytes containing sulfur and halide is in the form of Li_6PS_5X (X = Cl, Br, or I), which exhibits an argyrodite structure, analog of Ag_8GeS_6 [42]. This type of solid electrolytes presents a high ionic conductivity of 7×10^{-3} S cm^{-1} and a good chemical and electrochemical stability. The effect of halides on the ionic conductivity was studied and the results indicated the ones containing Cl and Br are much higher than the one with I, which cannot be explained by the bottleneck size or bonding between Li^+ ions and halides [43]. The possible reasons given by Wagermaker et al. are that the distribution of halides and sulfurs is not uniform and higher conductivity for the ones containing Br and Cl is attributed to the preference at the sites of 4a and 4c with a certain ratio [44].

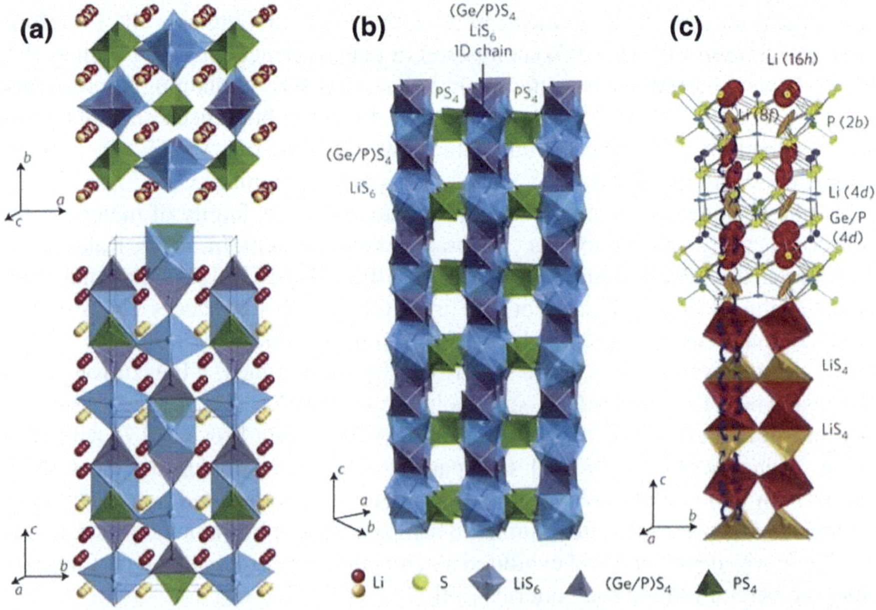

Fig. 5.3 Crystalline structure of LGPS and Li^+ transport pathway in LGPS. **a** Crystalline structure of LGPS viewed from c and a axis; **b** transport path constructed by LiS_6 and $(Ge/P)S_4$ polyhedral; **c** Li hoping between 16 h sites and 8f sites. Reproduced with permission from Ref. [1], Copyright © 2011 Nature publishing group

5.1.3.2 NASICON-Structured Electrolytes: Grain Boundary Effects

NASICON-typed solid electrolytes gained lot of attentions for their high ionic conductivity and good thermal stability. Goodenough and Hong first reported a polycrystalline material—$Na_{1+x}Zr_2Si_xP_{3-x}O_{12}$ in 1976 [21]. This material presents the rhombohedral structure in a space group of *R3c* and has a suitable channel size for migration of Na^+ ions. Inspired by this study, Li-containing NASICON-structured materials such as $LiGe_2(PO_4)_3$, $LiTi_2(PO_4)_3$, $LiZr_2(PO_4)_3$ and $LiHf_2(PO_4)_3$ were developed. In these materials, octahedron MO_6 and tetrahedron PO_4 are linked by the shared corner to form an $M_2(PO_4)_3$ framework and Li^+ ions move in the three-dimensional cavities in the structure. The bottleneck sizes and the activation energy for bulk conductivity change with the tetravalent ions in NASICON structures. Figure 5.4a shows the relationship between activation energy and unit cell volume. Surprisingly the minimum activation energy (around 0.3 eV) is observed in the sample with suitable volume size (1.31 nm^3) not in the biggest in $LiHf_2(PO_4)_3$. The optimized suitable volume is identical with that of $LiTi_2(PO_4)_3$. It is easy to understand that the one with a smaller tunnel size presents higher activation energy because the bottleneck size is too narrow to freely migrate. On the other side, in the one with bigger tunnel size the strong interaction between Li^+ ions and O^{2-} plays the key role, resulting in high activation energy. Furthermore, the tetravalent ions in NASICON structure can be substituted with other aliovalent ions. In the system of $Li_{1+x}Al_xTi_{2-x}(PO_4)_3$, the substitution of Ti^{4+} with Al^{3+} leads to a high active Li^+ ions concentration. The highest conductivity can be obtained in the case of $Li_{1.3}Al_{0.3}Ti_{1.7}(PO_4)_3$ and can reach up to 3×10^{-3} S cm^{-1} [24]. Perovskite-structured oxide materials was attempted to be designed as Li^+ ionic conductors and the famous system is $Li_{3x}La_{2/3-x}\square_{1/3-2x}TiO_3$ where $x = 1/6$ gives the highest ionic conductivity. The substitution of Li^+ with La^{3+} ions creates numerous vacancies and Li^+ ions can migrate through these vacancies [22]. But the increased conductivity in NASICON after doping with Al^{3+} ions is believed to be attributed to the decreased grain boundary resistance rather than increasing the ionic mobility since the activation energy for bulk conductivity keeps stable around 0.3 eV regardless with doping [24]. In NASICON-typed conductors, tuning the grain boundary resistance is crucial for the improvement of ionic conductivity. Additives, such as lithium compounds Li_2O, Li_3PO_4, and $LiBO_3$, which cannot form solid solution with NASICON electrolytes were used to increase the density of ceramic electrolytes, which only play binders or solders to facilitate the formation of dense ceramics [45–47]. Figure 5.4b gives the experimental data about the effects of Li_2O additive on the activation energy for bulk and grain boundary resistance [45]. It is notable that the content of additives needs to be optimized as well for different systems.

5.1.3.3 Li-Rich Garnet

Garnet-structured electrolytes were developed after 2000 and have raised numerous attentions owing to their high chemical stability against metallic lithium and high ionic conductivity [23, 48]. The garnet structure in formula of $A_3B_3C_2O_{12}$ contains a

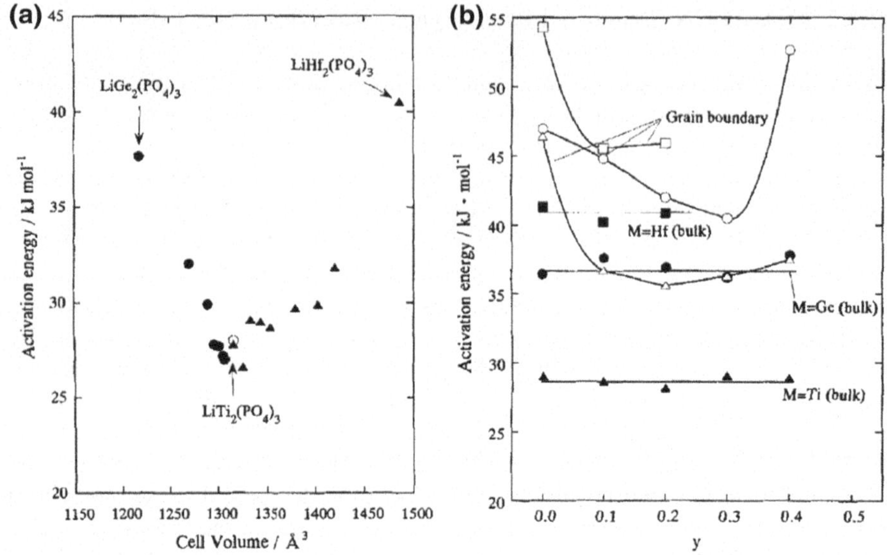

Fig. 5.4 **a** The relationship between unit cell and activation energy in solid-solution $LiM_xTi_{2-x}(PO_4)_3$, M = Ge^{4+}, Hf^{4+}. **b** The changes of bulk and grain boundary activation energy with additive level (Li_2O). Reproduced with permission from Ref. [47], Copyright © 1996 Wiley-VCH

$B_3C_2O_{12}$ framework with B cations in dodecahedral sites and C cations in octahedral sites. The framework contains a 3D-connected interstitial space consisting of the tetrahedral 24d-A sites. The stoichiometric garnet such as $Li_3Nd_3W_2O_{12}$ shows a very low Li^+ conductivity because of the fully occupied 24d sites. This situation was changed with the discovery of Li-rich garnet $Li_5La_3M_2O_{12}$ (M = Nb, Ta) in 2003 and the ionic conductivity for them can reach 10^{-6} S cm^{-1} [48]. A 3D Li^+ ion transport pathway composed of energetically equivalent and partially occupied sites was supported by a bond-valence analysis. Further, La^{3+} ions are partially substituted with bivalent alkali earth Ba^{2+}, Sr^{2+}, and Ca^{2+} to produce the garnet containing 6 Li^+ ions and the one containing Ba^{2+} garnet $Li_6BaLa_2Ta_2O_{12}$ exhibited a high ionic conductivity of 4×10^{-5} S cm^{-1} and an activation energy of 0.4 eV due to the biggest ionic radius of Ba^{2+} among alkali earth elements [49]. In 2007, a garnet containing 7 Li^+ ions $Li_7La_3Zr_2O_{12}$ was developed by Weppner et al. and presented a bulk ionic conductivity of 7.74×10^{-4} S cm^{-1}, which flourished the study on garnet-structured Li^+ ionic conductors [23, 50, 51]. With increasing Li^+ content in garnet, the Li^+ occupancy at tetrahedral sites slightly decreases and the ones at octahedral sites increase, as shown in Fig. 5.5 [50]. The study from Janek et al. indicated that the Al-doping on purpose or not can stabilize the cubic phase and the one without Al doping presents tetragonal phase with a low ionic conductivity of 2×10^{-6} S cm^{-1}. Further optimizations such as doping, sintering temperature, atmosphere and pressure were carried out and the overall ionic conductivity for garnet containing 7 Li^+ ions can be higher than 10^{-3} S cm^{-1} [52, 53].

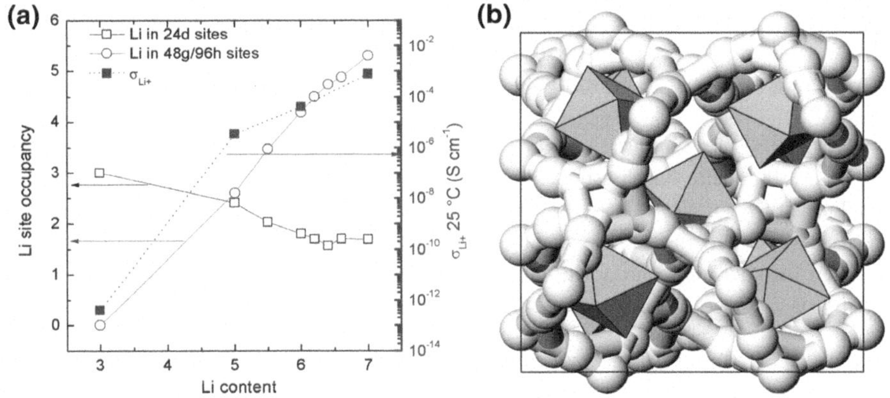

Fig. 5.5 **a** Li^+ occupied sites in garnet typed electrolytes with different Li^+ content [53]; **b** Crystalline structure of LLZO and the Li transport pathway [51]. The octahedrons and balls represent the ZrO_6 and lithium, respectively. **a** was reproduced with permission from Ref. [50], Copyright © 2014 Royal Society of Chemistry and **b** was reproduced with permission from Ref. [51], Copyright 2011 © American Chemical Society

5.1.3.4 Hydride and Halide-Based Solid-State Electrolytes

Hydrides were widely studied as materials for the storage of hydrogen and can be used as fast ionic conductors as well. The famous one is $LiBH_4$, which undergoes a phase transition from orthorhombic structure to hexagonal structure at the temperature of 100 °C. The $LiBH_4$ with hexagonal phase exhibits a high ionic conductivity of 2×10^{-3} S cm^{-1}, hence stabilization of this phase is the primary topic in this field [54]. Substitution of complex ions $[BH_4]^-$ (0.205 nm) with iodide ions I^- (0.211 nm) and confining $LiBH_4$ into ordered mesoporous silica were used to stabilize hexagonal phase at room temperature. For instance, solid solution $Li_4(BH_4)_3I$ exhibits as high ionic conductivity as 2×10^{-5} S cm^{-1} at 300 K [55]. Complex hydrides such as $Li_2(BH_4)(NH_2)$ and $Li_4(BH_4)(NH_2)_3$ were studied as well and the ionic conductivity can reach 2×10^{-4} S cm^{-1} with a very low activation energy of 0.26 eV for the later one [56]. Due to the limited stable window, hydride-based electrolytes are normally used for cathode with a low potential, such as sulfur, TiS_2 [57].

Recently, Li-rich antiperovskite Li_3OX (X = Cl or Br) was developed as well and the ionic conductivity for these solid electrolytes can be 2×10^{-3} S cm^{-1} for $Li_3OCl_{0.5}Br_{0.5}$ [58]. A low three-atom hop mechanism involving Li interstitial dumbbells was regarded as the transport mechanism through a computational simulation but it may decompose into Li_2O_2, LiCl and $LiClO_4$ at the potential of 2.5 V [59]. Later on amorphous phase Li_3OCl and doped $Li_{2.99}Ba_{0.005}OCl$ were developed by Braga and Goodenough et al., and these superionic conductors present an ionic conductivity of higher than 10^{-2} S cm^{-1} at room temperature and low activation energy of 0.06 eV [60–62]. It is believed that the high ionic conductivity of this materials is derived from the thermal excitation of quasi-rigid unit modes of the cluster ions and the exceptionally large channel size created by mixing large cluster ions together with the elementary ions [63]. But there are controversial comments about the batteries based on this kind of solid-state electrolytes, such as ever-increasing capacity [64, 65].

Overall speaking, charge carrier concentrations and mobility play the key role in improving the ionic conductivity. Doping or substitution is a widely used approach to achieve this but a patient seeking for the suitable ions and optimized amounts are needed. Except from the bulk conductivity, the grain boundary resistance needs to be taken into account and additives such as lithium salts are needed to increase the density of ceramic or decrease the interfacial energy barrier.

5.1.4 Thermal and Electrochemical Stability of Solid-State Electrolytes

The practical application of solid electrolytes in batteries is strongly dependent on the chemical and electrochemical stability of solid-state electrolytes and these properties of solid-state electrolytes will be elaborated in this section.

The thermal stability involving two parameters—glass transition temperature and melting point is crucial for polymer electrolytes, which not only determines the ionic conductivity but also the safety issue. As mentioned above, decreasing the crystallinity of polymers and lowering the glass transition temperature are the main strategies to enhance the ionic conductivity at room temperature. These properties of polymers can be obtained from the thermal analysis (differential scanning calorimetry and thermogravimetry) and those parameters for different polymers are shown in Table 5.2. The electrochemical stability and compatibility with anode or cathode materials are other essential factors. According to the report from Croce et al., it is recognized that metallic lithium can undergo a passivation process with contact with PAN gel electrolytes and this passivation layer can grow with the time. Since the component in PAN gel electrolytes is complicated, this stability test can be affected by water content, plasticizers and anions of lithium salts [66]. For the anodic stability, the electrochemical stability of PAN is located in the range of 4.3–4.8 V [67]. PMMA-based electrolytes present higher transference number of up to 0.7 than PEO-based electrolytes and the anodic stability higher than 4.5 V versus Li^+/Li as well [68]. The PVDF-based gel electrolytes containing EC and PC as plasticizers and $LiN(SO_2CF_3)_2$ as lithium salts show an anodic stability up to 4.5 V on stainless steel electrode, however the critical issue is the stability against lithium, where LiF can be formed after the reaction between fluorinated polymers and metallic lithium [69]. It is notable that the stability results can vary much dependent on the measured methods, component of plasticizers, electrodes etc.

Heat treatment is requisite for the preparation of inorganic solid electrolytes and the thermal stability of special phase with high ionic conductivity is crucial. The typical cases are AgI and $LiBH_4$, which have been illustrated in the above part. For sulfide-based solid-state electrolytes, it is well known that they can react with moisture in atmosphere and release very harmful gas—H_2S [70]. Recent study showed

Table 5.2 Thermal properties of polymer hosts

Polymer host	T_g/(°C)	Melting point/(°C)
PEO	− 64	65
PAN	125	317
PMMA	105	Amorphous
PVDF	− 40	171

that the solid-electrolytes based on Li_2S–SnS_2 can be stable in air [71]. Besides, the recent study about the phase stability has shown that the $Li_7P_3S_{11}$ can decompose at the temperature of higher than 280 °C under inert gas and more harmful gas can be produced when it is oxidized in atmosphere [72]. The ones containing P and S all possess this potential crisis. For the super ionic conductor—$Li_{10}GeP_2S_{12}$, it can be partially decomposed into $Li_4P_2S_6$ at the temperature of 600 °C in the inert gas [40]. The oxide-based electrolytes are recognized to be safe and stable in a quite wide temperature range. Because of the high mobility of Li^+ ion in super ionic conductors, the exchange reaction between Li^+ and proton in atmosphere can be found in many electrode materials and solid-state electrolytes. This effect can be ignored in some cases, such as in spinel $Li_4Ti_5O_{12}$ anode. On the other side, the effect can be fatal, for instance LLZO. After exchange reaction between Li^+ and proton, lithium hydroxide and a further product of lithium carbonate are formed, which induce giant interfacial/surface resistance (much larger than the bulk resistance) [73]. Careful storage and processing of garnet LLZO can avoid this effect and further surface modifications with interfacial layer are the effective approaches to deal with this issue as well [74, 75].

The electrochemical stability of solid electrolytes determines the practical voltage of a battery and has been widely studied experimentally and theoretically. Unfortunately, most inorganic solid electrolytes are not stable with metallic lithium and similarly with in aprotic electrolytes an interfacial layer can be formed except from the garnet LLZO. The spontaneously formed interfacial layer can be divided into two categories—ionic conducting layer and mixed (electronic and ionic) conducting layer. The ionic conducting interfacial layers are normally observed in amorphous solid electrolytes such as LiPON and Li_2S:P_2S_5 system. The reduced products after contact with metallic lithium for these samples are mainly Li_3N, Li_3P and Li_2S etc., which can present reasonable Li^+ ionic conductivity. Hence, thin film battery with LiPON as solid electrolytes and metallic lithium as anode can be successfully obtained and operated with quite long life time. For Li_2S:P_2S_5, the interfacial resistance can be the main barrier for the Li^+ transfer between electrolytes and metallic lithium for the low ionic conductivity of Li_2S and Li_3P. For the ceramic electrolytes containing metal ions, the stability of them against lithium is mainly dependent on the redox potential of metal ions. An easy assessment can be conducted based on the enthalpy of the reduction of metal oxides by lithium and the results indicate that only the oxides of Sc, Y, Ca, Be, Mg and La are stable with metallic lithium. As a consequence, the solid electrolytes, such as LGPS, LATP, LAGP and LLTO can be reduced by lithium to induce an interphase including Ge^{x+}, Ti^{3+}, Ta^{4+} and Al^0, which presents high electronic conductivity. Because of this, this interphase is not self-limited and grows very fast with time, eventually resulting in short circuit of the battery [76]. The specific electrochemical window of different electrolytes is normally measured by linear sweep voltammetry (LSV) or cyclic voltammetry (CV). But in some cases, the results obtained from linear sweep voltammetry can be wider than the real situation due to the slow rate of the interfacial reaction. For instance, the electrochemical window for LGPS from CV is from 0 to 5 V versus Li^+/Li, but the real situation is that the LGPS can be reduced in catholic process and oxidized during anodic process, presenting a quite narrow stable window (1.8–2.2 V) [1, 77, 78]. Another trustful approach to obtain the electrochemical stable window is theoretical calculation and lot of work including the compatibility with electrode materials have been completed recently [79, 80]. The detailed results of several solid electrolytes are shown in Table 5.1.

5.1.5 Nanoscale Modification at Solid/Solid Interface

Except from the interface among ceramic electrolytes—grain boundary, the interfaces between solid electrolytes and electrodes can be divided into two categories: the one with blocking electrode, another with non-blocking electrode. For non-blocking electrode interface (including most electrode materials/electrolytes) such as in Fig. 5.6b, the redox pairs Li/Li^+ (the oxidation of Li to Li^+ and the reduction of Li^+ to Li) could soon reach equilibrium given a potential, while in blocking electrode interface only an electrostatic equilibrium (accumulation of electron and cations on the two sides of the interface) is achieved, i.e. no steady current flows. In the later one, the electrostatic equilibrium can be broken when the potential is beyond the electrochemical window of solid electrolytes. As a consequence, the solid electrolytes get oxidized or reduced. The measurements (alternative or direct current perturbations) on these interfaces can be applied to determine the bulk conductivity, grain boundary conductivity, electronic conductivity and interfacial resistance. It is worth noting that the interfaces are not as pure as shown in Fig. 5.6 in the real system. As mentioned above, most solid state electrolytes possess limited stable electrochemical window, therefore varied interphases can be formed between SSE and electrolytes. Hence, there are some buffer layers either in situ formed or inserted on purpose between solid-electrolytes and electrodes.

The chemical and electrochemical stability between solid-state electrolytes and metallic lithium has been illustrated above and the interface has been elaborated. With respect to the different interfaces, modification strategies are listed as following. For solid-state electrolytes stable against metallic lithium, such as LLZO, the interfacial barrier between solid-state electrolytes and LLZO is mainly derived from the impurity phases, such as Li_2O on the surface of lithium or Li_2CO_3 on the surface of LLZO. The modification for this interface is mainly to realize the intimate contact between LLZO and metallic lithium and remove the interfacial barrier.

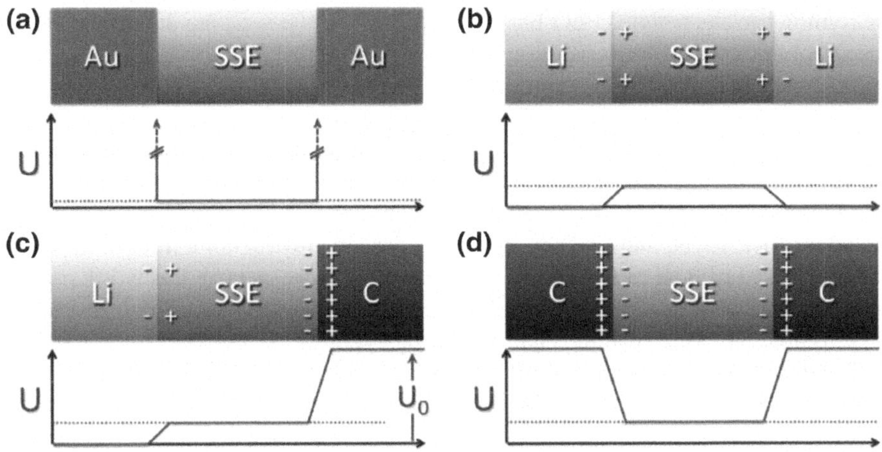

Fig. 5.6 Interfaces lying in solid-state batteries. C stands for cathode. **a** Double blocking electrode systems, which can be used to determine ionic conductivity and electronic conductivity. It is notable that Au electrode is not really blocking electrode for Au can react with Li to form alloys. **b**, **d** Symmetrical non-blocking electrodes, which can be used to determine the interfacial resistances between Li or cathode with SSEs. **c** Solid-state lithium battery. The potential drops at the above-mentioned interfaces are shown below responding systems. Reproduced with permission from Ref. [85], Copyright 2015 © American Chemical Society

Careful processing and storage of LLZO can prevent the production of Li_2CO_3 on the surface and significant reduction of interfacial resistance, as illustrated by Sakamoto et al., where the prepared LLZO pallet can present ignorable interfacial resistance with metallic lithium after suitable polishing and calcination [74]. Moreover, more and more modifications were used to improve the wettability between metallic lithium and SSEs and further to realize the intimate contact. According to the literature, the intimate contact between LLZO and Li can be achieved via inserting functional layers with the thickness of several nanometers, including some oxides such as Al_2O_3, ZnO, and metals such as Mg, Ge, and Al [75, 81, 82]. These functional layers not only can decrease the interfacial resistance between LLZO and metallic lithium, but also prevent the propagation of lithium dendrites in LLZO during the electrodeposition of lithium to some extent. For solid-state electrolytes, which are not stable with lithium and mixed conducting interphases are formed at the interface, such as LATP, LAGP, LLTO, and LGPS, a modification layer was utilized mainly to prevent the direct contact with metallic lithium. This layer should be stable (whether kinetically or thermally) against metallic lithium and possess very high ionic conductivity and low interfacial resistance at the two newly occurred interfaces. Although LLZO layer prepared by sputtering was used as protection layer for LATP electrolytes, the new issues caused by LLZO as mentioned above is still a challenge [83]. LiPON and PEO-based polymer electrolytes were regarded to be effective protection layer for LATP or other solid electrolytes. PAN-based polymer electrolytes were recognized to be unstable with metallic lithium but can stand for higher potential, hence an asymmetric electrolyte (PEO-based polymer electrolytes facing anode and PAN-based polymer electrolytes facing cathode) was designed and successfully applied in solid-state metallic lithium batteries [84]. For the rest solid electrolytes, which is unstable with lithium but an ionic conducting dominant interphase is formed, such as LiPON, Li_3PS_4, Li_3PO_4, and $LiZr_2(PO_4)_3$, the in situ formed interphase plays the key role for the electrochemical performance of the battery. The famous case is LiPON, which is unstable with metallic lithium but highly conductive interphases including Li_3P, Li_3N, and Li_2O are formed, consequently metallic lithium thin film batteries based LiPON can work for more than 10,000 cycles [16]. It is worth paying attention that most interphases would induce interface issue due to high resistance and unstable interface (volume changes during cycling).

The interfacial resistance in solid-state lithium batteries could be originated from the anode part and cathode part, but more and more study suggested that the interfacial resistance in cathode/SSE is dominating the power density [85]. The potential drop from cathode to anode can be several voltages and it can be observed that the significant drop occurs at the interface between cathode and SSE, which would induce more fundamental difficulties to optimize the interface, as shown in Fig. 5.6c. The potential drop in lithium ion batteries based on liquid electrolytes can be screened by several approaches, such as concentration polarization, dipole change of solvents, and electrochemical double layer, but in solid-state electrolytes, only mobile ions or holes accumulation or depletion work. For inorganic solid electrolytes, an interlayer playing the role to screen the potential drop is crucially important. In order to achieve this, polymer electrolytes were widely applied to screen the potential drops. The recent study has shown that the stability of solid batteries with polymer/ceramic/polymer structure is much better than the one without this structure, which proves the successful strategies to alleviate the interface issue [86].

5.1.6 Electrochemical Characters of Solid Electrolytes

To satisfy the future demands from electric vehicles and energy storage, the solid electrolytes should meet the following requirements:

(1) Ionic conductivity at working temperature should be higher than 10^{-4} S cm^{-1} and meanwhile the electronic conductivity should be lower than 10^{-10} S cm^{-1}.
(2) The electrochemical stable window should be as large as possible (0–6 V vs Li^+/Li).
(3) Good compatibility with electrode materials.
(4) Good thermal stability and mechanical properties.
(5) Low cost and facile preparation.

The abovementioned electrochemical properties can be measured via some electrochemical methods, such as electrochemical impedance spectroscopy, cyclic or linear voltammetry. The following part will give some short and practical introduction about these methods. In order to obtain the crystalline structure and other physico-chemical properties, many other methods such as X-ray-based scattering and adsorption spectroscopy, light-based scattering and absorption spectroscopy, neutron-based scattering, solid-state NMR can be applied and will not be elaborated in this chapter. The readers interested in these techniques can find them in the literatures [87].

Electrochemical impedance spectroscopy (EIS) was applying an alternative voltage/current with a low amplitude to one system and then recording the response of this system at different frequency (normally from 10^{-2} to 10^7 Hz). After applying an alternating voltage on a system, both resistive and reactive components responses with the function of frequency are recorded and represent the different features of the given system. The different components could be distinguished because the characteristic relaxation time or time constant can be differentiated [88]. Here our attention is focused on ceramic materials and discussion about their EIS responses. For ceramic samples, they normally possess bulk and grain boundary regions and the impedance responses of these regions give two semicircles in Nyquist plot (a plot in form of imaginary (capacitive) against real (resistive) impedance) seeing Fig. 5.7a. The distinction about these two semicircles is very important to understand their physical meanings. It is well-known that a semicircle in Nyquist plot represents a parallel resistor-capacitor (RC) circuit. The issue is assigning RC circuits to different regions (bulk part or grain boundary part). The assignment is based on the magnitudes of the capacitances and the corresponding interpretation for RC with different capacitance value is listed in Table 5.3 [88]. In case of that the range of capacitance for grain boundary and bulk component is overlapped, they can be distinguished with the temperature-dependent impedances due to their different responses to temperature. The bulk and grain boundary resistance can be distinguished in Fig. 5.7. Furthermore, the temperature-dependent EIS measurement can provide the relationship between conductivity and temperature and the activation energy for bulk and grain boundary can be obtained as well according to Eq. 5.6. In NASICON-structured ceramic electrolytes, the bulk and grain boundary can be observed as well and through this distinction the different strategies were applied to enhance the total conductivity as mentioned above [45].

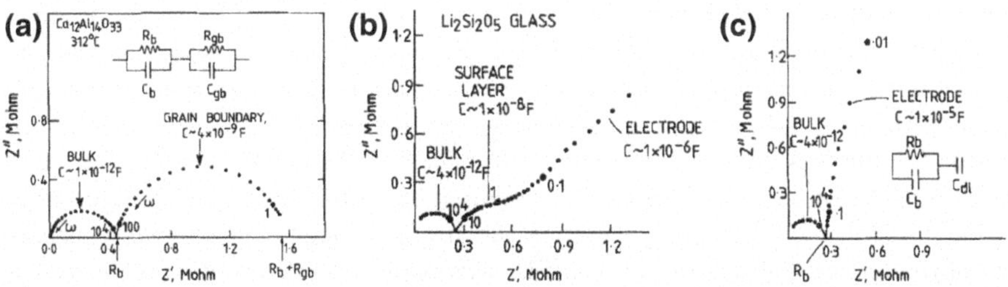

Fig. 5.7 **a** Impedance spectroscopy of oxygen ion conductor $Ca_{12}Al_{14}O_{33}$. The bulk resistance and grain boundary resistance can be clearly distinguished. Impedance spectra of $Li_2Si_2O_5$ with (**b**) and without (**c**) surface layer. Reproduced with permission from Ref. [88], Copyright 1990 © Wiley VCH

Table 5.3 Possible interpretations on impedance spectra. Reproduced with permission from Ref. [88], Copyright 1990 © Wiiley VCH

Capacitance/F	Possible interpretation
10^{-12}	Bulk
10^{-11}	Minor, second phase
10^{-11}–10^{-8}	Grain boundary
10^{-10}–10^{-9}	Bulk ferroelectric
10^{-9}–10^{-7}	Surface layer
10^{-7}–10^{-5}	Sample-electrode interface
10^{-4}	Electrochemical reaction

Surface layers can be observed in ceramic materials, when they are sensitive to the atmosphere. For instance, garnet-structured Li^+ conductor is sensitive to moisture and lithium carbonate can be found on the surface of the ceramic pallet. In this case, the surface layer would result in a big interfacial resistance and the distinction from surface layer and bulk materials can provide novel approach to improve the practical application of this material [73]. The case shown in Fig. 5.7b is lithium silicate glass $Li_2Si_2O_5$, which can form a hydrated or carbonate surface layer during cooling of the melt, similar with ceramic garnet. In the impedance spectroscopy an additional poorly-resolved semicircle represent this layer, which provides a visible proof and method to study the interfacial resistance caused by the formation of surface layer [88]. The interface between ceramic and electrodes is important since it can tell the nature of the conducting species within the ceramic, such as ions or electrons. The interface between ceramic and different electrodes (blocking or non-blocking electrodes) could be represented by the low frequency spike or no spike in impedance spectroscopy. The low frequency spike stands for the charge build-up at the blocking metal electrodes, while the conducting species are electrons due to no impedance barrier for the charge transfer in non-blocking electrode. It is notable that the semicircle for bulk or grain boundary resistance cannot be observed in polymer electrolytes or in ceramic at very high temperature.

The widely-used method to determine the transference number in polymer electrolyte was potentiostatic polarization, developed by Bruce et al. in 1987 [89]. In detail, a small potential perturbation was applied to a battery system composed of polymer electrolytes and two non-blocking electrodes. The response after the perturbation is a current through the system and the current decreases with time until a steady state. Given the very small potential perturbation ~10 mV, assumption of no redox of anions was given and hence the current at the steady state was only carried by the cations. In this case, the transference number of cations can be easily determined by the initial current divided by steady-state current. Since the interfacial resistance can attribute to the total current, the EIS was applied to determine the resistance before and after potentiostatic polarization. After contracting the contribution from the interfacial resistance, the transference number can be deduced:

$$t_+ = I_{ss}(\Delta V - I_0 R_0)/I_0(\Delta V - I_{ss} R_{ss}) \quad (5.9)$$

where I_{ss} is the steady state current, I_0 is the initial current, ΔV the applied potential, R_{ss} and R_0 are the electrode resistances after and before the polarization, respectively.

Besides, galvanostatic polarization, electromotive force method and nuclear magnetic resonance also can be used to determine the transference number. A comparative study about these methods to determine the transference number was given by Gores et al. in their publication [28]. The authors suggest that each method has their own restriction and the selection of the method is dependent on the measured system. For instance, the results from the pulse field gradient NMR cannot distinguish the pair ions and free ions, while galvanostatic polarization includes three measurements and is very time-consuming.

In the ceramic electrolytes, electronic conductivity is a crucial factor because an electrolyte with high electronic conductivity would result in serious current leakage and strong self-discharge of a battery. The determination of electronic conductivity in a mixed conductor was mainly based the method proposed by Hebb and Wagner (Hebb-Wagner method) [29, 30]. In the pristine work (the determination of electronic conductivity of silver sulfide), the suppression of ionic conductivity can be achieved by using blocking electrode such as platinum or graphite. And the potentiostatic polarization was applied and the steady-state current was regarded to be mainly carried by electrons. From the Ohm's law, the electronic conductivity can be obtained. Similarly, the pure ionic conductivity can be obtained using electron-blocking electrode, such as silver iodide electrode. Silver iodide is well known for the Ag^+ conductor but very poor electronic conductor.

5.2 Electrode Structures for Solid-State Batteries

5.2.1 Cathode Structures for Solid-State Batteries

As mentioned above, with more attention on the development of super ionic conductors, nowadays the Li^+ ionic conductivity can be as high as 25 mS cm^{-1} and may be further enhanced. This value has surpassed the one of conventional liquid electrolytes and due to these significant achievements, more and more study focuses are switched to design and optimize solid-state batteries. In solid-state batteries, two major concerns are interfacial issue between solid-state electrolytes and electrode materials and kinetic process. The above section has mentioned that the interfacial issues in cathode part could be more serious than the one in the anode and some strategies have been proposed from the view of solid electrolytes, such as using sandwich-structured electrolytes. It is worth noting that modification of cathode is another effective and more operative strategy to alleviate the interfacial issues.

Furthermore, the solid electrolytes have very poor permeability compared to liquid electrolytes, the contact between electrolytes and electrode materials is not as homogeneous as the one in liquid electrolytes, and therefore the kinetic process (ions and electron transport inside the electrode materials and between electrode materials and solid-state electrolytes) is crucial for the performance of solid-state batteries. Given the diffusion coefficient D, the characteristic constant time for diffusion t can be determined by $t = L^2/D$, where L is the diffusion length. Nanomaterials are regarded as the most powerful and effective strategy to significantly decrease the diffusion length of ion and electron and improve the mixing between nanomaterials and solid-state electrolytes. Maier et al. have showed that the power density in all-solid-state batteries/capacitors (the charge or discharge rate can be as high as 1240 C) can be significantly enhanced by creating suitable interfaces between highly conductive ionic conductors and electronic conductors [5]. Moreover, the nanostructuring could improve the homogenous mixing between solid electrolytes and electrode materials. For instance, researchers have prepared core/shell structure with Li_2S as the core and then Li_3PS_4 in situ formed on the surface Li_2S to achieve the homogenous mixing between cathode and solid electrolytes and the all-solid-state Li–S batteries based on the special design present excellent electrochemical performance [90].

It has been shown that nanomaterials and nanostructuring can be the effective strategies to improve the electrochemical performance of all-solid-state batteries. In this section, several cathode materials, such as sulfur and Li_2S for Li–S batteries, transitional metal oxides, such as $LiCoO_2$ and NCM for lithium batteries were exemplified to illustrate the practical application of nanomaterials and nanostructuring in solid batteries. For organic materials as cathode for all-solid-state battery, the dissolution of these organic cathodes can be totally avoided and a long and stable cycling can be achieved. For instance, an all-solid-state lithium organic battery with pillar[5]quinone ($C_{35}H_{20}O_{10}$) as cathode and poly(methacrylate) (PMA)/poly(ethylene glycol) (PEG)-$LiClO_4$-3 wt% SiO_2 as solid electrolytes exhibits an average operation voltage of ~2.6 V and a high initial capacity of 418 mA h g^{-1} with a stable cyclability (94.7% capacity retention after 50 cycles at 0.2 C rate) [91].

5.2.1.1 Sulfur Cathode

Sulfur was regarded as a promising cathode for the low cost, high abundance and very high theoretical energy density and the major challenge for Li–S batteries is the shuttle effect of the intermediate products—polysulfides. Polysulfides can be dissolved in organic solvents and migrate between cathode and anode, further results in poor lifetime [92]. Theoretically, solid-state electrolytes in lithium-sulfur batteries can eventually overcome the polysulfide shuttle effect and realize a long cycle life. Sulfur and the final discharging product-Li_2S both possess poor electronic and ionic conductivity, and therefore how to optimize the kinetic process is crucial for solid Li–S batteries.

Nanomaterials and nanostructures were widely applied into this field. Ultra-small amorphous sulfur was coated on electronic conductive reduced graphene oxide (rGO) to form rGO@S composites. The rGO@S composites were mixed with superionic conductive $Li_{10}GeP_2S_{12}$ and acetylene black to achieve a high ionic and electronic conductivity. As the $Li_{10}GeP_2S_{12}$ electrolyte is not stable with lithium, a 75%Li_2S-24%P_2S_5-1%P_2O_5 electrolyte layer was put between the lithium anode and $Li_{10}GeP_2S_{12}$ electrolyte [93]. This design provides fast ionic and electronic transport pathways in cathode and offers an excellent electrochemical performance of all-solid-state Li–S batteries (more than 1500 mA h g^{-1} at 0.05 C at 60 °C). The typical all-solid-state Li–S batteries are illustrated in Fig. 5.8. Apart from the rGO, other carbon materials, such as mesoporous carbon, carbon nanotube and carbon nanofibers, were also used to construct the electronic pathways [94]. The composite electrode of sulfur and acetylene black with sizes of 1–10 nm could be prepared via a gas/solid mixing, method and close contact between sulfur nanoparticles and conductive carbon could be simultaneously achieved through this method [95]. A reversible capacity of 900 mA h g^{-1} at a current density of 0.013 mA cm^{-2} was achieved with the thio-LISICON electrolyte. However, when thio-LISICON electrolytes were mixed with carbon, LGPS-carbon composites could serve as both a cathode and an anode, as the LGPS will be oxidized to P_2S_5, S, and GeS_2 at high potentials and be reduced to $Li_{15}Ge_4$, Li_3P and Li_2S at a low potential [78]. This makes the assessment of the function of carbon difficult in solid-state batteries and the interface between carbon and solid state electrolytes should be taken into account for the practical design of solid-state batteries.

The dissolution and preparation of solid electrolytes in organic solvents offers a promising opportunity to obtain homogenous mixture between electrolytes and cathode materials. Li_2S with several nanometers, Li_6PS_5Cl and PVP were dissolved in ethanol to form a homogeneous solution. After precipitation and carbonization process, the Li_2S-Li_6PS_5Cl-C cathode was obtained and can deliver a capacity of 830 mA h g^{-1} for 60 cycles at 50 mAg^{-1} at room temperature [96]. $Li_7P_3S_{11}$ electrolytes can be anchored on rGO@CoS based on the preparation in solution, and hence the composite cathode rGO@CoS@$Li_7P_3S_{11}$ provides a fast ionic and electronic pathway. The solid-state-batteries employing the cathode and lithium as anode exhibit excellent rate capability and cycling stability, delivering a discharge capacity of 421 mA h g^{-1} at 1.27 mA cm^{-2} after 1000 cycles at room temperature [97].

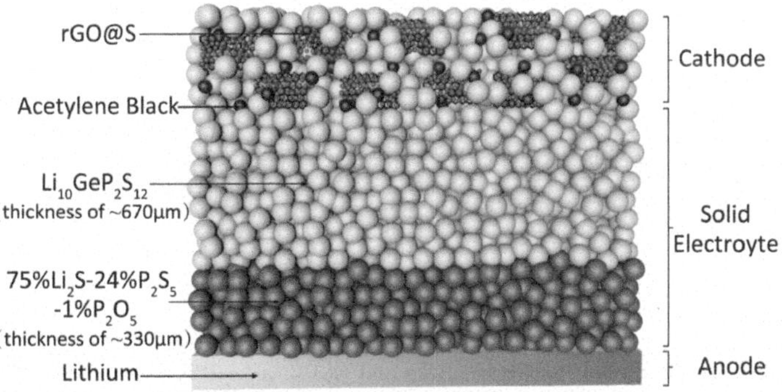

Fig. 5.8 Typical construction of an all-solid-state lithium–sulfur battery. The cathode is composed of solid-state electrolytes, conductive additives and active materials. In order to handle the instability of solid electrolyte against with lithium, a buffer layer was inserted. Reproduced with permission from Ref. [93], Copyright © 2017 Wiley VCH

It is worth noting that the interaction between sulfide-based electrolytes and sulfur provides an opportunity to incorporate them in a molecular level, seeing Fig. 5.9 [94]. It has been proved that Li_3PS_4 can react with sulfur to form lithium polysulfidophosphates, which can reversibly transfer into Li_3PS_4 and Li_2S [98]. Although the polysulfidophosphates present a lower ionic conductivity than Li_3PS_4, the low charge-transfer resistance and short distance between cathode and electrolytes facilitate the electrochemical reaction. Beneficial from the nanostructure, the all-solid-state Li–S batteries exhibit very high sulfur utilization, stability and rate performance (the initial capacity of 1400 mA h g^{-1}, 86% capacity retention after 300 cycles and more than 800 mA h g^{-1} at 1 C). It is believed that sulfur can react with P_2S_5 to form amorphous P_2S_{5+x} via high energy ball milling. The amorphous P_2S_{5+x} can into transfer nanosized, homogeneously distributed Li_2S and Li_3PS_4 during discharge process [99]. This process can significantly improve electrochemical performance of Li–S batteries.

Hydride-based electrolytes have limited electrochemical stable window and cannot be used as electrolytes for transitional oxide cathodes. Several attempts about the application of hydride-based electrolytes in Li–S batteries have been made. A bulk-type all-solid-state battery comprised of TiS_2 cathode, $LiBH_4$ electrolyte, and Li anode was assembled and operated at 393 K at 0.2 C. Chemical and electrochemical oxidation of $LiBH_4$ electrolyte occurred during the initial charge process. A new phase of $Li_2B_{12}H_{12}$ was precipitated at the interface between $LiBH_4$ and TiS_2. The new interphase has a high Li^+ conductivity and superior stability at the high potential to $LiBH_4$, thus could be acting as a stable interphase that enables ultra-long discharge-charge cycles [100]. Orimo et al. introduced the atom deficiencies for

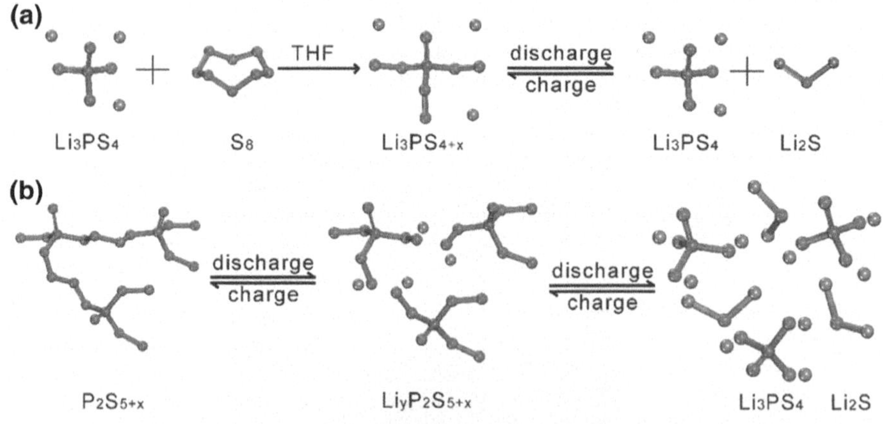

Fig. 5.9 Molecular interaction between sulfur and SSEs. **a** Li_3PS_4 can interact with sulfur in THF and accommodate sulfur to form lithium polysulfidophosphates. The polysulfidophates can be reversibly reduced into Li_3PS_4 and Li_2S during discharging and charging. **b** P_2S_5 with sulfur can form P_2S_{5+x} during the ball milling process. This can be transferred into Li_3PS_4 and Li_2S during the discharging. Reproduced with permission from Ref. [94], Copyright © 2018 Wiley-VCH

electrolytes to enhance the ionic conductivity of $Li_2B_{12}H_{12}$ (2.0×10^{-5} S cm^{-1} at 30 °C), and the solid-state battery consisting of Ti_2S cathode, a ball-milled $Li_2B_{12}H_{12}$ and Li metal as anode could operate at 80 °C [101].

Oxide-based solid electrolytes possess a wider electrochemical window than sulfide-based solid electrolytes and therefore the side reactions with cathodes and conductive additives could be ignored in Li–S batteries. The main drawback for oxide-based electrolytes is the mechanical property, which makes designing and fabricating batteries difficult. A bilayer-structured solid-state garnet electrolyte was designed for the solid-state Li–S batteries. The 3D solid-state electrolyte framework containing a dense-porous bilayer structure was designed and carbon nanotube (CNT) was coated into porous layer to form continuous Li^+/electron pathways. The thick porous layer could host the S cathode and liquid electrolyte, and the thin dense layer blocks polysulfide diffusion and impedes Li dendrite propagation. The bilayer garnet structure can accommodate the volume expansion of S electrode during the cycling as well [102]. This nanostructured design offers an excellent electrochemical performance for solid Li–S batteries. Furthermore, oxide electrolytes were widely used as fillers in polymer electrolytes to enhance the ionic conductivity and improve the mechanical properties and these modified polymer electrolytes present better electrochemical properties and compatibility with electrode materials. The LLZO nanoparticle-filled poly(ethylene oxide) electrolyte with a lower interface resistance shows improved ionic conductivity, and the sulfur composite cathode based on LLZO@C can deliver an capacity of >900 mA h g^{-1} at 37 °C [103].

Compared to liquid electrolytes, polymer-based electrolytes can prevent the shuttle effect of Li–S batteries to a certain extent at a reasonable operative temperature. For instance, the acidized carbon nanotube can induce in situ polymerization of DOL/DME solvents in commercially used liquid electrolytes. The soft and sticky solid electrolyte with the thickness of 100 nm presents a good ionic conductivity at room temperature and moreover prevents the migration of polysulfides through the separator, therefore enhancing cycling performance of Li–S batteries. The Li–S battery based on the selective and in situ formed membrane showed an initial specific capacity of 683 mA h g^{-1} at 1675 mA g^{-1} [104]. The in situ formed solid or gel electrolytes provide a new strategy for designing and fabricating solid-batteries.

Overall speaking, different kinds of solid electrolytes including sulfide-based, oxide-based, hydride-based and polymer-based electrolytes can be applied into Li–S batteries and the construction of a fast ionic and electronic pathway in cathode is crucial for the electrochemical performance of solid Li–S batteries. Nanomaterials and nanostructure design offer a promising opportunity to improve the performance and finally to realize the practical application of solid Li–S batteries.

5.2.1.2 Oxygen or Air Cathodes

Lithium–oxygen or lithium-air batteries are the most promising energy storage and conversion system with a high energy density. Since there is nitrogen, CO_2 and moisture in ambient air, the lithium-air batteries face more challenge than lithium oxygen batteries. The first attempt on polymer electrolyte-based rechargeable lithium –oxygen battery was reported in 1996, where PAN polymer electrolyte membrane serves both as electrolytes and separators. With cobalt phthalocyanine catalyzed Chevron carbon as cathode, the solid Li–O_2 can deliver a charge capacity of 112 mA h g^{-1} with the cutoff potential of 4.1 V given the discharge capacity is 100 mA h g^{-1} [105]. Generally speaking, there are some special requirements for solid electrolytes in Li–O_2 batteries, such as good stability against metallic lithium and oxygen at the same time, sufficient Li^+ conductivity, O_2 solubility and diffusion, and poor volatility to avoid evaporation. PEO-based polymer electrolytes were applied to Li–O_2 battery, which can present a comparable discharge capacity to glyme-based Li–O_2 cells and stable cycle for 40 cycles at a current density of 0.2 mA cm^{-2} [106]. Inorganic solid electrolytes such as LAGP were introduced into Li–O_2 battery by Kumar et al., which were composed of metallic lithium anode, a composite cathode with 25 wt% of carbon and 75 wt% of LAGP, LAGP membrane coated by PEO electrolytes on both sides [107]. And then the solid lithium-air battery based on LAGP was further developed by Zhou et al., where the discharge and charge capacities at a current density of 500 mAg^{-1} with the voltage range of 2.0–4.2 V are about 1700 mA h g^{-1} and 900 mA h g^{-1}, respectively [108]. Although solid electrolytes can alleviate the lithium dendrite issue to some extent and improve the safety issue, the main challenge in Li–O_2 battery is the oxidation and reduction of oxygen. It is believed that the reduction and oxidation go through different pathway and this difference induces the observed separation of the charge and discharge voltages (large polarization). Further singlet O_2 is formed on the oxidation of Li_2O_2, whereas reduction process involves triplet state and the singlet-to-triplet O_2 transition is spin forbidden and consequently kinetically not favorable, resulting in a big overpotential between charge and discharge process [109]. Furthermore, the interfacial resistance derived from the insufficient contact between electrode and electrolytes and the formation of Li_2CO_3 from the reaction of Li_2O_2 and C are other major challenges [110].

5.2.1.3 Transitional Metal Oxides

Compared with sulfur or sulfide-based cathodes, transitional metal oxides, such as $LiCoO_2$, NCM and NCA possess higher redox potential, which would induce more serious interface issues. These issues include continuously increased interfacial resistance caused by the decomposition of solid electrolytes, contact loss between active materials and solid electrolytes, and this would result in continuous capacity fade during cycling, Fig. 5.10a [111]. As a consequence, the modification of these cathode materials is regarded as an effective and practical strategy applied in solid batteries.

When a high-voltage cathode contacts with sulfide electrolytes, such as the $Li_{1-x}CoO_2/Li_{3.25}Ge_{0.25}P_{0.75}S_4$ system, the potential of the cathode tends to lower the Li^+ concentration on the electrolyte side, resulting in lithium depletion at the interface (space charge layer). Furthermore, due to limited electrochemical stable

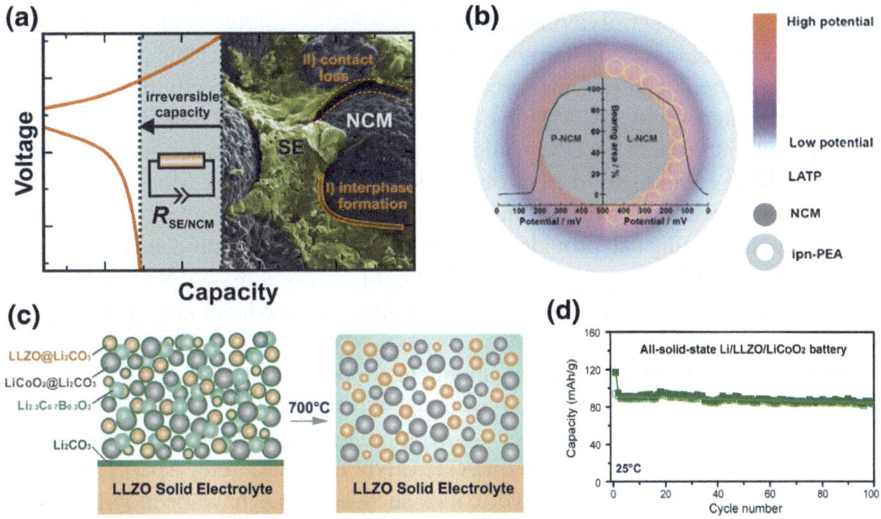

Fig. 5.10 **a** Interfacial issue in cathode part in all-solid-state batteries. Volume changes during cycling induce contact loss and instability between cathode materials and solid electrolytes leads to interphases. These two parts contribute to capacity loss. **b** Coating used to alleviate the interphase formation via mitigating potential drop at interface (pure NCM and LATP modified NCM). **c** All-in-one optimization of cathode interfaces to achieve successful work of an all-solid-state battery at room temperature. Additive of $Li_{2.3}C_{0.7}B_{0.3}O_3$ was used as a flux to achieve a good interface. **d** The battery presents a specific capacity of around 100 mA h g^{-1} and stable cycling. **a** was reproduced with permission from Ref. [111], Copyright © 2017, American Chemical Society; **b** was reproduced with permission from Ref. [113], Copyright © 2017, American Chemical Society; **c** and **d** were reproduced with permission from Ref. [118], Copyright © 2018, Elsevier

window sulfide-based electrolytes can be oxidized by the cathode materials and the space charge layer could worsen this process. A coating layer with several nanometers on cathode materials, such as $LiNbO_3$, $LiTaO_3$, and $Li_4Ti_5O_{12}$ can prevent the direct contact between cathode and electrolytes, and the amorphous phase could alleviate the effect from the space charge layer [20]. Furthermore, these coatings present suitable ionic conductivity, which can decrease the charge transfer resistance and improve the kinetic process. Hence all-solid-state lithium ion batteries with spinel $Li_4Ti_5O_{12}$ as anode, superionic conductor–$Li_{9.54}Si_{1.74}P_{1.44}S_{11.7}Cl_{0.3}$ as solid electrolytes (25 mS cm^{-1} at room temperature), and $LiNbO_3$ coated $LiCoO_2$ as cathode can present a specific capacity of around 120 mA h g^{-1} at 1.8 C and more than 80 mA h g^{-1} at 60 C at the operative temperature of 25 °C, which is even better than the batteries based on conversional liquid electrolytes. This work proved the advantage and feasibility of all-solid-state LIBs and encouraged more attentions on this field.

The coating strategy is not only applied in sulfide-based electrolytes, but also for polymer-based electrolytes. For polymer electrolytes, the coating such as Li_3PO_4 and LATP can prevent the oxidation decomposition on the solid polymer electrolyte and further promote to fabricate a high-voltage and high-capacity all-solid-state batteries based on a polymer electrolyte/inorganic composite electrolyte [112]. As mentioned above, one coating layer can harmonize the interfaces and buffer the potential drop between the cathode and polymer electrolyte, which can be directly observed from atomic force microscopy (AFM). One case is LATP modified $LiNi_{0.6}Co_{0.2}Mn_{0.2}O_2$ prototype cathode, where LATP can mitigate polarization and accelerate ion transport dynamics at the cathode-solid electrolyte boundary (Fig. 5.10b). More importantly, atomic force microscopy with the unique sensitivity to interfacial potential fluctuation verifies that the coating can provide a sustain-released potential drop at the interfaces and enhance the interface dynamics and stability. This work directly proves the function of coating on alleviating the effect of space charge layer and improving the dynamic process [113].

The widely-used coating materials should have special features, such as high ionic conductivity, low electronic conductivity and a wide electrochemical stable window. Hence some superionic conductors have been regarded as the most promising candidates, such as the abovementioned LATP, $LiNbO_3$, $LiTaO_3$ etc. Apart from these ionic conductors, stable and inactive oxides such as Al_2O_3 have been widely used as coating materials. For instance, Al_2O_3 with controlled thickness can be coated on the surface of $LiCoO_2$ via an atom layer deposition and the electrochemical results suggest that the Al_2O_3 layer can suppress the development of interfacial resistance at $LiCoO_2/Li_{3.15}Ge_{0.15}P_{0.85}S_4$ interface, and reduce the formation of interfacial layer which causes capacity fade during cycling [114]. Since Al_2O_3 possesses very poor ionic and electronic conductivity, the thicker Al_2O_3 layer would bring about new issues such as high charge-transfer resistance and sluggish kinetic process. The controllable growth of these stable and inactive oxides is crucial for the electrochemical performance [114]. Furthermore, a model system ($LiCoO_2$/interlayer/LATP-GC) was established to clarify the charge transfer reaction at the electrode-solid electrolyte interface and confirm the function of oxides layer. A depth-resolved X-ray absorption spectroscopy (DR-XAS) confirmed the electrode-solid electrolyte interface with a space resolution of ~7 nm. Co_3O_4 could be formed at the interface, even with the introduction of the NbO_2 layer. However, with the NbO_2 layer, a mixture phase of $LiNbO_3$ and $LiCoO_2$ was formed and restricted the abrupt change in Co–O bond during delithiation. The interface could also mitigate the stress at the interface during lithiation/delithiation [115].

The interfaces in cathode include the ones between cathode materials and solid electrolytes, cathode materials and conductive additives, and solid electrolytes and conductive additives. The interface between cathode materials and solid electrolytes is the predominant issue for the electrochemical performance of solid batteries, which can be alleviated by coating cathode materials. The interface between solid electrolytes and conductive carbon possesses effects on cyclic stability as well, especially for sulfide-based electrolytes [116]. A single coating strategy may not solve all the above mentioned interface issues and interface engineering through adding additives was widely applied. Due to the limitation of electrochemical window of LLZO, it cannot directly contact with oxide-based cathode. An irreversible electrochemical decomposition at the voltage of higher than 3.0 V versus Li^+/Li was identified in a clean cathode interface, and the decomposition was able to be avoided by a surface modification of Li_3BO_3 [117]. The surface of LLZO and $LiCoO_2$ is not always clean and Li_2CO_3 is easily formed as mentioned above. A solder of $Li_{2.3}C_{0.7}B_{0.3}O_3$ can melt and react with Li_2CO_3 to form a high Li^+ conducting $Li_{2.3-x}C_{0.7+x}B_{0.3-x}O_3$ interphase at 700 °C. As shown in Fig. 5.10c, after this treatment, the interface can be optimized and an all-solid-state battery can work at room temperature with a specific capacity of 80 mA h g^{-1} [118].

Except for the attention on avoiding the decomposition of solid electrolytes and formation of interphases between cathode and solid electrolytes, the property of interphases is crucial for electrochemical performance of solid batteries. The well-known example is an all-solid-state thin film battery, which is composed of metallic lithium as anode, lithium phosphorus oxynitride (LiPON) as SSEs, and the spinel $LiNi_{0.5}Mn_{1.5}O_4$ as cathode. This thin film battery exhibits a high working voltage of 4.8 V and can work stably for 10,000 cycles, superior to those based on liquid electrolytes [16]. As mentioned above, LiPON possesses very narrow electrochemical stable window and the excellent electrochemical performance is mainly derived from the formed interphases at both anode and cathode part. It can be deduced that the interphases in solid batteries in some cases can work as the solid electrolyte interphase (SEI) film in liquid electrolytes. The interphases enable the successful work of a high-voltage lithium metal battery based polymer electrolytes, and one example is bacterial cellulose supported poly(methyl vinyl ether-*alt*-maleic anhydride) as polymer electrolytes. In the anode part, the polymer can participate in the formation of the SEI film on the lithium metal anode and the formed SEI is very uniformly distributed and keeps unchanged, hence the current density for the lithium plating/striping in a symmetric battery can reach up to 9.5 mA cm^{-2}. For the cathode part, $LiCoO_2$ was charged to 4.45 V versus Li^+/Li with the capacity of 160 mA h g^{-1} after 700 cycles based on this electrolyte. It has shown that this polymer can facilitate a stable CEI formation [119]. These two cases show the great importance to understand the interphase at the cathode part for the practical application.

To achieve the high active-material-utilization in electrodes with high active-material loadings, Wakayama et al. described a self-assembled block copolymer-template synthesis of a three-dimensional bicontinuous nanocomposite electrode material consisting of $LiCoO_2$ and $Li_7La_3Zr_2O_{12}$. When the active-material content was 89.4%, 98% of the active material was utilized, which can be ascribed to its unique bicontinuous nanostructure consisting of the nanosized mixture of cathode materials and electrolytes [120]. An brilliant idea was proposed by Guo et al. to handle the interfacial issue, which is upgrading the conventional aprotic electrolytes to quasi-solid-state electrolytes , see Fig. 5.11a. More importantly, the catalyst for ring-open reaction is the commonly-used salt—$LiPF_6$, which will not result in any impurity in the electrochemical system [121]. Furthermore, the design of current collectors can facilitate the utilization of active materials. An aluminum current collector with microsized channel arrays can be prepared by laser etching and then the cathode materials and electrolytes were filtered into these channels. The liquid electrolytes can in situ polymerized under the assistance of UV light and gradiently polymerized electrolytes with solid stiff surface on the anode side and ductile inclusions on the cathode can be formed due to the limitation of the permeability of UV light (Fig. 5.11b). The gradiently descending degree of polymerization (DP) on the cathode side was used to improve the interface contact, while the high degree of DP on the Li metal anode was used to provide mechanical stability to inhibit Li dendrites. This can be confirmed by fluorescence molecular method. The high-rate quasi-solid-state Li metal batteries was well realized at room temperature based on the gradiently polymerized solid electrolyte concept [122]. Furthermore, the solution processing and preparation of sulfide-based electrolytes provide a new strategy to improve the kinetic process [123, 124]. For instance, solid-state electrolytes—LiI-Li_4SnS_4 can be homogeneously mixed with conductive carbon and $LiCoO_2$ in methanol and then the solid-electrolyte-coated cathodes can be obtained. This cathode presents very high rate capabilities (90 mA h g^{-1} at 2 C). Furthermore, the preparation of composite cathode in traditional liquid electrolytes can be transplanted for preparation of cathode for solid batteries, such as doctor blade method, see Fig. 5.11c [123].

In summary, how to optimize the kinetic process and manipulate the interphases in cathode part determines the rate performance and cyclic stability of a solid battery. Nanomaterials and nanotechnology such as coating play key role to achieve this. Furthermore, more attention should be paid to the formed interphases, including their components, functions and manipulation of their components and functions.

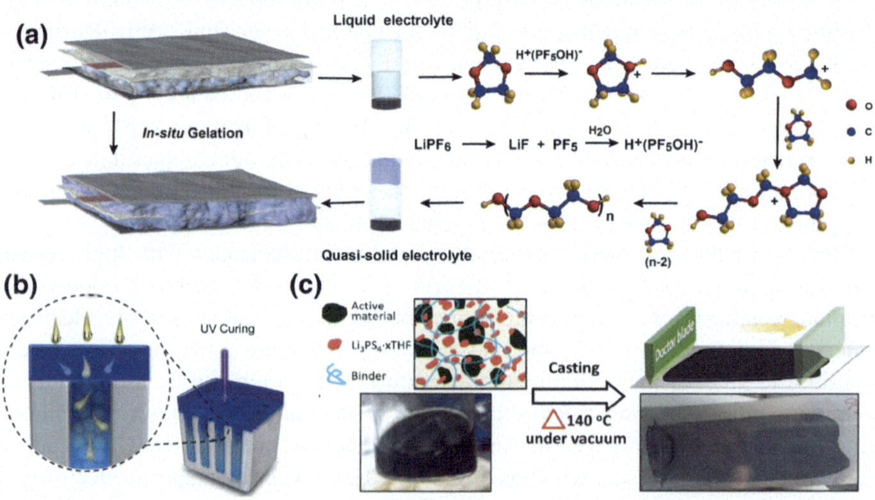

Fig. 5.11 Strategies to improve the kinetic process at cathode part. **a** In situ upgrading conventional aprotic electrolytes to quasi-solid electrolytes. **b** A gradient polymerization and 3D current collector are combined to achieve good electrochemical performance. **c** Preparation of composite cathode for solid-state batteries based on solution-processable solid-state electrolytes. **a** was reproduced with permission from Ref. [121], Copyright © 2018, American Association for the Advancement of Science; **b** was reproduced with permission from Ref. [122], Copyright © 2018, American Chemical Society; **c** and **d** were reproduced with permission from Ref. [120], Copyright © 2018, the Royal Society of Chemistry

5.2.2 Anode Materials for Solid-State Batteries

Similar with lithium batteries based on liquid electrolytes, graphite, silicon/carbon composite, and $Li_4Ti_5O_{12}$ etc. can be applied in solid batteries as well. In order to improve the energy density, metallic lithium with a theoretical capacity (3860 mA h g^{-1}) is regarded as the most promising candidate for the next generation batteries. In conventional liquid electrolytes, lithium plating and striping would induce several serious issues, such as safety issue and poor Coulombic efficiency, and therefore metallic lithium anode was abandoned at the birth of lithium ion batteries. The development of solid-state electrolytes revived metallic lithium anode and the reason lies in the following three aspects: the solid-state electrolytes are much safer than flammable liquid electrolytes; dendrites may be suppressed by solid electrolytes due to the high mechanical properties and unit transference number; Coulombic efficiency can be enhanced due to no interphase or SEI formation. Hence solid-state lithium batteries present giant advantages and valid feasibility. This section focuses on lithium plating/striping in solid electrolytes, and further the successful applications of nanomaterials and nanostructures in metallic lithium anodes are also elaborated.

5.2.2.1 Polymer-Based Electrolytes

The compatibility between polymer and metallic lithium is one of the major issues in polymer electrolytes-based lithium metal batteries. The charge transfer resistance and formation/conversion of the passivation layer on the Li electrode could influence the cycling life, Coulombic efficiency, and energy efficiency of the polymer electrolytes based lithium metal batteries. The electrochemical impedance spectroscopy (EIS) technique is widely used to monitor the change of the interfacial resistance which continuously increases when Li metal contacts with polymer electrolytes [67]. Other techniques such as in situ attenuated total reflection-Fourier transform infrared spectroscopy (ATR-FTIR) and XPS have widely been used to investigate the polymer-lithium electrode interfaces as well. It revealed that the deposited metallic Li reacted with PEO, which resulted in the formation of an irreversible cleavage of the ether functionality and generated alkoxides. Furthermore, reduction reactions of lithium salt and/or water (trace) with lithium are other side reactions. Lithium alkoxides, Li_2CO_3 and $ROCO_2Li$ have been detected as major reaction products. In the case of the gel electrolytes including PC or other conventional liquid electrolytes, the surface chemistry of the Li anode is dominated by the reduction of alkyl carbonate solvent to $ROCO_2Li$.

To improve the cyclability of Li metal anode in polymer electrolytes, a stable and low-resistance interphase layer needs to be formed. Many strategies have been used to improve the interfacial compatibility from either lithium anode side or the polymer electrolytes side. For Li anode, functional artificial SEI is normally built up to achieve a stable interface and uniform lithium plating and stripping. The surface modification of polymer electrolytes is a relatively easier strategy compared to the super active Li metal electrode. A variety of self-assembly monolayers in the forms of R–(CH_2CH_2O–H could enhance the lithium anode-electrolyte interfaces, where R–(CH_2CH_2O)–H could form a well-ordered continuous layer that avoid the direct contact of lithium with the polymer electrolytes, thus preventing reactions on the interface.

According to an assessment from Newton and Monroe, lithium dendrites can be suppressed when the modulus of electrolytes is several times higher than metallic lithium [125, 126]. Adding plasticizers such as EC and PC can significantly increase the ionic conductivity but worsen the mechanical properties of polymer electrolytes. Hence crosslinking and blending of polymers would be the better choices to achieve both goals at the same time. For instance, Guo et al. have developed a novel polymer electrolyte via a photo-polymerizing branched acrylate in interpenetrating poly(ether-acrylate) (ipn-PEA) network in the PEO based polymer electrolytes (Fig. 5.12a) [127]. The obtained polymer electrolyte possesses very high ionic conductivity (0.22 mS cm^{-1}) at room temperature and good mechanical strength (ca. 12 GPa). Beneficial from the high ionic conductivity, strong mechanical strength, and good stability with metallic lithium, a symmetric battery (Li/ipn-PEA/Li) can work at 0.5 mA cm^{-2} and capacity of 2 mA h cm^{-2} without the propagation of lithium dendrites [127]. Furthermore, a high transference number can significantly improve the cycling performance of Li plating/striping. Hence a single ion triblock copolymer was developed and the transference number is 0.85 (Fig. 5.12b) [14]. Aside from the high transference number, the high ionic conductivity (10^{-5} S cm^{-1} at 60 °C), excellent mechanical properties, and electrochemical stability (up to 5 V vs Li^+/Li) of this copolymer enable the successful work of solid-state batteries [14].

Furthermore, inorganic fillers can improve the ionic conductivity and the mechanical properties as well. For instance, a composite polymer electrolyte, composed of 3D ordered bicontinuous conducting $Li_{14}Al_{0.4}Ge_{1.6}(PO_4)_3$ (LAGP) ceramic and insulating epoxy polymer microchannels was reported [128]. The polymer channels endow the mechanical properties improvement (in particular mitigation of the ceramic brittleness) compared with that of the ceramic alone. As the electrical and mechanical properties depend on the microarchitecture, the gyroid filled with epoxy giving the best combination of conductivity and mechanical properties, and the mechanical properties of the electrolyte exhibit a 28% higher compressive failure strain and up to five times the flexural failure strain of a LAGP pellet before rupture [128]. A 3D garnet fiber network was constructed by electrospinning and then PEO based electrolytes were filtered inside to form a composite electrolyte (Fig. 5.12c) [129]. The electrolyte membrane can effectively block dendrites in a symmetric Li/electrolyte/Li cell during repeated lithium stripping/plating at room temperature with a current density of 0.5 mA cm^{-2} for over 300 h due to the improved ionic conductivity and mechanical properties [129]. The additive of inorganic fillers in polymer electrolytes can improve the transference number as well. According to the sand time model, increasing the transference number can prolong the sand time for exhausting the ions at the surface, at which time the lithium dendrite starts to grow [130]. The composite electrolytes can have a transference number as high of as 0.58, as reported in the literature, which is two to three times higher than that in the liquid electrolytes [131].

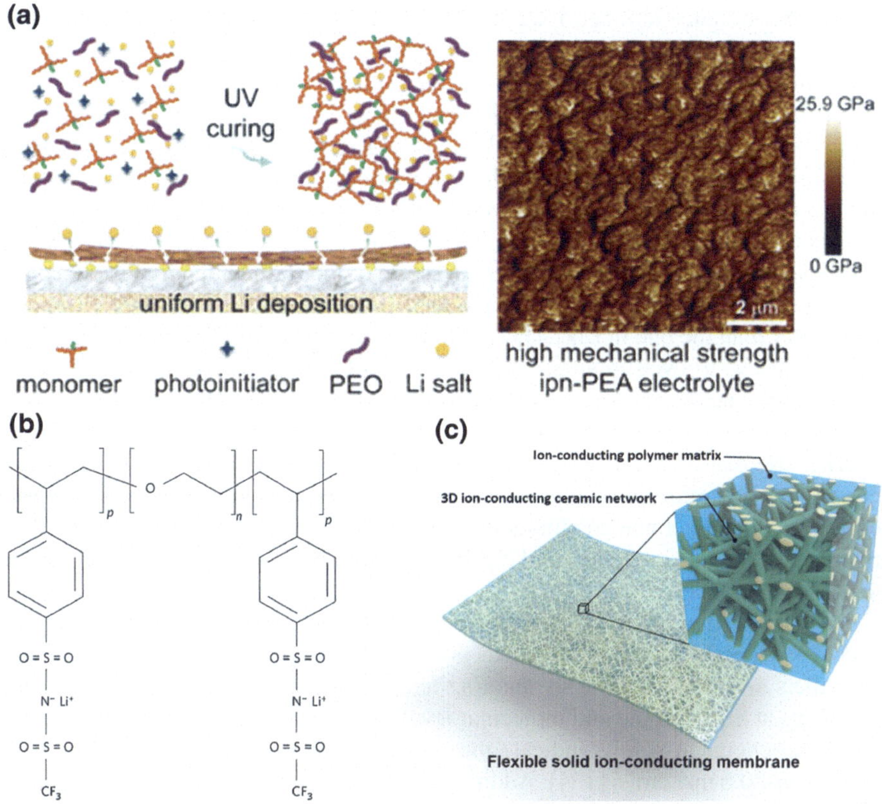

Fig. 5.12 Several strategies to handle the dendrite issue in polymer electrolytes. **a** Chemically blending to increase the mechanical strength and ionic conductivity for stable Li plating/striping [127]. **b** Single ion BAB triblock copolymer electrolytes [14]. **c** 3D LLZO fiber networks as fillers for polymer electrolytes [129]. **a** was reproduced with permission from Ref. [127], Copyright © 2016, American Chemical Society; **b** was reproduced with permission from Ref. [14], Copyright © 2013, Springer Nature Publishing; **c** was reproduced with permission from Ref. [129], Copyright © 2016, National Academy of Sciences

Numerous efforts have been paid to improve the stability of lithium plating and striping through increasing the ionic conductivity, improving the mechanical properties, and enhancing the transference number, but the standing maximum current density is still lower than 1 mA cm^{-2} in polymer electrolytes. New insights into lithium plating and striping in polymer electrolytes should be to further dig and new strategies and approaches should be given.

5.2.2.2 Inorganic Solid Electrolytes

Compared with polymer electrolytes, solid inorganic electrolytes meet both requirements for suppression of lithium dendrites including their high mechanical properties and unit transference number. Except from the inorganic solid electrolytes instable with metallic lithium to form mixed conducting interphases, such as LATP, LAGP, LLTO, and LGPS, the rest inorganic solid electrolytes, such as LLZO, $LiZr_2(PO_4)_3$, and Li_3PS_4 have been tested in solid lithium batteries. In contrast to the expectation, the propagation of lithium dendrites in inorganic solid electrolytes has been verified in more and more studies. Moreover, compared with the solid electrolytes which is not stable against metallic lithium to form ionic conducting interphase, the stable solid-state electrolytes against lithium could promote the growth of lithium dendrites, which is ascribed to a larger curvature of the formed dendrite and the less consumption of dendrites [132]. Consequently, the short circuit in LLZO happens much faster than the one in $LiZr_2(PO_4)_3$. It has been shown that lithium propagation has been observed even in single crystalline LLZO [133]. These results indicated that the failure mechanism is Griffith-like and does not totally follow the shear modulus criterion. According to the literature, the formation and propagation of lithium dendrites are mainly derived from the defects lying in solid electrolytes, and these defects could be microscopic level or molecular level, such as voids, cracks, grain boundaries, and local nonstoichiometry in ceramic. A critical current is required to propagate lithium through the solid electrolytes, which is significantly dependent on the processing of ceramics.

The garnet pallets were prepared by the hot pressing at the temperature ranging from 1100 to 1300 °C and these pallets possess very high density (more than 99%). The pallets are transparent at the treating temperature of higher than 1200 °C and the defects at microscopic and molecular level can be eliminated to the great extent. The electrochemical results indicate the critical current density can reach up to a maximum of 0.6 mA cm^{-2} [134]. Additionally, the temperature and pressure applied during cycles can alert great effect on the propagation of lithium dendrites. The critical current density can increase from 0.05 to 3 mA cm^{-2} when the temperature increases from 30 to 130 °C [135]. The evaluated temperature can improve the interface between garnet and metallic and ionic conductivity of solid electrolytes. Furthermore, it is believed that the effects of defects on heterogeneous conductivity can be alleviated to some extent. Last but not the least, an interface either self-formed or inserted on purpose would be an effective way to alleviate this issue as well. For instance, the functional thin layers such as Al, Ge, Si, Al_2O_3, SnO_2 were used to achieve the intimate contact between garnet and metallic lithium and to decrease the interfacial resistance.

These strategies can improve the critical current density of given solid-state electrolytes to certain extent, which cannot meet the normal requirement for lithium anode. This provokes more studies on the detailed mechanism of lithium propagation in solid-state batteries. One of explanations is the pristine defects of solid-state electrolytes, therefore lithium can propagate through the voids or even create cracks on solid-state electrolytes (Fig. 5.13a) [133, 136]. A critical and novel mechanism on the lithium propagation into solid-state electrolytes was revealed via neutron depth profile. The results indicated that metallic lithium can directly form in solid state electrolytes and the reason is ascribed to the high electronic conductivity of solid-state electrolytes. The researchers stress that lowering electronic conductivity is more critical and straightforward approach to alleviate dendrite issue in solid-state

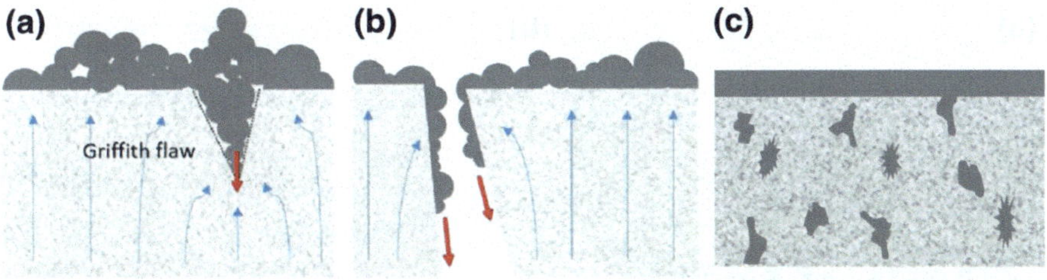

Fig. 5.13 Different modes of lithium propagation. In conventional liquid electrolytes, critical current density is mainly dependent on the transference number and plating capacity. In inorganic solid-state electrolytes, lithium propagates along the pre-existing defects (**b**) and/or creates cracks like Griffith flaw (**a**). **c** Lithium directly formed inside SSEs due to high electronic conductivity. **a** and **b** were reproduced with permission from Ref. [136], Copyright © 2018, Electrochemical Society; **c** was reproduced with permission from Ref. [137], Copyright © 2019 Springer Nature Publishing

electrolytes (Fig. 5.13c) [137]. Recently, an in situ formed LiF-rich solid electrolyte interphase between sulfide based electrolytes and lithium anode was established. A high concentrated liquid electrolyte was used as transferred materials and then the liquid phase was removed at vacuum at 120 °C. Beneficial from the ultra-low electronic conductivity and high stability of this interphase, the sulfide-based solid-state electrolytes can stand a current density as high as >2 mA cm^{-2}. More importantly, the Coulombic efficiency during Li plating/striping can reach up to more than 98% [138].

Since the critical current density is mainly dependent on solid electrolytes, whose physicochemical properties are hardly tuned, increasing the areal contact surface is another choice to alleviate lithium propagation. Nanomaterials and nanostructures used as lithium host have been widely applied in liquid electrolyte system, but rarely in solid electrolyte system. A porous/dense/porous garnet electrolyte was designed and molten lithium was penetrated into one porous layer as a host for metallic lithium anode (Fig. 5.14) [139]. The dense layer can prevent the lithium growing through and reaching cathode part, leading to a short circuit. The porous structure can enhance the areal surface area and lower local current density, and increase the tolerant applied current density. In this special design, the critical current density for lithium plating/striping can reach up to 2 mA cm^{-2} at room temperature, which is much higher than those without this structure [139].

In summary, the improved mechanical properties and transference number from liquid electrolytes, to polymer electrolytes, and to inorganic solid electrolytes, should improve the stability of lithium plating/striping, but the experimental data give the contrast results. The behind reason is recognized the homogeneity of lithium ions distribution. From the view of lithium part, the quasi-solid electrolytes or batteries represent the practical future.

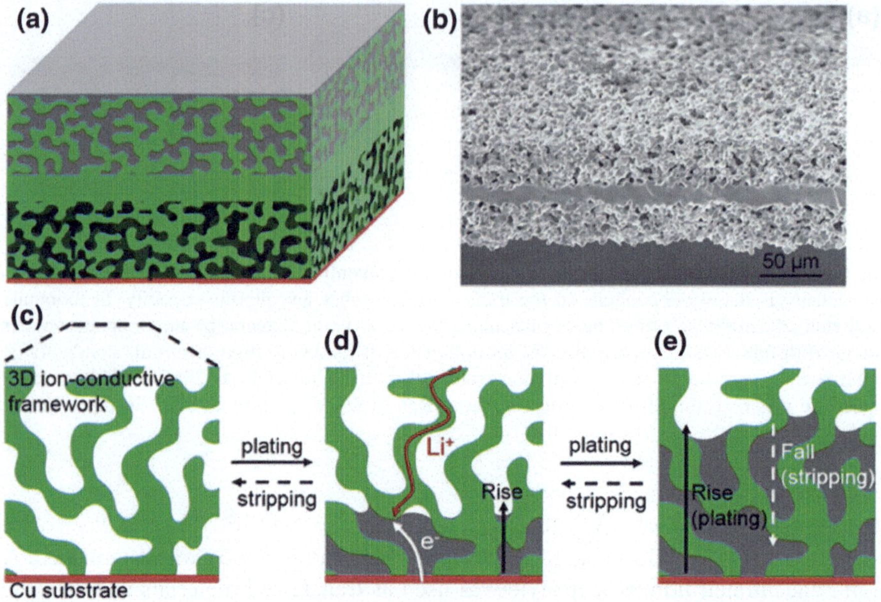

Fig. 5.14 Nanostructure applied in lithium anode part. **a** Schematic of 3D conductive host lithium anode. **c**, **d**, **e** 2D local schematics for lithium plating and stripping in this nanostructured host. **b** SEM image. Reproduced with permission from Ref. [139], Copyright © 2018 National Academy of Sciences

5.2.3 *Other Types of Solid-State Batteries*

Due to the rich abundance of Na, sodium batteries obtain more and more attentions for the application in large-scale energy storage. The cost, safety and cycling stability of sodium batteries are the dominating concerned factors. The improved safety and higher energy density beneficial from the use of solid electrolytes provoke the study on solid sodium batteries. The category of Na^+ solid electrolytes is similar with Li^+ solid electrolytes, and the ionic conductive mechanism is similar as well, which is mentioned in the above section. The development of solid electrolytes revives the old system Na–S batteries. High temperature Na–S battery was constructed in 1970s based on β-alumina solid electrolytes. The high temperature and safety issue are the big challenges. Sulfide-based solid electrolytes provide an opportunity of room temperature solid Na–S battery. Bulk-type all-solid-state Na/S cells with the Na_3PS_4 glass-ceramic electrolyte exhibited a specific capacity of 1100 mA h g^{-1} and no large capacity degradation at room temperature. Meanwhile, the utilization of sulfur was ~2 times higher than that of high-temperature-operating Na/sintered β″-alumina/S battery [140]. As mentioned above, cost issue is one of the most concerns for sodium battery, hence $Na_2MnFe(CN)_6$ earns much more attentions for its low cost. The dissolution of $Na_2MnFe(CN)_6$ in aprotic electrolytes makes the cycling stability worse.

The solid-state electrolytes can eventually eliminate this issue and offer an opportunity for long lifetime. The NASICON-typed electrolyte—$Na_3Zr_2Si_2PO_{12}$ presents a high ionic conductivity of 1.2×10^{-3} S cm^{-1} at 60 °C and the electrochemical performance of an all-solid-state sodium battery based on this electrolyte is comparable with the one based on aprotic electrolytes. Further, compared with the battery based on liquid electrolytes, the thin thickness and simple components of the SEI in the solid-state batteries offer a long and stable operation of sodium batteries [141]. The recent survey on solid-state sodium battery can referred in the literature [142]. Solid potassium batteries were attempted as well. Recently Goodenough et al. reported a solid potassium battery with a PMMA polymer-gel electrolyte and a polyaniline cathode, and this battery can deliver a specific capacity of 130 mA h g^{-1} and very good cycling stability [143].

5.3 Key Points in Construction of Solid-State Batteries

5.3.1 Temperature

Many aspects should be considered to construct the solid-state batteries with a reasonable electrochemical performance and temperature is a substantially important factor among these factors. For instance, high temperature can lower the interfacial resistance and significantly enhance the critical current density for lithium anode. Moreover, most polymer solid electrolytes exhibit reasonable ionic conductivity (more than 10^{-4} S cm^{-1}) only at high temperature, therefore, many solid-state batteries worked at a high temperature in a wide range of 55–100 °C. An all-solid-state lithium battery can be constructed through a rapid thermal annealing method to melt the cathode (V_2O_5) and form a continuous contact between cathode and solid electrolytes. This Li/garnet/V_2O_5 battery cannot cycle at room temperature very well because of the low diffusivity of Li^+ in V_2O_5, the solid-state-batteries were cycled at 100 °C, and the specific discharge capacity of the V_2O_5 is 150 mA h g^{-1} [144]. Recently, a molten lithium battery was built up with LLZO as solid-state electrolytes, and a molten Sn–Pb or Bi–Pb alloy cathode was used as cathode at 240 °C. This battery can stably cycle for more than one month at a suitable current density and very high Coulombic efficiency (99.98%). The specific power of this battery can reach 175 mW cm^{-2} with coupling with Bi–Pb cathode, and hence this battery offers prospects for grid energy storage [145]. High temperature can improve the electrochemical performance but would be a serious issue concerning about the safety risk and maintaining high environmental temperature. The electrochemical performance of solid batteries at low temperature is a more serious issue as the environmental temperature in winter is normally low. Unfortunately, there is no report about the electrochemical performance of solid batteries at −20 °C. It is still uncertain that whether a solid battery should be designed to work at –20 °C or a system to keep the environmental temperature around battery at 40 °C should be explored. As mentioned above, numerous solid electrolytes or quasi-solid electrolytes with ionic conductivity of more than 10^{-4} S cm^{-1} at room temperature have been studied, meanwhile the concept about thermostat system for solid batteries has been proposed as well. The temperature issue could be controlled via a combination of developments from these two parts.

5.3.2 *Strain and Stress*

Mechanical properties of solid electrolytes are very crucial for electrochemical performance of solid-state batteries. The Young's modulus and formability at powder compaction could be used for the evaluation of the solid electrolytes. For instance, the addition of lithium halides into sulfide-based electrolytes would decrease the Young's modulus of the electrolytes, and batteries using Si electrodes and glass electrolytes with lithium halides exhibited a larger capacity compared to that without lithium halides [146]. Furthermore, insertion and extraction of Li^+ in the electrodes are accompanied by severe volume changes, leading to significant pressures within the battery system, which is more critical in the case of solid-state batteries than liquid batteries. The stress and strain in the electrodes can cause particle cracking, electrode fracture and battery deformation. In a solid-state battery, an unfavorable combination of electrode materials and solid electrolytes with different mechanical properties, e.g. hardness and elasticity, may be detrimental to the battery performance [147]. There are several methods to evaluate or monitor the stress changes in solid-state batteries (Fig. 5.15) [147]. In situ X-ray diffraction could be used to obtain the crystallographic volume change of electrode materials during cycling. The volumetric changes can be calculated from the unit cell changes. Secondly the changes of the open circuit voltage with the different given pressure can provide information about the reaction volume. The reaction volume was defined by the difference of the partial molar volumes of lithium in both electrodes. Thirdly, the battery pressure can be monitored at different state of charge, which can be linked with crystallographic volume change of electrode materials through the concept of the reaction volume.

Typically, solid-state batteries may experience severe mechanical degradation due to the volume changes of the active electrode materials. The cathode materials such as $LiCoO_2$ (LCO) and $Li[Ni_{1\text{-}x\text{-}y}Co_xMn_y]O_2$ (NCM) exhibit reversible volume changes. $LiCoO_2$ is a rare exception for the unit cell volume of $LiCoO_2$ increases upon delithiation to $Li_{0.5}CoO_2$ by roughly +2%. With the further increase of lithium ions extracted from LCO, there would be a decrease of overall unit cell volume as the repulsion between oxygen layers is smaller. The unit cell volume for NCM materials decreases over the whole range of delithiated compositions, and the higher nickel-content in NCM and deep delithiation will bring about a more severe volume change. The volumetric shrinking of $LiNi_{0.6}Co_{0.2}Mn_{0.2}$ is only 2 and 5% with delithiated to $Li_{0.4}Ni_{0.6}Co_{0.2}Mn_{0.2}$ and $Li_{0.2}Ni_{0.6}Co_{0.2}Mn_{0.2}$, respectively. Such volume change can induce interfacial issues such the loss contact between cathode and electrolytes, resulting in continuous capacity fading. $LiFePO_4$ exhibits a two-phase

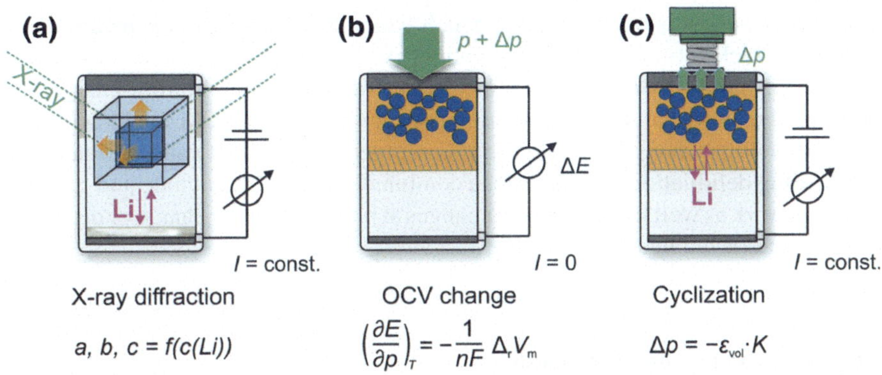

Fig. 5.15 Schematic of the different experiments related to volume changes in electrode materials. **a** Measurement of the lattice parameters (**a**, **b** and **c**) of crystalline active materials by in or ex situ X-ray diffraction as function of SOC (state of charge), i.e. of lithium concentration. **b** Measurement of the OCV (E) as function of pressure p and evaluation of the reaction volume $\Delta_r V_m$ of the cell reaction, with n being the number of transferred electrons and F denoting the Faraday constant. **c** Monitoring the pressure change Δp of an operating cell, which in a first approximation can be described by the product of volumetric strain ε_{vol} and the bulk modulus K for small values of the strain. Reproduced with permission from Ref. [147], Copyright © 2018 Royal Society of Chemistry

transition with a volume strain of ~−7% from $LiFePO_4$ to $FePO_4$. For anode materials, such as silicon, the volumetric expansion can be up to 400% during lithiation $Li_{4.4}Si$. For graphite, it undergoes a volumetric change of ~13% with full lithiation to Li_6C. $Li_4Ti_5O_{12}$ is known to be a zero-strain material with a negligible volumetric strain of −0.2% upon delithiation [147]. The volume change for metallic lithium is 100% during plating and striping. The stress induced by anode is the dominating contribution in solid battery, especially for silicon or lithium anode. It is believed that the major challenge of increasing the capacity in thin film battery (Li/LiPON/LCO) is the stress induced by too thick lithium anode.

The chemical stability of the solid-state electrolytes against Li metal strongly affects the mechanical integrity. When $Li_{1.5}Al_{0.5}Ge_{1.5}(PO_4)_3$ contacts with Li metal, a mixed ionic and electronic conductor formed on the interface. The formed interphases accompanying with residual C, Li_2O, Li_2O_2, and Li_2CO_3 lead to the mechanical deterioration of the SSE because it causes pulverization of the LAGP pellet with repeated micro crack formation at the anode side [148].

Plasticity of the solid-electrolyte should be considered in future studies, as the component needs to compensate volume changes of electrode materials. The Yound's modulus for sulfide-based electrolytes is in the range of 20–40 GPa, and the value for NASICON and garnet based locate in the range of 120–140 GPa, and 150–160 GPa, respectively [147]. From the viewpoint of alleviation of battery strain, the sulfide based electrolytes is preferred. Furthermore, the design of electrode could offer help to this issue. For instance, porous structure in electrode and particle size could compensate the stress induced by electrodes. Adding flexible components such as polymers is another strategy, which can reduce local stress generation and counteract crack formation or delamination. Additionally, reasonable selection of electrode materials could achieve a low mean volume change during charge and discharge. For instance, the combination between LCO and NCM could achieve a low mean volume change because the volume increases for LCO and decreases for NCM during delithiation. Meanwhile, a combination of NCM cathode and graphite anode can work as well for the volume changes at both part during charge or discharge can be counteracted to certain extent.

5.4 Summary and Outlook

The fast development of super ionic conductors and the demands in high energy and good safety of batteries flourish the studies of solid-state batteries. The thin film battery based LiPON has been commercialized and successfully applied in some microsystems. The solid-state battery based on PEO and $LiFePO_4$ has been applied in electric vehicles by the France's Bollore group. Furthermore, more and more governments and corporations enlarge the investment in the study on solid batteries and believe that the commercialization solid batteries with high energy density and high safety may be achieved in the near future. In spite of the bright future, there are some fundamental and practical issues, listed as followed:

(1) Insufficient ionic conductivity. Although there are some solid electrolytes (sulfide-based electrolytes) with comparable ionic conductivity with aprotic electrolytes at room temperature, the value for most of them locates in the range of 10^{-3}–10^{-5} S cm^{-1} (10^{-4}–10^{-3} S cm^{-1} for garnet and NASICON, 10^{-5}–10^{-4} for polymer based electrolytes).
(2) Interfacial issues. The interface issues include the interfacial compatibility between electrode and electrolytes, the ion and electron transport in composite electrode.
(3) Metallic lithium anode. The most attractive anode brings about the most difficult challenge, including lithium dendrites, large volume changes, and low Coulombic efficiency.
(4) Operative condition. The effects from internal stains and environmental temperature should be taken into account.
(5) Cost and safety.

Nanomaterials and nanotechnology can be applied in solid-state batteries to handle some of the challenges, including insufficient ionic conductivity, interfacial issues, and metallic lithium anode, as summarized in this section. The challenges from operative conditions, cost and safety should be handled in combination with industrial design and productions.

References

1. Kamaya, N., Homma, K., Yamakawa, Y., et al. (2011). A lithium superionic conductor. *Nature Materials, 10,* 682–686.
2. Hull, S. (2004). Superionics: Crystal structures and conduction processes. *Reports on Progress in Physics, 67,* 1233–1314.
3. Owens, B. B., & Argue, G. R. (1967). High conductivity solid electrolytes-MAg_4I_5. *Science, 157,* 308–310.
4. Bradley, J. N., & Greene, P. D. (1967). Solids with high ionic conductivity in group 1 halide systems. *Transactions of the Faraday Society, 63,* 424–430.
5. Chen, C.-C., Fu, L., & Maier, J. (2016). Synergistic, ultrafast mass storage and removal in artificial mixed conductors. *Nature, 536,* 159–164.
6. Whitting, M. S., & Huggins, R. A. (1971). Measurement of sodium ion transport in beta alumina using reversible solid electrodes. *Journal of Chemical Physics, 54,* 414–416.
7. Whittingham, M. S., & Huggins, R. A. (1971). Transport properties of silver beta-alumina. *Journal of the Electrochemical Society, 118,* 1–6.
8. Bachman, J. C., Muy, S., Grimaud, A., et al. (2016). Inorganic solid-state electrolytes for lithium batteries: Mechanisms and properties governing ion conduction. *Chemical Reviews, 116,* 140–162.
9. Fenton, D. E., Parker, J. M., & Wright, P. V. (1973). Complexes of alkali-metal ions with poly(ethylene oxide). *Polymer, 14,* 589–589.
10. Armand, M. (1994). The history of polymer electrolytes. *Solid State Ionics, 69,* 309–319.
11. Meyer, W. H. (1998). Polymer electrolytes for lithium-ion batteries. *Advanced Materials, 10,* 439–448.
12. Croce, F., Appetecchi, G. B., Persi, L., et al. (1998). Nanocomposite polymer electrolytes for lithium batteries. *Nature, 394,* 456–458.
13. Chazalviel, J. N. (1990). Electrochemical aspects of the generation of ramifield metallic electrodeposits. *Physical Review A, 42,* 7355–7367.
14. Bouchet, R., Maria, S., Meziane, R., et al. (2013). Single-ion BAB triblock copolymers as highly efficient electrolytes for lithium-metal batteries. *Nature Materials, 12,* 452–457.
15. Seino, Y., Ota, T., Takada, K., et al. (2014). A sulphide lithium super ion conductor is superior to liquid ion conductors for use in rechargeable batteries. *Energy & Environmental Science, 7,* 627–631.
16. Li, J., Ma, C., Chi, M., et al. (2015). Solid electrolyte: The key for high-voltage lithium batteries. *Advanced Energy Materials, 5,* 1401408.
17. Bates, J. B., Dudney, N. J., Gruzalski, G. R., et al. (1992). Electrical properties of amorphous lithium electrolyte thin films. *Solid State Ionics, 53–56,* 647–654.
18. Kanno, R., Hata, T., Kawamoto, Y., et al. (2000). Synthesis of a new lithium ionic conductor, thio-LISICON-lithium germanium sulfide system. *Solid State Ionics, 130,* 97–104.
19. Hong, H. Y. P. (1978). Crystal structure and ionic conductivity of $Li_{14}Zn(GeO_4)_4$ and other new Li^+ superionic conductors. *Materials Research Bulletin, 13,* 117–124.
20. Kato, Y., Hori, S., Saito, T., et al. (2016). High-power all-solid-state batteries using sulfide superionic conductors. *Nature Energy, 1,* 16030.
21. Goodenough, J. B., Hong, H. Y. P., & Kafalas, J. A. (1976). Fast Na +-ion transport in skeleton structures. *Materials Research Bulletin, 11,* 203–220.
22. Inaguma, Y., Liquan, C., Itoh, M., et al. (1993). High ionic conductivity in lithium lanthanum titanate. *Solid State Communications, 86,* 689–693.
23. Murugan, R., Thangadurai, V., & Weppner, W. (2007). Fast lithium ion conduction in garnet-type $Li_7La_3Zr_2O_{12}$. *Angewandte Chemie International Edition, 46,* 7778–7781.
24. Aono, H., Sugimoto, E., Sadaoka, Y., et al. (1990). Ionic conductivity of solid electrolytes based on lithium titanium phosphate. *Journal of the Electrochemical Society, 137,* 1023–1027.
25. Blanchard, D., Nale, A., Sveinbjörnsson, D., et al. (2015). Nanoconfined LiBH4 as a fast lithium ion conductor. *Advanced Functional Materials, 25,* 184–192.

26. Boukamp, B. A., & Huggins, R. A. (1978). Fast ionic conductivity in lithium nitride. *Materials Research Bulletin, 13,* 23–32.
27. He, X., Zhu, Y., & Mo, Y. (2017). Origin of fast ion diffusion in super-ionic conductors. *Nature Communications, 8,* 15893.
28. Zugmann, S., Fleischmann, M., Amereller, M., et al. (2011). Measurement of transference numbers for lithium ion electrolytes via four different methods, a comparative study. *Electrochimica Acta, 56,* 3926–3933.
29. Hebb, M. H. (1952). Electrical conductivity of silver sulfide. *The Journal of Chemical Physics, 20,* 185–190.
30. Wagner, C. (1953). Investigations on silver sulfide. *The Journal of Chemical Physics, 21,* 1819–1827.
31. Sata, N., Eberman, K., Eberl, K., et al. (2000). Mesoscopic fast ion conduction in nanometre-scale planar heterostructures. *Nature, 408,* 946–949.
32. Guo, Y.-G., Hu, Y.-S., Lee, J.-S., et al. (2006). High-performance rechargeable all-solid-state silver battery based on superionic AgI nanoplates. *Electrochemistry Communications, 8,* 1179–1184.
33. Maekawa, H., Iwatani, T., Shen, H., et al. (2008). Enhanced lithium ion conduction and the size effect on interfacial phase in Li_2ZnI_4–mesoporous alumina composite electrolyte. *Solid State Ionics, 178,* 1637–1641.
34. Maekawa, H., Tanaka, R., Sato, T., et al. (2004). Size-dependent ionic conductivity observed for ordered mesoporous alumina-LiI composite. *Solid State Ionics, 175,* 281–285.
35. Guo, X. (2011). Can we achieve significantly higher ionic conductivity in nanostructured zirconia? *Scripta Materialia, 65,* 96–101.
36. Hu, Y. W., Raistrick, I. D., & Huggins, R. A. (1977). Ionic Conductivity of lithium orthosilicate—lithium phosphate solid solutions. *Journal of the Electrochemical Society, 124,* 1240–1242.
37. Rodger, A. R., Kuwano, J., & West, A. R. (1985). Li+ ion conducting γ solid solutions in the systems Li_4XO_4-Li_3YO_4: X = Si, Ge, Ti; Y = P, As, V; Li_4XO_4-$LiZO_2$: Z = Al, Ga, Cr and Li_4GeO_4-Li_2CaGeO_4. *Solid State Ionics, 15,* 185–198.
38. Kanno, R., & Maruyama, M. (2001). Lithium ionic conductor thio-LISICON—The Li2S-GeS2-P2S5 system. *Journal of the Electrochemical Society, 148,* A742–A746.
39. Takada, K., Inada, T., Kajiyama, A., et al. (2003). Solid-state lithium battery with graphite anode. *Solid State Ionics, 158,* 269–274.
40. Weber, D. A., Senyshyn, A., Weldert, K. S., et al. (2016). Structural insights and 3D diffusion pathways within the lithium superionic conductor $Li_{10}GeP_2S_{12}$. *Chemistry of Materials, 28,* 5905–5915.
41. Bron, P., Johansson, S., Zick, K., et al. (2013). $Li_{10}SnP_2S_{12}$: An affordable lithium superionic conductor. *Journal of the American Chemical Society, 135,* 15694–15697.
42. Deiseroth, H.-J., Kong, S.-T., Eckert, H., et al. (2008). Li_6PS_5X: A class of crystalline Li-rich solids with an unusually high Li^+ mobility. *Angewandte Chemie International Edition, 47,* 755–758.
43. Kraft, M. A., Culver, S. P., Calderon, M., et al. (2017). Influence of lattice polarizability on the ionic conductivity in the lithium superionic argyrodites Li_6PS_5X (X = Cl, Br, I). *Journal of the American Chemical Society, 139,* 10909–10918.
44. Yu, C., Ganapathy, S., de Klerk, N. J. J., et al. (2016). Unravelling Li-Ion Transport from picoseconds to seconds: Bulk versus interfaces in an argyrodite Li_6PS_5Cl–Li_2S All-solid-state Li-Ion battery. *Journal of the American Chemical Society, 138,* 11192–11201.
45. Aono, H., Sugimoto, E., Sadaoka, Y., et al. (1993). The electrical properties of ceramic electrolytes for LiM x Ti2 − x (PO 4) 3+ yLi2 O, M = Ge, Sn, Hf, and Zr systems. *Journal of the Electrochemical Society, 140,* 1827–1833.
46. Aono, H., Sugimoto, E., Sadaoka, Y., et al. (1991). Electrical property and sinterability of $LiTi_2(PO_4)_3$ mixed with lithium salt (Li_3PO_4 or Li_3BO_3). *Solid State Ionics, 47,* 257–264.
47. Adachi, G.-Y., Imanaka, N., & Aono, H. (1996). Fast $Li^{\oplus}$ conducting ceramic electrolytes. *Advanced Materials*, 8, 127–135.

48. Thangadurai, V., Kaack, H., & Weppner, W. J. F. (2003). Novel fast lithium ion conduction in garnet-type $Li_5La_3M_2O_{12}$ (M = Nb, Ta). *Journal of the American Ceramic Society, 86,* 437–440.
49. Awaka, J., Kijima, N., Takahashi, Y., et al. (2009). Synthesis and crystallographic studies of garnet-related lithium-ion conductors $Li_6CaLa_2Ta_2O_{12}$ and $Li_6BaLa_2Ta_2O_{12}$. *Solid State Ionics, 180,* 602–606.
50. Thangadurai, V., Narayanan, S., & Pinzaru, D. (2014). Garnet-type solid-state fast Li ion conductors for Li batteries: Critical review. *Chemical Society Reviews*, 43, 4714–4727.
51. Geiger, C. A., Alekseev, E., Lazic, B., et al. (2011). Crystal chemistry and stability of "Li7La3Zr2O12" Garnet: A fast lithium-ion conductor. *Inorganic Chemistry*, 50, 1089–1097.
52. Bernuy-Lopez, C., Manalastas, W., del Amo, J. M. L., et al. (2014). Atmosphere controlled processing of Ga-substituted garnets for high Li-Ion conductivity ceramics. *Chemistry of Materials, 26,* 3610–3617.
53. Li, Y., Han, J.-T., Wang, C.-A., et al. (2012). Optimizing Li+ conductivity in a garnet framework. *Journal of Materials Chemistry, 22,* 15357–15361.
54. Unemoto, A., Matsuo, M., & Orimo, S.-I. (2014). Complex hydrides for electrochemical energy storage. *Advanced Functional Materials, 24,* 2267–2279.
55. Maekawa, H., Matsuo, M., Takamura, H., et al. (2009). Halide-stabilized LiBH4, a room-temperature lithium fast-ion conductor. *Journal of the American Chemical Society, 131,* 894–895.
56. Matsuo, M., Remhof, A., Martelli, P., et al. (2009). Complex hydrides with (BH4)− and (NH2)−anions as new lithium fast-ion conductors. *Journal of the American Chemical Society, 131,* 16389–16391.
57. Friedrichs, O., Remhof, A., Hwang, S. J., et al. (2010). Role of Li2B12H12 for the formation and decomposition of $LiBH_4$. *Chemistry of Materials, 22,* 3265–3268.
58. Zhao, Y., & Daemen, L. L. (2012). Superionic conductivity in lithium-rich anti-perovskites. *Journal of the American Chemical Society, 134,* 15042–15047.
59. Emly, A., Kioupakis, E., & Van der Ven, A. (2013). Phase stability and transport mechanisms in antiperovskite Li_3OCl and Li_3OBr superionic conductors. *Chemistry of Materials, 25,* 4663–4670.
60. Braga, M. H., Murchison, A. J., Ferreira, J. A., et al. (2016). Glass-amorphous alkali-ion solid electrolytes and their performance in symmetrical cells. *Energy & Environmental Science, 9,* 948–954.
61. Braga, M. H., Ferreira, J. A., Stockhausen, V., et al. (2014). Novel Li3ClO based glasses with superionic properties for lithium batteries. *Journal of Materials Chemistry A, 2,* 5470–5480.
62. Braga, M. H., Grundish, N. S., Murchison, A. J., et al. (2017). Alternative strategy for a safe rechargeable battery. *Energy & Environmental Science, 10,* 331–336.
63. Fang, H., & Jena, P. (2017). Li-rich antiperovskite superionic conductors based on cluster ions. *Proceedings of the National Academy of Sciences, 114,* 11046–11051.
64. Steingart, D. A., & Viswanathan, V. (2018). Comment on "Alternative strategy for a safe rechargeable battery" by M. H. Braga, N. S. Grundish, A. J. Murchison, & J. B. Goodenough. (2017). *Energy & Environmental Science, 10,* 331–336. *Energy & Environmental Science, 11,* 221–222.
65. Braga, M. H., Subramaniyam, C. M., Murchison, A. J., et al. (2018). Nontraditional, safe, high voltage rechargeable cells of long cycle life. *Journal of the American Chemical Society, 140,* 6343–6352.
66. Groce, F., Gerace, F., Dautzemberg, G., et al. (1994). Synthesis and characterization of highly conducting gel electrolytes. *Electrochimica Acta, 39,* 2187–2194.
67. Dautzenberg, G., Croce, F., Passerini, S., et al. (1994). Characterization of PAN-based gel electrolytes. electrochemical stability and lithium cyclability. *Chemistry of Materials, 6,* 538–542.
68. Appetecchi, G. B., Croce, F., & Scrosati, B. (1995). Kinetics and stability of the lithium electrode in poly(methylmethacrylate)-based gel electrolytes. *Electrochimica Acta, 40,* 991–997.
69. Choe, H. S., Giaccai, J., Alamgir, M., et al. (1995). Preparation and characterization of poly(vinyl sulfone)- and poly(vinylidene fluoride)-based electrolytes. *Electrochimica Acta, 40,* 2289–2293.

70. Muramatsu, H., Hayashi, A., Ohtomo, T., et al. (2011). Structural change of $Li_2S–P_2S_5$ sulfide solid electrolytes in the atmosphere. *Solid State Ionics, 182,* 116–119.
71. Sahu, G., Lin, Z., Li, J., et al. (2014). Air-stable, high-conduction solid electrolytes of arsenic-substituted Li_4SnS_4. *Energy & Environmental Science, 7,* 1053–1058.
72. Busche, M. R., Weber, D. A., Schneider, Y., et al. (2016). In situ monitoring of fast Li-Ion conductor $Li_7P_3S_{11}$ crystallization inside a hot-press setup. *Chemistry of Materials, 28,* 6152–6165.
73. Buschmann, H., Dolle, J., Berendts, S., et al. (2011). Structure and dynamics of the fast lithium ion conductor "$Li_7La_3Zr_2O_{12}$". *Physical Chemistry Chemical Physics, 13,* 19378–19392.
74. Sharafi, A., Kazyak, E., Davis, A. L., et al. (2017). Surface chemistry mechanism of ultra-low interfacial resistance in the solid-state electrolyte $Li_7La_3Zr_2O_{12}$. *Chemistry of Materials, 29,* 7961–7968.
75. Han, X., Gong, Y., Fu, K., et al. (2016). Negating interfacial impedance in garnet-based solid-state Li metal batteries. *Nature Materials, 16,* 572–579.
76. Hartmann, P., Leichtweiss, T., Busche, M. R., et al. (2013). Degradation of NASICON-type materials in contact with lithium metal: Formation of mixed conducting interphases (MCI) on solid electrolytes. *The Journal of Physical Chemistry C, 117,* 21064–21074.
77. Han, F., Zhu, Y., He, X., et al. (2016). Electrochemical Stability of $Li_{10}GeP_2S_{12}$ and $Li_7La_3Zr_2O_{12}$ Solid Electrolytes. *Advanced Energy Materials, 6,* 1501590.
78. Han, F., Gao, T., Zhu, Y., et al. (2015). A battery made from a single material. *Advanced Materials, 27,* 3473–3483.
79. Tian, Y., Shi, T., Richards, W. D., et al. (2017). Compatibility issues between electrodes and electrolytes in solid-state batteries. *Energy & Environmental Science, 10,* 1150–1166.
80. Richards, W. D., Miara, L. J., Wang, Y., et al. (2016). Interface stability in solid-state batteries. *Chemistry of Materials, 28,* 266–273.
81. Luo, W., Gong, Y., Zhu, Y., et al. (2017). Reducing interfacial resistance between garnet-structured solid-state electrolyte and Li-metal anode by a germanium layer. *Advanced Materials, 29,* 1606042.
82. Fu, K., Gong, Y., Fu, Z., et al. (2017). Transient behavior of the metal interface in lithium metal-garnet batteries. *Angewandte Chemie International Edition, 56,* 14942–14947.
83. Reinacher, J., Berendts, S., & Janek, J. (2014). Preparation and electrical properties of garnet-type $Li_6BaLa_2Ta_2O_{12}$ lithium solid electrolyte thin films prepared by pulsed laser deposition. *Solid State Ionics, 258,* 1–7.
84. Duan, H., Yin, Y.-X., Shi, Y., et al. (2018). Dendrite-free Li-metal battery enabled by a thin asymmetric solid electrolyte with engineered layers. *Journal of the American Chemical Society, 140,* 82–85.
85. Luntz, A. C., Voss, J., & Reuter, K. (2015). Interfacial challenges in solid-state Li ion batteries. *Journal of Physical Chemistry Letters, 6,* 4599–4604.
86. Zhou, W., Wang, S., Li, Y., et al. (2016). Plating a dendrite-free lithium anode with a polymer/ceramic/polymer sandwich electrolyte. *Journal of the American Chemical Society, 138,* 9385–9388.
87. Grey, C. P., & Tarascon, J. M. (2016). Sustainability and in situ monitoring in battery development. *Nature Materials, 16,* 45–56.
88. Irvine, J. T. S., Sinclair, D. C., & West, A. R. (1990). Electroceramics: Characterization by impedance spectroscopy. *Advanced Materials, 2,* 132–138.
89. Evans, J., Vincent, C. A., & Bruce, P. G. (1987). Electrochemical measurement of transference numbers in polymer electrolytes. *Polymer, 28,* 2324–2328.
90. Lin, Z., Liu, Z., Dudney, N. J., et al. (2013). Lithium superionic sulfide cathode for all-solid lithium-sulfur batteries. *ACS Nano, 7,* 2829–2833.
91. Zhu, Z., Hong, M., Guo, D., et al. (2014). All-solid-state lithium organic battery with composite polymer electrolyte and pillar[5]quinone cathode. *Journal of the American Chemical Society, 136,* 16461–16464.
92. Yin, Y.-X., Xin, S., Guo, Y.-G., et al. (2013). Lithium-sulfur batteries: Electrochemistry, materials, and prospects. *Angewandte Chemie International Edition, 52,* 13186–13200.

93. Yao, X., Huang, N., Han, F., et al. (2017). High-performance all-solid-state lithium–sulfur batteries enabled by amorphous sulfur-coated reduced graphene oxide cathodes. *Advanced Energy Materials, 7,* 1602923.
94. Yue, J., Yan, M., Yin, Y. -X., et al. Progress of the interface design in all-solid-state Li–S batteries. *Advanced Functional Materials* 1707533.
95. Kobayashi, T., Imade, Y., Shishihara, D., et al. (2008). All solid-state battery with sulfur electrode and thio-LISICON electrolyte. *Journal of Power Sources, 182,* 621–625.
96. Han, F., Yue, J., Fan, X., et al. (2016). High-performance all-solid-state lithium–sulfur battery enabled by a mixed-conductive Li2S nanocomposite. *Nano Letters, 16,* 4521–4527.
97. Yao, X., Liu, D., Wang, C., et al. (2016). High-energy all-solid-state lithium batteries with ultralong cycle life. *Nano Letters, 16,* 7148–7154.
98. Lin, Z., Liu, Z., Fu, W., et al. (2013). Lithium polysulfidophosphates: A family of lithium-conducting sulfur-rich compounds for lithium–sulfur batteries. *Angewandte Chemie International Edition, 52,* 7460–7463.
99. Tanibata, N., Tsukasaki, H., Deguchi, M., et al. (2017). A novel discharge-charge mechanism of a S-P_2S_5 composite electrode without electrolytes in all-solid-state Li/S batteries. *Journal of Materials Chemistry A, 5,* 11224–11228.
100. Unemoto, A., Ikeshoji, T., Yasaku, S., et al. (2015). Stable interface formation between TiS_2 and $LiBH_4$ in bulk-type all-solid-state lithium batteries. *Chemistry of Materials, 27,* 5407–5416.
101. Kim, S., Toyama, N., Oguchi, H., et al. (2018). Fast lithium-ion conduction in atom-deficient closo-type complex hydride solid electrolytes. *Chemistry of Materials, 30,* 386–391.
102. Fu, K., Gong, Y., Hitz, G. T., et al. (2017). Three-dimensional bilayer garnet solid electrolyte based high energy density lithium metal-sulfur batteries. *Energy & Environmental Science, 10,* 1568–1575.
103. Tao, X., Liu, Y., Liu, W., et al. (2017). Solid-state lithium–sulfur batteries operated at 37 °C with composites of nanostructured $Li_7La_3Zr_2O_{12}$/carbon foam and polymer. *Nano Letters, 17,* 2967–2972.
104. Xu, G., Kushima, A., Yuan, J., et al. (2017). Ad hoc solid electrolyte on acidized carbon nanotube paper improves cycle life of lithium–sulfur batteries. *Energy & Environmental Science, 10,* 2544–2551.
105. Abraham, K. M., & Jiang, Z. (1996). A polymer electrolyte-based rechargeable lithium/oxygen battery. *Journal of the Electrochemical Society, 143,* 1–5.
106. Hassoun, J., Croce, F., Armand, M., et al. (2011). Investigation of the O_2 electrochemistry in a polymer electrolyte solid-state cell. *Angewandte Chemie International Edition, 50,* 2999–3002.
107. Kumar, B., Kumar, J., Leese, R., et al. (2010). A solid-state, rechargeable, long cycle life lithium-air battery. *Journal of the Electrochemical Society, 157,* A50–A54.
108. Kitaura, H., & Zhou, H. (2012). Electrochemical performance and reaction mechanism of all-solid-state lithium–air batteries composed of lithium, $Li_{1+x}Al_yGe_{2-y}(PO_4)_3$ solid electrolyte and carbon nanotube air electrode. *Energy & Environmental Science, 5,* 9077–9084.
109. Bruce, P. G., Freunberger, S. A., Hardwick, L. J., et al. (2011). Li–O_2 and Li–S batteries with high energy storage. *Nature Materials, 11,* 19–29.
110. Li, F., Kitaura, H., & Zhou, H. (2013). The pursuit of rechargeable solid-state Li–air batteries. *Energy & Environmental Science, 6,* 2302–2311.
111. Koerver, R., Aygün, I., Leichtweiß, T., et al. (2017). Capacity fade in solid-state batteries: Interphase formation and chemomechanical processes in nickel-rich layered oxide cathodes and lithium thiophosphate solid electrolytes. *Chemistry of Materials, 29,* 5574–5582.
112. Miyashiro, H., Kobayashi, Y., Seki, S., et al. (2005). Fabrication of all-solid-state lithium polymer secondary batteries using Al_2O_3-coated $LiCoO_2$. *Chemistry of Materials, 17,* 5603–5605.
113. Liang, J.-Y., Zeng, X.-X., Zhang, X.-D., et al. (2018). Mitigating interfacial potential drop of cathode-solid electrolyte via ionic conductor layer to enhance interface dynamics for solid batteries. *Journal of the American Chemical Society, 140,* 6767–6770.

114. Woo, J. H., Trevey, J. E., Cavanagh, A. S., et al. (2012). Nanoscale interface modification of $LiCoO_2$ by Al_2O_3 atomic layer deposition for solid-state Li batteries. *Journal of the Electrochemical Society, 159,* A1120–A1124.
115. Okumura, T., Nakatsutsumi, T., Ina, T., et al. (2011). Depth-resolved X-ray absorption spectroscopic study on nanoscale observation of the electrode–solid electrolyte interface for all solid state lithium ion batteries. *Journal of Materials Chemistry, 21,* 10051–10060.
116. Zhang, W., Leichtweiß, T., Culver, S. P., et al. (2017). The detrimental effects of carbon additives in $Li_{10}GeP_2S_{12}$-based solid-state batteries. *ACS Applied Materials & Interfaces, 9,* 35888–35896.
117. Park, K., Yu, B.-C., Jung, J.-W., et al. (2016). Electrochemical nature of the cathode interface for a solid-state lithium-ion battery: Interface between $LiCoO_2$ and garnet-$Li_7La_3Zr_2O_{12}$. *Chemistry of Materials, 28,* 8051–8059.
118. Han, F., Yue, J., Chen, C., et al. (2018). interphase engineering enabled all-ceramic lithium battery. *Joule, 2,* 497–508.
119. Dong, T., Zhang, J., Xu, G., et al. (2018). A multifunctional polymer electrolyte enables ultra-long cycle-life in a high-voltage lithium metal battery. *Energy & Environmental Science, 11,* 1197–1203.
120. Wakayama, H., Yonekura, H., & Kawai, Y. (2016). Three-dimensional bicontinuous nanocomposite from a self-assembled block copolymer for a high-capacity all-solid-state lithium battery cathode. *Chemistry of Materials, 28,* 4453–4459.
121. Liu, F. -Q., Wang, W. -P., Yin, Y. -X., et al. (2018). Upgrading traditional liquid electrolyte via in situ gelation for future lithium metal batteries. *Science Advances, 4,* eaat5383.
122. Dong, W., Zeng, X.-X., Zhang, X.-D., et al. (2018). Gradiently polymerized solid electrolyte meets with micro-/nanostructured cathode array. *ACS Applied Materials & Interfaces, 10,* 18005–18011.
123. Oh, D. Y., Kim, D. H., Jung, S. H., et al. (2017). Single-step wet-chemical fabrication of sheet-type electrodes from solid-electrolyte precursors for all-solid-state lithium-ion batteries. *Journal of Materials Chemistry A, 5,* 20771–20779.
124. Park, K. H., Oh, D. Y., Choi, Y. E., et al. (2016). Solution-processable glass LiI-Li_4SnS_4 superionic conductors for all-solid-state Li-Ion batteries. *Advanced Materials, 28,* 1874–1883.
125. Monroe, C., & Newman, J. (2005). The impact of elastic deformation on deposition kinetics at lithium/polymer interfaces. *Journal of the Electrochemical Society, 152,* A396–A404.
126. Monroe, C., & Newman, J. (2004). The effect of interfacial deformation on electrodeposition kinetics. *Journal of the Electrochemical Society, 151,* A880–A886.
127. Zeng, X.-X., Yin, Y.-X., Li, N.-W., et al. (2016). Reshaping lithium plating/stripping behavior via bifunctional polymer electrolyte for room-temperature solid Li metal batteries. *Journal of the American Chemical Society, 138,* 15825–15828.
128. Zekoll, S., Marriner-Edwards, C., Hekselman, A. K. O., et al. (2018). Hybrid electrolytes with 3D bicontinuous ordered ceramic and polymer microchannels for all-solid-state batteries. *Energy & Environmental Science, 11,* 185–201.
129. Fu, K., Gong, Y., Dai, J., et al. (2016). Flexible, solid-state, ion-conducting membrane with 3D garnet nanofiber networks for lithium batteries. *Proceedings of the National Academy of Sciences, 113,* 7094–7099.
130. Bai, P., Li, J., Brushett, F. R., et al. (2016). Transition of lithium growth mechanisms in liquid electrolytes. *Energy & Environmental Science*, 9, 3221–3229.
131. Zhao, C.-Z., Zhang, X.-Q., Cheng, X.-B., et al. (2017). An anion-immobilized composite electrolyte for dendrite-free lithium metal anodes. *Proceedings of the National Academy of Sciences, 114,* 11069–11074.
132. Wang, S., Xu, H., Li, W., et al. (2018). Interfacial chemistry in solid-state batteries: Formation of interphase and its consequences. *Journal of the American Chemical Society, 140,* 250–257.
133. Porz, L., Swamy, T., Sheldon, B. W., et al. (2017). Mechanism of lithium metal penetration through inorganic solid electrolytes. *Advanced Energy Materials, 7,* 1701003.
134. Sharafi, A., Haslam, C. G., Kerns, R. D., et al. (2017). Controlling and correlating the effect of grain size with the mechanical and electrochemical properties of $Li_7La_3Zr_2O_{12}$ solid-state electrolyte. *Journal of Materials Chemistry A, 5,* 21491–21504.

135. Sharafi, A., Meyer, H. M., Nanda, J., et al. (2016). Characterizing the Li–$Li_7La_3Zr_2O_{12}$ interface stability and kinetics as a function of temperature and current density. *Journal of Power Sources, 302,* 135–139.
136. Kerman, K., Luntz, A., Viswanathan, V., et al. (2017). Review-practical challenges hindering the development of solid state Li ion batteries. *Journal of the Electrochemical Society*, 164, A1731–A1744.
137. Han, F., Westover, A. S., Yue, J., et al. (2019). High electronic conductivity as the origin of lithium dendrite formation within solid electrolytes. *Nature Energy, 4,* 187–196.
138. Fan, X., Ji, X., Han, F., et al. (2018). Fluorinated solid electrolyte interphase enables highly reversible solid-state Li metal battery. *Science Advances, 4,* eaau9245.
139. Yang, C., Zhang, L., Liu, B., et al. (2018). Continuous plating/stripping behavior of solid-state lithium metal anode in a 3D ion-conductive framework. *Proceedings of the National Academy of Sciences, 115,* 3770–3775.
140. Tanibata, N., Deguchi, M., Hayashi, A., et al. (2017). All-solid-state Na/S batteries with a Na_3PS_4 electrolyte operating at room temperature. *Chemistry of Materials, 29,* 5232–5238.
141. Gao, H., Xin, S., Xue, L., et al. (2018). Stabilizing a high-energy-density rechargeable sodium battery with a solid electrolyte. *Chem, 4,* 833–844.
142. Zhao, C., Liu, L., Qi, X., et al. (2018). Solid-state sodium batteries. *Advanced Energy Materials, 8,* 1703012.
143. Gao, H., Xue, L., Xin, S., et al. (2018). A high-energy-density potassium battery with a polymer-gel electrolyte and a polyaniline cathode. *Angewandte Chemie International Edition, 57,* 5449–5453.
144. Liu, B., Fu, K., Gong, Y., et al. (2017). Rapid thermal annealing of cathode-garnet interface toward high-temperature solid state batteries. *Nano Letters, 17,* 4917–4923.
145. Jin, Y., Liu, K., Lang, J., et al. (2018). An intermediate temperature garnet-type solid electrolyte-based molten lithium battery for grid energy storage. *Nature Energy, 3,* 732–738.
146. Kato, A., Yamamoto, M., Sakuda, A., et al. (2018). Mechanical properties of Li_2S–P_2S_5 glasses with lithium halides and application in all-solid-state batteries. *ACS Applied Energy Materials, 1,* 1002–1007.
147. Koerver, R., Zhang, W., de Biasi, L., et al. (2018). Chemo-mechanical expansion of lithium electrode materials—On the route to mechanically optimized all-solid-state batteries. *Energy & Environmental Science, 11,* 2142–2158.
148. Chung, H., & Kang, B. (2017). Mechanical and thermal failure induced by contact between a $Li_{1.5}Al_{0.5}Ge_{1.5}(PO_4)_3$ solid electrolyte and Li metal in an all solid-state Li cell. *Chemistry of Materials, 29,* 8611–8619.

Chapter 6
Nanostructures and Nanomaterials for Sodium Batteries

Peng-Fei Wang, Yu-Bin Niu and Yu-Guo Guo

Abstract As a kind of post-Li-ion batteries (LIBs), the advantage of "high-quality and low-cost" reflected by Na-ion batteries (NIBs) in large-scale energy storage has attracted more and more attention and recognition. However, larger ionic radius and slower kinetic behavior of Na make it impossible to compete with more mature LIBs in terms of energy density and power density. As far as the electrode materials themselves are concerned, there is still a great potential to be tapped. For this reason, the nanoscale structural design also plays an important role in improving the overall performance of NIBs. In this chapter, we systematically summarize and discuss the research progress of nanostructured strategies for cathode and anode materials of NIBs, and prospect for their future development as well as highlight the impact of nanostructure on the electrochemical performance of the batteries.

Abbreviation

ABF	Annular bright field
AAO	Anodic aluminum oxide
ALD	Atomic layer deposition
CNT	Carbon nanotubes
CVD	Chemical vapor deposition
CTAB	Cetyltrimethyl ammonium bromide
ESD	Electrostatic spray deposition
G	Graphene
HRTEM	High-resolution transmission electron microscope
HAADF	High-angle annular dark field
LIBs	Li-ion batteries

P.-F. Wang
University of Maryland at College Park, College Park, Maryland 20742, USA

Y.-B. Niu · Y.-G. Guo (✉)
Institute of Chemistry, Chinese Academy of Sciences, Beijing 100190, People's Republic of China
e-mail: ygguo@iccas.ac.cn

Y.-G. Guo, *Nanostructures and Nanomaterials for Batteries*,
https://doi.org/10.1007/978-981-13-6233-0_6

MWCNTsTs	Multi-walled carbon nanotubes
MLD	Molecular layer deposition
SBA-13	Mesoporous silicon
NIBs	Na-ion batteries
DMF	*N*,*N*-dimethylformamide
NMP	*N*-methyl-2-pyrrolidone
CMK-3	Ordered mesoporous carbon
1D	One-dimensional
PVP	Polyvinyl pyrrolidone
PEO	Polyethylene oxide
PAN	Polyacrylonitrile
PVA	Polyvinyl alcohol
PVdF	Polyvinylidene fluoride
PVD	Physical vapor deposition
P123	Polyethylene oxide-polypropylene oxide-polyethylene oxide
rGO	Reduced graphene oxide
3D	Three-dimensional
TM	Transition metal
TEM	Transmission electron microscope
SHE	Standard hydrogen electrode
SEM	Scanning electron microscope
SAED	Selected area electron diffraction
STEM	Scanning transmission electron microscopy
SDS	Sodium dodecyl sulfate
SWCNTs	Single walled carbon nanotubes
XPS	X-ray photoelectron spectroscopy
0D	Zero-dimensional

6.1 Sodium Ion—An Important Charge Carrier for Energy Storage

Since graphite//$LiCoO_2$ cell commercialized in 1991 for the first time, rechargeable LIBs have stepped into our daily life as secondary energy storage device. It is well acknowledged that lithium is widely existed in the Earth's crust but is not regarded as an abundant element [1, 2]. In this context, we have to reconsider the sustainability of lithium element. According to the newest report, the relative abundance of lithium in the Earth's crust is limited to be only 20 ppm. Indeed, the raw materials cost such as the price of Li_2CO_3 is 20–30 times more expensive than corresponding Na equivalents. What is worse, lithium resources are mainly distributed in South America, leading to that the production of LIBs relies on the import of lithium from South America. In comparison, as one of the most abundant elements in the Earth's crust, sodium resources are unlimited everywhere especially for the infinite Na in

Table 6.1 Comparison of physical properties for "lithium" and "sodium" as charge carriers for rechargeable batteries

	Li	Na
Cation radius (pm)	76	102
Atomic weight (g mol^{-1})	6.94	23.00
E^0 versus SHE (V)	−3.04	−2.71
Coordination preference	Octahedral or tetrahedral	Octahedral or prismatic
Melting point (°C)	180.5	97.7
Distribution	70% in South America	Everywhere
Abundance (mg kg^{-1})	20	23,600
Price (US$ per ton)	5800	250–300

the ocean. In addition, sodium is the second lightest and smallest alkali metal next to lithium as shown in Table 6.1 and shows similar electrochemical properties with Li. On the basis of this context, the material's abundance and similar electrochemical properties are simple and clear reason as to why sodium ions are attractive as charge carriers for rechargeable batteries.

6.2 Differences and Similarities Between LIBs and NIBs

6.2.1 A Brief Summary of the History of LIBs and NIBs

In the early 1980s, Li^+/Na^+ ions as charge carriers for electrochemical energy storage at ambient temperature were almost studied at the same time. Early study was mainly focused on electrochemical Li/Na insertion into TiS_2 [3, 4]. Since the first report of the superior electrode performance of $LiCoO_2$ in 1980, intensive attention has been concentrated only on Li insertion materials, and studies on NIBs for energy storage once almost disappeared. Besides, another important reason was the absence of suitable anode materials for NIBs in a long period of time. In the mid-1980s, carbonaceous materials such as disordered carbons and graphite have been found to be promising candidates for Li insertion hosts. The founding of the graphite negative electrode material for LIBs, which delivers high reversible capacity (theoretically 372 mA h g^{-1}) with a low and flat operating potential of 0.1–0.2 V versus Li^+/Li, further accelerating the development of LIBs. Unfortunately, graphite cannot be used as an insertion host of sodium ions.

In 2000, Stevens and Dahn found that hard carbon could deliver a high reversible capacity of 300 mA h g^{-1} in a Na cell, which is close to that for lithium insertion in graphitic carbon. Even though its cyclability was insufficient for battery applications at that time, hard carbon has then been extensively studied as a potential anode electrode material candidate for NIBs. Another important founding is that Okada and coworkers [5] reported that $NaFeO_2$ is electrochemically active in Na cells based on the unique Fe^{3+}/Fe^{4+} redox couple, which is an important subject to realize the high-energy and cost-effective NIBs in the future. This two important discovery become the turning point in the research for NIBs and research interest for sodium ions has completely attractively changed since then.

6.2.2 *Working Principles*

Following the "rock-chair" working principles of existing LIBs, an intercalation sodium battery could consist of an anode made of a high-capacity, dense Na-metal alloy (e.g. Na metal, Na_4Sn or Na_xC) and a high-capacity cathode because sodium metal has, at 97.72 °C, a lower melting point than lithium (180.5 °C), room temperature poses a risk for Na metal electrodes. Generally speaking, a NIB consists of two sodium insertion materials, cathode and anode electrodes, which are electronically separated by electrolyte (in general, electrolyte salts dissolved in aprotic polar solvents) as a pure ionic conductor. As a "rocking-chair" type battery, the charge-discharge reactions of NIBs take place through a reversible Na^+-intercalation mechanism, while Na ions move off from the cathode host, pass through the electrolyte and then insert into the anode host during charge, and vice versa during discharge (Fig. 6.1). Meanwhile, electrons move from anode to cathode during charge or from cathode to anode during discharge through the external circuit, thus storing or delivering electric energy in a direct electrochemical manner.

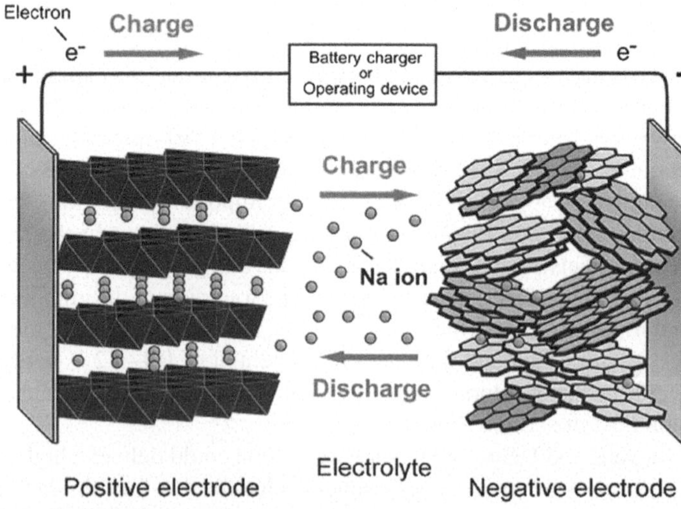

Fig. 6.1 Working principles of Na-ion batteries. Wherein, figure was reproduced with permission from Ref. [1], Copyright © 2014 American Chemical Society

6.2.3 *Size Effects of Large Na Ions*

The size effects of large Na ions make a large difference between Li and Na batteries. Firstly, due to the heavier atomic weight (23 g mol^{-1} vs. 7 g mol^{-1}) and larger radius (1.02 Å vs. 0.76 Å) of Na^+ than that of Li^+, there is no doubt that the large Na ions will make a decrease in both the theoretical gravimetric and volumetric reversible capacity when compared with Li insertion electrodes. For example, layered $LiCoO_2$ and $NaCoO_2$ oxides have the same crystal structure. Assuming that one-electron Co^{3+}/Co^{4+} redox is likely to occur, the difference in the calculated theoretical reversible capacity becomes smaller (274 and 235 mA h g^{-1} for $LiCoO_2$ and $NaCoO_2$, respectively). Similarly, because the molar volume of Na (39.3 $Å^3$ per Na atom) is much larger than that of Li (21.3 $Å^3$ per Li atom), the volumetric capacity of Na metal becomes much smaller. Although facing the unsatisfactory energy density, NIBs are expected to be still the competitive battery system for large scale energy storage considering the original target of NIBs battery technology is based on the Na ions, not Na metal. The capacity sacrifice can be potentially reduced.

Secondly, many Na based compounds exhibit enlarged interlayer distance and multiple phase transitions due to the relatively bulky Na ions, resulting in different electrochemistry properties. These Na-based layered materials can be categorized into two main groups: P2 type or O3 type according to the surrounding Na^+ environment and the number of unique oxide layer stacking, which was first specified by Delmas et al. [6] Symbols of "P" or "O" represents a prismatic or octahedral coordination environment of Na ions, and the "2" or "3" suggests the number of transition metal layers with different kinds of O stacking in a single cell unit. In general, both P2 and O3 phases commonly suffer from a series of phase transitions involving different stacking sequences of the oxide layers during the electrochemical cycling. P2 phase usually transforms to O2 phase due to gliding ($\pi/3$ rotation) of some TMO_6 octahedra sheets if a certain amount of Na^+ are extracted, resulting in a significant crystal structure contraction and decreased interlayer distance (Fig. 6.2a–b). Instead, in the O3-type phase Na^+ are originally stabilized at edge-shared octahedral sites with TMO_6 octahedra as shown in Fig. 6.2c. When Na ions are partly extracted from the O3-type phase, Na^+ at prismatic sites become energetically stable, associated with the formation of vacancies, similar to the P2-type phase. Then the formation of spacious prismatic sites is achieved by the gliding of TMO_2 sheets without breaking TM-O bonds. As a result, the typical oxygen stacking changes from "AB CA BC" to "AB BC CA", and this new phase is classified as a P3-type phase (Fig. 6.2d) [7].

Lastly, the larger ionic radii of Na^+ is found to be a weak solvation energy in polar solvents, which has been evidenced by systematic and theoretical studies on Li^+, Na^+, and Mg^{2+} together with different aprotic polar solvents. Since the desolvation energy highly influences the kinetics of alkali-ion insertion processes at the electrolyte interface, the relatively low energy for desolvation is an important finding to design high-power batteries. Small Li ions have a relatively high charge density around ions when compared with Na ions as the same monovalent ions. Li ions are, therefore, energetically stabilized by accepting/sharing more electrons from/with the

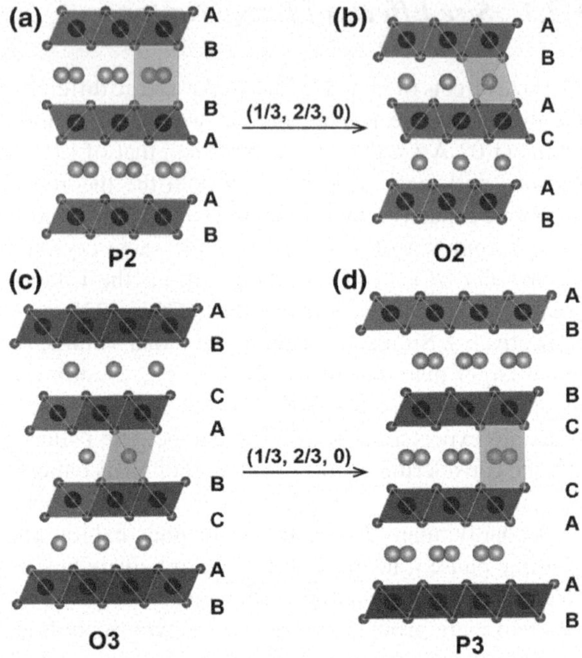

Fig. 6.2 A schematic of the crystal structures for layered Na-containing oxides (Na_xTMO_2) for **a** P2-type, **b** O2-type, **c** O3-type, and **d** P3-type stackings. The blue and yellow balls represent the transition metal and Na ions in the O-type frameworks, respectively. Wherein, **a**–**d** were reproduced with permission from Ref. [7], Copyright © 2018 John Wiley & Sons

solvated polar molecules, namely, Li^+ is classified as a relatively strong Lewis acid. As a result, a relatively large energy for the desolvation process is required for Li^+ compared with Na^+. Similarly, results of first-principles calculation suggest that the activation energy of Na^+ diffusion is relatively small for $NaCoO_2$ when compared with $LiCoO_2$. The high ion conductivity of Na^+-based electrolyte is also beneficial to increase the battery performance compared to Li^+ electrolyte. A comparative study of the molar conductivity of $NaClO_4$ and $LiClO_4$ has revealed a relatively low viscosity of $NaClO_4$ solutions with aprotic solvents, and conductivity is also high (10–20%) in comparison to $LiClO_4$ solutions. These facts also probably originate from the size difference between Na and Li in relation to the solvation energy and solvated states of ions in aprotic polar solvents.

6.3 Nanostructures and Nanomaterials for Cathodes of NIBs

Currently, the application of cathodes for NIBs has several challenges including low ion mobility, low electrical conductivity, and severe structural degradation during the long-term cycling. How to improve the electrochemical performance of various cathode materials has become the main bottleneck in the development of high-performance cathodes for NIBs.

6.3.1 Surface Coating at Nanoscale

Typical polyanion-type materials and organics are commonly subject to low electrical conductivities due to the intrinsically isolating nature of polyanions. The construction of nanocomposites with conductive carbon networks and nanostructured active particles demonstrate effective to address these issues. As one of the most promising cathodes, NASICON-based electrode material $Na_3V_2(PO_4)_3$ (NVP) shows high Na^+ ion conductivity, small volume changes upon Na^+ intercalation/deintercalation, moderate potential plateau (3.4 V), high theoretical energy density (400 Wh kg^{-1}), high thermal stability, facile reaction mechanism, and rich structural diversity [8–10]. The 3D framework structures of NVP (*R*-3*c* space group) are composed of corner-shared VO_6 octahedra and PO_4 tetrahedra. The transport behavior is consisting of two different types of Na^+, including one Na^+ at M1 sites and the other two Na^+ at M2 sites (Fig. 6.3a). However, the cycling stability and rate capability of electrode material NVP appeared unsatisfactory because of the intrinsic poor electronic conductivity of the phosphate framework [11, 12]. To solve this problem, many efforts have been devoted to seeking for electronically conductive medium to improve the overall transport in the past few years [13–15]. Owing to high surface area, high electrical conductivity, good chemical stability and low cost, various carbon materials as conductive substrates have been demonstrated an efficient strategy to improve electronic conductivity, aiming at obtaining better electrochemical performance. As to the carbon materials, amorphous carbon, multi-walled carbon nanotubes (MWCNTsTs), mesoporous ordered carbon frameworks, graphene, interconnected carbon frameworks, and N-doped carbon have been confirmed to improve the electrochemical performance during long-term cycles at high current densities [16–20].

Amorphous carbon: amorphous carbon can be formed after many organic polymers and surfactants decomposition in inert atmosphere, which frequently used to control the NVP size. Chen and co-workers reported that ascorbic acid and polyethylene glycol 400 can not only provide the uniform surface coating of conductive carbon, but also efficiently control the particle growth (Fig. 6.3b) [21]. In general, as displayed in Fig. 6.3c, the electrochemical reaction of Na^+ can only occur on the outside part of the large particles due to limited diffusion length and the electronic transfer is blocked by the low conductivity, leading to poor battery performance. The core-shell structure was composed of a 40 nm NVP core and a few nanometers amorphous carbon shell after optimization, resulting in the excellent cycle life and rate capability. The enhanced electrochemical performance is attributed to the synergistic effect of small particle size and uniform carbon coating, which can make full utilization of the active material and the migration of both electrons and ions.

Carbon nanotubes: MWCNTsTs is known to excellent electrical conductivity, mechanical property, and good thermal stability, which is also regarded as one of most fascinating conductive mediums for NVP. Yang et al. proposed the MWCNTsTs modified NVP/C composite via a microwave solid-state route to improve the rate performance and cycling stability of NVP/C (Fig. 6.3d) [22]. As shown in Fig. 6.4e, f, the electrodes delivered specific capacities of 146.5, 142.3, 135.2, 126.7,

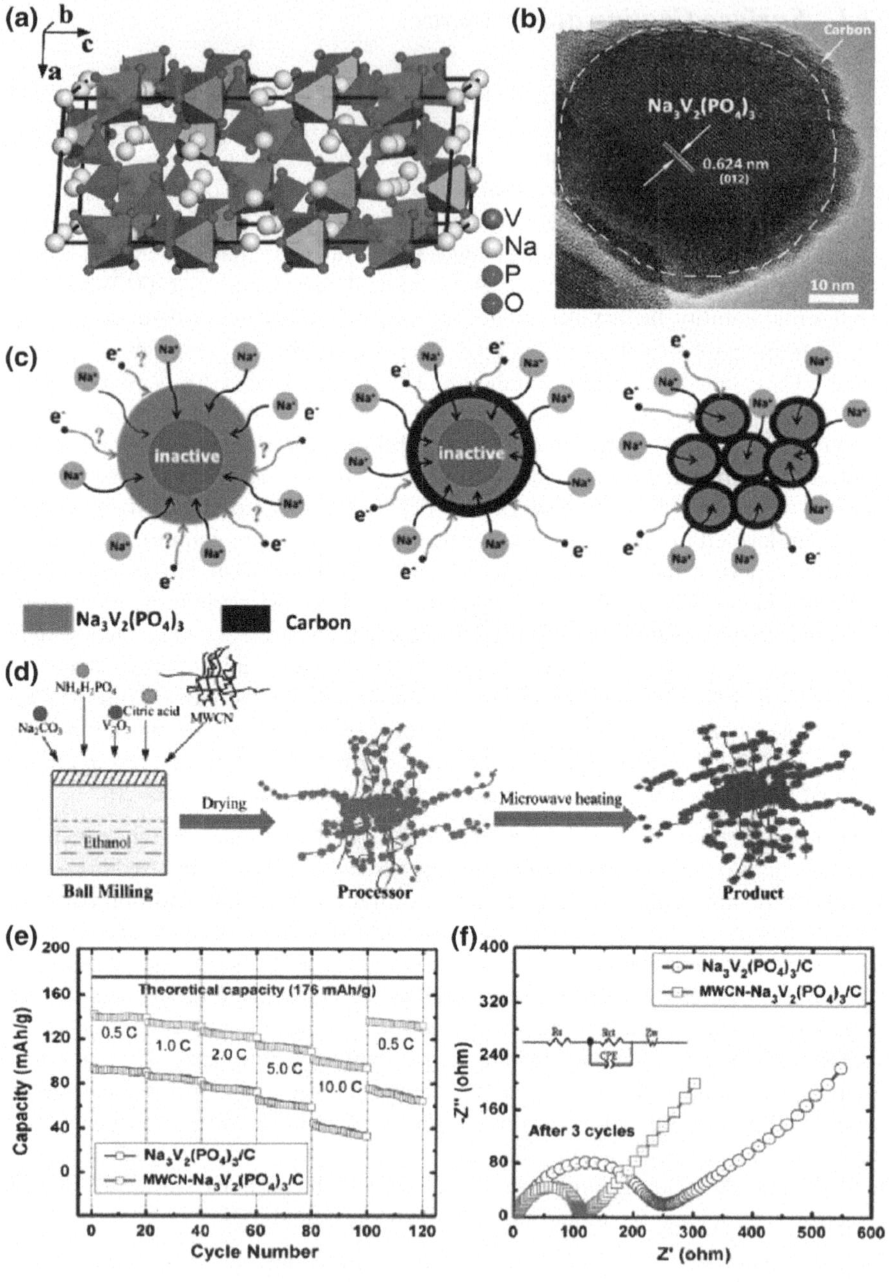
(a)
b
c
a
V
Na
P
O
(b)
Carbon
$Na_3V_2(PO_4)_3$
0.624 nm
(012)
10 nm
(c)
inactive
inactive
$Na_3V_2(PO_4)_3$
Carbon
(d)
$NH_4H_2PO_4$
Na_2CO_3
V_2O_5
Citric acid
MWCN
Ethanol
Ball Milling
Drying
Processor
Microwave heating
Product
(e)
Theoretical capacity (176 mAh/g)
0.5 C
1.0 C
2.0 C
5.0 C
10.0 C
0.5 C
Capacity (mAh/g)
Cycle Number
$Na_3V_2(PO_4)_3/C$
MWCN-$Na_3V_2(PO_4)_3/C$
(f)
$Na_3V_2(PO_4)_3/C$
MWCN-$Na_3V_2(PO_4)_3/C$
After 3 cycles
-Z'' (ohm)
Z' (ohm)

◂**Fig. 6.3** **a** Crystal structure of NASICON $Na_3V_2(PO_4)_3$. **b** HRTEM image of the obtained nano NVP@C. **c** Schematic illustration of electron conduction and Na^+ transport within NVP, NVP/C, and nano NVP@C. **d** Schematic illustration of the synthesis procedure of MWCNTs modified $Na_3V_2(PO_4)_3$/C sample. **e** Rate performances of the MWCNTs modified samples. **f** Electrochemical impedance spectroscopy profiles of the pristine and MWCNTs modified $Na_3V_2(PO_4)_3$/C sample. Wherein, **a–c** were reproduced with permission from Ref. [21], Copyright © 2014 Royal Society of Chemistry. **d–f** Were reproduced with permission from Ref. [22], Copyright © 2015 Elsevier

114.6, and 101.8 mA h g^{-1} at 0.1, 0.5, 1, 2, 5, and 10 C in the voltage range of 2.5–4.6 V, respectively. Meanwhile, the charge-transfer resistance of NVP/C is obviously reduced after MWCNTsT were added, indicating that MWCNTs plays a key role in improving the electronic conductivity, rate capability, and cycling retention of NVP/C electrode.

The long-term technologies have the potential to deliver significantly higher gravimetric (volumetric) energy densities or show obvious cost advantage relative to the state-of-art LIBs. However, the commercial success of these techniques is still not guaranteed due to the concern of their long-term operation, cost and safety. Active materials, such as metal anodes and conversion cathodes with new redox chemistry bypassing the limitation of intercalation mechanism of the current LIBs, bring new opportunities to boost the energy density, but severe challenges as well [23, 24]. Basically, these emerging post-LIBs show two major problems: irreversible phase transitions of active materials and/or unstable electrode/electrolyte interphases. As a result, innovations in structure and interphase stabilization strategy to improve the operation stability of component materials should be the focus of future research. Besides, the working principles and cell construction of many emerging post-LIBs techniques are quite different from what is typically employed in commercial LIBs, which requires creative engineering techniques for fabricating cells to facilitate their practical use [25, 26]. In spite of the electrode materials, other cell components, such as current collectors, binders, conductive additives, electrolytes, and separators, also require optimization for various techniques. The ultimate goal of the post-LIBs is to use Li metal anode accompanied with solid electrolytes. In this section, we will give a general outlook on the fundamental issues and show how to gain better understanding on the behavior of the electrodes and electrode/electrolyte interphases with nanotechnologies. Some potential solutions enabled by nanostructures and nanomaterials will be discussed too.

Graphene: Graphene is a flat monolayer of carbon atoms tightly packed into a honeycomb lattice, which has attracted immense interest and applications in many research fields because of its good thermal stability, superior electronic conductivity, and high specific surface area. As plotted in Fig. 6.4a, a novel 3D NVP@C@rGO cathode is developed with hierarchically porous structures via a convenient freeze-drying method. This cathode displays an ultralong-life cycling stability (64% capacity retention after 10,000 cycles at 100 C) in Fig. 6.4b and outstanding rate performance (91 mA h g^{-1} at 80 C and 86 mA h g^{-1} at 100 C), which can be attribute to high electrical conductivity, bicontinuous ion/electron pathways, and a large electrode-

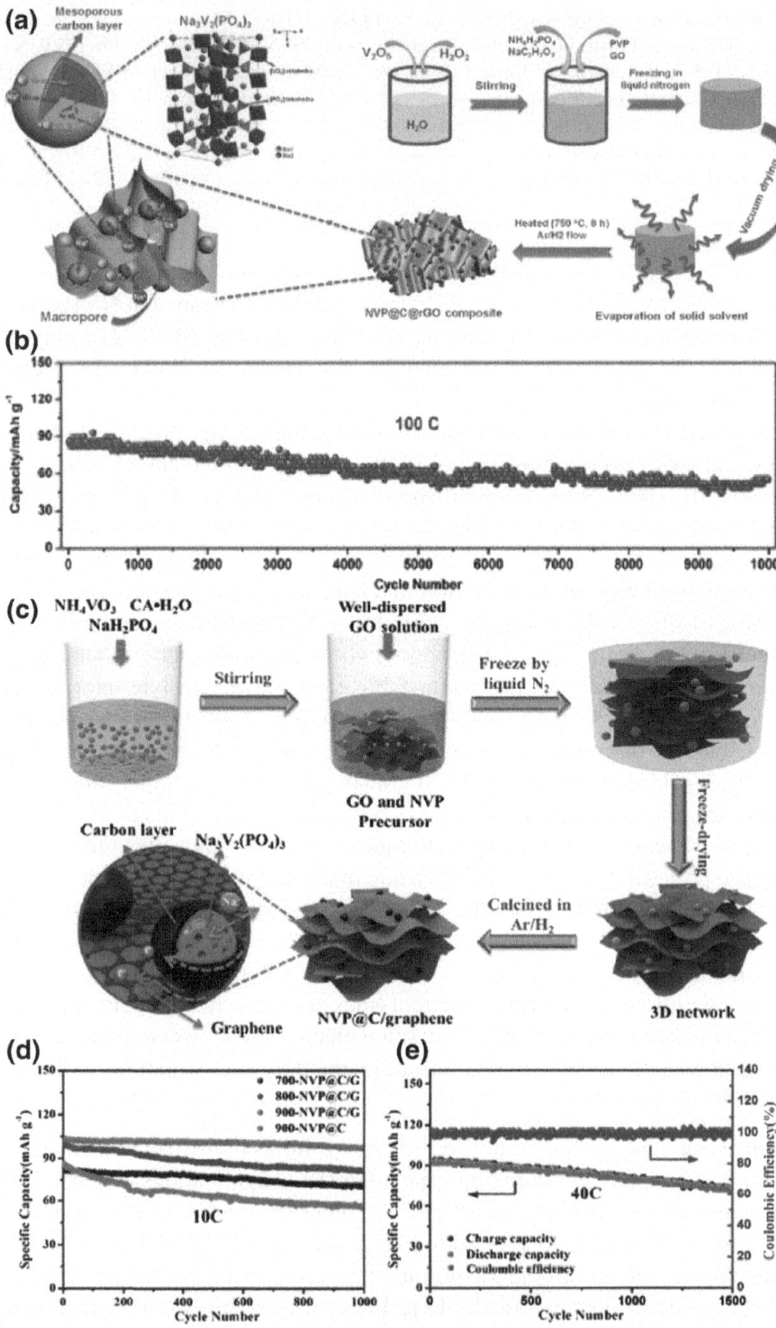
(a)
Mesoporous carbon layer
$Na_3V_2(PO_4)_3$
V_2O_5
H_2O_2
H_2O
Stirring
$NH_4H_2PO_4$
PVP
GO
Freezing in liquid nitrogen
Vacuum drying
Heated (750 °C, 8 h) Ar/H2 flow
Macropore
NVP@C@rGO composite
Evaporation of solid solvent
(b)
Capacity/mAh g^{-1}
100 C
Cycle Number
(c)
NH_4VO_3 CA•H_2O
NaH_2PO_4
Stirring
Well-dispersed GO solution
Freeze by liquid N_2
GO and NVP Precursor
Freeze-drying
Carbon layer
$Na_3V_2(PO_4)_3$
Calcined in Ar/H_2
Graphene
NVP@C/graphene
3D network
(d)
700-NVP@C/G
800-NVP@C/G
900-NVP@C/G
900-NVP@C
10C
Specific Capacity(mAh g^{-1})
Cycle Number
(e)
40C
Charge capacity
Discharge capacity
Coulombic efficiency
Specific Capacity(mAh g^{-1})
Coulombic Efficiency(%)
Cycle Number

◂**Fig. 6.4** **a** Schematic of 3D hierarchical meso- and macroporous NVP@C@rGO cathode with pathways for both electrons and sodium ions (upper right corner: crystal structure of NASICON-type $Na_3V_2(PO_4)_3$ and schematic of freeze-drying-assisted method to prepare NVP@C@rGO composite. **b** Long-term cycle life of NVP@C@rGO for 10,000 cycles at 100 C. **c** Schematic diagram of the synthesis process of 3D porous NVP@C/G. **d** Cyclic performance of 900-NVP@C, 700-NVP@C/G, 800-NVP@C/G, and 900-NVP@C/G at 10 C. **e** Ultra-long cyclic performance of 900-NVP@C/G at 40 C. Wherein, **a**, **b** were reproduced with permission from Ref. [27], Copyright © 2015 John Wiley & Sons. **c–e** Were reproduced with permission from Ref. [28], Copyright © 2016 Royal Society of Chemistry

electrolyte contact area for rapid Na^+ and electron transport [27]. It is worth noting that Wang and co-workers also proposed a simple freeze-drying-assisted thermal treatment method to obtain porous NVP@C nanocomposites enwrapped in a 3D graphene network (Fig. 6.4c) [28]. Owing to the enhancement of the electronic conductivity by the 3D conductive graphene support network and the carbon-coated structure, the novel electrode displays a superior rate capability (76 mA h g^{-1} at 60 C) and ultra-long cyclability at different rates (82% capacity retention for 1500 cycles at 40 C) in Fig. 6.4d, e.

Ordered mesoporous carbon: As one of the most well-known member of highly ordered mesoporous carbon materials family, CMK-3 has been developed in many energy areas due to the high surface area, uniform porosity, well-interconnected framework, and excellent electrical conductivity [29, 30]. These advantages make CMK-3 an ideal material to confine the carbon-coated nanosized NVP particles in the 3D porous structure via a simple nanocasting technique. As shown in Fig. 6.5a, Yu and co-workers developed a simple, effective method to synthesize NVP@C@CMK-3 nanocomposite through the impregnation of NVP@C nanoparticles into conductive and interconnected carbon host (CMK-3) [31]. Benefitting from the advantages of a 3D network for electron transfer and many open pores for electrolyte easy access, the electrode delivers superior rate capability (81 mA h g^{-1} at 30 C) and ultralong cyclability (78 mA h g^{-1} at 5 C after 2000 cycles) in Fig. 6.5b, c, suggesting that the novel "double carbon coating" design may offer new insights into the rational design of high-performance cathode materials for NIBs.

3D interconnected carbon: With the purpose of combining the advantages of the high-energy density of batteries with the high-power density of supercapacitors, developing a simple and generalizable approach to construct 3D self-supported tricontinuous electrode architecture is highly desired. In this regard, free-standing and binder-free features can be realized by 3D interconnected carbon frameworks, which can be prepared via various methods such as electrostatic spray deposition (ESD), spray-drying, and filtration-freezing-drying techniques. Yu and co-workers proposed a rather general strategy to prepare 3D self-supported tricontinuous cathode material via a facile ESD technique (Fig. 6.5d) [32]. An interconnected 3D tricontinuous NVP:rGO-CNT cathode deposited on the current collectors without any conductive additives or binders is successfully fabricated. Benefitting from the advantages of homogenously distributed NVP nanoparticles and ultra-fast charge/ionic trans-

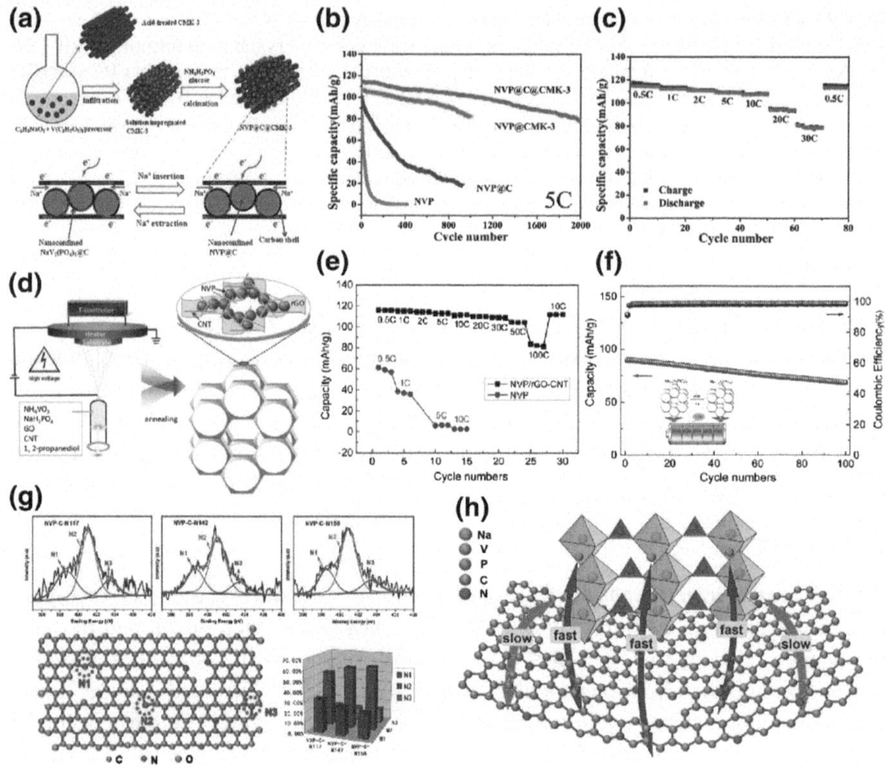

Fig. 6.5 **a** Schematically illustrates the experimental procedures for fabrication NVP@C@CMK-3 cathode materials and the reversibility of electrochemical reactions for the NVP@C nanoconfined within a single carbon pore. **b** Cycle performances of the NVP, NVP@C, NVP@CMK-3, and NVP@C@CMK-3 at 5 C. **c** Rate performance of the NVP@C@CMK-3 electrode. **d** Schematic illustration of ESD technique to fabricate the interpenetrating 3D tricontinuous NVP:rGO-CNT cathode. **e** Rate capabilities of 3D porous NVP and interpenetrating 3D tricontinuous NVP:rGO-CNT as cathode materials. **f** Cycling performance of symmetric full sodium battery at 10 C for 100 cycles. **g** High-resolution XPS spectra of N1 s for the NVP-C-N samples, schematic structure of the binding conditions of N in a carbon lattice, and the histogram for ratio of different N species in various NVP-C-N samples; **h** A schematic illustration of sodium ion storage mechanism in the NVP-C-N composite electrode. Wherein, **a–c** were reproduced with permission from Ref. [31], Copyright © 2015 John Wiley & Sons. **d–f** Were reproduced with permission from Ref. [32], Copyright © 2016 John Wiley & Sons. **g**, **h** Were reproduced with permission from Ref. [33], Copyright © 2015 John Wiley & Sons

port networks, this electrode displays a specific capacity of 82 mA h g^{-1} (around 70% of its theoretical capacity) at 100 C (charge and discharge in 36 s), an outstanding capacity retention of 96% after 2000 cycles at 10 C, and remarkable Na-ion full battery performance (Fig. 6.5e, f). It is worth noting that the excellent performance is achieved without conductive additives and binders. Meanwhile, this unique strategy has the potential to be a milestone for precisely designing high-energy density and high-power density cathode materials.

N-doped carbon: Besides the mentioned traditional carbon materials, N-doped carbon exhibits much higher electrical conductivity because of three types of binding configurations. These common pyridinic N, pyrrolic N, and quaternary N could create atomic defects and active sites for sodium storage. For example, Wang et al. initially prepared a N-doped carbon coated NVP cathode material of NIBs via sol-gel method and three types of nitrogen species are normally obtained in the carbon layer [33]. When used as a cathode material for sodium storage, this NVP/C+N electrode displays outstanding rate capacity of 84.3 mA h g^{-1} at 5 C and an excellent capacity retention of 93% is maintained even after 100 cycles, which can be attributed to the enhancement of the electronic conductivity and extrinsic defects with active sites of the suitable N-doped carbon coating (Fig. 6.5g, h). In short, it can be summarized that the surficial carbon coating nanotechnology could indeed modify the chemical and physical properties of polyanionic materials and be readily extended to application of other electrode materials with poor electronic conductivity for NIBs.

6.3.2 Cationic Substitution in Crystal Framework

As another promising cathode candidate, layered Na_xMO_2 exhibits more complex phase transitions during the removal of Na ions from alkali ion layers compared with its Li analogues because of larger Na^+ ions coupled with charge ordering and the ordering arrangement between Na^+ and vacancies at different Na contents. Cationic substitution including Li doping, Mg doping, Cu doping, and Ti doping etc. seems effective to suppress or reduce these irreversible phase transitions, improving the electrochemical performance of layered oxides.

Li doping: Unlike Li-based layered oxides, Li and TM occupy the alternating octahedral layers. Li ions and TM ions in P2-type oxides are located in TM sites, the presence of monovalent Li ions in the TM layers allows more Na ions to reside in the prismatic sites, stabilizing the overall charge balance of the compound. Consequently, more Na ions remain in the compound upon charge, the P2 structure is retained in the high voltage region, and the P2-O2 phase transformation is delayed [34, 35]. The incorporation of Li into the O3-type could mitigate the Jahn-Teller distortion [36], refrain unfavorable high-voltage phase transition [37] and tend to bond with the F^- anion and coordinate with solvent molecules [38], which could help to prevent the loss of electroactive TM cations from the cathode and contribute to a stable electrolyte-cathode interphase. Interestingly, Li substitution for transition metals also enhances the contribution of oxide ion in the redox in O3-type or P3-type, which is similar to "Li-rich" materials such as $Li[Li_{1/3}Mn_{2/3}]O_2$ in Li batteries [39, 40]. Despite, Li-substituted layered Na_xTMO_2 delivers large reversible capacity and excellent rate performance due to stabilization of the layered structure by migration of Li from the slab to interslab space and the pillar effect in the interslab space.

Cu/Mg/Zn doping: Substitution of divalent Cu/Mg/Zn elements transfers the well-known phase transition from P2-O2 to P2-OP4 by stabilizing the reversible OP4 phase at the end of charge in P2 type oxides and reduce huge volume change [41–43]. Unlike Mg^{2+} ([Ne]) and Zn^{2+} ([Ar]3d^{10}), Cu^{2+} ([Ar]3d^9) is a Jahn-Teller active ion with electron configuration ($t_{2g}^6e_g^3$) and induces lattice distortion. Noting that Cu^{2+}/Cu^{3+} redox couple is reported active in Cu-substituted O′3-$NaCu_{0.67}Sb_{0.33}O_2$ [44] and P2-$Na_{2/3}Cu_{1/3}Mn_{2/3}O_2$ [41], etc. Moreover, Cu^{2+} doping improves water-resistant property of O3 and P2 type materials [45, 46]. Mg/Zn doping commonly suppresses the P2-O2 phase transition observed in the undoped composition during charging (during Na removal) and improves Na-ion mobility [47, 48]. Furthermore, contribution of additional capacity is provided through charge compensation by oxygen redox chemistry and some oxygen loss in the Mg-substitution compound [49]. Oxygen redox occurs in O2p orbitals that interact with alkali ions in the TM and alkali-ion layers (that is, oxygen redox occurs in compounds containing Li^+-O(2p)-Li^+ interactions). Mg^{2+} resides in the TM layers rather than alkali-metal ions, which demonstrates that excess alkali-metal ions are not required to activate oxygen redox The extraction of alkali ions from the alkali and TM layers in the alkali-rich compounds results in severely under-bonded oxygen, which promotes oxygen loss, whereas Mg^{2+} remains in $Na_{2/3}[Mg_{0.28}Mn_{0.72}]O_2$, which stabilizes oxygen [50].

Al doping: Unlike divalent Mg/Zn/Cu metals strongly suppress the P2-O2 phase transition and slight amount of Na is remained in the structure, Al-substituted P2 type sample also shows P2-O2 phase transition by large volume shrinkage but good cycle stability [51]. It is found that trivalent Al metals slightly suppress the shrinkage of interslab distance, while resulting in slightly smaller capacity. AlO_6 octahedra probably attracts Na ion for charge neutrality and slightly remained Na ion would play a role as pillars between slabs to maintain the large interslab distance of P2 layer. In comparison, MgO_6 octahedra strongly attracts more Na ions than AlO_6 for charge compensation, and more amount of Na remained in the interslab space would also act as pillars to suppress shrinkage of the interslab distance. These assumptions reasonably explain mechanism of suppressed shrinkage of interslab distance. Despite, the volume change does not reasonably explain tendency of the cycle stability for doped samples because small amount of Al doping suppresses degradation of particle surface and results in good cycle and rate performances [52].

Ti doping: Ti substitution in P2-type compounds could not only effectively decrease the site energy difference between Na_e and Na_f, thus suppressing the Na-vacancy ordering, which could refrain the formation of strongly ordered intermediate phases, and promote complete solid-solution behavior over a wider range of Na content, but also refrain the transition meta slabs glides and suppress the P2-O2 phase transition in the high voltage region, which is beneficial to reduce the volume change upon charge-discharge, leading to the good cyclability [53, 54]. The introduction of Ti in O3-type oxides could enlarges interslab distance, restrain the unfavorable and irreversible multiphase transformation above 3 V and retain a reversible O3-P3 phase transition with a P3 solid solution zone over a wide capacity range [55, 56]. So the phase boundary movement for several two-phase regions is largely reduced thus improved cycling stability and rate capability are both achieved. In total, Ti doping in layered framework makes the voltage profile transition from stepwise to smooth. In addition, tetravalent Ti might modify the orbital hybridization between oxygen 2p and transition metal 3d orbitals and elevate the redox potential of layered transition-metal oxides in Na batteries. Note that although higher Ti content leads to the smoother voltage curves and the higher average charge and discharge voltage, voltage hysteresis also increases.

Sn/Sb/Te/Bi doping: Heavier chalcogenides elements (Bi, Sb, Te) incorporation surprisingly generates a characteristic "honeycomb" ordered layered materials [57–59]. In this characteristic structure, every XO_6 octahedron is surrounded by six MO_6 edge-sharing octahedron forming the superstructure lattice and the Na ions are hosted in-between the M_2XO_6 slabs. $Na_3[M^{2+}]_2[X^{5+}]O_6$ (M = Ni, Cu, Mg, Zn; X = Bi, Sb) possesses an O′3 stacking sequence and generally ascribe a typical honeycomb structure that is generated by the 2:1 M^{2+}/Sb^{5+} cationic ordering to the metal layers (each SbO_6 octahedron is surrounded by six MO_6). Pure $Na_3Cu_2SbO_6$, $Na_3Mg_2SbO_6$, and $Na_3Zn_2SbO_6$ demonstrate electrochemically inactive because Cu^{2+}, Mg^{2+}, and Zn^{2+} cannot be further oxidized to higher valence states. $Na_3Ni_2SbO_6$ is the only layered antimonite showing good electrochemical performance: a high capacity (117 mA h g^{-1}), high voltage (two large discharge plateaus at 3.7 and 3.3 V) as well as a high recharge rate due to the reversible insertion/extraction of two Na, owing to the Ni^{2+}/Ni^{3+} redox couple when used as positive electrode materials for NIBs [60]. Similarly, $Na_2M_2TeO_6$ (M = Ni, Co, Zn, Mg) represent two different superstructures of the known layered P2-type and, despite complete ordering of the M^{2+} and Te^{6+} in the rigid lattice, have an apparently disordered sodium arrangement and high sodium ion conductivity. The conductor of typical $Na_2Mg_2TeO_6$ exhibits high conductivity of 2.3×10^{-4} S cm^{-1} at room temperature and a large electrochemical window of ~4.2 V (versus Na^+/Na), showing potential for use in all-solid-state NIBs [61]. Sn^{4+} ([Kr] $4d^{10}$) does not possess single d electrons and is thus unable to interact with O through its d orbitals. Reduced orbital hybridization between the oxygen 2p and nickel 3d orbitals within the metallic layers increases Ni–O bond ionicity. An abnormal O3-type sodium insertion material, $Na_{0.7}Ni_{0.35}Sn_{0.65}O2$, displays a high redox potential of 3.7 V versus Na^+/Na (based on the Ni^{2+}/Ni^{3+} couple). It can be summarized that later elements chemical substitutions having (d^0) and (d^{10}) configuration shells commonly obtains unique ordered or abnormal layered structure and provides a chemical approach to raise cell voltage by virtue of increased M–O bond iconicity [62, 63].

6.3.3 Structure Design and Optimization

Multiple structure design is also employed as an effective approach to improving the electrochemical performance of cathode materials owing to facilitation of the reaction dynamics. In order to integrate the advantages and conquer the disadvantages of cooperative component, fabricating composite structure or a hierarchical columnar with varied crystal structure and unique electrochemical behaviors were reported for NIBs [64]. The multiphase positive electrode materials, especially for the P2/P3, P2/O3, P3/O3, P2/P3/O3, P2/tunnel and O3 gradient composites, are the major extensively studied topic in the research of biphasic structures.

P2/P3 composites: As discussed above, the P-type structure is a unique phase in NIBs due to the larger radius of sodium ions than that of lithium ions. Na ion occupy a wider triangular prism sites, which will be beneficial for the diffusing of sodium. Typical P2 phase owns ABBA oxygen stacking including two TMO_6 octahedra per unit cell and shows two different sites Na cations, either sharing faces with two TMO_6 (Na_f) octahedra of adjacent slabs or sharing edges with six TMO_6 octahedra (Na_e). In P3 structure, the stacking sequence of the oxygen layer is ABBCCA, the three TMO_2 sheets per unit cell had prisms sharing one face with one TMO_6 octahedron and three edges with TMO_6 octahedra of the next layer, resulting in only one sodium site. From the strategy of synthesizing, the calcining temperature will induce P3-P2 phase transition due to the sodium layers slipped as annealing temperature rise and high energy is required to break/reform the TM–O bonds. Inspired by the characteristics in synthetic of this P-type structure, many researchers focus on P2/P3 multiphase compounds to realize preeminent electrochemical performance. The influence of annealing temperature on structure and electrochemical performance of P-type $Na_xNi_{0.22}Co_{0.11}Mn_{0.66}O_2$ materials was systematically studied as shown in Fig. 6.6a–c [65]. The cathodes synthesized at 700 °C is assigned to P3 phase with space of R3m, while the materials synthesized at 800 °C are indexable as P2 phase with space of P63/mmc. Interestingly, the samples annealed at 750 °C belong to the multiphase of P2-type and P3-type. Moreover, this layered P2/P3-$Na_xNi_{0.22}Co_{0.11}Mn_{0.66}O_2$ cathode can be processed in open air without major changes at high sodium content ($x > 0.33$) [66]. Chen et al. integrated the synthesis condition of P2 into P3 layered materials and obtained a P3/P2-$Na_{0.66}Co_{0.5}Mn_{0.5}O_2$ cathode, which exhibited a high discharge capacity of 156.1 mA h g^{-1} at 1 C in the voltage range of 1.5–4.3 V [67]. A stable layered P3/P2-$Na_{0.5}Ni_{0.2}Co_{0.15}Mn_{0.65}O_2$ intergrowth nanocomposite with long cycle performance in full cell system was recently reported (Fig. 6.6d–f) [68]. It must be pointed out that the poor initial Coulombic efficiency and the relatively low operate voltage, as well as relatively toxic elements exists in those works should be further developed to realize the large-scale energy storage in NIBs.

P2/O3 composites: In P2-type layered compounds, the trigonal prismatic sites in a Na layer are face-sharing each other and Na-ion diffuses directly between Na site, which led to a high-rate performance of P2-type cathodes. In comparison, the intervention of intermediate sites in the sodium-ion migration needs to overcome a high energy barrier and it undergoes a complex phase transition in O3-type

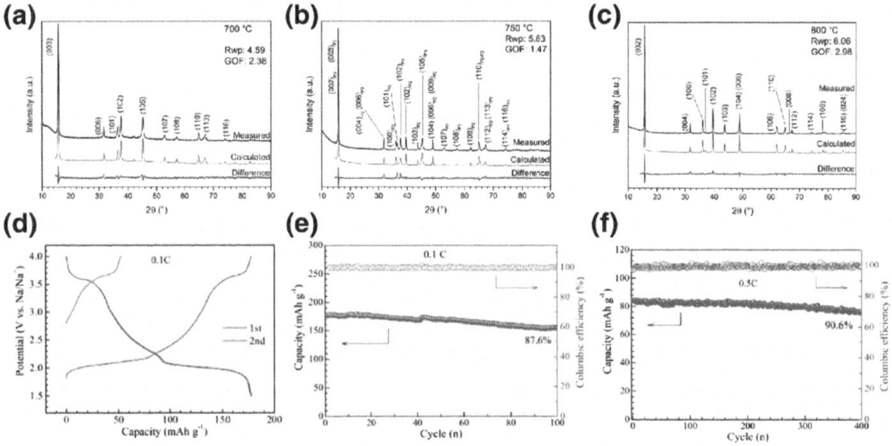

Fig. 6.6 X-ray diffraction pattern and Rietveld refinement of Na $_x$Ni$_{0.22}$Co$_{0.11}$Mn$_{0.66}$O$_2$ at **a** 700, **b** 750, **c** 800 °C. **d** The initial two charge/discharge curves and **e** cycle-life of P3/P2-Na$_{0.5}$Ni$_{0.2}$Co$_{0.15}$Mn$_{0.65}$O$_2$ at 0.1 C in the range of 1.5–4.0 V. **f** The cycle performance of the P3/P2-Na$_{0.5}$Ni$_{0.2}$Co$_{0.15}$Mn$_{0.65}$O$_2$//hard carbon sodium-ion full battery in the potential range of 1.5–4.0 V. Wherein, **a–c** were reproduced with permission from Ref. [65], Copyright © 2014 Royal Society of Chemistry. **d–f** Were reproduced with permission from Ref. [68], Copyright © 2018 Royal Society of Chemistry

structure, which will unfavorable for the electrochemical performance. However, O3-type structure exhibits much more reversible capacity than its P2 polytype in the same operate voltage range in layered oxides. Generally speaking, P2 and O3 correspond, respectively, to high and low sodium content in materials. Therefore, researchers have paid much attention to biphasic structures with O3 and P2 phases by controlling synthesis condition to combine the respective advantages, recently. Lee et al. first reported P2/O3-Na$_{1-x}$Li$_x$Ni$_{0.5}$Mn$_{0.5}$O$_{2+d}$ ($0 \leq x \leq 1$) intergrowth cathode and investigated the generation of each of structures as shown in Fig. 6.7a [69]. Upon Li substitution for Na in Na$_{1-x}$Li$_x$Ni$_{0.5}$Mn$_{0.5}$O$_{2+d}$, Na-O3 peaks fade away and Na-P2, Li-O3 peaks forms gradually. Although a smooth curve of charge/discharge was observed, the O3-O3″ phase transition still influenced the cycling performance due to O3 phase played the majority in the composites. Guo and co-workers designed P2/O3 Na$_{0.66}$Li$_{0.18}$Mn$_{0.71}$Ni$_{0.21}$Co$_{0.08}$O$_{2+d}$ layered oxide cathodes with P2 type structure as the major phase [70]. The SAED patterns of O3 and P2 phase are shown in Fig. 6.7b, c. the bright spots of the O3 phase projected along the [010] direction accords to a typical diffraction pattern for O3 structure lattice. And the bright spots projected along [71] direction can be assigned to characteristic reflections of P2 structure lattice. Figure 6.7d–h reveal the detailed atomic arrangements of P2 phase marked in red frames and O3 phase marked in blue frames. Furthermore, it shows that this P2/O3-Na$_{0.66}$Li$_{0.18}$Mn$_{0.71}$Ni$_{0.21}$Co$_{0.08}$O$_{2+d}$ provides large reversible discharge capacity of 200 mA h g^{-1} with the voltage range of 1.5–4.5 V and the energy density is calculated as 640 Wh kg^{-1}. Both Co-free layered P2/O3 biphasic

$Na_{0.67}Mn_{0.55}Ni_{0.25}Ti_{0.2-x}Li_xO_2$ materials [13] and P2/O3-$Na_{0.78}Ni_{0.2}Fe_{0.38}Mn_{0.42}O_2$ hybrid structures [72] are proposed. The designed P2/O3-materials delivers good cycle performance with reversible P2/O3-P2/O3/P3-P2/P3-P2/O′3 phase transition. All of those work show that layered intergrowth structures are a promising strategy to explore for novel layered electrodes and investigate the hybrid phase with O3/P2 to achieve a balance among capacity, cycling stability, and rate capability is necessary.

Na-P3/Li-O′3 composites: As mentioned in front part, the P-type structure is a unique structure in NIBs due to the radius of sodium ions is larger than that of Li ions and Na ion occupy a wider triangular prism site, thus partial Li ions incorporated into P-type structure may form the O3 phase. Therefore, Huang et al. designed a novel Na-P3/Li-O′3-$Na_{0.57}Li_{0.17}Mn_{0.67}Ni_{0.16}Co_{0.17}O_2$ heteroepitaxial nanostructure via an in situ composition modulation route [73]. The in situ formation of Li_2MnO_3 (Li-O′3) changes the crystallographic and chemical features of the neighboring Na-P3 layered matrix significantly and leads to the alternating Na-P3/Li-O′3 heteroepitaxial nanostructure, which greatly enhance the structural durability of the Na-P3 cathode material by impeding the transformation of the P3 structure into the unstable P3″ structure during charging. This alternating heteroepitaxial nanostructure delivers an high reversible capacity of ~210 mA h g^{-1} between 1.5 and 4.5 V versus Na^+/Na, much improved cycling stability and excellent electrode kinetics compared with P3-$Na_{0.57}Li_{0.17}Mn_{0.67}Ni_{0.16}Co_{0.17}O_2$. Further investigation into the interactive mechanism and optimal distributions of the Na-P3 and Li-O′3 phases in the composite structure is further needed.

P2/P3/O3 composites: As mentioned above, many researches have proved that P-type and O-type multiphase compounds exhibited better electrochemical performance than pure phase, unfortunately, the chosen voltage range of these works is too wide to enable Na-ion cells with reasonable voltages. Keller et al. synthesized a P3/P2/O3-type $Na_{0.76}Mn_{0.5}Ni_{0.3}Fe_{0.1}Mg_{0.1}O_2$, which shows a superior capacity retention of about 91.3% and the highest discharge capacities in the 2nd and 50th cycles than P2 or P3 phase in a limited 2.5–4.3 V [74]. This enables improved electrochemical performance due to P-type and O-type phases might stabilize each other as their *c* axis parameters exhibit opposing changes upon Na^+ extraction/insertion and phase transition rather than simple mixing pure P-type and O-type phases. Moreover, it might reduce the strain inside the electrode and stabilize the cycling behavior of the active material.

Gradient structure: It is important to create unique morphologies that endow high-capacity, long-term cyclability and high energy density to overcome the limited capacity of O3 structure. Hwang et al. reported a concept of the nanorod-structured full concentration gradient materials and designed a novel structure, which have the different chemical constituent with $NaNi_{0.75}Co_{0.02}Mn_{0.23}O_2$ inner while $NaNi_{0.58}Co_{0.06}Mn_{0.36}O_2$ outer of the structure and $NaNi_{0.75}Co_{0.08}Mn_{0.17}O_2$ inner to $NaNi_{0.58}Co_{0.14}Mn_{0.28}O_2$ in the outer end, respectively [75, 76]. This work of their group exhibited that the sample with high Ni composition inner offered high reversible capacity and the surface material can enhance the stability in electrochem-

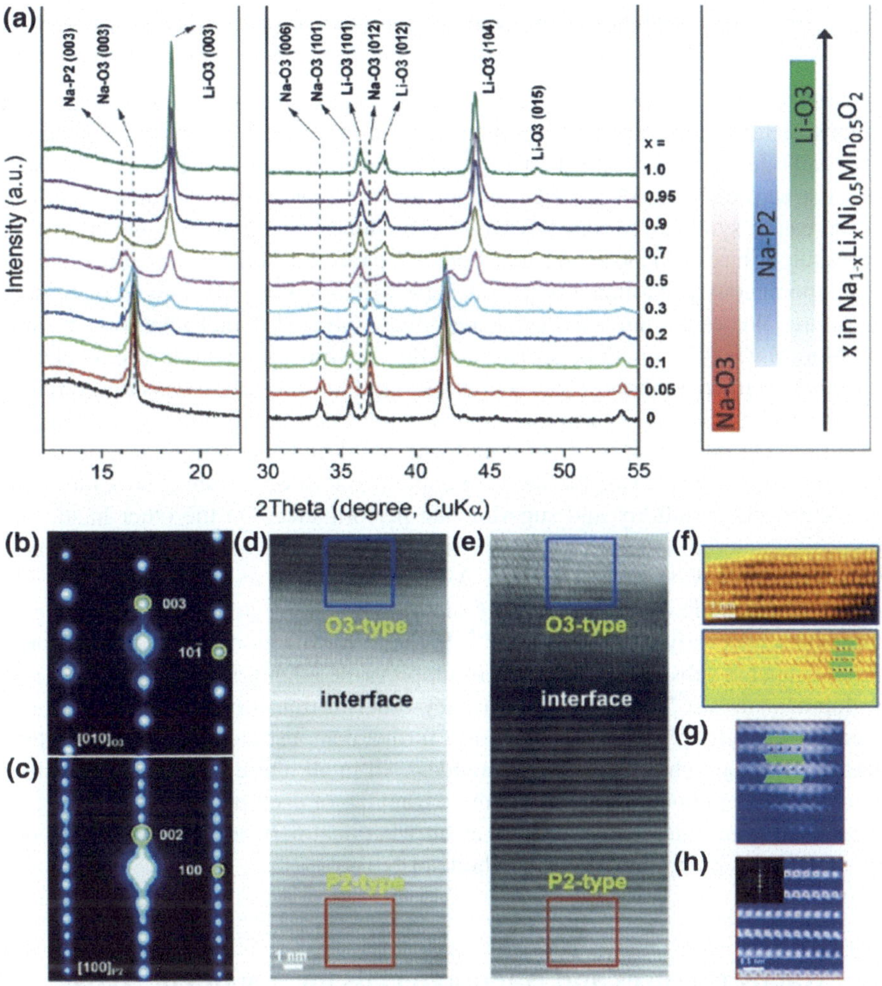

Fig. 6.7 X-ray diffraction patterns of P2/O3-$Na_{1-x}Li_xNi_{0.5}Mn_{0.5}O_{2+d}$ ($0 \leq x \leq 1$) with different Li content. The SAED patterns of the P2/O3-$Na_{0.66}Li_{0.18}Mn_{0.71}Ni_{0.21}Co_{0.08}O_{2+d}$ composite: **b** O3-type structure and **c** P2-type structure. The STEM images: **d** HADDF and **e** ABF images of P2/O3-$Na_{0.66}Li_{0.18}Mn_{0.71}Ni_{0.21}Co_{0.08}O_{2+d}$ composite, the blue and red rectangle represent O3 structure and P2 structure areas. The enlarged images corresponding to the rectangle areas are shown in (**f**, **g**, **h**). Wherein, **a** was reproduced with permission from Ref. [69], Copyright © 2014 John Wiley & Sons. **b–h** Were reproduced with permission from Ref. [70], Copyright © 2015 John Wiley & Sons

ical cycling. The researches of the gradient materials provide a guidance to design cathode with high Ni composition, as well as the high capacity and long cycle life.

Layered/tunnel composites: Recently, hybrid materials combining layered and spinel structures have attracted significant attention for their promising application in the cathodes of rechargeable batteries. As for NIBs, sodium manganese oxide with tunnel structure ($Na_{0.44}MnO_2$, *Pbam*) can substitute for spinel structure because it can be easily synthesized and has larger diffusion channels for Na ion transport. Moreover, it has been observed that both P2 phase and tunnel structures can be contemporaneously stabilized when the ratio of Na ions to Mn ions is increased. Therefore, Wu et al. and Xiao et al. reported a layered-tunnel intergrowth cathode $Na_{0.6}MnO_2$ [77, 78]. The structure of this intergrowth cathode was is clearly confirmed by micro scanning electron microscopy, high-resolution transmission electron microscopy as well as scanning transmission electron microscopy with atomic resolution. On one hand, the hybrid material could integrate advantages of the P2 layered structure with available high Na content and those of tunnel structure with excellent cycling stability and superior rate performance. On the other hand, this composite material undergoes a highly reversible P2/tunnel-OP4/tunnel structural evolution over a wide capacity range. As a consequence, the composite electrode delivers a reversible discharge capacity of 198.2 mA h g^{-1} at 0.2 C rate, leading to a high-energy density of 520.4 Wh kg^{-1}. In the following work, Chen et al. adopted partial copper to substitute manganese to suppress the Jahn-Teller effect of Mn^{3+} and the doping of the Cu^{2+} improved Na^{+} vacancy-ordering transition and contribute more reversible capacity [79]. All of these works further demonstrated that heterostructure material is a good choice for NIBs cathodes. All in all, these results indicated that the multiphase cathode can combine the advantage of each pure phase at a certain degree to realize much more superior electrochemical performance, which smooth the way for designing more useful cathode of NIBs.

6.4 Nanostructures and Nanomaterials for Anodes of NIBs

A systematic review of current cathode materials for NIBs is provided in the previous section. It can be observed that the cathode material has developed rapidly both in terms of rate capability and cycle durability, and why the information on the practical use of NIBs is still rarely reported, which is mainly the limitation of the anode material. Although most anode materials have much higher capacity performance than that of the cathode materials, it is not negligible that their cycle stability, rate performance, and initial Coulombic efficiency are not ideal. Nanostructuring of materials is an effective method to address the above issues, but there are also some problems, such as increased synthesis costs, lower initial Coulombic efficiency, and lower volumetric energy density. Therefore, how to apply nanostructured technology more effectively in the construction of the anode materials is a problem that researchers need to focus on next. The next part of this chapter will mainly review the exist-

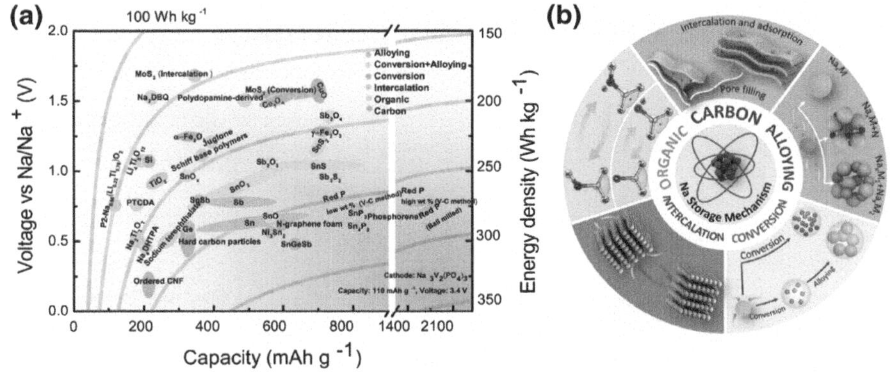

Fig. 6.8 Current reports on anode materials (**a**) and the corresponding sodium storage mechanisms (**b**) for NIBs. Wherein, **a**, **b** were reproduced with permission from Ref. [80], Copyright © 2017 Elsevier

ing system of anode materials for NIBs, and provide a comprehensive overview of strategies for nanostructured construction and performance improvement.

6.4.1 Typical Anode Materials of NIBs

The current anode materials for NIBs can be mainly divided into the following five categories: carbonaceous materials, titanium-based materials, elements/alloys, chalcogenides, and organic materials, as shown in Fig. 6.8. Carbonaceous materials mentioned here generally include graphite-based carbon, soft carbon, and hard carbon. These materials typically exhibit lower operation voltage, but lower Coulombic efficiency and capacity, moderate cycle performance, poor rate performance, and worrying safety greatly limit their widespread use. Titanium-based materials are a class of anode electrode materials that have been studied earlier beyond carbonaceous materials, and mainly contain polymorphic TiO_2, $Na_2Ti_nO_{2n+1}(2 \leq n \leq 9)$, $Li_4Ti_5O_{12}$, $NaTi_2(PO_4)_3$, and transition metal titanate oxides. The working voltage range of these materials is generally between 0.5 and 1 V with various advantages such as low cost, non-toxicity, and structural stability. However, in the ointment, their specific capacity and initial Coulombic efficiency are low.

On the way to pursue the capacity performance of materials, material scientists seem to have found better candidate materials, namely elements/alloys and chalcogenides, which can address very high capacity, but larger volume change has become their fatal weakness. Indeed, anode materials for NIBs are not limited to inorganic systems. There are some prominent roles in organic systems that are worth exploring. Organic electrode materials are also called green electrode materials. As their name suggests, they generally have the advantages of low cost and environmental protection of the synthetic process. Low initial Coulombic efficiency and poor cycle

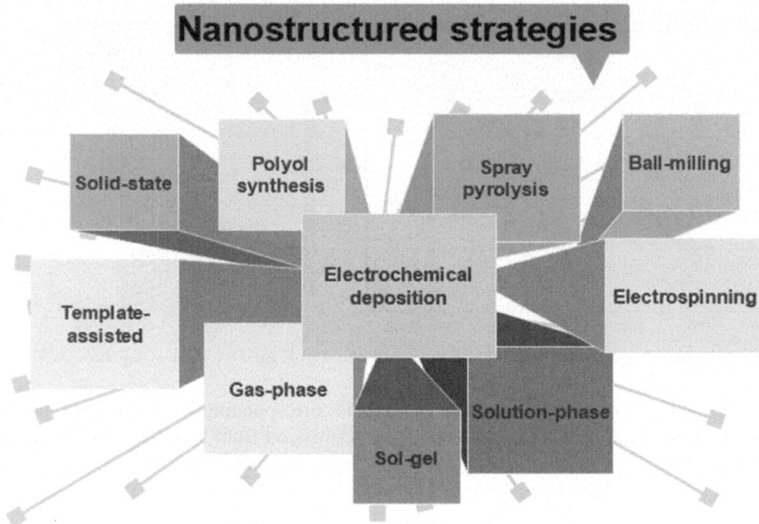

Fig. 6.9 Schematic description of typical nanostructured strategies

performance are also the main difficulties they currently face. From the above brief introduction, it can be seen that all anode materials, like cathode materials, have their own advantages, and it is difficult to mask their inherent disadvantages. How to package them is the key to determine their worth or not, and as far as the current research results are concerned, nanotechnology is a very effective tool. Therefore, the following will discuss in detail the current synthesis strategies of nanostructure and the application of nanostructured materials in NIBs.

6.4.2 Synthesis Strategies for Constructing Nanostructured Anode Materials

Changing or improving synthetic methods is one of the simpler strategies for constructing different nanostructures. At present, the synthesis strategies can be described by using "*let a hundred flowers blossom and a hundred schools of thought contend*", including solid-state reaction, high-energy ball milling method, sol-gel route, solution-phase method, gas-phase strategy, electrospinning method, electrochemical deposition, template method, spray drying, and polyol synthesis method (Fig. 6.9). These technologies will be introduced one by one below.

Solid-state reaction: Solid-state reaction is one of the simplest and most efficient methods to synthesize inorganic materials. When this method is used, it is only necessary to mix the various raw materials uniformly, and the desired product can be obtained by heat treatment at a relatively high temperature for a certain period of time in air, oxygen or an inert atmosphere. The materials synthesized by this method generally have larger particle sizes, irregular morphology and lower specific surface area. At present, this method is mainly employed to synthesize cathode electrode materials; however, its use in the synthesis of anode electrode materials is rarely reported.

High-energy ball milling: High-energy ball milling is one way in which a direct synthesis of a material is achieved in a closed container using collisions between tiny balls and precursors formed at a high speed to induce chemical reactions [80, 81]. Many alloys or composite anode materials can be obtained by this route, such as Sn_5SbP_3/C [82], Sb_2Te_3/graphite [83] and P/C composite [84]. The advantage of this method is that it is simple and even large-scale production can be achieved. The materials synthesized by this method can reach the nanometer level, but have no regular morphology.

Sol-gel process: Compared with the above two methods, the main advantage of the sol-gel route is that the size of the synthesized material particles is smaller and more uniform. Its general synthesis process mainly includes two steps: gelation and heat treatment. Commonly used gelling agents are citric acid, tartaric acid, etc. They mainly act to chelate metal ions, which is why the material prepared by this method has a uniform size. It should be noted that the disadvantage of this method is also obvious, that is, the synthesis process is too time-consuming [85].

Solution-phase synthesis: The solution-phase synthesis mentioned here mainly refers to the hydrothermal/solvothermal method. This method is a senior materials engineer and has been hired to construct a number of anode materials with special morphologies. The main mechanism of such method is to change the growth and nucleation of particles in the solution by adjusting the temperature and pressure to achieve the purpose of constructing a variety of nanostructures. Filling degree, reaction time, reaction temperature, solution concentration, solvent properties, and reagent type are the key factors that affecting the morphology of the final product. Therefore, it is necessary to conduct a series of control studies on the above parameters in order to obtain a good nanostructured material. Compared with the cathode materials, this method exhibits greater generality and compatibility with the anode materials and many anode materials can be synthesized by a one-pot method, such as metal, chalcogenide, etc. However, due to the application of high temperature and high pressure, safety has become a major obstacle to limiting its large-scale practical application.

Electrospinning: Electrospinning is widely used in the preparation of cathode or anode nanofibers. Simple electrospinning device can be composed of three parts: high voltage DC power supply, injection system and collecting substrate. Its working principle is to use electrostatic force to pull the viscous precursor solution pushed out of the injection system onto the opposite collector under high voltage to obtain the nanofiber precursor of the desired material, and then a final product with a special morphology can be obtained undergo proper annealing. Among them, the viscous precursor solution is generally obtained by adding a high molecular polymer to the solution. The commonly used polymers include polyvinyl pyrrolidone (PVP), polyethylene oxide (PEO), polyacrylonitrile (PAN), polyvinyl alcohol (PVA), and polyvinylidene fluoride (PVdF). In light of the different solubility of the reagents, the solvents that can be selected are also diversified, such as H_2O, *N,N*-dimethylformamide (DMF), *N*-methyl-2-pyrrolidone (NMP) and ethanol are all conventional candidates. It should be noted that although electrospinning is very effective for constructing nanostructures, the results are influenced by many factors such as polymer molecular weight, solvent type, solution concentration, voltage magnitude, spray distance, injection rate, ambient humidity, and heating/cooling rate, that is, the various parameters mentioned above have to be faced in order to obtain the desired structures [86–88].

Gas-phase route: The gas-phase route is a technology that designs materials from the atomic or molecular level, and mainly includes physical/chemical vapor deposition (PVD/CVD), atomic layer deposition (ALD), and molecular layer deposition (MLD). This method is widely used to construct nanostructured materials with different dimensions. In the field of anode materials, many metal/carbon and oxide composites with unique structures and excellent electrochemical performance are synthesized with the help of this technique.

Electrochemical deposition: Electrochemical deposition is a technique for constructing diverse nanostructures. This method has two major advantages. First, compared with electrospinning, its parameters are relatively simple to adjust. Current and deposition time are two of its key parameters. Second, the material deposited on the conductive substrate (such as Ni foam, stainless steel substrate, Ti foil, Cu foil, etc.) can be used directly as binder- and additive-free electrodes. In general, this method is mainly employed to prepare anode materials such as metals, alloys, and oxides. It should also be mentioned that the early electrodeposition was mainly performed directly on the conductive substrates, so it was difficult to obtain materials with special morphologies. With the development of the template-assisted method, many researchers [89–91] have combined such two methods together and achieved some advanced structures.

Template-assisted method: Template-assisted method is one of the most widely spread strategies for synthesizing nanostructured materials. Its advantage lies in the controllable morphology and size. Whether it is used in the cathode or the anode system, it has a gratifying application, and it can even be said that it is a gospel for cathode materials that are difficult to tailor morphology. At present, template types employed by the template-assisted method can be classified into hard template and soft template. Among them, the hard template can be further subdivided into limited domain template and oriented template. Specifically, commonly used anodic aluminum oxide membranes (AAO), mesoporous silicon (SBA-13), and mesoporous carbon (CMK-3) are confined templates, while graphene and carbon nanotubes belong to the category of oriented templates. As for soft template-assisted method, it mainly uses surfactants including cetyltrimethyl ammonium bromide (CTAB), polyethylene oxide-polypropylene oxide-polyethylene oxide (P123), and sodium dodecyl sulfate (SDS) as templating agents. During the synthesis process, in situ synthesis of templates is achieved through the assembly of surfactants. Another difference between the two template-assisted methods is that the former requires additional *etc*hing, but the latter does not.

Spray pyrolysis process: In terms of synthetic electrode materials, spray pyrolysis can be regarded as a powerful technique. Its basic principle is that the precursor solution is first atomized by an atomizer, and then the mist enters a certain length of high-temperature closed reaction zone with the aid of a carrier gas to perform a series of physical and chemical reactions to obtain the desired product, wherein, the above physical and chemical processes include solvent vaporization, solute precipitation/diffusion/decomposition, and particles growth/agglomeration. In the synthesis process, the parameters such as atomization process, gas flow rate, reaction temperature, and reaction zone length have a profound influence on the morphology of the final product. The biggest benefits of this method are controllable parameters and short synthesis period. It should be pointed out that subsequent heat treatment is required to exhibit better electrochemical performance for most materials. At present, many nanostructures can be constructed by this method, including dense, hollow, layered, porous, core-shell structures, nanowires, and nanosheets [92].

Polyol synthesis: Another fascinating path to build nanostructured electrode materials is polyol synthesis that can also be called self-propagating combustion. A typical synthesis procedure is as follows: First, various reagents are firstly dissolved in a polyol, and then the resulting solution is poured into a pre-heated plate, and ignition is initiated with an open flame to induce the occurrence of a self-propagating combustion reaction. The reaction can be completed in a few minutes, so it is a very expeditious synthesis method. Specifically, since the preheating is generally carried out near the boiling point of the solvent, the preheating temperature is also different depending on the selected solvent. Polyols play a variety of roles in the synthesis process, which can be utilized as both a fuel and a carbon source [93]. The material synthesized by this method is generally in the form of nanoparticles. In order to increase the crystallinity of the as-prepared material, it is necessary to perform appropriate heat treatment. At present, the anode materials prepared by this method mainly include $NaTi_2(PO_4)_3$ [94] and $Na_2Ti_6O_{13}$ [95].

In short, strategies for constructing nanostructured electrode materials have been extensively developed. Objectively, not every nanostructure is beneficial for improving the electrochemical performance of electrode materials. In the following section, the effect of nanostructures on the electrochemical performance of anode electrode materials will be comprehensively discussed.

6.4.3 Anode Materials with Diversified Nanostructures

6.4.3.1 Nanostructured Carbonaceous Materials

As we know, graphite has been successfully used as an anode material for LIBs, but it is disappointing that when it is used as an anode for NIBs, it exhibits poor electrochemical activity, mainly because the sodium-graphite intercalation compounds formed during discharge are not stable [96]. To this end, the researchers [97] treated the graphite by oxidation-reduction method to obtain expanded graphite with a large interlayer distance, which increased the reversible capacity of graphite from ~35 to ~284 mA h g^{-1} and showed better cycle stability. However, its initial Coulombic efficiency is still low (~49.53%) due to the presence of oxygen-containing functional groups. On this basis, graphene derivatives have attracted considerable attention, such as reduced graphene oxide (rGO), graphene foam, nitrogen-doped graphene and so on. It has been reported that rGO can exhibit a reversible capacity of around 450 mA h g^{-1}, which is mainly attributed to the adsorption property of graphene surface oxygen-containing functional groups, but as described above, its initial Coulombic efficiency is also low [98]. In addition to material design to improve the sodium storage properties of graphite-based carbon materials, electrolyte optimization and selection are also good strategies. For example, replacing the ester-based electrolyte with an ether-based electrolyte, which is mainly due to the existence of solvent co-intercalation phenomena [99, 100]. It should be noted that this strategy also shows good compatibility with other carbon materials.

Compared with graphitized carbon materials, non-graphite carbon materials have received more attention, which can be divided into soft carbon and hard carbon. Their main difference is that the former is more easily graphitized than the latter at high temperatures, and the degree of disorder in the former is significantly lower than that of the latter. From the charge-discharge profiles, the soft carbon shows only the slope characteristics in the all electrochemical window range, while the hard carbon shows two characteristics of the potential slope up 0.2 V and the potential plateau below 0.2 V. That is, soft carbon has a higher operating potential than hard carbon, which somewhat reduces the potential safety hazard caused by the formation of sodium dendrites. Regarding the electrochemical performance, soft carbon has better rate capability, while hard carbon shows better capacity performance. Although they all have own advantages, but their initial Coulombic efficiencies are still low, which is a common issue for carbonaceous-based anode materials. Doeff et al. first reported on the sodium storage behavior of soft carbon anode with a capacity of about 90 mA h g^{-1} prepared from petroleum coke in 1993 [71]. Since then, in order to improve the electrochemical performance of soft carbon, the material scientists' ideas are mainly from the perspective of precursor selection and structural modification. Currently reported precursors mainly include pitch, plastics, tar, anthracite, etc. In terms of structural modification, one approach is elemental doping, and the other is nanostructured design. Undoubtedly, they play a more or less role in improving the electrochemical performance of soft carbon. Mesoporous soft carbon as an anode material for NIBs prepared by Song et al. through a simple nano-$CaCO_3$ template method delivers a reversible capacity of 331 mA h g^{-1} at 30 mA g^{-1}, and retains a capacity of 103 mA h g^{-1} at 500 mA g^{-1} after 3000 cycles [101].

Frankly, among the most widely studied carbonaceous-based anode material for NIBs is hard carbon. This is mainly due to the fact that hard carbon exhibits a lower operating potential, which is a great advantage for increasing the energy density of the battery. However, as mentioned earlier, the main bottlenecks faced by hard carbon are rate capability, cycle stability, and safety issues under high current rates. At present, in addition to strategies based on binders, electrolytes, precursors, etc., nanostructures have also been proposed extensively to solve the above issues. Liu et al. prepared hard carbon nanoparticles by pyrolysis of a polyaniline precursor with discharge capacities of about 200 and 45 mA h g^{-1} at current densities of 50 and 2500 mA g^{-1}, respectively. After 500 cycles, the capacity retention is 70%, all of which are attributed to an optimum spacing of 0.366 nm between the graphitic layers and the nano particular size resulting in a low-barrier Na^+ cation insertion [102]. A amorphous monodispersed hard carbon microsphere was synthesized using sucrose as the precursor. The carbonization temperature was found has significant effect of the microstructure and electrochemical properties of the obtained product. When the carbonization temperature was 1600 °C, the obtained product exhibits the highest plateau capacity of 220 mA h g^{-1} and excellent cycle stability with a capacity retention of 93% after 100 cycles. Interestingly, its initial Coulombic efficiency can be increased from 50 to 80% by coating soft carbon [103]. In addition, hard carbon microtubes were also obtained through the carbonization of cotton. When being used for sodium storage, they show the highest reversible capacity of 315 mA h g^{-1} for the sample carbonized at 1300 °C. The results further indicate that the carbonization temperature has a crucial influence on the microstructure and electrochemical properties of hard carbon [104]. Besides, many nanostructured hard carbons have also been prepared and exhibit excellent electrochemical properties such as hollow carbon nanowires [105], hollow macroporous carbon particles [106], banana peel pseudographite [107], and low-tortuosity mesoporous wood carbon [108].

6.4.3.2 Nanostructured Titanium-Based Materials

Titanium-based materials have received wide attention in the research field of anode materials for NIBs due to their low cost, non-toxicity, high safety, and low redox potential. The titanium-based anode materials reported so far are mainly of the following types: TiO_2, $Na_2Ti_nO_{2n+1}(2 \leq n \leq 9)$, $Li_4Ti_5O_{12}$, and $NaTi_2(PO_4)_3$. In the following, these systems will be discussed in detail from the perspective of storage mechanisms and nanostructures.

TiO_2 includes various crystal forms such as amorphous, anatase, rutile, brookite, hollandite and bronze, all of which are in dilemma of poor electronic conductivity, which results in their poor rate capability. Just like carbonaceous-based materials, nanostructured tactics have been introduced into titanium-based systems to improve their electrochemical performance, including morphology engineering, heteroatom doping, and carbon coating. Bi et al. [109] used electrochemical anodization technique to obtain self-organized amorphous TiO_2 nanotube arrays with good electrochemical performance on different Ti substrates (Fig. 6.10a). The porous Ti foam-based material exhibited superior electrochemical performance compared with that of the conventional Ti foil, which is mainly attributed to higher surface area and porosity. In addition to using Ti as a conductive substrate, Cu and Ni can also be used to construct electrode materials those do not require conductive additives and binders which to some extent shorten the diffusion path of ions and increase the transport speed of electrons. Yang et al. [110]. prepared anatase TiO_2 nanocubes with encircled (100) facets through hydrothermal method, which is more favorable for sodium ion attachment, showing an excellent rate capability and good cycling stability even after 1000 cycles, however, with a low initial Coulombic efficiency of ~38.3% (Fig. 6.10b). It exhibits a high reversible specific capacity of 242.7 mA h g^{-1} at 0.5 C and maintains a considerable capacity of 115.9 mA h g^{-1} at 20 C. In addition, it was found to have great cycle stability in which a capacity of ~95.5% can be retained even after 3000 cycles at 25 C. All of these come from the incorporation of nitrogen and the uniform carbon coating, resulting in fast electron transfer. Interestingly, nanostructured black anatase TiO_2 with oxygen vacancies was obtained via hydrothermal method using $NaBH_4$ as a reductant. Without any treatment, it displays good rate performance and cycle stability. Due to the higher intrinsic electronic conductivity, Na insertion kinetics are effectively improved, showing a relatively high initial Coulombic efficiency [111]. Nanoporous rutile TiO_2 was designed with a large surface area, providing high accessibility to electrolytes and more active sites. This large surface makes electron and ion transport faster, effectively increasing capacity and reaction kinetics (Fig. 6.10c) [112].

The most effective and widely used way to increase the electronic conductivity of an electrode material is to combine it with carbon materials. Wang et al. [113] used atomic layer deposition technique to grow a layer of amorphous TiO_2 on carbon nanotubes and carbon fiber paper substrates. The composite exhibits excellent electrochemical performance even beyond the performance of anatase TiO_2 having an open structure. 3D petal-like graphene-rich wrapped rutile TiO_2 nanoneedles were studied for NIB anode materials, showing a high capacity of 245.3 mA h g^{-1} at 0.25 C, and 94.4% capacity retention was obtained at 10 C after 4000 cycles (Fig. 6.10d) [114]. Doping is an excellent strategy for improving the intrinsic electronic conductivity of a material. Yang et al. [115] developed N-doped TiO_2-B nanorods utilizing hydrothermal treatment, ion exchange and subsequent low temperature calcination process, which exhibit higher electrochemical performance in terms of specific capacity, cycling stability and rate capability than those of anatase TiO_2 powder raw materials and TiO_2(B) nanorods without N-doping.

Titanates should be the most promising anode materials for NIBs in titanium-based materials because they are easy to synthesize, have moderate operating voltage and loose sodium ion migration channels. Currently reported titanates are mainly $Na_2Ti_3O_7$, $Na_2Ti_6O_{13}$, $Na_2Ti_7O_{15}$, etc., which can be described by the general formula $Na_2Ti_nO_{2n+1}$ ($2 \leq n \leq 9$). The main problems they face are poor electronic conductivity, low initial Coulombic efficiency, and structural instability, leading to

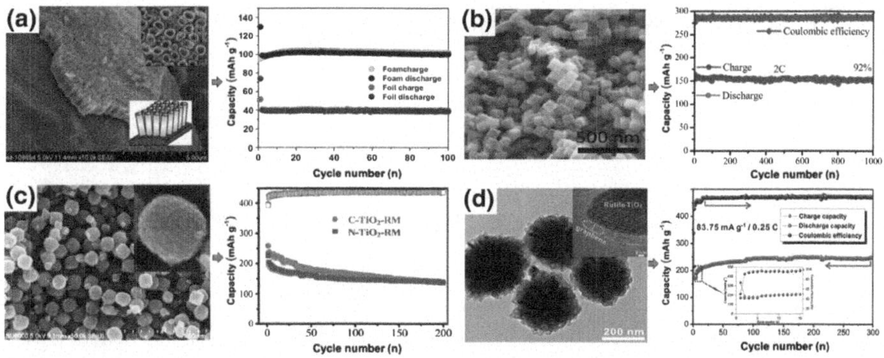

Fig. 6.10 **a** SEM image of amorphous TiO_2 nanotube arrays on the Ti foam and the corresponding cycle performance at a current density of 10 mA cm^{-2}. **b** SEM image of anatase TiO_2 nanocubes and the corresponding cycle performance at 2 C. **c** SEM image of nanoporous rutile TiO_2 mesocrystals and the corresponding cycle performance at 0.5 A g^{-1}. **d** TEM image of graphene-rich wrapped petal-like rutile TiO_2 and the corresponding cycle performance at 0.25 C. Wherein, **a** was reproduced with permission from Ref. [109], Copyright © 2013 Elsevier. **b** Was reproduced with permission from Ref. [110], Copyright © 2015 Royal Society of Chemistry. **c** Was reproduced with permission from Ref. [112], Copyright © 2015 Royal Society of Chemistry. **d** Was reproduced with permission from Ref. [114], Copyright © 2016 John Wiley & Sons

their poor rate capability and cycle stability. In order to improve their electrochemical performance, many efforts have been adopted, mainly including morphology engineering, element doping, and carbon coating or compounding. Li et al. [116] successfully obtained $Na_2Ti_7O_{15}$ nanotubes on a Ti mesh substrate by hydrothermal method and used it directly as a binder-free anode for NIBs (Fig. 6.11a). This material exhibits high reversible capacity (258 mA h g^{-1} at 50 mA g^{-1} for the 2nd cycle, 108 mA h g^{-1} at 1.5 A g^{-1}), and excellent rate capability. Even after 200 cycles at a high current density of 1.0 A g^{-1}, it still displays a capacity of 130 mA h g^{-1} and a capacity retention as high as 96%. These outstanding electrochemical properties can be attributed to the penetration of the electrolyte, the shortening of the ion diffusion path, and the increase in electronic conductivity. In addition, Ni et al. [117] used the same method to prepare $Na_2Ti_3O_7$ nanotube arrays on Ti substrate, and investigated the effects of TiO_2 coating realized by ALD and S doping on the array electrode materials (Fig. 6.11b). The composite delivers high reversible capacity of 221 mA h g^{-1} and exhibits a superior cycle efficiency and rate capability, retaining 78 mA h g^{-1} at 10 C (1770 mA g^{-1}) over 10,000 cycles. Li et al. [118]. synthesized ultralong $Na_2Ti_3O_7$ nanowires with a length of more than 100 mm and a width of 100–200 nm were grown on a carbon cloth substrate by a one-pot hydrothermal method (Fig. 6.11c). As a binder-free flexible electrode for NIBs, it demonstrates a high specific capacity of 170.3 mA h g^{-1} (close to 100% of the theoretical capacity), excellent cyclability with a capacity retention of over 96% after 200 cycles at 2 C. In addition to the above nanostructures, many nanostructures have also been reported, such as nanosheets [119] (Fig. 6.11d), microflowers [120] (Fig. 6.11e), nanorods

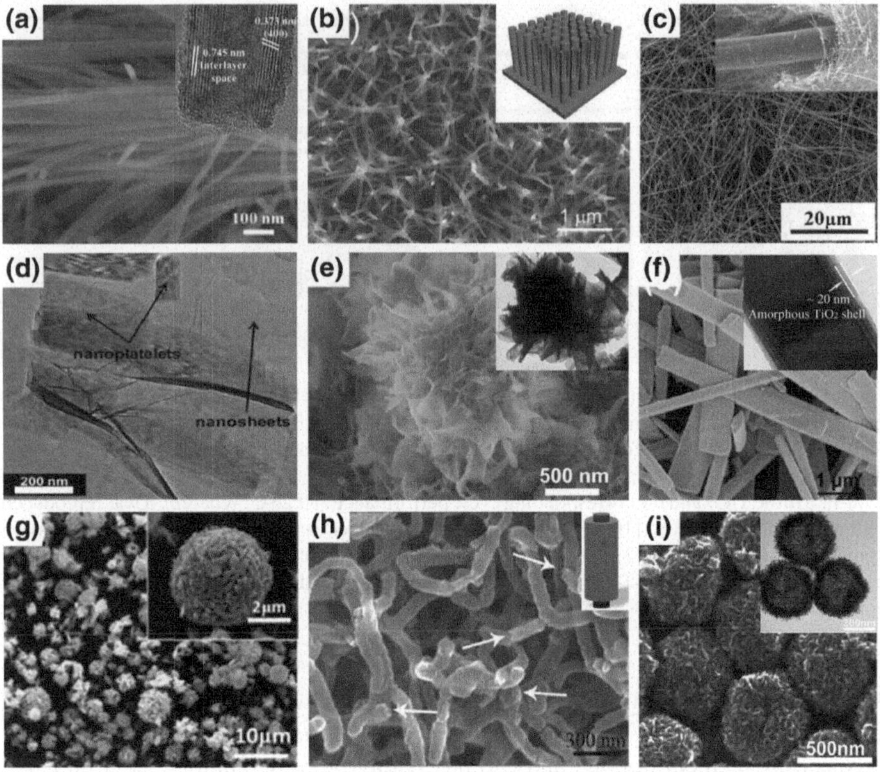

Fig. 6.11 SEM or TEM images of $Na_2Ti_7O_{15}$ nanotubes on a Ti net substrate (**a**), S-doped $Na_2Ti_3O_7$ nanotube arrays grown on Ti foil (**b**), ultra-long $Na_2Ti_3O_7$ nanowires@carbon cloth composite (**c**), $Na_2Ti_3O_7$ nanoplatelets (**d**), $Na_2Ti_3O_7$ nanosheets-formed three-dimensional microflowers (**e**), $Na_2Ti_6O_{13}$@TiO_2 core-shell nanorods (**f**), $Na_2Ti_3O_7$ microspheres (**g**), $Na_2Ti_3O_7$@CNT coaxial nanocables (**h**), and $Na_2Ti_3O_7$@N-Doped carbon hollow spheres (**i**), respectively. Wherein, **a** was reproduced with permission from Ref. [116], Copyright © 2015 Royal Society of Chemistry. **b** Was reproduced with permission from Ref. [117], Copyright © 2016 John Wiley & Sons. **c** Was reproduced with permission from Ref. [118], Copyright © 2016 Royal Society of Chemistry. **d** Was reproduced with permission from Ref. [119], Copyright © 2017 American Chemical Society. **e** Was reproduced with permission from Ref. [120], Copyright © 2017 American Chemical Society. **f** Was reproduced with permission from Ref. [121], Copyright © 2018 Elsevier. **g** Was reproduced with permission from Ref. [122], Copyright © 2014 Elsevier. **h** Was reproduced with permission from Ref. [123], Copyright © 2015 Royal Society of Chemistry. **i** Was reproduced with permission from Ref. [125], Copyright © 2017 John Wiley & Sons

[121] (Fig. 6.11f), spheres [122] (Fig. 6.11g), nanocables [123] (Fig. 6.11h) and hollow spheres [124, 125] (Fig. 6.11i). In general, after nanostructuring, the coherence of electron transfer and the path of ion migration will be improved to varying degrees and thus to achieve the purpose of improving their electrochemical performance.

$Li_4Ti_5O_{12}$ is mainly used as an anode material for LIBs, but according to related researches show that it can also be used as one for NIBs. It can deliver a reversible capacity of around 145 mA h g^{-1} with a 0.9 V average voltage. It has been reported that the sodium storage activity of $Li_4Ti_5O_{12}$ is related to particle size. Only by making it have a small particle size, $Li_4Ti_5O_{12}$ can exhibit good electrochemical activity. Therefore, to make $Li_4Ti_5O_{12}$ better used in NIBs, nanosizing or nanostructuring is particularly important, which can also solve the issue of poor kinetics. Tian et al. [126] fabricated a nanocomposite composed of crystalline $Li_4Ti_5O_{12}$ nanosheets and multi-walled carbon nanotubes via a facile heteroassembly process and followed thermal treatment. As for sodium storage, the anode delivers a reversible capacity of 82 mA h g^{-1} at 5 C and high stable capacity of 99 mA h g^{-1} at 1 C after 200 cycles. Furthermore, Yun et al. [127] reported a B-doped carbon-coated porous $Li_4Ti_5O_{12}$ using a facile one-pot wet-chemical method. As an anode for NIBs, it can display a high specific capacity of 148 mA h g^{-1} at 0.5 C and 98.4 mA h g^{-1} at 10 C with a great capacity retention of about 90% at 1 and 10 C after 300 cycles, which is mainly due to the fact that the carbon coating improves the electronic conductivity of the material, while the B doping further reduces the diffusion barrier of Na ions.

$NaTi_2(PO_4)_3$ is a NASICON structured compound, and thus exhibits good sodium ion conductivity and good stability due to the presence of a polyanion framework structure. However, since TiO_6 octahedrons are separated by PO_4 tetrahedrons, its electronic conductivity is poor, which in turn makes the material exhibit poor cycle stability and rate capability when used as an anode material for NIBs. In view of this, many efforts focused on morphology design, carbon coating or compositing, and doping have been successively proposed. Wu et al. [128] developed a self-assembly route to construct a novel architecture consisting of nanosized porous $NaTi_2(PO_4)_3$ particles embedded in microsized 3D graphene network (Fig. 6.12a). Such architecture synergistically combines the advantages of 3D graphene networks and 0D porous nanoparticles. It greatly increases the electron/ion transport kinetics and assures the integrity of electrode structure, resulting in attractive electrochemical performance, which is reflected in high rate capability (112 mA h g^{-1} at 1 C, 105 mA h g^{-1} at 5 C, 96 mA h g^{-1} at 10 C, 67 mA h g^{-1} at 50 C), long cycle life (capacity retention of 80% after 1000 cycles at 10 C), and high initial Coulombic efficiency (> 79%). Guo et al. [129] synthesized uniform $NaTi_2(PO_4)_3$ nanocubes with size of about 500 nm by means of a low-temperature solid-phase method (Fig. 6.12b). After co-coating with carbon and rutile TiO_2, the resulting composite exhibits excellent electrochemical performance. At 10 C, it can deliver a high rate capability of 83.5 mA h g^{-1} and an ultrastable property of 89.3% capacity retention after 10,000 cycles. Zhang et al. [130] used solvothermal method in combination with subsequent calcination to prepare $NaTi_2(PO_4)_3$ nanoparticles that were homogeneously embedded in a mesoporous CMK-3 matrix (Fig. 6.12c). Compared with pure $NaTi_2(PO_4)_3$, the composite exhibits higher capacity, better rate performance, and longer cycle stability. In addition to complexing with carbon materials, many efforts have also focused on building the morphology of $NaTi_2(PO_4)_3$. Yang et al. [131] synthesized four interesting porous $NaTi_2(PO_4)_3$ quasi-nanocubes via a one-pot solvothermal method with different organic precursors (ethylene glycol, triethylene glycol, n-propanol and n-butyl alcohol). The as-synthesized products deliver excellent high rate performance and cycle stability as NIB anodes. After 10,000 cycles at 10 C, 75.5%

of the initial capacity is retained (Fig. 6.12d–g). Even at very high rate of 100 C, these $NaTi_2(PO_4)_3$ quasi-nanocubes can still deliver considerable reversible capacities after deep charging/discharging for 15,000 cycles. However, since the working voltage of $NaTi_2(PO_4)_3$ is about 2 V, its application prospect in non-aqueous NIBs is far less than that in aqueous systems.

6.4.3.3 Nanostructured Alloy Materials

Elements and alloys have been considered as anode electrode materials with the highest theoretical capacity for NIBs, which has great advantages in reducing the cost and weight of a battery. At present, electrode materials of this type has been extensively studied, including Si, Ge, Sn, P, SnP, and SbGe. Since these materials store energy mainly by alloying with sodium, they inevitably undergo huge volume expansion, which is fatal to the cycle life of the battery. Nanostructured materials exhibit exhilarating positive effect in suppressing structural collapse caused by volume expansion.

Red P has a suitable operating voltage, which is currently reported as the anode material with the highest theoretical specific capacity (2600 mA h g^{-1}), but lower electrical conductivity and larger volume expansion have become a stumbling block to its practical application. To this end, optimization strategies such as doping, cladding, and nanocrystallization have been successively proposed. Zhu et al. [132] used a modified vaporization-condensation method to achieve uniformly distributed red P between tangled SWCNTs bundles. Thanks to the presence of SWCNTs, the electrical conductivity and mechanical strength of the composite are greatly improved, which is also significantly enhances its sodium storage properties. As a result, the composite exhibits high capacities of ~700 and ~300 mA h g^{-1} at 50 and 2000 mA g^{-1}, respectively, along with 80% capacity retention after 2000 cycles. Using the same method, Yu et al. [133] evaporated red phosphorus into microporous carbon YP-80F with confined space. Electrochemical tests showed that the resulting composite exhibits better electrochemical performance than that of P@CNT composite (Fig. 6.13a). In order to further increase the diffusion rate of sodium ions, reducing the particle size of the active material is an effective means. Liu et al. [134] obtained red P quantum dots on the surface of rGO by vaporization-condensation process. The tiny red P particles kept the network structure of rGO, which not only improves the electronic/ionic conductivity of the composite, but also simplifies the process of preparing flexible electrode. The fabricated highly flexible electrode can provide a charge capacity of 1211 mA h g^{-1} at the initial cycle and maintain 914 mA h g^{-1} after 300 cycles at a current density of 1593.9 mA g^{-1}. Even at high current densities of ~31.8 and ~47.8 A g^{-1}, the electrode also achieves capacities of 510.6 and 135.3 mA h g^{-1}, respectively (Fig. 6.13b). In addition to surface attachment strategies, Liu et al. [135] also used electrospinning technique to achieve uniform encapsulation of red P nanoparticles in N-doped carbon nanofibers (Fig. 6.13c). This not only greatly increases the electrical conductivity between red P particles, but also shortens the diffusion path of sodium ions. Thanks

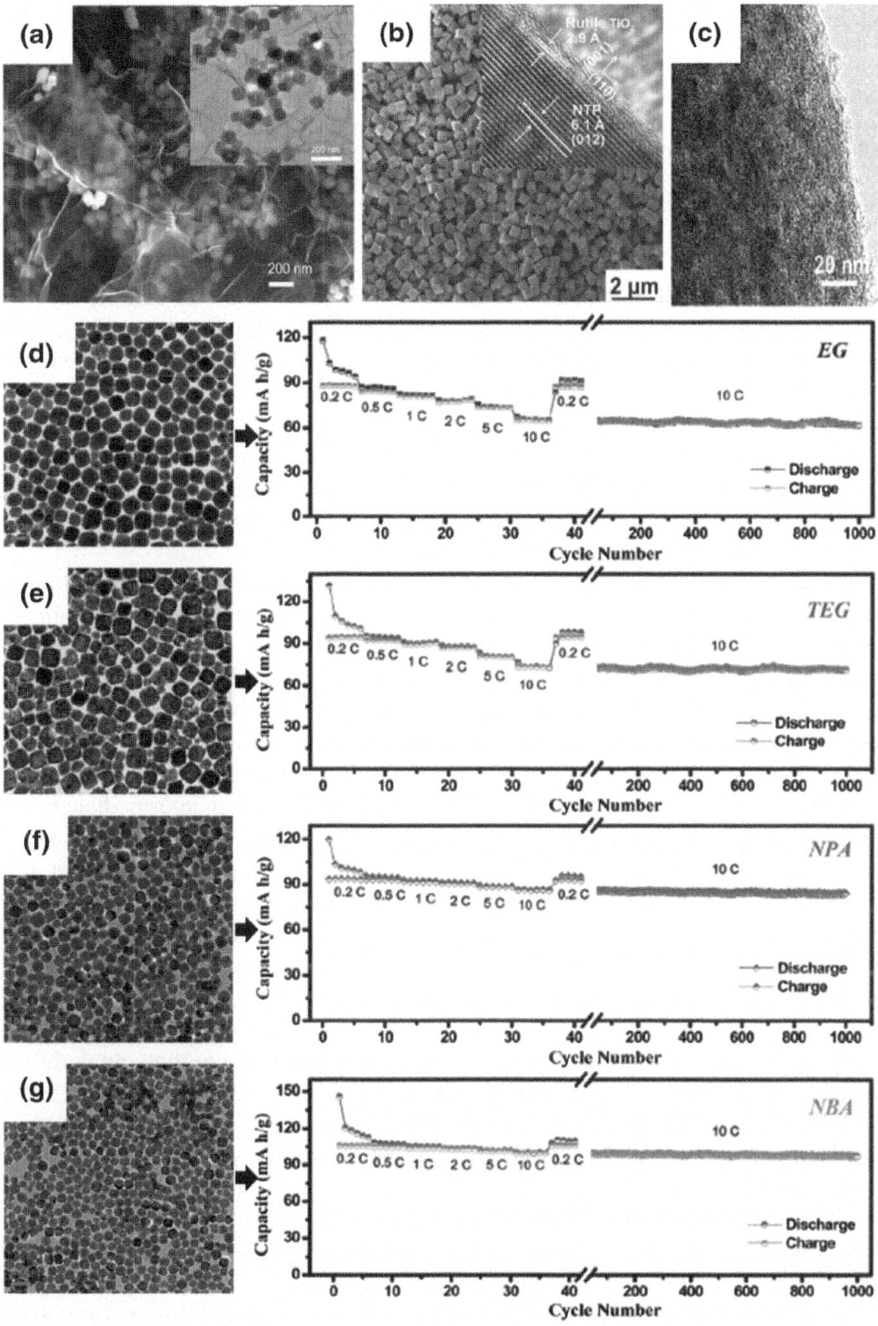
(a)
200 nm
200 nm
(b)
Rutile TiO2
2.9 Å
(001)
(110)
NTP
6.1 Å
(012)
2 μm
(c)
20 nm
(d)
EG
Capacity (mA h/g)
0.2 C
0.5 C
1 C
2 C
5 C
10 C
0.2 C
10 C
Discharge
Charge
Cycle Number
(e)
TEG
Capacity (mA h/g)
0.2 C
0.5 C
1 C
2 C
5 C
10 C
0.2 C
10 C
Discharge
Charge
Cycle Number
(f)
NPA
Capacity (mA h/g)
0.2 C
0.5 C
1 C
2 C
5 C
10 C
0.2 C
10 C
Discharge
Charge
Cycle Number
(g)
NBA
Capacity (mA h/g)
0.2 C
0.5 C
1 C
2 C
5 C
10 C
0.2 C
10 C
Discharge
Charge
Cycle Number

◂**Fig. 6.12** **a** SEM and TEM images of porous NASICON-type $NaTi_2(PO_4)_3$ particles embedded in microsized 3D graphene network. **b** SEM and TEM images of rutile TiO_2-coated $NaTi_2(PO_4)_3$ nanocubes. **c** TEM image of $NaTi_2(PO_4)_3$/CMK-3 nanocomposite. **d–g** TEM images and rate performances of the four $NaTi_2(PO_4)_3$ nanocubes electrodes. Wherein, **a** was reproduced with permission from Ref. [128], Copyright © 2015 American Chemical Society. **b** Was reproduced with permission from Ref. [129], Copyright © 2015 John Wiley & Sons. **c** Was reproduced with permission from Ref. [130], Copyright © 2014 Royal Society of Chemistry. **d–g** Were reproduced with permission from Ref. [131], Copyright © 2015 Royal Society of Chemistry

to this unique design nanostructure, the resulting composite can exhibit specific capacities of 1308 and 637 mA h g^{-1} at 200 and 5000 mA g^{-1}, respectively, with an excellent capacity retention of ~81% after 1000 cycles. Indeed, individual red P with special morphologies can also be constructed without skeleton structure. Based on the gas-bubble-directed hollow-nanosphere-formation mechanism, Zhou et al. [136] used solvothermal method to synthesize porous red P hollow nanospheres. The as-prepared material can reveal a capacity of 1364.7 mA h g^{-1} at 2 C and excellent long cycling performance, which can be attributed to the effective suppression of volume expansion and particle pulverization (Fig. 6.13d).

Similar nanostructured tactics are also widely used in other element and alloy systems. In order to solve poor conductivity and large volume change of Si anode, Liu et al. [137] employed electrospinning technique to construct a bamboo-rattle like nanostructure, yolk-shell carbon/silicon nanobeads embedded in 1D carbon nanofibers composite. This unique structure effectively relieves the problem of volume expansion and exhibits excellent electrochemical performance (Fig. 6.14a). Liu et al. [138] used viral scaffolds on a stainless steel substrate to prepare a core-shell 3D Sn nanoforest array with a carbon coating as a shell, Sn as an intermediate layer, and Ni as an inner core, which can deliver an initial high capacity of ~722 mA h g^{-1} and a Coulombic efficiency of ~78% along with 405 mA h g^{-1} retained after 150 cycles toward Na-ion storage (Fig. 6.14b). Its outstanding performance is mainly due to the unique design of the 3D structure. The interdigital spaces between individual rods effectively relieves the volume expansion and stress changes during charge and discharge, while additional carbon coating further inhibits Sn aggregation and improves electrical conductivity. Liu et al. [139] synthesized hollow Sb@C yolk-shell nanospheres through a nanoconfined galvanic replacement route that can deliver a capacity of 280 mA h g^{-1} at 1000 mA g^{-1} after 200 cycles (Fig. 6.14c). Huang et al. [140] prepared Sn_4P_3 nanospheres with multi-shell structure using a white phosphorus solution-based solvothermal phosphidation route using commercially available Sn nanospheres (Fig. 6.14d). This architecture provides a shorter sodium ion diffusion path and more buffer space for volume expansion, thereby benefiting from this, the resulting material exhibits excellent electrochemical performance in terms of reversibility, rate capability and cycle stability. Liang et al. [141] adopted the AAO-based template method to construct Sb-Ni nanoarrays (Fig. 6.14e). Such unique design not only ensures fast conduction of electrons, but also ensures good structural stability. When used as an anode material, the nanoarrays exhibit high capacity retention rate and excellent rate capability.

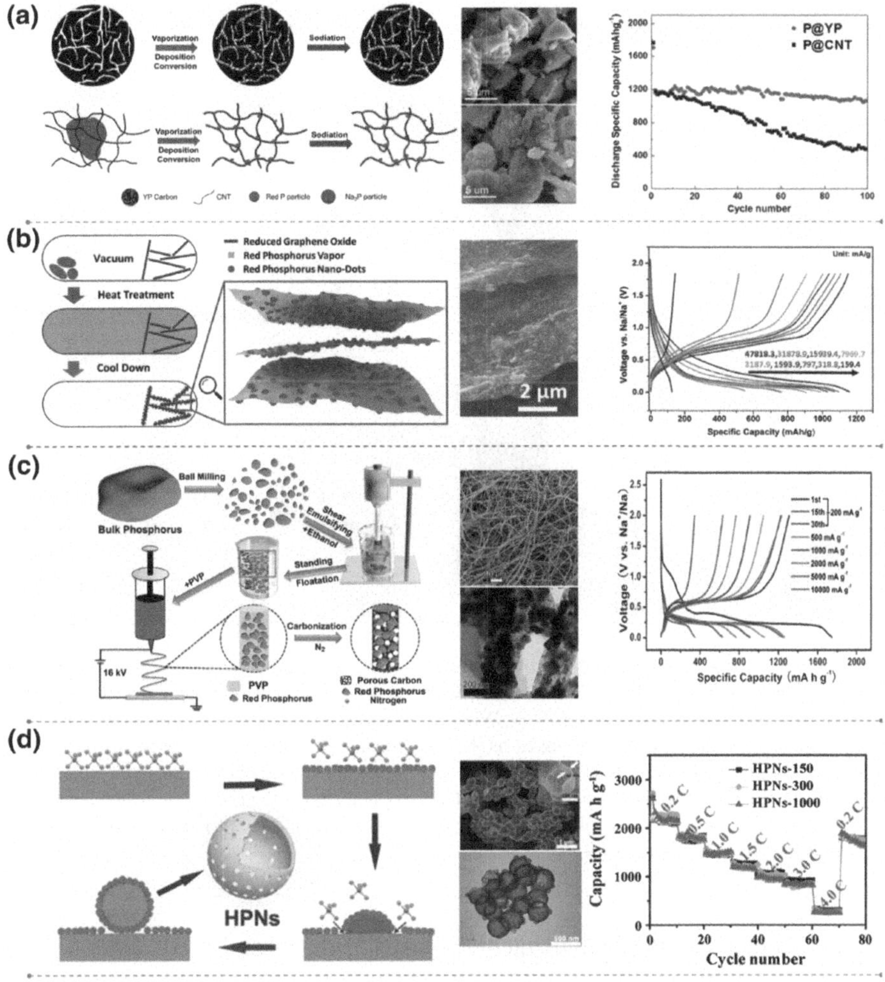

Fig. 6.13 **a** Schematic illustration of preparation process and sodiation, SEM images and cycling performance within 100 cycles at 0.1 C of P@YP-80F and P@CNT composites. **b** Schematic description of synthesis route, SEM images and rate capability of P@rGO. **c** Schematic illustration of the preparation process, SEM images and rate capability of red P@C nanofibers. **d** Schematics illustrate the proposed formation mechanism, SEM/TEM images, and rate performance with different average diameters of porous hollow red-phosphorus nanospheres. Wherein, **a** was reproduced with permission from Ref. [133], Copyright © 2017 Elsevier. **b** Was reproduced with permission from Ref. [134], Copyright © 2017 American Chemical Society. **c** Was reproduced with permission from Ref. [135], Copyright © 2017 Elsevier. **d** Was reproduced with permission from Ref. [136], Copyright © 2017 John Wiley & Sons

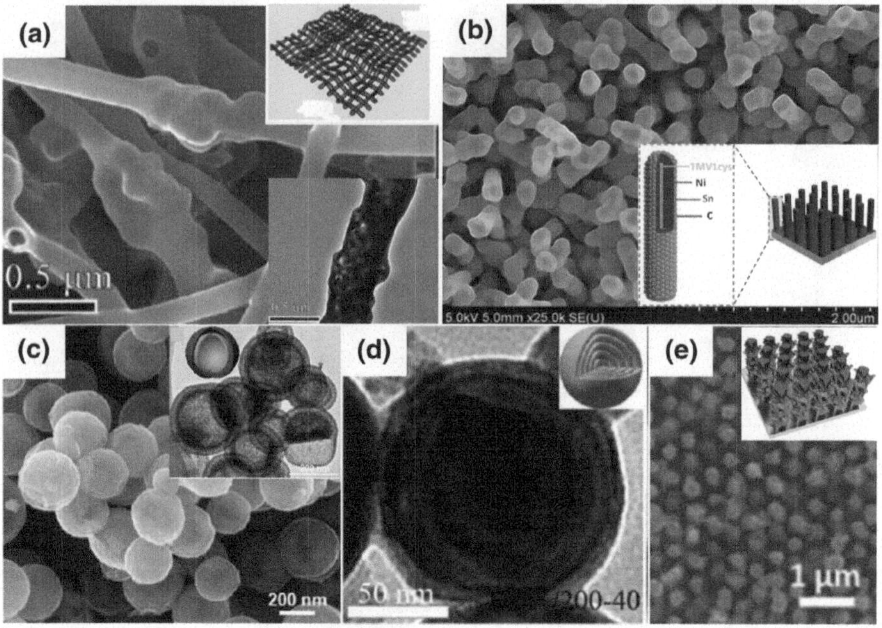

Fig. 6.14 SEM and TEM images of bamboo-rattle-like Si/C composites (**a**), Sn-coated viral nanoforests (**b**), hollow Sb@C yolk-shell spheres (**c**), multi-shell Sn_4P_3 nanospheres (**d**) and hierarchical Sb-Ni nanoarrays (**e**), respectively. Wherein, **a** was reproduced with permission from Ref. [137], Copyright © 2017 John Wiley & Sons. **b** Was reproduced with permission from Ref. [138], Copyright © 2013 American Chemical Society. **c** Was reproduced with permission from Ref. [139], Copyright © 2017 American Chemical Society. **d** Was reproduced with permission from Ref. [140], Copyright © 2017 Royal Society of Chemistry. **e** Was reproduced with permission from Ref. [141], Copyright © 2017 Springer Nature

6.4.3.4 Nanostructured Chalcogenide-Based Materials

Chalcogenides are generic names for oxides, sulfides, and selenides. As anode materials, such materials mainly follow two reaction mechanisms, namely, conversion reaction mechanism and conversion-alloying reaction mechanism. The materials belonging to the former are generally transition metal compounds that cannot form alloys with sodium, such as Fe_3O_4, Co_3O_4, Ni_3S_2, etc., while other compounds usually follow the latter reaction mechanism, such as SnO_2, SnO, SnS_2, Sb_2O_3, Sb_2Se_3 and so on. Since these materials will undergo structural changes and volume expansion during charge and discharge, their cycle stability and rate capability have yet to be improved. To this end, various nanostructured strategies have been proposed successively for different electrode materials.

As mentioned earlier, reducing particle size is an effective trick for releasing stress and increasing ionic conductivity. Wang et al. [142] synthesized a SnO_2@reduced graphene oxide nanocomposite by hydrothermal method, wherein the size of SnO_2 is

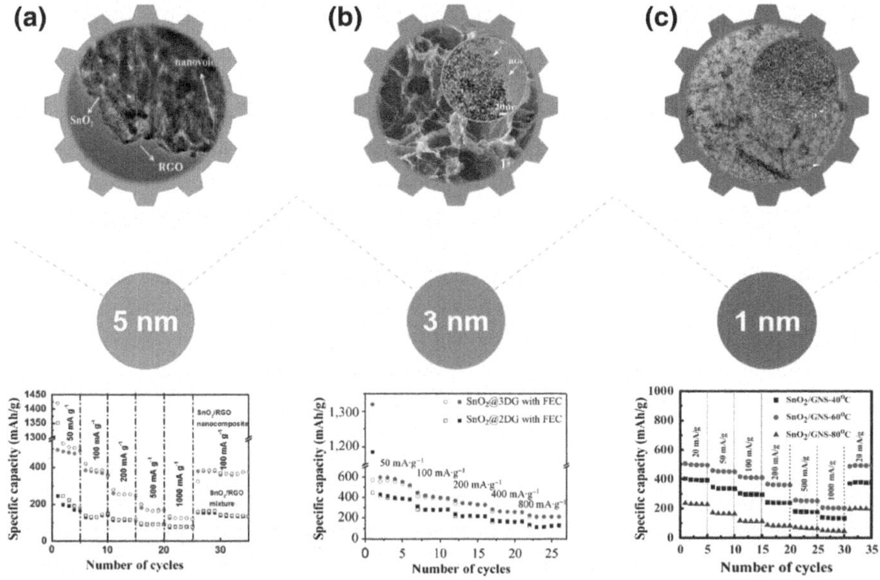

Fig. 6.15 SEM, TEM images and the corresponding rate performances of ultrafine SnO_2 nanoparticles (~5 nm) loading onto reduced graphene oxide (**a**), ultrasmall SnO_2 nanoparticles (~3 nm) embedded in three dimensional graphene (**b**) and SnO_2 particles (~1 nm) between graphene nanosheets (**c**), respectively. Wherein, **a** was reproduced with permission from Ref. [142], Copyright © 2014 Royal Society of Chemistry. **b** Was reproduced with permission from Ref. [143], Copyright © 2017 Springer Nature. **c** Was reproduced with permission from Ref. [144], Copyright © 2016 Elsevier

about 5 nm. As an anode material, the resulting composite is maintained at 330 mA h g^{-1} after 150 cycles and the capacity retention is approximately 81%. When the current density increases to 2 C, the corresponding capacity is maintained at around 120 mA h g^{-1} (Fig. 6.15a). Chen et al. [143] first constructed a 3D structure of reduced graphene oxide using the ice template method, and then hydrothermally grown SnO_2 nanoparticles with a particle size of about 3 nm on this structure. The prepared composite exhibits high cycle stability and rate performance. At current densities of 100 and 800 mA g^{-1}, its reversible capacities reach 440 and 210 mA h g^{-1}, respectively (Fig. 6.15b). In addition, the composite obtained by Chang et al. [144]. under supercritical CO_2 fluid conditions exhibits outstanding electrochemical performance, which is mainly attributed to the SnO_2 particle size as low as about 1 nm (Fig. 6.15c).

Certainly, in addition to the uniform dispersion of SnO_2 nanoparticles on the surface of graphene, many special structures have also been commended to improve electrochemical performance, such as SnO_2@carbon nanofibers [145, 146], SnO_2@Carbon cloth composite [147], SnO_2@polyaniline core-shell hollow spheres [148], etc. Ma et al. [145] constructed oxygen-vacancy-containing SnO_{2-x}@carbon nanofibers based on a flowable sulfur template and electrospinning technique with high reversible capacity, ultralong cyclic stability, and excellent rate capability for advanced NIBs (Fig. 6.16a). The discharge capacities of 634 and 447 mA h g^{-1} at 0.1 and 2 A g^{-1} were found to remain after 300 and 2500 cycles, respectively. Liu et al. [146] obtained SnO_2@void@carbon nanofibers composite having a void structure by filling the porous SnO_2 nanowires synthesized by a modified polyethylene glycol-directed method with SiO_2 and carbon and then *etc*hing the SiO_2 with NaOH (Fig. 6.16b). With the help of this reasonable structure, the composite exhibits a high capacity of 308 mA h g^{-1} at a current density of 200 mA g^{-1} after 100 cycles. Furthermore, Liu et al. [147] used a hydrothermal method to construct a composite structure of SnO_2-coated carbon cloth (Fig. 6.16c). The soft nature of the carbon cloth and the nanocrystal structure of SnO_2 layers can effectively accommodate the volume change associated with the sodiation process. After being coated with carbon or Al_2O_3, the electrochemical properties of the resulting material were significantly improved. The carbon-coated SnO_2@carbon cloth core-shell structure anode achieved 501 and 144 mA h g^{-1} capacity at 0.1 and 30 C, respectively. Meanwhile, a specific capacity of 375 mA h g^{-1} after 100 deep cycles with an 80% retention is achieved by Al_2O_3-coated SnO_2@carbon cloth anode. Zhao et al. [148] achieved the improvement of electrochemical performance of SnO_2 hollow spheres by coating polyaniline using two-step hydrothermal method and thermal treatment (Fig. 6.16d), that is, the reversible capacity still be maintained at 213.5 mA h g^{-1} after 400 cycles at 300 mA g^{-1}. In addition to the typical SnO_2 system, similar nanostructured tactics have also been used in other chalcogenide systems, such as SnO_x [149] (Fig. 6.16e), Sb_2S_3 [150] (Fig. 6.16f) and SnS [151, 152] (Fig. 6.16g–i).

6.4.3.5 Nanostructured Organic Materials

Apart from inorganic anode materials, organic anode materials have received extensive attention in recent years due to their abundant resources, low cost, and environmental friendliness. Up to now, the reported organic anode materials can be broadly classified into four categories, namely, quinones, carboxylates, anhydrides, and Schiff bases. However, as electrode materials, their low electronic conductivity and high solubility tend to result in poor electrochemical cycle stability and rate performance. For this reason, similar strategies such as morphology design, surface coating and conductive compounding were also introduced into the system in order to make up for their deficiencies.

Elizabeth et al. [153] studied the effect of Ketjen black on the electrochemical performance of polymeric Schiff bases synthesized by polycondensation reaction. When the Ketjen black content reaches 50%, the reversible capacity of the resulting composite can reach 360 mA h g^{-1} at 0.1 C, even if the current density increases to 1 C, it still retains a capacity of 190 mA h g^{-1}, however, high carbon content inevitably leads to too low initial Coulombic efficiency. Using spray-drying method, Wu et al. [154] prepared $Na_2C_6H_2O_4$ spheres sandwiched by carbon nanotubes (Fig. 6.16j). As an anode material, it exhibits good capacity, rate, and cycle stability. At 0.1 C, the reversible capacity is 259 mA h g^{-1} with an initial Coulombic efficiency of 88%.

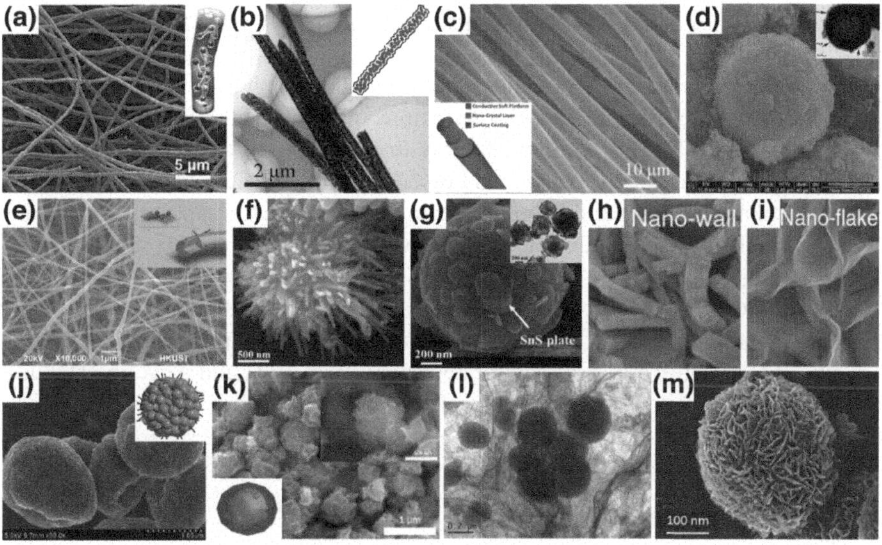

Fig. 6.16 SEM and TEM images of robust SnO_{2-x} nanoparticles impregnated carbon nanofibers (**a**), yolk-shell structured SnO_2@void@C porous nanowires (**b**), SnO_2 coated carbon cloth composite (**c**), Core-shell structured SnO_2 hollow spheres-polyaniline composite (**d**), ultrafine amorphous SnO_x embedded in carbon nanofiber/carbon nanotube composites (**e**), $Bi_{0.94}Sb_{1.06}S_3$ nanorod clusters (**f**), SnS-C microspheres (**g**), SnS nanowall arrays (**h**) and SnS nanosheet arrays (**i**) directly grown on a graphene foam backbone, carbon nanotubes-sandwiched $Na_2C_6H_2O_4$ spheres (**j**), graphene oxide-wrapped croconic acid disodium salt composite (**k**), graphene-wrapped $Na_2C_{12}H_6O_4$ nanoflowers (**l**) and pristine $Na_2C_{12}H_6O_4$ nanoflowers (**m**), respectively. Wherein, **a** was reproduced with permission from Ref. [145], Copyright © 2018 John Wiley & Sons. **b** Was reproduced with permission from Ref. [146], Copyright © 2016 Elsevier. **c** Was reproduced with permission from Ref. [147], Copyright © 2015 Elsevier. **d** Was reproduced with permission from Ref. [148], Copyright © 2015 Royal Society of Chemistry. **e** Was reproduced with permission from Ref. [149], Copyright © 2015 John Wiley & Sons. **f** Was reproduced with permission from Ref. [150], Copyright © 2015 American Chemical Society. **g** Was reproduced with permission from Ref. [151], Copyright © 2015 Springer Nature. **h, i** Were reproduced with permission from Ref. [152], Copyright © 2016 Springer Nature. **j** Was reproduced with permission from Ref. [154], Copyright © 2015 Royal Society of Chemistry. **k** Was reproduced with permission from Ref. [156], Copyright © 2014 Elsevier. **l–m** Were reproduced with permission from Ref. [157], Copyright © 2016 John Wiley & Sons

Even at 7 C, it still maintains a capacity of 142 mA h g^{-1}. Meanwhile, its reversible capacity is still kept at 191 mA h g^{-1} at 1 C after 30 cycles. In addition to the simple compounding strategies, coating and morphology engineering strategies have also shown good attractive effects in improving the electrochemical properties of organic materials. Zhao et al. [155] used ALD to coat different thicknesses of Al_2O_3 protective layer on the surface of $Na_2C_8H_4O_4$/Ketjen black electrode. Electrochemical tests showed that the electrochemical performance of the electrode was significantly improved after coating in terms of initial Coulombic efficiency, rate capability, and cycling performance. The uncoated electrode presented an initial charge capacity of 258 mA h g^{-1}, which slowly faded to 192 mA h g^{-1} after 50 cycles, with a capacity retention of 74.4%, while the electrodes with coating thicknesses of 20 and 50 nm exhibit initial charge capacities of 282 and 255 mA h g^{-1}, which decayed to 225 and 211 mA h g^{-1}, with capacity retentions of 79.8% and 82.7%, respectively. Similarly, Luo et al. [156] employed spray-drying route to achieve the construction of graphene-wrapped croconic acid disodium salt nanostructures (Fig. 6.16k). This particular structure exhibits good electrochemical performance and maintains good morphology stability during cycling. Moreover, Deng et al. [157] prepared graphene-encapsulated $Na_2C_{12}H_6O_4$ nanoflowers through a precipitation-ultrasonic process (Fig. 6.16l, m). Due to its high interfacial surface area, abundant electronic and ionic transport channels, the as-prepared material demonstrates a high reversible capacity of > 220 mA h g^{-1}, a strong rate capability at 10 C and a cycling stability with 92% capacity retention over 100 cycles.

6.5 Summary and Outlook

In summary, the main problems faced by sodium ion electrode materials can be summarized as structural instability and poor ion/electron conductivity. Depending on the type of material, rational design of different strategies or nanostructures is an effective means to solve the above problems. Although many nanostructured tactics have been proposed at present, research on a certain material is still not systematic enough. In addition, from the practical application point of view, how to design cost-effective nanostructured materials for NIBs and realized them in large-scale synthesis should be the direction in the future.

References

1. Yabuuchi, N., Kubota, K., Dahbi, M., et al. (2014). Research development on sodium-ion batteries. *Chemical Reviews, 114*(23), 11636–11682.
2. Hwang, J.-Y., Myung, S.-T., & Sun, Y.-K. (2017). Sodium-ion batteries: Present and future. *Chemical Society Reviews, 46*(12), 3529–3614.
3. Whittingham, M. S. (1978). Chemistry of intercalation compounds: Metal guests in chalcogenide hosts. *Progress in Solid State Chemistry, 12*(1), 41–99.
4. Newman, G. H., & Klemann, L. P. (1980). Ambient temperature cycling of an Na-TiS_2 Cell. *Journal of the Electrochemical Society, 127*(10), 2097–2099.
5. Takeda, Y., Nakahara, K., Nishijima, M., et al. (1994). Sodium deintercalation from sodium iron oxide. *Materials Research Bulletin, 29*(6), 659–666.
6. Delmas, C., Fouassier, C., & Hagenmuller, P. (1980). Structural classification and properties of the layered oxides. *Physica B+C, 99*, 81–85.
7. Wang, P.-F., You, Y., Yin, Y.-X., et al. (2018). Layered oxide cathodes for sodium-ion batteries: Phase transition, air stability, and performance. *Advanced Energy Materials, 8*(8), 1701912.
8. Jiang, Y., Zhou, X., Li, D., et al. (2018). Highly reversible Na storage in $Na_3V_2(PO_4)_3$ by optimizing nanostructure and rational surface engineering. *Advanced Energy Materials, 8*(6), 1800068.

9. Chen, S., Wu, C., Shen, L., et al. (2017). Challenges and perspectives for NASICON-type electrode materials for advanced sodium-ion batteries. *Advanced Materials, 29*(48), 1700431.
10. Xu, Y., Wei, Q., Xu, C., et al. (2016). Layer-by-layer $Na_3V_2(PO_4)_3$ embedded in reduced graphene oxide as superior rate and ultralong-life sodium-ion battery cathode. *Advanced Energy Materials, 6*(14), 1600389.
11. Jian, Z., Han, W., Lu, X., et al. (2013). Superior electrochemical performance and storage mechanism of $Na_3V_2(PO_4)_3$cathode for room-temperature sodium-ion batteries. *Advanced Energy Materials, 3*(2), 156–160.
12. Wei, T., Yang, G., & Wang, C. (2017). Bottom-up assembly of strongly-coupled $Na_3V_2(PO_4)_3$/C into hierarchically porous hollow nanospheres for high-rate and -stable Na-ion storage. *Nano Energy, 39,* 363–370.
13. Ren, W., Zheng, Z., Xu, C., et al. (2016). Self-sacrificed synthesis of three-dimensional $Na_3V_2(PO_4)_3$ nanofiber network for high-rate sodium-ion full batteries. *Nano Energy, 25,* 145–153.
14. Guo, D., Qin, J., Yin, Z., et al. (2018). Achieving high mass loading of $Na_3V_2(PO_4)_3$@carbon on carbon cloth by constructing three-dimensional network between carbon fibers for ultralong cycle-life and ultrahigh rate sodium-ion batteries. *Nano Energy, 45,* 136–147.
15. Wang, X., Niu, C., Meng, J., et al. (2015). Novel $K_3V_2(PO_4)_3$/C bundled nanowires as superior sodium-ion battery electrode with ultrahigh cycling stability. *Advanced Energy Materials, 5*(17), 1500716.
16. Zhai, Y., Dou, Y., Zhao, D., et al. (2011). Carbon materials for chemical capacitive energy storage. *Advanced Materials, 23*(42), 4828–4850.
17. Li, S., Dong, Y., Xu, L., et al. (2014). Effect of carbon matrix dimensions on the electrochemical properties of $Na_3V_2(PO_4)_3$ nanograins for high-performance symmetric sodium-ion batteries. *Advanced Materials, 26*(21), 3545–3553.
18. Zhu, C., Song, K., van Aken, P. A., et al. (2014). Carbon-coated $Na_3V_2(PO_4)_3$ embedded in porous carbon matrix: An ultrafast Na-storage cathode with the potential of outperforming Li cathodes. *Nano Letters, 14*(4), 2175–2180.
19. Fang, Y., Xiao, L., Ai, X., et al. (2015). Hierarchical carbon framework wrapped $Na_3V_2(PO_4)_3$ as a superior high-rate and extended lifespan cathode for sodium-ion batteries. *Advanced Materials, 27*(39), 5895–5900.
20. Raccichini, R., Varzi, A., Passerini, S., et al. (2015). The role of graphene for electrochemical energy storage. *Nature Materials, 14*(3), 271–279.
21. Duan, W., Zhu, Z., Li, H., et al. (2014). $Na_3V_2(PO_4)_3$@C core-shell nanocomposites for rechargeable sodium-ion batteries. *Journal of Materials Chemistry A, 2*(23), 8668–8675.
22. Chen, M., Kou, K., Tu, M., et al. (2015). Fabrication of multi-walled carbon nanotubes modified $Na_3V_2(PO_4)_3$/C and its application to high-rate lithium-ion batteries cathode. *Solid State Ionics, 274,* 24–28.
23. Lin, D., Liu, Y., & Cui, Y. (2017). Reviving the lithium metal anode for high-energy batteries. *Nature Nanotechnology, 12,* 194.
24. Xin, S., Chang, Z., Zhang, X., et al. (2017). Progress of rechargeable lithium metal batteries based on conversion reactions. *National Science Review, 4*(1), 54–70.
25. Manthiram, A., Yu, X., & Wang, S. (2017). Lithium battery chemistries enabled by solid-state electrolytes. *Nature Reviews Materials, 2,* 16103.
26. Kerman, K., Luntz, A., Viswanathan, V., et al. (2017). Review-practical challenges hindering the development of solid state Li ion batteries. *Journal of the Electrochemical Society, 164*(7), A1731–A1744.
27. Rui, X., Sun, W., Wu, C., et al. (2015). An advanced sodium-ion battery composed of carbon coated $Na_3V_2(PO_4)_3$ in a porous graphene network. *Advanced Materials, 27*(42), 6670–6676.
28. Fang, J., Wang, S., Li, Z., et al. (2016). Porous $Na_3V_2(PO_4)_3$@C nanoparticles enwrapped in three-dimensional graphene for high performance sodium-ion batteries. *Journal of Materials Chemistry A, 4*(4), 1180–1185.
29. Ji, X., Lee, K. T., & Nazar, L. F. (2009). A highly ordered nanostructured carbon-sulphur cathode for lithium-sulphur batteries. *Nature Materials, 8*(6), 500–506.

30. Schuster, J., He, G., Mandlmeier, B., et al. (2012). Spherical ordered mesoporous carbon nanoparticles with high porosity for lithium-sulfur batteries. *Angewandte Chemie International Edition, 51*(15), 3591–3595.
31. Jiang, Y., Yang, Z., Li, W., et al. (2015). Nanoconfined carbon-coated $Na_3V_2(PO_4)_3$ particles in mesoporous carbon enabling ultralong cycle life for sodium-ion batteries. *Advanced Energy Materials, 5*(10), 1402104.
32. Zhu, C., Kopold, P., van Aken, P. A., et al. (2016). High power-high energy sodium battery based on threefold interpenetrating network. *Advanced Materials, 28*(12), 2409–2416.
33. Shen, W., Wang, C., Xu, Q., et al. (2015). Nitrogen-doping-induced defects of a carbon coating layer facilitate Na-storage in electrode materials. *Advanced Energy Materials, 5*(1), 1400982.
34. Xu, J., Lee, D. H., Clément, R. J., et al. (2014). Identifying the critical role of Li substitution in P2-$Na_x[Li_yNi_zMn_{1-y-z}]O_2$ ($0 < x, y, z < 1$) intercalation cathode materials for high-energy Na-ion batteries. *Chemistry of Materials, 26*(2), 1260–1269.
35. Clément, R. J., Xu, J., Middlemiss, D. S., et al. (2017). Direct evidence for high Na^+ mobility and high voltage structural processes in P2-$Na_x[Li_yNi_zMn_{1-y-z}]O_2$ ($x, y, z \leq 1$) cathodes from solid-state NMR and DFT calculations. *Journal of Materials Chemistry A, 5*(8), 4129–4143.
36. Zheng, S., Zhong, G., McDonald, M. J., et al. (2016). Exploring the working mechanism of Li^+ in O3-type $NaLi_{0.1}Ni_{0.35}Mn_{0.55}O_2$ cathode materials for rechargeable Na-ion batteries. *Journal of Materials Chemistry A, 4*(23), 9054–9062.
37. Oh, S.-M., Myung, S.-T., Hwang, J.-Y., et al. (2014). High capacity O3-type $Na[Li_{0.05}(Ni_{0.25}Fe_{0.25}Mn_{0.5})_{0.95}]O_2$ cathode for sodium ion batteries. *Chemistry of Materials, 26*(21): 6165–6171.
38. You, Y., Xin, S., Asl, H. Y., et al. (2018). Insights into the improved high-voltage performance of Li-incorporated layered oxide cathodes for sodium-ion batteries. *Chem, 4*(9), 2124–2139.
39. Rong, X., Liu, J., Hu, E., et al. (2018). Structure-induced reversible anionic redox activity in Na layered oxide cathode. *Joule, 2*(1), 125–140.
40. Du, K., Zhu, J., Hu, G., et al. (2016). Exploring reversible oxidation of oxygen in a manganese oxide. *Energy & Environmental Science, 9*(8), 2575–2577.
41. Kang, W., Zhang, Z., Lee, P.-K., et al. (2015). Copper substituted P2-type $Na_{0.67}Cu_xMn_{1-x}O_2$: A stable high-power sodium-ion battery cathode. *Journal of Materials Chemistry A, 3*(45), 22846–22852.
42. Wang, P.-F., You, Y., Yin, Y.-X., et al. (2016). Suppressing the P2-O2 phase transition of $Na_{0.67}Mn_{0.67}Ni_{0.33}O_2$ by magnesium substitution for improved sodium-ion batteries. *Angewandte Chemie International Edition,* 55(26), 7445–7449.
43. Wu, X., Guo, J., Wang, D., et al. (2015). P2-type $Na_{0.66}Ni_{0.33-x}Zn_xMn_{0.67}O_2$ as new high-voltage cathode materials for sodium-ion batteries. *Journal of Power Sources, 281,* 18–26.
44. Xu, S.-Y., Wu, X.-Y., Li, Y.-M., et al. (2014). Novel copper redox-based cathode materials for room-temperature sodium-ion batteries. *Chinese Physics B, 23*(11), 118202.
45. Li, Y., Yang, Z., Xu, S., et al. (2015). Air-stable copper-based P2-$Na_{7/9}Cu_{2/9}Fe_{1/9}Mn_{2/3}O_2$ as a new positive electrode material for sodium-ion batteries. *Advanced Science, 2*(6), 1500031.
46. Yao, H.-R., Wang, P.-F., Gong, Y., et al. (2017). Designing air-stable O3-type cathode materials by combined structure modulation for Na-ion batteries. *Journal of the American Chemical Society, 139*(25), 8440–8443.
47. Singh, G., Tapia-Ruiz, N., Lopez del Amo, J. M., et al. (2016). High voltage Mg-doped $Na_{0.67}Ni_{0.3-x}Mg_xMn_{0.7}O_2$ ($x = 0.05, 0.1$) Na-ion cathodes with enhanced stability and rate capability. *Chemistry of Materials, 28*(14), 5087–5094.
48. Wu, X., Xu, G. L., Zhong, G., et al. (2016). Insights into the effects of zinc doping on structural phase transition of P2-type sodium nickel manganese oxide cathodes for high-energy sodium ion batteries. *ACS Applied Materials & Interfaces, 8*(34), 22227–22237.
49. Yabuuchi, N., Hara, R., Kubota, K., et al. (2014). A new electrode material for rechargeable sodium batteries: P2-type $Na_{2/3}[Mg_{0.28}Mn_{0.72}]O_2$ with anomalously high reversible capacity. *Journal of Materials Chemistry A, 2*(40), 16851–16855.
50. Maitra, U., House, R. A., Somerville, J. W., et al. (2018). Oxygen redox chemistry without excess alkali-metal ions in $Na_{2/3}[Mg_{0.28}Mn_{0.72}]O_2$. *Nature Chemistry, 10*(3), 288–295.

51. Zhang, X. H., Pang, W. L., Wan, F., et al. (2016). P2-$Na_{2/3}Ni_{1/3}Mn_{5/9}Al_{1/9}O_2$ microparticles as superior cathode material for sodium-ion batteries: Enhanced properties and mechanism via graphene connection. *ACS Applied Materials & Interfaces, 8*(32), 20650–20659.
52. Pang, W.-L., Zhang, X.-H., Guo, J.-Z., et al. (2017). P2-type $Na_{2/3}Mn_{1-x}Al_xO_2$ cathode material for sodium-ion batteries: Al-doped enhanced electrochemical properties and studies on the electrode kinetics. *Journal of Power Sources, 356,* 80–88.
53. Yoshida, H., Yabuuchi, N., Kubota, K., et al. (2014). P2-type $Na_{2/3}Ni_{1/3}Mn_{2/3-x}Ti_xO_2$ as a new positive electrode for higher energy Na-ion batteries. *Chemical Communications, 50*(28), 3677–3680.
54. Wang, P.-F., Yao, H.-R., Liu, X.-Y., et al. (2018). Na^+/vacancy disordering promises high-rate Na-ion batteries. *Science Advances, 4*(3), eaar6018.
55. Wang, P.-F., Yao, H.-R., Liu, X.-Y., et al. (2017). Ti-substituted $NaNi_{0.5}Mn_{0.5-x}Ti_xO_2$ cathodes with reversible O3-P3 phase transition for high-performance sodium-ion batteries. *Advanced Materials, 29*(19), 1700210.
56. Zheng, L., & Obrovac, M. N. (2017). Investigation of O3-type $Na_{0.9}Ni_{0.45}Mn_xTi_{0.55-x}O_2$ ($0 \leq x \leq 0.55$) as positive electrode materials for sodium-ion batteries. *Electrochimica Acta, 233,* 284–291.
57. Evstigneeva, M. A., Nalbandyan, V. B., Petrenko, A. A., et al. (2011). A new family of fast sodium ion conductors: $Na_2M_2TeO_6$ (M = Ni Co, Zn, Mg). *Chemistry of Materials, 23*(5), 1174–1181.
58. Seibel, E. M., Roudebush, J. H., Wu, H., et al. (2013). Structure and magnetic properties of the α-$NaFeO_2$-type honeycomb compound $Na_3Ni_2BiO_6$. *Inorganic Chemistry, 52*(23), 13605–13611.
59. Schmidt, W., Berthelot, R., Sleight, A. W., et al. (2013). Solid solution studies of layered honeycomb-ordered phases O3-$Na_3M_2SbO_6$ (M = Cu, Mg, Ni, Zn). *Journal of Solid State Chemistry, 201,* 178–185.
60. Yuan, D., Liang, X., Wu, L., et al. (2014). A honeycomb-layered $Na_3Ni_2SbO_6$: A high-rate and cycle-stable cathode for sodium-ion batteries. *Advanced Materials, 26*(36), 6301–6306.
61. Li, Y., Deng, Z., Peng, J., et al. (2018). New P2-type honeycomb-layered sodium-ion conductor: $Na_2Mg_2TeO_6$. *ACS Applied Materials & Interfaces, 10*(18), 15760–15766.
62. Wang, P. F., Xin, H., Zuo, T. T., et al. (2018). An abnormal 3.7 volt O3-type sodium-ion battery cathode. *Angewandte Chemie International Edition, 57*(27), 8178–8183.
63. Sathiya, M., Jacquet, Q., Doublet, M.-L., et al. (2018). A chemical approach to raise cell voltage and suppress phase transition in O3 sodium layered oxide electrodes. *Advanced Energy Materials, 8*(11), 1702599.
64. Kubota, K., Kumakura, S., Yoda, Y., et al. (2018). Electrochemistry and solid-state chemistry of $NaMeO_2$ (Me = 3d transition metals). *Advanced Energy Materials, 8*(17), 1703415.
65. Chagas, L. G., Buchholz, D., Vaalma, C., et al. (2014). P-type $Na_xNi_{0.22}Co_{0.11}Mn_{0.66}O_2$ materials: linking synthesis with structure and electrochemical performance. *Journal of Materials Chemistry A, 2*(47), 20263–20270.
66. Buchholz, D., Chagas, L. G., Vaalma, C., et al. (2014). Water sensitivity of layered P2/P3-$Na_xNi_{0.22}Co_{0.11}Mn_{0.66}O_2$ cathode material. *Journal of Materials Chemistry A,* 2(33), 13415–13421.
67. Chen, X., Zhou, X., Hu, M., et al. (2015). Stable layered P3/P2 $Na_{0.66}Co_{0.5}Mn_{0.5}O_2$ cathode materials for sodium-ion batteries. *Journal of Materials Chemistry A, 3*(41), 20708–20714.
68. Hou, P., Yin, J., Lu, X., et al. (2018). A stable layered P3/P2 and spinel intergrowth nanocomposite as a long-life and high-rate cathode for sodium-ion batteries. *Nanoscale, 10*(14), 6671–6677.
69. Lee, E., Lu, J., Ren, Y., et al. (2014). Layered P2/O3 intergrowth cathode: Toward high power Na-ion batteries. *Advanced Energy Materials, 4*(17), 1400458.
70. Guo, S., Liu, P., Yu, H., et al. (2015). A layered P2- and O3-type composite as a high-energy cathode for rechargeable sodium-ion batteries. *Angewandte Chemie International Edition, 54*(20), 5894–5899.

71. Doeff, M. M., Ma, Y., Visco, S. J., et al. (1993). Electrochemical insertion of sodium into carbon. *Journal of the Electrochemical Society, 140*(12), L169–L170.
72. Qi, X., Liu, L., Song, N., et al. (2017). Design and comparative study of O3/P2 hybrid structures for room temperature sodium-ion batteries. *ACS Applied Materials Interfaces, 9*(46), 40215–40223.
73. Huang, Q., Liu, J. T., Zhang, L., et al. (2018). Tailoring alternating heteroepitaxial nanostructures in Na-ion layered oxide cathodes via an in-situ composition modulation route. *Nano Energy, 44,* 336–344.
74. Keller, M., Buchholz, D., & Passerini, S. (2016). Layered Na-ion cathodes with outstanding performance resulting from the synergetic effect of mixed P- and O-type phases. *Advanced Energy Materials, 6*(3), 1501555.
75. Hwang, J. Y., Oh, S. M., Myung, S. T., et al. (2015). Radially aligned hierarchical columnar structure as a cathode material for high energy density sodium-ion batteries. *Nature Communications, 6,* 6865.
76. Hwang, J.-Y., Myung, S.-T., Yoon, C. S., et al. (2016). Novel cathode materials for Na-ion batteries composed of spoke-like nanorods of $Na[Ni_{0.61}Co_{0.12}Mn_{0.27}]O_2$ assembled in spherical secondary particles. *Advanced Functional Materials, 26*(44), 8083–8093.
77. Wu, Z. G., Li, J. T., Zhong, Y. J., et al. (2017). Mn-based cathode with synergetic layered-tunnel hybrid structures and their enhanced electrochemical performance in sodium ion batteries. *ACS Applied Materials & Interfaces, 9*(25), 21267–21275.
78. Xiao, Y., Wang, P.-F., Yin, Y.-X., et al. (2018). A layered-tunnel intergrowth structure for high-performance sodium-ion oxide cathode. *Advanced Energy Materials, 8*(22), 1800492.
79. Chen, T.-R., Sheng, T., Wu, Z.-G., et al. (2018). Cu^{2+} dual-doped layer-tunnel hybrid $Na_{0.6}Mn_{1-x}Cu_xO_2$ as a cathode of sodium-ion battery with enhanced structure stability, electrochemical property, and air stability. *ACS Applied Materials & Interfaces, 10*(12), 10147–10156.
80. Cui, J., Yao, S., & Kim, J.-K. (2017). Recent progress in rational design of anode materials for high-performance Na-ion batteries. *Energy Storage Materials, 7,* 64–114.
81. James, S. L., Adams, C. J., Bolm, C., et al. (2012). Mechanochemistry: opportunities for new and cleaner synthesis. *Chemical Society Reviews, 41*(1), 413–447.
82. Zhang, W., Mao, J., Pang, W. K., et al. (2017). Large-scale synthesis of ternary Sn_5SbP_3/C composite by ball milling for superior stable sodium-ion battery anode. *Electrochimica Acta, 235,* 107–113.
83. Yang, Z., Sun, J., Ni, Y., et al. (2017). Facile synthesis and in situ transmission electron microscopy investigation of a highly stable Sb_2Te_3/C nanocomposite for sodium-ion batteries. *Energy Storage Materials, 9,* 214–220.
84. Ramireddy, T., Xing, T., Rahman, M. M., et al. (2015). Phosphorus-carbon nanocomposite anodes for lithium-ion and sodium-ion batteries. *Journal of Materials Chemistry A, 3*(10), 5572–5584.
85. Feinle, A., Elsaesser, M. S., & Husing, N. (2016). Sol-gel synthesis of monolithic materials with hierarchical porosity. *Chemical Society Reviews, 45,* 3377–3399.
86. Jung, J.-W., Lee, C.-L., Yu, S., et al. (2016). Electrospun nanofibers as a platform for advanced secondary batteries: a comprehensive review. *Journal of Materials Chemistry A, 4*(3), 703–750.
87. Li, D., & Xia, Y. (2004). Electrospinning of nanofibers: Reinventing the wheel? *Advanced Materials, 16*(14), 1151–1170.
88. Niu, C., Meng, J., Wang, X., et al. (2015). General synthesis of complex nanotubes by gradient electrospinning and controlled pyrolysis. *Nature Communications, 6,* 7402.
89. Liang, L., Xu, Y., Wang, C., et al. (2015). Large-scale highly ordered Sb nanorod array anodes with high capacity and rate capability for sodium-ion batteries. *Energy & Environmental Science, 8*(10), 2954–2962.
90. Nam, D.-H., Hong, K.-S., Lim, S.-J., et al. (2014). Electrochemical synthesis of a three-dimensional porous Sb/Cu_2Sb anode for Na-ion batteries. *Journal of Power Sources, 247,* 423–427.

91. Xu, Y., Zhou, M., Zhang, C., et al. (2017). Oxygen vacancies: Effective strategy to boost sodium storage of amorphous electrode materials. *Nano Energy, 38,* 304–312.
92. Zhu, Y., Choi, S. H., Fan, X., et al. (2017). Recent progress on spray pyrolysis for high performance electrode materials in lithium and sodium rechargeable batteries. *Advanced Energy Materials, 7*(7), 1601578.
93. Gim, J., Mathew, V., Lim, J., et al. (2012). Pyro-synthesis of functional nanocrystals. *Scientific Reports, 2,* 1–6.
94. Niu, Y., Xu, M., Guo, C., et al. (2016). Pyro-synthesis of a nanostructured $NaTi_2(PO_4)_3/C$ with a novel lower voltage plateau for rechargeable sodium-ion batteries. *Journal of Colloid and Interface Science, 474,* 88–92.
95. Ghosh, S., Mani, A. D., Kishore, B., et al. (2017). Autocombustion synthesis of nanostructured $Na_2Ti_6O_{13}$ negative insertion material for Na-ion batteries: Electrochemical and diffusion mechanism. *Journal of the Electrochemical Society, 164*(9), A1881–A1886.
96. Nobuhara, K., Nakayama, H., Nose, M., et al. (2013). First-principles study of alkali metal-graphite intercalation compounds. *Journal of Power Sources, 243,* 585–587.
97. Wen, Y., He, K., Zhu, Y., et al. (2014). Expanded graphite as superior anode for sodium-ion batteries. *Nature Communications, 5,* 4033.
98. Wan, J., Shen, F., Luo, W., et al. (2016). In situ transmission electron microscopy observation of sodiation-desodiation in a long cycle, high-capacity reduced graphene oxide sodium-ion battery anode. *Chemistry of Materials, 28*(18), 6528–6535.
99. Kim, H., Yoon, G., Lim, K., et al. (2016). A comparative study of graphite electrodes using the co-intercalation phenomenon for rechargeable Li, Na and K batteries. *Chemical Communications, 52*(85), 12618–12621.
100. Zhang, J., Wang, D.-W., Lv, W., et al. (2017). Achieving superb sodium storage performance on carbon anodes through an ether-derived solid electrolyte interphase. *Energy & Environmental Science, 10*(1), 370–376.
101. Cao, B., Liu, H., Xu, B., et al. (2016). Mesoporous soft carbon as an anode material for sodium ion batteries with superior rate and cycling performance. *Journal of Materials Chemistry A, 4*(17), 6472–6478.
102. Xiao, L., Cao, Y., Henderson, W. A., et al. (2016). Hard carbon nanoparticles as high-capacity, high-stability anodic materials for Na-ion batteries. *Nano Energy, 19,* 279–288.
103. Li, Y., Xu, S., Wu, X., et al. (2015). Amorphous monodispersed hard carbon micro-spherules derived from biomass as a high performance negative electrode material for sodium-ion batteries. *Journal of Materials Chemistry A, 3*(1), 71–77.
104. Li, Y., Hu, Y.-S., Titirici, M.-M., et al. (2016). Hard carbon microtubes made from renewable cotton as high-performance anode material for sodium-ion batteries. *Advanced Energy Materials, 6*(18), 1600659.
105. Cao, Y., Xiao, L., Sushko, M. L., et al. (2012). Sodium ion insertion in hollow carbon nanowires for battery applications. *Nano Letters, 12*(7), 3783–3787.
106. Ding, J., Wang, H., Li, Z., et al. (2013). Carbon nanosheet frameworks derived from peat moss as high performance sodium ion battery anodes. *ACS Nano, 7*(12), 11004–11015.
107. Lotfabad, E. M., Ding, J., Cui, K., et al. (2014). High-density sodium and lithium ion battery anodes from banana peels. *ACS Nano, 8*(7), 7115–7129.
108. Fei, S., Wei, L., Jiaqi, D., et al. (2016). Ultra-thick, low-tortuosity, and mesoporous wood carbon anode for high-performance sodium-ion batteries. *Advanced Energy Materials, 6*(14), 1600377.
109. Bi, Z., Paranthaman, M. P., Menchhofer, P. A., et al. (2013). Self-organized amorphous TiO_2 nanotube arrays on porous Ti foam for rechargeable lithium and sodium ion batteries. *Journal of Power Sources, 222,* 461–466.
110. Yang, X., Wang, C., Yang, Y., et al. (2015). Anatase TiO_2 nanocubes for fast and durable sodium ion battery anodes. *Journal of Materials Chemistry A, 3*(16), 8800–8807.
111. Chen, J., Ding, Z., Wang, C., et al. (2016). Black anatase titania with ultrafast sodium-storage performances stimulated by oxygen vacancies. *ACS Applied Materials & Interfaces, 8*(14), 9142–9151.

112. Hong, Z., Zhou, K., Zhang, J., et al. (2015). Facile synthesis of rutile TiO_2 mesocrystals with enhanced sodium storage properties. *Journal of Materials Chemistry A, 3*(33), 17412–17416.
113. Wang, H., Jia, G., Guo, Y., et al. (2016). Atomic layer deposition of amorphous TiO_2 on carbon nanotube networks and their superior Li and Na ion storage properties. *Advanced Materials Interfaces, 3*(21), 1600375.
114. Zhang, Y., Foster, C. W., Banks, C. E., et al. (2016). Graphene-rich wrapped petal-like rutile TiO_2 tuned by carbon dots for high-performance sodium storage. *Advanced Materials, 28*(42), 9391–9399.
115. Yang, Y., Liao, S., Shi, W., et al. (2017). Nitrogen-doped TiO_2(B) nanorods as high-performance anode materials for rechargeable sodium-ion batteries. *RSC Advances, 7*(18), 10885–10890.
116. Li, H., Fei, H., Liu, X., et al. (2015). In situ synthesis of $Na_2Ti_7O_{15}$ nanotubes on a Ti net substrate as a high performance anode for Na-ion batteries. *Chemical Communications, 51*(45), 9298–9300.
117. Ni, J., Fu, S., Wu, C., et al. (2016). Superior sodium storage in $Na_2Ti_3O_7$ Nanotube arrays through surface engineering. *Advanced Energy Materials, 6*(11), 1502568.
118. Li, Z., Shen, W., Wang, C., et al. (2016). Ultra-long $Na_2Ti_3O_7$ nanowires@carbon cloth as a binder-free flexible electrode with a large capacity and long lifetime for sodium-ion batteries. *Journal of Materials Chemistry A, 4*(43), 17111–17120.
119. Ko, J. S., Doan-Nguyen, V. V. T., Kim, H.-S., et al. (2017). $Na_2Ti_3O_7$ nanoplatelets and nanosheets derived from a modified exfoliation process for use as a high-capacity sodium-ion negative electrode. *ACS Applied Materials & Interfaces, 9*(2), 1416–1425.
120. Anwer, S., Huang, Y., Liu, J., et al. (2017). Nature-inspired $Na_2Ti_3O_7$ nanosheets-formed three-dimensional microflowers architecture as a high-performance anode material for rechargeable sodium-ion batteries. *ACS Applied Materials & Interfaces, 9*(13), 11669–11677.
121. Zhou, X., Zhong, D., Luo, H., et al. (2018). $Na_2Ti_6O_{13}$@TiO_2 core-shell nanorods with controllable mesoporous shells and their enhanced photocatalytic performance. *Applied Surface Science, 427,* 1183–1192.
122. Zou, W., Li, J., Deng, Q., et al. (2014). Microspherical $Na_2Ti_3O_7$ prepared by spray-drying method as anode material for sodium-ion battery. *Solid State Ionics, 262,* 192–196.
123. Dong, S., Shen, L., Li, H., et al. (2015). Pseudocapacitive behaviours of $Na_2Ti_3O_7$@CNT coaxial nanocables for high-performance sodium-ion capacitors. *Journal of Materials Chemistry A, 3*(42), 21277–21283.
124. Chen, S., Pang, Y., Liang, J., et al. (2018). Red blood cell-like hollow carbon sphere anchored ultrathin $Na_2Ti_3O_7$ nanosheets as long cycling and high rate-performance anodes for sodium-ion batteries. *Journal of Materials Chemistry A, 6,* 13164–13170.
125. Xie, F., Zhang, L., Su, D., et al. (2017). $Na_2Ti_3O_7$@N-doped carbon hollow spheres for sodium-ion batteries with excellent rate performance. *Advanced Materials, 29*(24), 1700989.
126. Tian, Y., Wu, Z. L., Xu, G. B., et al. (2017). Hetero-assembly of a $Li_4Ti_5O_{12}$ nanosheet and multi-walled carbon nanotube nanocomposite for high-performance lithium and sodium ion batteries. *RSC Advances, 7*(6), 3293–3301.
127. Yun, B.-N., Du, H. L., Hwang, J.-Y., et al. (2017). Improved electrochemical performance of boron-doped carbon-coated lithium titanate as an anode material for sodium-ion batteries. *Journal of Materials Chemistry A, 5*(6), 2802–2810.
128. Wu, C., Kopold, P., Ding, Y.-L., et al. (2015). Synthesizing porous $NaTi_2(PO_4)_3$ nanoparticles embedded in 3D graphene networks for high-rate and long cycle-life sodium electrodes. *ACS Nano, 9*(6), 6610–6618.
129. Yang, J., Wang, H., Hu, P., et al. (2015). A high-rate and ultralong-life sodium-ion battery based on $NaTi_2(PO_4)_3$ nanocubes with synergistic coating of carbon and rutile TiO_2. *Small, 11*(31), 3744–3749.
130. Pang, G., Nie, P., Yuan, C., et al. (2014). Mesoporous $NaTi_2(PO_4)_3$/CMK-3 nanohybrid as anode for long-life Na-ion batteries. *Journal of Materials Chemistry A, 2*(48), 20659–20666.
131. Yang, G., Song, H., Wu, M., et al. (2015). Porous $NaTi_2(PO_4)_3$ nanocubes: A high-rate nonaqueous sodium anode material with more than 10 000 cycle life. *Journal of Materials Chemistry A, 3*(36), 18718–18726.

132. Zhu, Y., Wen, Y., Fan, X., et al. (2015). Red phosphorus-single-walled carbon nanotube composite as a superior anode for sodium ion batteries. *ACS Nano, 9*(3), 3254–3264.
133. Yu, Z., Song, J., Wang, D., et al. (2017). Advanced anode for sodium-ion battery with promising long cycling stability achieved by tuning phosphorus-carbon nanostructures. *Nano Energy, 40,* 550–558.
134. Liu, Y., Zhang, A., Shen, C., et al. (2017). Red phosphorus nanodots on reduced graphene oxide as a flexible and ultra-fast anode for sodium-ion batteries. *ACS Nano, 11*(6), 5530–5537.
135. Liu, Y., Zhang, N., Liu, X., et al. (2017). Red phosphorus nanoparticles embedded in porous N-doped carbon nanofibers as high-performance anode for sodium-ion batteries. *Energy Storage Materials, 9,* 170–178.
136. Zhou, J., Liu, X., Cai, W., et al. (2017). Wet-chemical synthesis of hollow red-phosphorus nanospheres with porous shells as anodes for high-performance lithium-ion and sodium-ion batteries. *Advanced Materials, 29*(29), 1700214.
137. Zhang, L., Hu, X., Chen, C., et al. (2017). In operando mechanism analysis on nanocrystalline silicon anode material for reversible and ultrafast sodium storage. *Advanced Materials, 29*(5), 1604708.
138. Liu, Y., Xu, Y., Zhu, Y., et al. (2013). Tin-coated viral nanoforests as sodium-ion battery anodes. *ACS Nano, 7*(4), 3627–3634.
139. Liu, J., Yu, L., Wu, C., et al. (2017). New nanoconfined galvanic replacement synthesis of hollow Sb@C yolk-shell spheres constituting a stable anode for high-rate Li/Na-Ion batteries. *Nano Letters, 17*(3), 2034–2042.
140. Huang, S., Meng, C., Xiao, M., et al. (2017). Multi-shell tin phosphide nanospheres as high performance anode material for a sodium ion battery. *Sustainable Energy & Fuels, 1*(9), 1944–1949.
141. Liang, L., Xu, Y., Wen, L., et al. (2017). Hierarchical Sb-Ni nanoarrays as robust binder-free anodes for high-performance sodium-ion half and full cells. *Nano Research, 10*(9), 3189–3201.
142. Wang, Y.-X., Lim, Y.-G., Park, M.-S., et al. (2014). Ultrafine SnO_2 nanoparticle loading onto reduced graphene oxide as anodes for sodium-ion batteries with superior rate and cycling performances. *Journal of Materials Chemistry A, 2*(2), 529–534.
143. Pei, L., Jin, Q., Zhu, Z., et al. (2015). Ice-templated preparation and sodium storage of ultra-small SnO_2 nanoparticles embedded in three-dimensional graphene. *Nano Research, 8*(1), 184–192.
144. Patra, J., Chen, H.-C., Yang, C.-H., et al. (2016). High dispersion of 1-nm SnO_2 particles between graphene nanosheets constructed using supercritical CO_2 fluid for sodium-ion battery anodes. *Nano Energy, 28,* 124–134.
145. Ma, D., Li, Y., Mi, H., et al. (2018). Robust SnO_{2-x} nanoparticle-impregnated carbon nanofibers with outstanding electrochemical performance for advanced sodium-ion batteries. *Angewandte Chemie International Edition, 57*(29), 8901–8905.
146. Li, H. Z., Yang, L. Y., Liu, J., et al. (2016). Improved electrochemical performance of yolk-shell structured SnO_2@void@C porous nanowires as anode for lithium and sodium batteries. *Journal of Power Sources, 324,* 780–787.
147. Liu, Y., Fang, X., Ge, M., et al. (2015). SnO_2 coated carbon cloth with surface modification as Na-ion battery anode. *Nano Energy, 16,* 399–407.
148. Zhao, X., Zhang, Z., Yang, F., et al. (2015). Core-shell structured SnO_2 hollow spheres-polyaniline composite as an anode for sodium-ion batteries. *RSC Advances, 5*(40), 31465–31471.
149. Zhang, B., Huang, J., & Kim, J.-K. (2015). Ultrafine amorphous SnO_x embedded in carbon nanofiber/carbon nanotube composites for Li-ion and Na-ion batteries. *Advanced Functional Materials, 25*(32), 5222–5228.
150. Zhao, Y., & Manthiram, A. (2015). $Bi_{0.94}Sb_{1.06}S_3$ Nanorod cluster anodes for sodium-ion batteries: Enhanced reversibility by the synergistic effect of the Bi_2S_3-Sb_2S_3 solid solution. *Chemistry of Materials,* 27(17), 6139–6145.

151. Choi, S. H., & Kang, Y. C. (2015). Aerosol-assisted rapid synthesis of SnS-C composite microspheres as anode material for Na-ion batteries. *Nano Research, 8*(5), 1595–1603.
152. Chao, D., Zhu, C., Yang, P., et al. (2016). Array of nanosheets render ultrafast and high-capacity Na-ion storage by tunable pseudocapacitance. *Nature Communications, 7,* 12122.
153. Elizabeth, C. M., Javier, C. G., & Michel, A. (2014). Polymeric schiff bases as low-voltage redox centers for sodium-ion batteries. *Angewandte Chemie International Edition, 53*(21), 5341–5345.
154. Wu, X., Ma, J., Ma, Q., et al. (2015). A spray drying approach for the synthesis of a $Na_2C_6H_2O_4$/CNT nanocomposite anode for sodium-ion batteries. *Journal of Materials Chemistry A, 3*(25), 13193–13197.
155. Zhao, L., Zhao, J., Hu, Y.-S., et al. (2012). Disodium terephthalate ($Na_2C_8H_4O_4$) as high performance anode material for low-cost room-temperature sodium-ion battery. *Advanced Energy Materials, 2*(8), 962–965.
156. Luo, C., Zhu, Y., Xu, Y., et al. (2014). Graphene oxide wrapped croconic acid disodium salt for sodium ion battery electrodes. *Journal of Power Sources, 250,* 372–378.
157. Deng, W., Qian, J., Cao, Y., et al. (2016). Graphene-wrapped $Na_2C_{12}H_6O_4$ nanoflowers as high performance anodes for sodium-ion batteries. *Small, 12*(5), 583–587.

Chapter 7
Traditional Nanostructures and Nanomaterials in Batteries

Xing-Long Wu, Jin-Zhi Guo and Yu-Guo Guo

Abstract Traditional nanostructures and nanomaterials, such as conductive additives, separators and current collectors also play a key role in batteries although they are minority components. In this section, we summarize the research progress of traditional nanostructures and nanomaterials in rechargeable batteries, discussing the effects of these nano components on the electrochemical properties of batteries.

Abbreviations

CNF	Carbon nanofiber
CNT	Carbon nanotube
LIB	Lithium ion battery
AB	Acetylene black
SP	Super P
KB	Ketjen black
SWCNT	Single-walled carbon nanotube
MWCNT	Multi-walled carbon nanotube
VCF	Vapor deposit carbon fiber
CVD	Chemical vapor deposition
MC	Mesoporous carbon
1D	One-dimensional
VGCF	Vapor-grown carbon fiber
PE	Polyethylene
PP	Polypropylene
GPE	Gel polymer electrolyte
PMMA NP	Poly(methyl methacrylate) nanoparticle array

X.-L. Wu · J.-Z. Guo
Northeast Normal University, Changchun, Jilin 130024, People's Republic of China

Y.-G. Guo (✉)
Institute of Chemistry, Chinese Academy of Sciences, Beijing 100190, People's Republic of China
e-mail: ygguo@iccas.ac.cn

Y.-G. Guo, *Nanostructures and Nanomaterials for Batteries*,
https://doi.org/10.1007/978-981-13-6233-0_7

PVDF-HFP	Poly(vinylidene-fluoride-co-hexafluoropropylene)
PEGDMA	Poly(ethylene glycol) dimethacrylate
PVDF-co-CTFE	Polyvinylidene-fluoride-co-chlorotrifluoroethylene
PEO	Poly(ethylene oxide)
PI	Polyimide
ANF	Aramid nanofiber
PMIA	Poly(m-phenylene isophthalamide)
PVA	Poly(vinyl alchol)
PPESK	Poly(phthalazinone ether sulfone ketone)
PMP	Polymethylpentene
PAEK	Poly(arylene ether ketone)
PLA	Poly(lactic acid)
GF/PI	Glass fiber/polyimide
MMT	Montmorillonite
[EMIm]TFSI	1-ethyl-3-methylimidazolium bis(trifluoromethanesulfonyl)imide
GNH	Graphene-carbon nanotube hybrid
Cu/CC	Cu-coated carbon cloth
3DDC	3D porous dendritic current collector
PET	Polyethylene terephthalate
PVD	Physical vapour deposition
SACNT	Super-aligned carbon nanotube
Al	Aluminum
Cu	Copper

7.1 Introduction of Traditional Nanostructures and Nanomaterials in Batteries

In batteries, there are also some traditional components, such as conductive additives, separators and current collectors that can improve the performance of batteries by using nanostructures and nanomaterials. Although they are minority in batteries, these components play a key role on the electrochemical cell performance. Conductive additives can construct a conductive percolation network to increase and keep the electronic conductivity of electrode, enabling it charge and discharge faster. In addition, conductive additives absorb and retain electrolyte, allowing an intimate contact between the lithium ions and active materials. Carbon nanomaterials generally include carbon black, carbon nanofibers (CNFs), carbon nanotubes (CNTs) and graphene, which all have superior properties such as low weight, high chemical inertia and high specific surface area. They are all useful conductive additives for batteries [1–3]. The separator with nanometer sized pores plays a crucial role in preventing the physical contact between the anode and cathode, simultaneously serving as an electrolyte reservoir to enable ion transporting between the electrodes [4]. Even though the separator itself does not take part in any cell reactions directly, its structure and properties can greatly influence the battery performance [5]. The current collector plays a vital bridge function in supporting active materials, binders, conductive additives, and electronically connecting overall structure with the external circuit, since minimizing the overall resistance of the electrochemical system is a basic principle for achieving better electrochemical performance [6, 7]. Therefore, the nanostructure feature for the interface characteristic of current collector has a significant effect on improving the specific capacity, rate performance, and long cycle life.

7.2 Conductive Additives for Batteries

Conductive additive plays an important role in batteries because many electrode materials usually have low electrical conductivity. It constructs a conductive percolation network for electrons transport [1]. In batteries, the electrochemical reactions need the electronics and lithium ions reached the surface of the active material at the same time, so the electronics can participate in the electrochemical reaction and the performance of active materials can be realized in good ways [8–11]. In addition, conductive additive can absorb and retain a part of electrolyte, allowing an intimate contact between the lithium ions and active materials. Although the percentage is low, general quality percentage is less than 5%, but the crystallinity of conductive agent, such as form and content to a large extent influence on the performance of the battery in charge and discharge processes. Conductive additives improve the performance of lithium ion batteries (LIBs) mainly in the following three aspects: (i) improved the electronic conductivity, the contacts between the active material particles, active materials and collectors; [12] (ii) improved the ionic conductivity, forming the pore of different size and shape between the active particles, which will make the active materials and electrolyte infiltrate fully; (iii) improved the compressibility of electrodes, accordingly, the volume energy density, bending, peek strength and so on have been improved.

Carbon materials are the most commonly used conductive agent in current LIBs, different types of carbon materials have been used as conductive additives, such as carbon black, CNTs, graphene, and so on [2, 13]. Carbon materials have many advantages, such as no toxicity and environmental protection, high corrosion resistance, low cost, low weight and batch yield. In addition, the mass density of carbon materials are low compared to the active materials, reducing the use amount of conductive additives can significantly increase the volume energy density of batteries. On the basis of their properties, in terms of surface area, particle size, electrolyte retention capability and intrinsic conductivity, certain, mostly empirical, rules have been developed for the selection of conductive additives. Therefore, we intend to discuss the effect of different conductive additives with nanostructured on the properties of electrode materials in the following section.

7.2.1 Carbon Black

Carbon black is the generic terms of the formation of molten polymer of the small particles of carbon and hydrocarbon pyrolysis products in the gas phase state. Carbon black is one of the most widely used conductive additive, mainly obtained from organic (natural gas, heavy oil, etc.) incomplete combustion and thermal decomposition, then improving its electrical conductivity and purity by high temperature treatment [14]. The characteristics of the carbon black is small particle size, specific surface area, particularly large conductive performance is very good, it can rise to protect liquid absorption in the batteries. Carbon black mainly includes, acetylene black (AB), Super P (SP), and Ketjen black (KB). They all have spherical primary particles. The conducting networks formed by them are dominated by the degree of inter-particle resistance which in turn depends on the degree of inter-particle contact.

The characteristic and dosage of carbon black in electrode play an important role in preparation for the best performance of the battery, but the amount of the carbon black is vary for different active materials. Liu et al. [15] investigated the effects of carbon black on the electrochemical performance of $LiCoO_2$ and $LiMn_2O_4$ cathode. While 2–5 wt% of carbon are adequate for the $LiMn_2O_4$ cathode to deliver good rate capability and maintain cyclability, the $LiCoO_2$ cathode requires more carbon (10 wt%) to achieve likewise. Moreover, the specific surface area of carbon black also can affect the performance of electrode, and the large specific surface area of carbon black can lead to the increase of side reaction in electrode, which results in charge loss, capacity decrease and life decay of batteries. In anode, the capacity loss will increase with the increase of specific surface area of electrode because of the formation of SEI. In the case of cathode, the large specific surface area of carbon black can increase the probability of electrolyte oxidation, especially, at charged state, and research found that it can accelerate the Mn dissolution of spinel $LiMn_2O_4$ [16–18]. This process is accompanied by the corrosion of carbon, which gives rise to severe capacity losses in the cycling.

Besides the dosage and own properties of conductive additives, the dispersing extent is an important factor to influence the electrochemical performance of electrode. Kuroda et al. [12] investigated the influence of the dispersion state of conductive additives on the properties of LIBs, and examined the charge/discharge properties of cathode prepared with suspensions of KB dispersed in an organic solvent. Discharge capacity, durability and resistance to high-current discharge of the cathode prepared from a well-dispersed suspension was higher than those of the cathode prepared with KB powder. Because KB particles with a well-dispersed suspension formed a conductive network uniformly in the cathode, despite the low amount of additive. Dominko et al. [19] studied the influence of carbon black distribution or arrangement in composite cathode on the cathode performance by using three types of cathode active materials: $LiMn_2O_4$, $LiCoO_2$, and $LiFePO_4$. To the active materials, carbon black is added in two different ways: (i) using a conventional mixing procedure (cathode active materials, carbon black and binder were directly mixed) and (ii) using a novel coating technology (gelatin pretreated the cathode active particles,

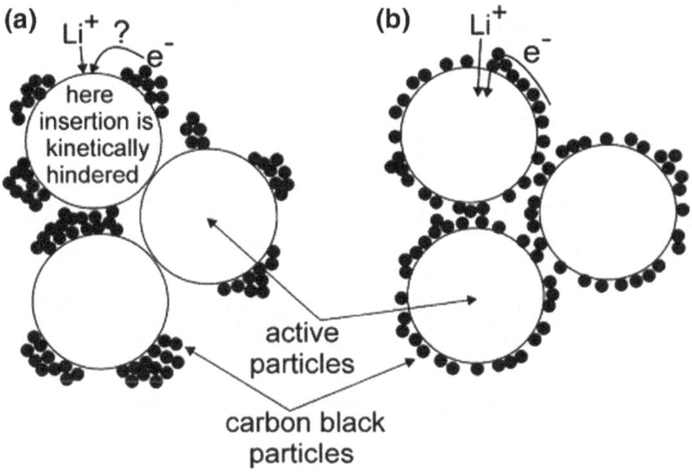

Fig. 7.1 A sketch indicating the possible impact of carbon black (electron conductor) on kinetics of Li^+/electron electrochemical insertion into active particles. Reproduced with the permission from Ref. [19], Copyright © 2003 Elsevier B.V

and then mixed with a dispersion of carbon black). Different technologies result in different distribution of carbon black around active particles. It is shown that the uniformity of carbon black distribution can affect the cathode kinetics, regardless of the types of active particles used. For example, in the case of $LiCoO_2$, even a much lower content (2 wt%) of uniformly distributed carbon black shows better kinetics than a higher content (10 wt%) of non-uniformly distributed carbon black. A simple model explaining the influence of carbon black distribution on cathode kinetics is presented in Fig. 7.1 [19]. In order for insertion to occur without excessive polarization, both the Li^+ and the electron have to be available at the same spot Fig. 7.1b by contrast, if a part of active material is not in contact with an electronic conductor Fig. 7.1a, the supply of electrons to this spot will be reflected in additional polarization, and thus, resulting in a lower reversible capacity.

7.2.2 Carbon Nanotubes

CNT is a one-dimensional (1D) tubular nanomaterial composed of single-layer or multi-layer graphite crimped, with the diameter of about 5 nm and the length of 10–20 μm. According to the graphite layers, the CNT can be divided into single-walled CNT (SWCNT) and multi-walled CNT (MWCNT). Adding a small amount of CNT can form an effective conducting network that fully connected with active materials, making for the capacity and cycle stability of the battery improved. CNT has high thermal conductivity that can benefit to heat transfer out generated from electrochemical reactions in battery, and also relieve internal polarization, improving the high/low temperature performance and security of battery [3, 20]. CNT has a high aspect ratio, reaching the same percolation threshold require the smaller amount of CNT than other conductive additives. Moreover, the CNT in the aspect of improving special capacity, energy density and cycle performance, also has the advantages.

As conductive additives, CNTs have the following advantages: [21, 22] (i) CNT with good electrical conductivity, fibrous structure can form the successive conductive network in the electrode active materials; (ii) CNT with the 1D structure can form abundant conductive contact points with the active material, following a "line-to-point" conducting mode, which effectively reduce the contact resistance between the electrode material particles, acting as wire in the conducting network; (iii) electrode plate has high toughness after adding CNTs due to its high flexibility and tensile strength, which can able to buffer material exfoliation caused by volume change during the process of charging and discharging, and then improve the cycle life of electrode; (iv) large specific area means CNTs can greatly improve the penetration ability of electrolyte in the electrode. Due to the above advantages, CNTs as conductive additives can effectively improve the overall performance of the batteries, and this makes the CNTs become the focus in the LIBs conductive additives and one of the most potential application direction.

In 2002, SWCNTs were used as a conductive additive in LIBs with the V_2O_5 as the active material [23]. By integrating the nanotubes as part of the sol-gel chemistry, uniform materials are prepared with controlled weight fractions and electrode loadings in the milligram per centimeter squared range. The similarity in morphology and dimensional scale for the SWCNTs and V_2O_5 ribbons leads to a very different electrode microstructure as compared to traditional intercalation electrodes. The V_2O_5/SWCNTs electrode retains high specific capacity at high discharge rate; this behavior is attributed to the fact that the electronically conducting network developed from the SWCNTs is more effective than one based on carbon black. Afterwards, the group of Li et al. [3] reported the MWCNTs as the conductive additives in LIBs with the $LiNi_{0.7}Co_{0.3}O_2$ as the active material, and compared with the AB conductive additives. The results show that $LiNi_{0.7}Co_{0.3}O_2$ particles were connected by MWCNTs to form a conductive network wiring, which was effective to improve reversible capacity and cycle performance. These studies indicate that CNTs have potential application as a conductive additive for high-power LIBs. The group of Sheem et al. [24] also indicated that an electrode containing MWCNT is superior to one containing carbon black, and the improved performance of the former electrode is due largely to the resilience of the MWCNT aggregates that form conductive bridges between particles of the active material. These resilient bridges maintain intimate contacts between the particles even when the composite expands during cycling. By contrast, similar but rigid bridges of carbon black are broken during cycling. Figure 7.2 is a schematic presentation of the two kinds of bridges in the composite cathodes. Similar results have been shown in the system of $LiFePO_4$, $LiCoO_2$, graphite [2, 25].

Guoping et al. [26] compared three conductive additives: AB, vapor deposit carbon fibers (VCF) and CNTs. The initial discharge capacities clearly show that CNTs are the most effective in enhancing charge/discharge rate performance of the composite cathode. They investigated other electrochemical performance (including cathode

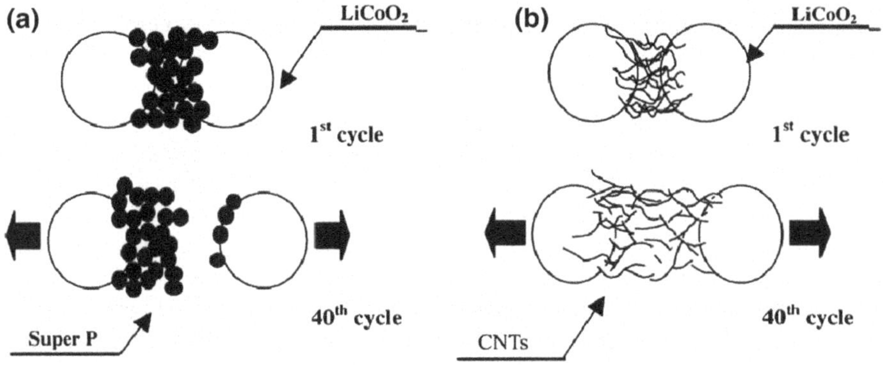

Fig. 7.2 **a** Schematic representation of changes in Super P loaded cathode; **b** schematic representation of changes in CNTs loaded cathode. Reproduced with the permission from Ref. [24], Copyright © 2006 Elsevier B.V

resistance, cyclic performance and discharge voltage profile) of the three composite cathodes. The results indicated that electrochemical performance of CNTs loaded cathode is the best. The reasons are as follows. The CNTs' wire-like shape, crystallinity and nano-size are all considered to be important factors. The wire-like shape favors forming continuous conductive network and then assures an efficient electronic transport throughout the cathode. The higher rapidly transfer electrons and lower percolation threshold in comparison with VCF. Moreover, the higher surface energy guarantees their close contact with the cathode active materials $LiCoO_2$.

However, CNTs as conductive additives of LIBs also have some problems, such as, the residual metal catalyst in the process of CNTs synthesis, can easily be oxidized and precipitated on the anode surface in charging and discharging process of high voltage, which lead to the micro short circuit in the internal of battery, serious self-discharge, or even safety accidents; the dispersity of CNTs is poor due to the strong van der Waals force between CNTs, which bring about the inferior electrical conductivity. Therefore, CNTs as conductive additives are applied in commercial LIBs that still have problems to be solved.

7.2.3 Graphene and Reduced Graphene Oxide (rGO)

Graphene is a new type of carbon nanomaterials, has a unique geometric structure characteristics and physical properties. Compared with CNTs, the study of graphene as a conductive additive is relatively few and the commonly used graphene in batteries is rGO. It has been reported that the graphene as conductive additives, to some extent, can also improve the performance of the battery [27]. In 2010, it was reported that graphene could be used as a conductive additive in LIBs [28], and then the researchers are beginning to pay attention to graphene as a conductive additive.

As conductive additives, graphene has the characteristics with gentle, thin and thick, and mainly has the following four advantages: (i) high electrical conductivity, and using a small amount of graphene can effectively reduce ohmic polarization of internal battery; (ii) 2D layer structure, compared with 0D carbon black particles and 1D CNTs, graphene and the active materials can realize the "plane-to-point" contact, has a lower conductive threshold, and it can construct a conductive network from the greater spatial span, realizing the long-range conductive in the entire electrode; (iii) ultra-thin characteristics, the graphene is a typical surface solid, compared with more than sp2 carbon layer of carbon black, conductive graphite and MWCNTs, and the all carbon atoms on graphene can be exposed to electron transfer, with a high utilization efficiency of atoms, so it can construct a complete conducting network in minimum usage, improving the energy density of batteries; (iv) high flexibility, can contract well with the active material, can buffer the volume expansion shrinkage in active material during the charge and discharge processes [24], and can inhibit rebound effect of the electrode, ensuring good cycle performance of batteries. Due to these advantages, based on graphene conductive material of LIBs can achieve compact build. The graphene conductive additives with the character of gentler, thinner and thicker exhibit a potential application prospect in batteries.

In addition to their own physical properties (good electronic conductivity) and structural characteristics (2D), the efficiency of graphene conductive additives are associated with particles unique contact mode between the graphene and the active materials. In 2010, the group of Yang propose the conducting network mechanism of graphene with flexible "plane-to-point" contact [28], shown in Fig. 7.3. As shown in Fig. 7.3, in graphene conductive network, the graphene and active materials follow a "plane-to-point" conducting mode while the carbon black and active materials follow a "point-to-point" mode. Compared to the carbon black, the graphene has higher conductive efficiency, therefore, it can achieve the whole electrode conductive threshold in less usage, the active materials show better electrochemical performance, and the energy density of the battery is improved. The research of Su et al. [27] shows that the effect of graphene as a conductive additive is closely related to its addition amount. The graphene can well form the conducting network under the smaller addition amount, and the effect is far better than that of carbon black [28]. However, in the case of a large amount, graphene with thick piece might hinder the Li^+ diffusion and lead to the lower ionic conduction of electrode. Zhang et al. [29] research the graphene as a conductive additive in $Li_4Ti_5O_{12}$ anode, and discusses its conductive threshold in detail. When the dosage of graphene is 5%, and the electrochemical properties of $Li_4Ti_5O_{12}$ is better than that of the 15% carbon black. Using the concept of the distance between the particles, Zhang et al. simulate that the graphene conductive threshold is 0.54%, lower than the carbon black one order of magnitude, showed good effect of graphene conductive additive from the perspective of quantitative. However, as the content of graphene continues to increase, more than 10%, the diffusion coefficient of lithium ion decreased due to the severe graphene agglomeration. The theoretical analysis also shows that the amount of graphene conductive additive is closely related to the size and layer of graphene, and the graphene as a conductive additive, should have larger size and thinner thickness.

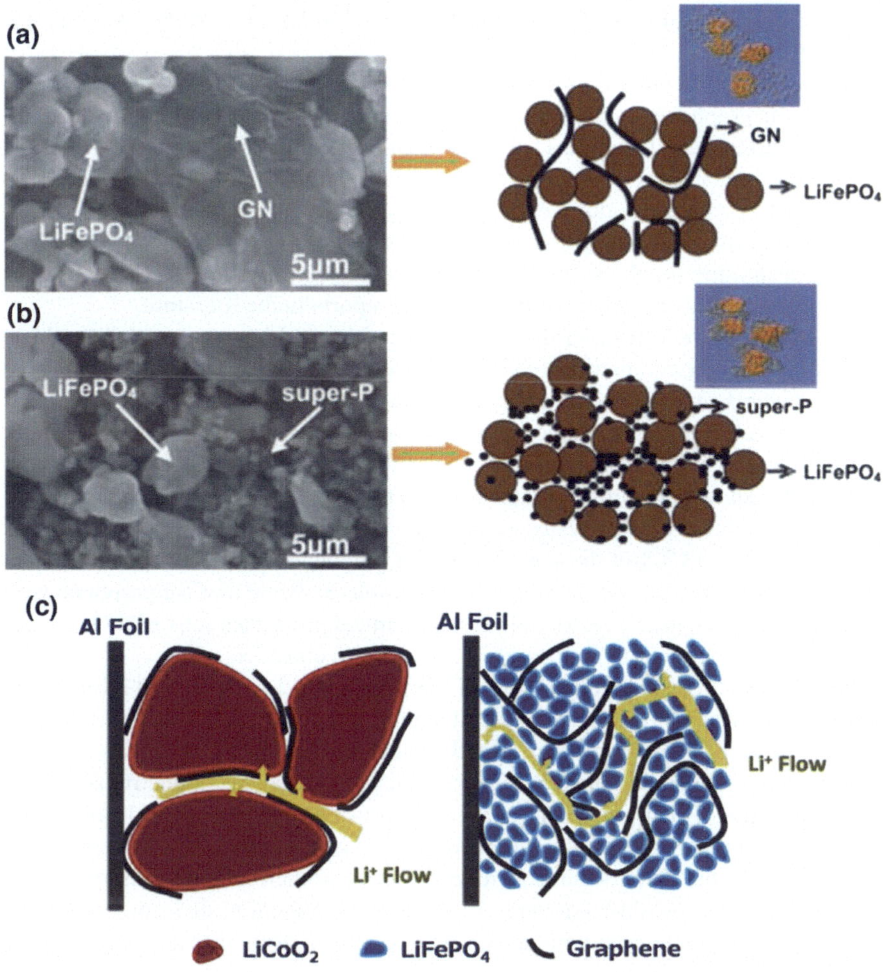

Fig. 7.3 Schematic representations of conducting mechanisms of GN and SP as conductive additives in C (**a** plane-to-point mode for GN case, **b** point-to-point mode for SP case); **c** schematics of the Li^+ transport paths in different size of cathode systems utilizing GN as the conductive additive. **a**, **b** was reproduced with the permission from Ref. [28], Copyright © 2010 Royal Society of Chemistry; **c** was reproduced with the permission from Ref. [37], Copyright © 2016 Elsevier

The use of graphene as conductive additives can largely reduce the dosage of conductive additives, thus effectively improve the volume energy density of LIBs. At present, the requirement for the volume energy density is far more than the mass energy density in the LIBs. Conductive additives can't provide capacity in the battery, but due to its smaller density, even if the small dosage will occupy a large electrodes space, to a great extent, reducing the volume energy density of the whole system. For example, the density of carbon black is 0.4 g cm^{-3}, generally is far less than the 2.0–2.3 g cm^{-3} of $LiFePO_4$ and 3.8–4.0 g cm^{-3} of $LiCoO_2$. In theory, every 1% reduction of carbon black is equivalent to increase about $LiFePO_4$ 5% of or 7–10% of $LiCoO_2$, can significantly improve the volume energy density of the whole system.

When the graphene prepared by different synthesis was used as conductive additives, their effects are different. The group of Bi [30] compared the conductive effect of graphene prepared by oxidation-reduction and chemical vapor deposition (CVD) method as the conductive additives in $LiFePO_4$ system. The results indicate that using CVD method for preparing graphene has a relatively good performance. Moreover, in the industry, the graphene prepared by liquid stripping method has been used as conductive additives in LIBs, but there are few reports in the literature.

In addition to the effects of different preparation methods, the size of graphene layers will affect the conduction. From the view of conductive threshold, the graphene with big layer is easier to produce the conducting network within the electrode. So, it is generally believed the graphene amount is less when graphene layer is larger under the same electrochemical properties. However, the group of Liu et al. [31] finding that the $LiFePO_4$ showed better performance when using the graphene with the diameter <5 μm as conductive additives (e.g. the discharge specific capacity of $LiFePO_4$ is 112 mA h g^{-1} at 5 C). But the same amount of graphene with the diameter >10 μm affect inversely, only has 79 mA h g^{-1}. This results indicate that for graphene with large-size, the influencing factors of electrochemical properties for $LiFePO_4$ are not just electronic conduction, ion transport also need to consider.

The charge and discharge reactions of LIBs need the electronics and lithium ions to reach the surface of the active material at the same time. Due to the barrier effect of graphene nanosheets to ion transport, the battery based on graphene as conductive additives has a slower ion transport than that of using traditional conductive additives, and polarization effect significantly increased inside the battery, so leading to a low capacity. However, with the rate increasing, the demand of Li^+ transport speed is increasing in the electrochemical reaction process, the rate-determining step by electronic conduction shift gradually to the ionic conduction, therefore, when the batteries use the graphene as conductive additives, the performance decreases rapidly in the large current densities.

Graphene nanosheet can cause the hindering effect to Li^+ transport, it has a little effect to the energy type batteries, but cannot be ignored in the power type batteries. The main factors that affect the process are the length of diffusion path and the tortuosity of transport path [32], and reflected to the electrochemical environment of the LIBs, corresponding to the thickness of the electrode and the size ratio of the active material particles and graphene, respectively [33, 34]. Studies have shown that graphene produce a small effect to lithium ion transport within the whole electrode under the electrode piece with thin thickness, however, in the thicker electrode, the effect is noteworthy [35]. In order to reduce the hindering effect of Li^+ transport lithium ion transport caused by the graphene, the graphene striping and surface porosity can decrease the Li^+ transport resistance [36]. Moreover, simulation results shown that the different size ratio of the graphene and active material can affect the tortuosity of electrode porosity, thus increasing the path of the Li^+ transport. When the size of graphene layer less than or equal to the active material, the Li^+

hindering effect caused by graphene can be neglected; however, when the former is significantly greater than the latter, the tortuosity of transport path is large. This mean that the size of the graphene was significantly less than the active materials when the graphene was used to power type batteries. In addition, the group of Yang also indicated this conclusion by experiment in the system of micro $LiCoO_2$ and nano $LiFePO_4$ [35, 37], and they put forward the model diagram of Li^+ transport in different size ratio of graphene and active materials (Fig. 7.3c). In the system of the $LiCoO_2$, the effect of graphene to lithium ion transport is not obvious in charge and discharge process, graphene not obvious influence on lithium ion transport behavior, while in $LiFePO_4$ system, graphene can cause the barrier effect to Li^+ transport [33], and the active material is the key factor to Li^+ transport path in the cathode.

Furthermore, the dispersity of the graphene is a key factor for the performance in LIBs using graphene as a conductive additive [38]. The role of conductive additive is to build the conductive network for electrons transport in the electrode. However, the structural advantage of the gentler, thinner and thicker is difficult to full play if the conductive additives cannot well disperse. The result is that the utilization rate of graphene greatly reduce, and the effective conductive network is difficult to build [39]. Due to the graphene with a big size, it can be influenced by van der Waals forces, easy to agglomerate together. Generally, ultrasonic pretreatment can significantly improve the effect of the dispersion of graphene conductive additives.

7.2.4 Multiple Conductive Additives

Compared with single conductive additives, multiple conductive additives can take advantages of two or more conductors. Therefore, multiple conductive additives usually show more superiority performance than single conductive additives, due to their synergistic effects. In fact, in the preparation processes of LIBs, also tend to use multiple conductive additive and build conducting networks in different scales of the electrode, utilizing the advantage of different conductive additive and improving the performance of batteries in a large extent [40–43]. Compared with the single conductive additive, the multiple conductive additive has a better effect, because the conductive additive with different sizes can build the conducting network in different scales of the electrode [44, 45].

Micro-sized graphite is easily dispersed with cathode active materials, but cannot form a valid conductive network with a small dosage. When nano-sized SP or like is also added, a valid conductive network is usually formed. Then the cathode has improved cyclic life and higher discharge capacity. This study had been done by Cheon et al. [41] and Hong et al. [46]. Zhang et al. [47] used binary conductive additives as conductive additives to improve the electrochemical performance of $LiCoO_2$ composite cathode, and binary conductive additives formed by sonication of mesoporous carbon (MC) and AB. The results showed that the electrochemical performance of a multiple conductive additives loaded $LiCoO_2$ composite cathode was better than that of a cathode loaded with AB. The possible mechanism is that the MC can adsorb and retain electrolyte solution, which allows an intimate contact between the Li^+ and the cathode active material $LiCoO_2$ due to its large mesoporous specific surface area. Figure 7.4a is a schematic view of the $LiCoO_2$ composite cathode loaded with multiple conductive additives.

As we all know, 1D carbon nanomaterials easily form a conductive network. But 1D carbon has less contact point with cathode active materials compared with particulate carbon. If particulate carbon and fiber-like carbon are mixed together to form multiple conductive additives, they will take advantages of the two kinds of conductors. Kang et al. [48] and Utsunomiya et al. [49] used mixtures of carbon black and vapor-grown carbon fiber (VGCF) as conductive additives, and obtained a high capacity at a higher current density than VGCF and AB themselves by improving the electric contact between $Li_{4/3}Ti_{5/3}O_4$ particles and a nickel current collector. Wang et al. [50] and Bian et al. [51] also studied the synergetic effect of binary conductive additives in the cathode of commercial $LiFePO_4$ and $Li_{1.18}Co_{0.15}Ni_{0.15}Mn_{0.52}O_2$. Results suggest that LIBs with a mixture of these two additives have an apparently improved performance in low and high rate discharge capacity compared with the LIBs with a single component additive with the same mass fraction, and construct an effective conductive network for electron transport. In addition, incorporation of carbon black and VGCFs can improve the electrochemical kinetics of the cathode material by retarding the harmful side reactions, promoting the charge transfer reactions and increasing the apparent lithium diffusion coefficient. As shown in Fig. 7.4b, the sketch of the composite electrode prepared by the CB/VGCFs binary conductive additive.

The contact modes between graphene and carbon black have better complementary effect, can establish the long-range and short-range conductive network at the same time inside the electrode. Although graphene conductive additives can build good conductive network under the less usage and greatly improve the conductivity of the electrode, which are specific to each active material particles, graphene nanosheets cannot fully cover the entire particle surface, which lead to the transport

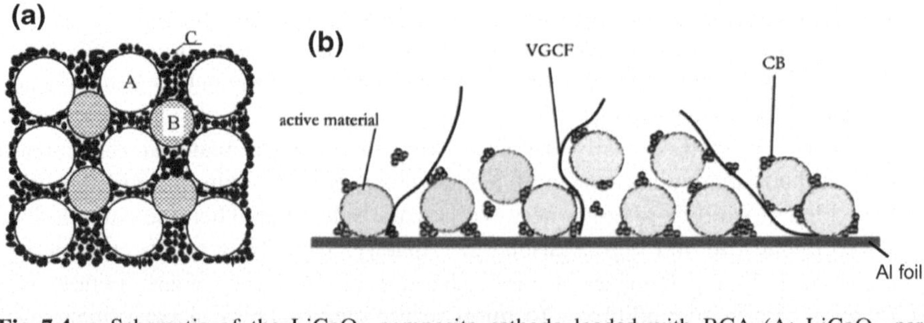

Fig. 7.4 a Schematic of the $LiCoO_2$ composite cathode loaded with BCA (A: $LiCoO_2$ particles, B: mesoporous carbon particles and C: acetylene black particles); **b** schematic of the $Li_{1.18}Ni_{0.15}Co_{0.15}Mn_{0.52}O_2$ electrode prepared by the CB/VGCFs binary conductive additive. **a** Was reproduced with the permission from Ref. [47], Copyright © 2009 Elsevier B.V.; **b** was reproduced with the permission from Ref. [51], Copyright © 2015 Elsevier B.V

lag of electron on the exposed surface [19]. However, if the graphene nanosheets completely coated active material particles, the electrochemical performance of the active material would greatly reduce due to the barrier effect of graphene nanosheets to ion transport [52]. Therefore, when using the graphene as conductive additive, other carbon materials with lower dimensions will be needed to solve the short-range conduction on the surface of particles [53, 54]. Carbon black is 0D carbon nanomaterials, can be evenly attached on the surface of active materials, improving the electron transport on the surface of active material particles [55, 56]. In $LiFePO_4$ and $LiCoO_2$, using multiple conductive additives can significantly reduce the polarization phenomenon in the battery, and compared to only use the graphene, graphene/carbon black multiple conductive additives can significantly reduce the needed dosage of graphene [27, 37]. Jiang et al. [57] found that the specific capacity and cycling performance of $LiMn_2O_4$ were obviously enhanced when graphene nanosheets co-existed with AB, which was attributed to the following two reasons: (i) the planar graphene nanosheets could bridge $LiMn_2O_4$ particles more effectively via a "plane-to-point" conducting mode; (ii) AB particles might serve as the fillings in the electrode and connect the isolated $LiMn_2O_4$ particles to graphene nanosheets through a "filling effect", thereby constructing a novel and more effective conducting network. In this way, the synergy effect between the "plane-to-point" conducting mode (due to graphene nanosheets) and the "filling" mode (due to AB) significantly decreased the charge-transfer resistance of the $LiMn_2O_4$-based electrode.

7.3 Separators for Batteries

The separator is placed between the anode and the cathode in a battery. The separator's properties play a critical role in obtaining optimum cell performance and the inherent safety of the cell. The main role of the separator is to prevent the physical contact between the anode and cathode in the batteries and to prevent it from internal short while serving as the electrolyte reservoir to enable ionic transport. Separators should possess a uniform pore structure (usually in the nanometer-scale) and exhibit a low electrical resistance or high ionic conductivity to achieve good performance. The separator property of a low electrical resistance has a pronounced effect in capacity, rate performance, fast charge capability, overall cell resistance, and cycle life [58]. Low electrical resistances can be achieved when separators absorb and retain big amounts of liquid electrolyte. Even though the separator does not involve directly in any cell reactions, its structure and properties can greatly influence the battery performance, including cycle life, safety, energy density, and power density, through influencing the cell kinetics. A wide variety of factors should be considered while selecting appropriate separators for use in LIBs. An ideal separator should possess high chemical/electrochemical and thermal stability to sustain a long period of operation. High wettability to the electrolyte and good permeability of the separator is also needed for efficient ion transport. The separator should have a proper thickness to achieve a balance between its mechanical properties and Li-ion transport

Table 7.1 General requirements for separators used in LIBs

Parameter	Requirement
Chemical and electrochemical stabilities	Stable for a long period of time
Wettability	Wet out quickly and completely
Mechanical property	>1000 kg cm^{-1} (98.06 MPa)
Thickness	20–25 μm
Pore size	<1 μm
Porosity	40–60%
Permeability (Gurley)	<0.025 s μm^{-1}
Dimensional stability	No curl up and lay flat
Thermal stability	<5% shrinkage after 60 min at 90 °C
Shutdown	effectively shut down the battery at elevated temperatures

properties. Table 7.1 summarizes the general requirements that should be considered for LIBs separators. In the following section, we summary the advanced separators for batteries, including the modification of the traditional separators with nanomaterials and novel nano-sized separators.

7.3.1 Modification of Traditional Separators with Nanomaterials for Batteries

The stability of the separator is directly correlated with the reliability of the battery. Therefore, it is critical to develop separators with high structural stability under working conditions and "self-protecting" separators that can shut down the battery under abnormal conditions. Traditional polyolefin separators, such as polyethylene (PE) and polypropylene (PP), with pore size in the range of micrometers have been commercialized and widely used in LIBs [59]. The performance, porosity, pore size, gas permeability, mechanical/thermal properties, and safety issues regarding commercial PP and PE monolayer separators have been studied by several groups [60–63]. These polyolefin separators are generally chemically stable, highly porous, and low cost. However, polyolefin separators also have inherent drawbacks, such as inferior electrolyte wettability, low thermal stability, poor electrolyte retention, and severe safety concerns, which limits the performance of the LIBs. Furthermore, a stretching process is usually involved in the production of conventional polyolefin separators to generate porous features. Application of the stretched separators at elevated temperatures normally leads to separator shrinkage resulting from the re-organization of the elongated polymer chains. This separator shrinkage may eventually cause battery shortage [64]. To solve these issues, functionalized polyolefin separators were proposed and various methods could be used to modify the structure of nanoporous polyolefin membrane separators. Nanoporous membranes consisting of multilayers of PP and PE have been designed as thermal shutdown separators with a safety advantage.

7.3.1.1 Modification of PE Separators

In the early 1960s, the study found that both dry and wet process can be made of PE microporous membrane, and thus began the study of PE separators. The pore diameter of PE separator is about 30–100 nm, and its porosity is about 30–50%. Celgard 2730 PE separator is made of melt stretching process, its excellent performance has been applied in the LIBs industry [4], its thickness is 20 μm, porosity is 43%, and the melting point is 135 °C. However, due to its low wettability and poor electrolyte retention as well as limited thermal stability. To improve these properties, various nanomaterials could be used to modify the structure of PE nanoporous membrane separators, such as polymer-coated, inorganic nanostructure-coated, inorganic material/polymer hybrid-coated. Table 7.2 shows the properties of modified PE nanoporous membrane separators.

Table 7.2 The properties of modified PE nanoporous membrane separators

Entry	Separator	Thickness (μm)	Electrolyte	Electrolyte uptake (%)	Ionic conductivity (mS cm^{-1})	References
1	PMMA NPs coated PE	25	$LiPF_6$-EC/DEC (1:1 v/v)	200	0.52	[70]
2	PVDF-HFP/PEGDMA coated PE	–	$LiClO_4$-EC/DEC (1:1 v/v)	125	0.38	[71]
3	PVDF-co-CTFE coated PE	19	$LiPF_6$-EC/DMC/EMC (1:1:1 v/v/v)	–	0.73	[72]
4	Al_2O_3 coated PE	23	–	70	1.12	[76]
5	ZrO_2 coated PE	–	$LiPF_6$-EC/EMC (3:7 v/v)	118	0.61	[77]
6	SiO_2 grafted PE	8	$LiPF_6$-EC/DMC/EMC (1:1:1 v/v/v)	–	0.45	[78]
7	TiO_2 grafted PE	8	$LiPF_6$-EC/DMC/DEC (1:1:1 v/v/v)	–	0.50	[79]
8	SiO_2@PMMA NPs coated PE	25	$LiPF_6$-EC/DMC/DEC (1:1:1 v/v/v)	89.5	1.08	[80]
9	ZrO_2/PAA coated PE	–	$LiPF_6$-EC/DMC/EMC (1:1:1 v/v/v)	325	0.51	[81]
10	PEDOT-PEG)/AlF_3 coated PE	–	$LiPF_6$-EC/DEC/EMC (3:2:5 v/v/v)	–	0.32	[82]

Many patents for PE-PP bilayer and PP-PE-PP trilayer membranes have been reported by separator manufacturers. In a multilayer structure, the PE layer melts and blocks the pathway of ions when the temperature is close to the melting temperature of PE. At the same time, the PP layer still retains its dimensional structure and mechanical strength, preventing the short circuit between two electrodes. Hence, such a multilayer structure can provide the safety assurance while maintaining sufficient mechanical strength. The shutdown characteristics of commercial trilayer membranes are often studied by measuring the impedance of batteries as a function of temperature. In addition to the shutdown function, it was also reported that multilayer membranes of PE and PP have higher puncture strength than the corresponding PE or PP monolayer membranes.

In spite of the properties of PE-PP and PP-PE-PP separators have been improved, they are still many defects. Gel polymer electrolyte (GPE), which is attractive for LIBs due to its high ionic conductivity and safety, has been studied most extensively [65–68]. However, GPE has not been commercialized yet because it cannot attain adequate mechanical properties to support the winding tension of anode/GPE/cathode in an assembly process without a separator. One of the ways to solve this problem is to use a mechanical support such as PE to prepare supported GPE [69]. Park et al. [70] developed a novel GPE-coated separator, which is based on introduction of close-packed poly(methyl methacrylate) nanoparticle arrays (PMMA NPs) onto a PE separator. Thanks to the ordered arrangement of the PMMA NPs, the ion transport in the composite separator was enhanced. As a result, the battery with the composite separator had a high rate capability (Table 7.2 entry 1). Sohn et al. [71] reported that poly(vinylidene-fluoride-*co*-hexafluoropropylene)/poly(ethylene glycol) dimethacrylate (PVDF-HFP/PEGDMA)-coated PE separators were prepared successfully by a dip coating at different humidity levels (0–50%), followed by electron beam irradiation. Nanoporous structures generated during dip-coating process at high humidity levels (above R.H. 25%), PEGDMA content, and electron beam irradiation dose were found to be important in determining liquid electrolyte uptakes, ion conductivity, and thermal shrinkage of the prepared separators. The highly porous separators showed an enhanced electrolyte uptake and ionic conductivity (Table 7.2 entry 2). In addition, Lee et al. [72] prepared nanofiber-coated membrane separators by electrospinning polyvinylidene-fluoride-co-chlorotrifluoroethylene (PVDF-co-CTFE) nanofibers onto PE nanoporous membrane substrate. The nanofiber coatings on the PE substrate formed an interconnected fibrous structure with average fiber diameters in the range of 129–134 nm. The nanofiber-coated membrane showed higher electrolyte uptake capacities than uncoated membrane. The ionic conductivity of liquid electrolyte soaked membrane was also improved by the coating of nanofibers on the membrane substrate. Moreover, the cells assembled with nanofiber-coated membrane exhibited lower interfacial resistance than those with uncoated membrane (Table 7.2 entry 3).

Besides polymer nanomaterials, metal oxide NPs were used to enhance the wettability and mechanical properties of the PE separator [73–75]. For example, Al_2O_3 NPs were commonly used for coating of polyolefin separators through a wet coating procedure (Fig. 7.5c). Coating of a thin Al_2O_3 layer on a PE separator can dramatically improve both the thermal stability and electrochemical performance of the separator [76]. The coating was reported to efficiently suppress the deformation of the separator at elevated temperatures and improve the battery performance (Table 7.2 entry 4). Kim et al. [77] fabricated ZrO_2 coatings on PE separators; the coated separator showed improved battery performance (Table 7.2 entry 5). Interestingly, it was observed that hydrated ZrO_2 is more favorable for the construction of the coating. The reason for this lies in the fact that the hydrated ZrO_2 can increase the nanopore content around the NPs, thus facilitating the ion transport within the separator. Coating of inorganic nanostructures on the PE separator was usually performed by coating of the as-synthesized inorganic NPs onto the separator. Commonly used polymer binders, such as PVDF, generally have low thermal stability and thus may affect the high temperature performance of the separator. While, binder-free coating of the inorganic nanostructures is beneficial for reducing the interfacial resistance between the nanoparticle coating and the substrate. Modification of the separator through the "grafting to" approach is a facile and feasible way to achieve this goal. Zhu et al. [78] reported electron-beam induced immobilization of vinylsilane species onto PE separators and subsequent grafting of SiO_2 onto the separator through a sol-gel process (Table 7.2 entry 6). This modification endows the separator with excellent thermal stability without sacrificing its ion transport capability and is therefore favorable for the coating modification of the separator. Compared to the coating with polymer binder, grafting of the NPs onto the polyolefin substrate is more beneficial to decrease the interfacial resistance of the separator. TiO_2 NPs were also grafted onto PE separators using an isopropyl trioleyl titanate linker through electron-beam induced grafting [79]. The nano-sized TiO_2 grafting significantly increased the hydrophilicity of the separator without influencing its morphology. As a result, the ionic conductivity of the separator increased and the battery performance was enhanced (Table 7.2 entry 7).

In addition, inorganic material/polymer hybrid coatings combine the benefits of both the inorganic and polymer phase. Therefore, hybrid coatings were expected to endow the separators with excellent mechanical stability at high temperatures and good affinity to the electrolyte. SiO_2@PMMA NPs were reported to enhance both the mechanical stability and the wettability of the separator [80]. The mechanical stability of the separator was enhanced owing to the presence of the SiO_2 core, while the wettability of the separator was improved because the PMMA shell has a higher affinity to the electrolyte compared to the pristine PE separator. The SiO_2@PMMA-coated PE separator showed both improved rate capability and safety performance compared to the PE separator (Table 7.2 entry 8). A ZrO_2/poly(acrylic acid) (PAA) coating was assembled onto a plasma-activated PE separator through a layer-by-layer process [81]. The ZrO_2/PAA coating was crosslinked with the PE substrate by carbodiimide crosslinking. The coating thickness could be tuned simply through changing the cycles of the layer-by-layer process (Table 7.2 entry 9). Generally, the composite separator exhibited better electrochemical performance and stability. Hybrid coatings that consisted of a functional polymer and inorganic NPs take the advantage of both structural moieties. For instance, a thin poly(3,4-ethylenedioxythiophene)-co-poly(ethylene glycol) (PEDOT-PEG)/AlF_3 hybrid coating was applied onto a PE

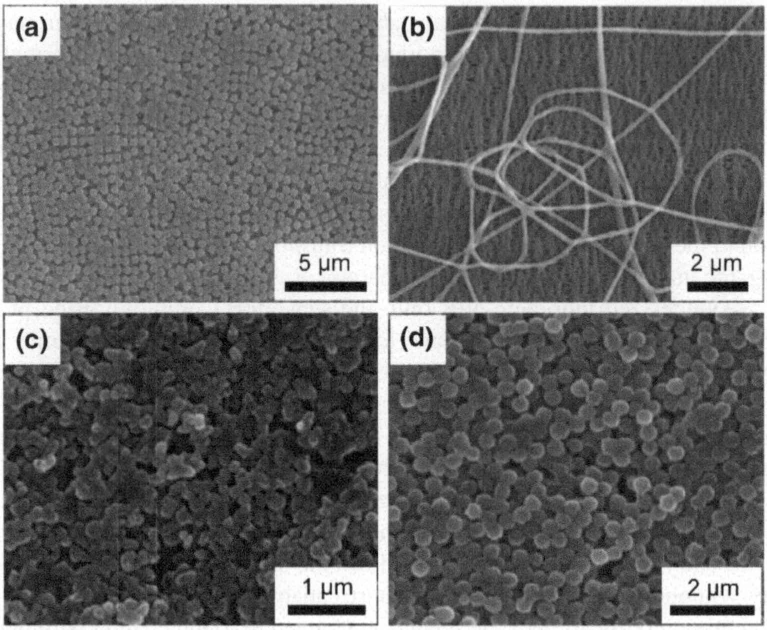

Fig. 7.5 Surface SEM images of (**a**) the closely packed PMMA-coated PE separator, **b** the electrospun PVDF-CTFE-coated PE separator, **c** an Al_2O_3-coated PE separator, and **d** the SiO_2@PMMA-coated PE separator. **a** Was reproduced with the permission from Ref. [70], Copyright © 2011 Elsevier B.V.; **b** was reproduced with the permission from Ref. [72], Copyright © 2014 Springer Nature; **c** was reproduced with the permission from Ref. [76], Copyright © 2014 Elsevier B.V.; **d** was reproduced with the permission from Ref. [80], Copyright © 2015 Elsevier B.V

separator [82]. The PEDOT-PEG phase in the coating is ionic conductive, which is beneficial for the electrochemical performance of the separator. The AlF_3 in the coating could enhance the thermal stability of the separator and stabilize the electrolyte. Owing to the synergistic effect of the two components, the coated PE separator showed suppressed thermal shrinkage and improved discharge capacity as well as enhanced cycling performance (Table 7.2 entry 10).

7.3.1.2 Modification of PP Separators

In the early 1970s, the PP nanoporous membrane was formed under tensile state through heat treatment process and specific directional tensile elastic modulus, and was applied to the production of Celgard 2400 PP nanoporous membrane, and its thickness is about 25 μm, porosity is 40%, the melting point is 165 °C [4]. The shock strength of PP is lower than PE at low temperature, but its ductility, tensile strength is superior to PE. So combining with the characteristics of PE and PP, PP-PE blending, PP/PE, PP/PE/PP composite membrane have been developed and applied to the production of LIBs. However, due to its low wetting capability and poor electrolyte retention. To improve these properties, various nanomaterials could be used to modify the structure of nanoporous PP membrane separators, such as polymer-coated, inorganic nanostructure-coated, Inorganic material/polymer hybrid-coated. Table 7.3 shows the properties of modified PP nanoporous membrane separators.

Physical coating is a simple method to modified PP separators. Lee et al. [72] reported that composite membrane separators were prepared by coating PVDF-co-CTFE nanofibers on the surface of PP using the electrospinning technique, due to the excellent electrochemical stability, good adhesion property, and high temperature resistance of PVDF-co-CTFE (Table 7.3 entry 1) [83–86]. Man et al. [87] present a new approach named hyperthermal hydrogen induced cross-linking technology to increase the electrolyte-affinity of PP separators by covalently cross-linking a thin layer of poly(ethylene oxide) (PEO) on surface-inert PP separators. PEO was selected as the surface modifier due to its higher polarity and good affinity with the electrolyte. As proved by solvent rinsing tests, contact angle measurements and Fourier transform infrared spectroscopy, a PEO coating was found firmly fixed on the separator surface, which results in significantly improved wetting with the electrolyte (Table 7.3 entry 2). Moreover, Hu et al. [83] reported that novel aramid nanofiber (ANF) coated PP separators for LIBs were successfully fabricated by dip-coating

Table 7.3 The properties of modified PP nanoporous membrane separators

Entry	Separator	Thickness (μm)	Electrolyte	Electrolyte uptake (%)	Ionic conductivity (mS cm^{-1})	References
1	PVDF-co-CTFE coated PP	18	$LiPF_6$-EC/DMC/EMC (1:1:1 v/v/v)	–	0.75	[72]
2	PEO coated PP	–	$LiClO_4$-EC/DMC (1:1 v/v)	–	–	[87]
3	Aramid NFs coated PP	–	$LiPF_6$-DMC/EC/PC (2:1:1 w/w/w)	124.5	0.69	[83]
4	Poly-A15C5 coated PP	–	$LiPF_6$-EC/DEC (1:2 v/v)	–	–	[88]
5	Al_2O_3/PVdF-HFP coated PP	26	$LiPF_6$-EC/DEC (1:1 v/v)	–	0.795	[89]
6	Al_2O_3/PEK coated PP	–	$LiPF_6$-EC/DEC (1:1 v/v)	380	1.35	[90]
7	SiO_2/PVDF coated PP	83	$LiPF_6$-EC/DEC (1:1 v/v)	646	0.747	[91]

a surface cationized PP separator in ANF dispersions. The thermal stabilities of the ANF-coated separators were significantly enhanced over those of the non-ANF-coating separators. The ANF-coated separators can provide better rate capacity and superior cycling performance over that of the pristine PP separator. More importantly, the cells fabricated with the ANF-coated separator can withstand high temperature shock in a hot oven test. In a word, coating porous polyolefin separators with thin ANF layers provides a new strategy to enhance the safety and reliability of LIBs (Table 7.3 entry 3). More interestingly, polymers bearing a macrocycle component were used to endow the commercially available polyolefin separator with barrier properties against Mn species. Such as, poly-[divinylbenzene-(vinylbenzyl-aza-15-crown-5)-vinyl-benzyl chloride] nanoparticles were synthesized and coated onto the polyolefin separator through filtration [88]. The modified separator effected a 39% reduction in capacity loss rate during cycling at 50 °C with 100% depth of discharge at C/5 rate, and was able to chelate Mn species and decrease the Mn species on the graphite electrode by ~50%. As a result of this effect, the battery with the modified separator showed slightly enhanced initial discharge capacity as well as significantly improved cycling stability at both ambient and elevated temperature (Table 7.3 entry 4).

Similar to the PE, metal oxide NPs were also used to enhance the wettability and mechanical properties of the PP separator. Coating of inorganic nanostructures on the polyolefin separators was usually performed by coating of the as-synthesized inorganic NPs onto the separator. Commonly used polymer binders, such as PVDF, generally have low thermal stability and thus may affect the high temperature performance of the separator. It was found surprisingly that Al_2O_3/PVDF-HFP composite coated PP has a lower dimensional stability at 150 °C than the pristine PP separator [89]. This finding has inspired the research of the effect of the polymer component of the coating on the thermal stability of the separator (Table 7.3 entry 5). Wang et al. [90] used phenolphthalein polyetherketone (PEK) as the binder instead of PVDF to prepare Al_2O_3-coated separator. The PEK has a high glass transition temperature up to 230 °C. The high temperature performance of the battery was significantly enhanced when PEK was used as the binder (Table 7.3 entry 6). Besides, SiO_2/PVDF hybrid NF coating was also applied on a PP non-woven separator to further enhance its performance [91]. The composite separator exhibited superior thermal stability and mechanical strength. Moreover, the wettability of the composite separator was high owing to the introduction of inorganic fillers in the composite (Table 7.3 entry 7).

In addition, coating of traditional separators with an additional barrier layer is a simple and straight forward approach to alleviate the shuttle effect in Li–S batteries. Nano-scale carbonaceous materials were also used to coat polyolefin separators Li–S batteries. Carbon-based coatings have been demonstrated to show excellent barrier properties against polysulfide species. Both physical trapping and chemical interactions with the polysulfide species can be realized with carbon coatings. Commercially available and low-cost SP carbon black was used to improve the battery performance of the PP separator. Li–S battery constructed with a SP-coated PP separator and S cathode showed superior discharge performance [92]. Additionally, the self-discharge of the Li–S battery can be well suppressed. Other commercial carbon materials, such as KB, have also been used to prepare the coated separator to suppress the shuttle effect in Li–S batteries [93]. The porous features of the carbon coating layer influences the battery performance of the separator. Balach et al. fabricated a MC and coated the MC onto a PP separator [94]. Ex situ XPS analysis of the MC coating after cycling demonstrated that LiPS species were trapped within the carbon coating. The composition of the MC coatings on the separator also influences its Li–S battery performance. It was shown that doping of the heteroatom in the carbon coating can promote the interaction between the polysulfide species and carbon coating, thus leading to an improved barrier properties [95–97]. To improve the polyolefin separator's barrier properties without affecting its ion conductivity, other forms of carbon coatings were developed for the modification of the polyolefin separator. Graphene and CNTs are emerging carbon materials with a number of outstanding physical and chemical properties [98–100].

7.3.2 New Separators for Batteries

Apart from the conventional polyolefin separators, the other polymers and inorganic materials can also serve as the separator for batteries. Separators based on polymers with high affinity to electrolytes have been suggested as potential alternatives to traditional separators. While, inorganic separators are more favorable than the polymer-based separators in terms of safety.

7.3.2.1 Separators Based on Other Polymers

Separators based on polymers with high affinity to electrolytes have been suggested as potential alternatives to traditional separators. PVDF can adsorb a large quantity of electrolyte, which is beneficial for the battery performance [101]. It was shown that an electrospun PVDF membrane with proper annealing possessed a lower interfacial resistance and higher discharge capacity than the commercial PP membrane (Fig. 7.6a, b, Table 7.4 entry 1) [102]. The annealing process can improve the mechanical properties of the PVDF separator as well as the battery performance. PVDF can be also copolymerized with hydrophilic polymers, such as HFP, to further increase its affinity to the electrolyte [103]. Despite its high affinity to electrolyte, PVDF separators generally have an intrinsically low tensile strength and thermal stability. On one hand, the low thermal stability of PVDF can be used for the creation of trilayer separators with thermal shut-down function [104]. On the other hand, PVDF can be composited with other materials to yield separators with high stability. Poly(m-phenylene isophthalamide) (PMIA), which shows both excellent mechanical property and thermal stability owing to the presence of benzene-amide chains in the molecule, was used to construct a tri-layer PVDF/PMIA/PVDF separator (Fig. 7.6c, d) [105]. Taking advantage of PMIA, the thermal stability of the composite separator was significantly enhanced, suppressing that for the Celgard 2320. By virtue of the PVDF phase, the composite separator had high ionic conductivity. Additionally, the

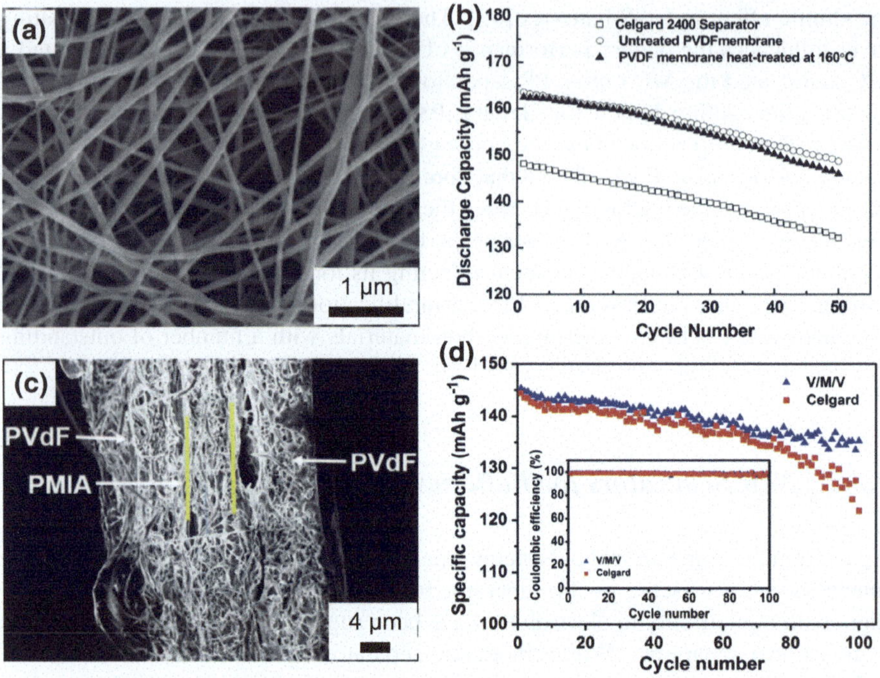

Fig. 7.6 **a** The SEM of electrospun PVDF fibrous membranes, **b** correspondingly cycle performance of Li/$LiFePO_4$ cells with different separators; **c** the FE-SEM image of the cross-section view of PVDF/PMIA/PVDF composite membranes, and **d** cycle performance of the cells assembled with the composite membranes and Celgard membrane at 0.2 C rate. **a**, **b** Were reproduced with the permission from Ref. [102], Copyright © 2013 Elsevier B.V.; **c**, **d** were reproduced with the permission from Ref. [105], Copyright © 2014 Royal Society of Chemistry

battery based on the PVDF/PMIA/PVDF separator exhibited better discharge capacity and cycling stability than the cell with the Celgard 2320 separator (Table 7.4 entry 2). To prevent undesired Mn diffusion, Banerjee et al. introduced a polymer component comprising iminodiacetic acid disodium salt into a porous PVDF-HFP separator [106]. The introduced polymer chelates Mn^{2+} and Mn^{3+} effectively and the separator with an optimal chelating polymer content could reduce the Mn species deposited onto the Li electrode by 90%. In addition, the polymer component helps to enhance the wettability/conductivity of the separator toward electrolytes. Owing to this synergistic effect, the composite half-cell with chelating polymer-modified PVDF-HFP separator and $LiMn_2O_4$ cathode exhibited higher first-cycle discharge capacity and much improved cycling performance at elevated temperatures than its counterpart with a pristine PVDF-HFP separator (Table 7.4 entry 3). Owing to its ease of preparation and relative low cost, this multi-functional separator holds great potential in practical battery applications.

Table 7.4 The properties of modified PP nanoporous membrane separators

Entry	Separator	Thickness (μm)	Electrolyte	Electrolyte uptake (%)	Ionic conductivity (mS cm^{-1})	References
1	PVDF fibrous	–	$LiPF_6$-EC/DMC (1:1 v/v)	380	1.35	[102]
2	PVDF/PMIA/PVDF	25	$LiPF_6$-EC/DMC/EMC (1:1:1 v/v/v)	–	0.81	[105]
3	P(VDF-HFP)	–	$LiPF_6$-EC/DMC (1:1 v/v)	413	3.1	[106]
4	PVA	28	$LiPF_6$-EC/DEC (1:1 v/v)	230	1.20	[107]
5	PPESK/PVDF/PPESK	182	$LiPF_6$-EC/DMC (1:1 v/v)	1079	5.94	[108]
6	PI@PVDF-HFP	35	$LiPF_6$-EC/DMC (1:1 v/v)	470	1.68	[110]
7	PAEK-PLA	35	$LiPF_6$-EC/DMC/DEC (1:1:1 v/v/v)	120	1.987	[111]
8	Cellulose nanofiber	10	$LiPF_6$-EC/DEC (1:1 v/v)	–	0.75	[112]

(continued)

Table 7.4 (continued)

Entry	Separator	Thickness (μm)	Electrolyte	Electrolyte uptake (%)	Ionic conductivity (mS cm^{-1})	References
9	IO separator	20	$LiPF_6$-EC/DEC (1:1 v/v)	–	1.0	[117]
10	Glass fiber	82	$LiPF_6$-EC/DEC/FEC (1:1:0.04 v/v/v)	–	–	[118]
11	GF/PI	46	$LiPF_6$-EC/DEC (1:1 v/v)	210	0.38	[119]
12	Al_2O_3 NWs	50	$LiPF_6$-EC/DMC/DEC (4:4:2 v/v/v)	190	1.7	[120]
13	boehmite NPs	22	$LiPF_6$-EC/DEC (3:7 v/v)	–	1.72	[121]
14	$MgAl_2O_4$ NPs	30–50	$LiPF_6$-EC/DEC (1:1 v/v)	81	1	[122]
15	$MgAl_2O_4$ NF	–	$LiPF_6$-EC/DEC (1:1 v/v)	415	2.80	[123]
16	MMT	70–80	$LiPF_6$-EC/DEC (1:1 v/v)	251	0.901	[124]

Poly(vinyl alchol) (PVA) is a highly hydrophilic polymer that adsorbs a large amount of electrolyte. A nanoporous PVA separator was prepared using non-solvent-induced phase separation [107]. The contact angle of the PVA separator toward the electrolyte is 8.58°, significantly lower 42.48° for the commercial PP separator. Additionally, the PVA separator possessed improved thermal stability relative to the PP separator. These features make the PVA separator a promising candidate for the high performance LIBs (Table 7.4 entry 4).

New separators based on robust polymer materials have been proposed to hold great promise for the development of LIB separators with enhanced safety properties. Poly(phthalazinone ether sulfone ketone) (PPESK), with a high glass transition temperature of ~300 °C, was also used to fabricate highly thermal stable separators (Table 7.4 entry 5) [108]. Polymethylphetene (PMP) has both high chemical stability and thermal stability. Nonwoven PMP separators were produced through an electrospinning process. Apart from its good mechanical properties, the as-obtained separator had a high electrolyte uptake and high conductivity. Polyimide (PI), featured by its high mechanical strength, good thermal resistance, and chemical stability, was used to prepare high performance separators for LIBs. For example, electrospun PI did not show any deformation after heating at 150 °C for 1 h [109]. Furthermore, the electronspun PI separator had a higher affinity toward the non-aqueous electrolyte compared to the conventional polyolefin separators. The battery with the electronspun PI separator showed a higher discharge capacity and rate capability than the battery using polyolefin separators. To improve the wettability of the PI-based separator, a PI@PVDF-HFP core-shell NF separator was developed (Table 7.4 entry 6) [110]. A coaxial electrospinning process was employed to prepare the PI@PVDF-HFP composite separator. The composite separator combined the higher thermal stability of PI and the excellent electrolyte affinity of PVDFHFP. As a result, the PI@PVDF-HFP composite separator showed improved battery performance compared with commercial polyolefin separators. Polymer separators derived from highly thermal stable polymers, such as poly (arylene ether ketone) (PAEK)/poly(lactic acid) (PLA) composite, were also reported (Table 7.4 entry 7) [111].

The use of abundant, eco-friendly and biodegradable cellulose material for the synthesis of separator has aroused great research interests. Using the homogenized cellulose nanofiber, separator was prepared through a vacuum filtering and pressing procedure [112]. Owing to the presence of closely packed polysaccharide chains in cellulose nanofiber, the cellulose nanofiber-based separator showed improved mechanical properties and porous properties compared to the separator based on macro/micro-sized cellulose. In addition, the permeability, wettability, ionic conductivity, and tensile strength of the resultant separator can be well balanced by controlling the solvent composition in the high pressure homogenization process. The cellulose nanofiber separator showed a high electrochemical stability at a voltage up to 4.5 V versus Li^+/Li and negligible open-circuit voltage drop after 70 h. The cellulose nanofiber separator-based battery also showed similar discharge capacity and cycling performance compared to the battery with a commercial PP/PE/PP separator (Table 7.4 entry 8).

A high performance cellulose separator was recently developed by force spinning of cellulose acetate, followed by hydrolysis in alkaline solutions [113]. Superior electrolyte uptake capability and thermal stability of the novel cellulose separator was demonstrated, thus securing their application in LIBs. A cellulose nanofiber-based separator prepared from a papermaking process also showed remarkable thermal stability and promising battery performance. A PE/methylcellulose composite separator was prepared by a thermally induced phase separation [114]. Incorporation of methylcellulose significantly enhanced the wettability of the composite separator, as revealed by the water contact angle measurement. As a result of the enhanced wettability, the separator exhibited high discharge capacity. However, the tensile strengthen of the composite separator was relatively low. An agarose/polyacrylonitrile composite separator was produced by a dual electrospinning process [115]. Polyacrylonitrile endowed the separator with high wettability, while the agarose could chelate transition metal ions such as Mn^{2+}. Such a combination effectively improved the battery performance. Bacterial cellulose also proved to be an ideal material for the construction of high performance separators for secondary batteries [116]. Owing to an abundance of hydroxyl groups, the bacterial cellulose separator had a high wettability. More importantly, the bacterial cellulose separator had a much higher stability at a higher temperature. Specifically, the novel separator had little shrinkage up to 180 °C, in sharp contrast to the pronounced deformation of Celgard 2325 separator under identical conditions.

Additionally, the morphology of the polymer separator has a great influence on its battery performance. Kim et al. [117] suggested that presence of well-ordered macroporosity in the separator significantly promotes its LIBs performance by facilitating ion transport within the separator (Table 7.4 entry 9). To produce a separator with highly ordered macroporosity, ethoxylated trimethylolpropane triacrylate, which showed good compatibility with silica template, was chosen as the monomer and nonwoven polyethylene terephthalate (PET) was used as the substrate. The battery assembled using this composite separator with tailored morphology showed superior discharge performance and cycling stability.

7.3.2.2 Separators Based on Inorganic Structures

Although the stability of polymer-based separators can be improved through appropriate functionalization, the polymeric nature of the separator still imposes safety concerns. From this perspective, inorganic separators are more favorable than the polymer-based separators in terms of safety. Separators based on pure inorganic materials have been shown. Low-cost glass fibers, which have been widely used as separators in batteries, were recognized to be potential separators for LIBs. Large-scale fabrication of weaved glass fiber separators were demonstrated by the group of Luo et al. [118]. The produced glass fiber separator showed greatly enhanced thermal stability and good mechanical strength compared to commercial PP (Table 7.4 entry 10). Good cell performance of the glass fiber separator was also shown. One of the major drawbacks of glass fiber as a separator is its large pore size (usually >1 μm).

The large pore size of the separator increases the risk of shortage of the battery. To ease this problem, the glass fiber must be thick, which is not favorable for highly efficient ion transport in the separator. To solve this problem, a glass fiber/polyimide (GF/PI) composite separator was developed through a paper making process [119]. The GF/PI separator has a low thickness of ~45 μm. The wettability, thermal stability, electrochemical stability, and battery performance as a result of the GF/PI was significantly improved compared to the commercial polyolefin separator (Table 7.4 entry 11). Al_2O_3-based pure inorganic separators were synthesized by vacuum filtration of the Al_2O_3 NWs derived from a hydrothermal process [120]. The Al_2O_3 separator has a high hydrophilicity and therefore can adsorb a high amount of electrolyte. Additionally, the high thermal stability of the Al_2O_3 separator guaranteed a higher safety performance at elevated temperature (Table 7.4 entry 12). Free standing inorganic separator based on boehmite NPs were also developed as highly thermally stable separators (Table 7.4 entry 13) [121].

Mixed oxides were also used for the development of inorganic separators for battery applications. $MgAl_2O_4$ NPs were synthesized and bonded with PVDF-HFP to produce a binder with a thickness of 30–50 μm [122]. The ratio of the binder in the separator was tuned in the range of 10–30% to investigate the effect of binder on the performance of the separator. The optimal separator exhibited good thermal stability and showed an initial discharge capacity of 80 mA h g^{-1} at 1C in a $Li/LiFePO_4$ cell (Table 7.4 entry 14). The $MgAl_2O_4$ NF separator was produced by electrospinning with PVDF-HFP as the binder. The NF separator also showed promising performances (Table 7.4 entry 15) [123].

A montmorillonite (MMT) separator was prepared by hot pressing with a PVDF-HFP binder [124]. The MMT separator had a high porosity of 40% and an electrolyte uptake of 251%. Although the discharge capacity and cycling performance of the MMT separator-based battery needs to be improved, preliminary battery performance tests show that the MMT separator exhibits high interface compatibility with the electrode (Table 7.4 entry 16).

The use of Li conductive NPs to construct the separator was expected to achieve a separator with high thermal stability and good Li conductivity. A ceramic separator was developed with $LiAlO_2$ and PVDF-HFP as the binder. The obtained separator, with an average thickness of 70–80 μm, showed high thermal stability and acceptable electrochemical performance [125]. Ferroelectric NPs, such as $BaTiO_3$, were also demonstrated for the preparation of inorganic separators with good conductivity [126].

7.4 Current Collectors for Batteries

Current collector is an essential component in rechargeable LIBs, which physically contacts with electrode materials including active materials, binders and conductive additives, and plays a key role in electronically connecting these electrode materials with external circuit [127, 128]. Regardless of multifarious electrode materials, aluminum (Al) foil and copper (Cu) foil are the most well-established current collectors for positive and negative electrode materials in LIBs, respectively, probably due to their high electric conductivity, moderate chemical and electrochemical stability, and more importantly, low cost, etc. [129]. However, both Al foil and Cu foil are susceptible to localized corrosions during long-term cycling [130–132]. Serious consequences can be readily perceived, including the increase of resistance, electric isolations of the active materials, the microshort-circuits induced by resultant Al/Cu fragments, the undesired parasitic side reactions, etc. [130, 133]. Therefore, the corrosions of Al/Cu foil are widely recognized as the nontrivial factor contributing to the degradation/failure of battery performance and reduced service life [131, 134]. Furthermore, the surface characteristics of current collectors are vital to improve the cycle life, gravimetric/volumetric energy densities, and power densities of LIBs, because the optimized surface structure could effectively reduce the additional charge transfer and transport resistances at the interface between current collector and electrode materials [135–139]. In this section, we will mainly discuss the effect of using nanomaterials modified on the surface of the traditional current collectors, and some nanostructured current collectors.

7.4.1 Surface Modification of Al Current Collector

The Al foil current collector is usually very rigid, which afford limited contacts and weak interaction with the electrode. And the anodic corrosions of Al foil are particularly prominent, considering the thermodynamic standard electrode potential of Al^{3+}/Al (1.38 V vs Li^{+}/Li) is much below the normal operating potential of most positive electrode [132]. As a result, the corrosions of Al foil not only deteriorate the long-term performances of current 4 V LIBs but also are generally believed to be a bottleneck for next-generation 5 V LIBs. Furthermore, whatever electrolyte systems are, the corrosion of Al foil as current collector is always a haunting issue plaguing the long-term performance and reducing the service life of LIBs. For the deployment of 5 V LIBs in the future, which is imperative among the battery community to increase the energy density, the case would be further exacerbated [127]. Here, we discuss the modifying methods of Al current collector, including solvent treatment, coating carbon nanomaterial and electrodepositing Al nanorods.

7.4.1.1 Solvent Treatment

The textured Al current collector can be obtained by chemical etching, which is a nontraditional machining process in which material removal is carried out by using a strong chemical solution, called etchant. Lecoeur et al. [139] reported a simple way to obtain a textured Al current collector, which consists of electrochemically oxidizing a bulk Al substrate, using an electrolytic bath composed of 1-ethyl-3-methylimidazolium bis(trifluoromethanesulfonyl)imide ([EMIm]TFSI) and $AlCl_3$ salt as an aggressive agent. This substrate, offering nanoneedles, was used as the current collector in LIBs using $LiFePO_4$ as an active material. Electrochemical tests indicated that $LiFePO_4$ electrodes display excellent capacity retention and high rate capability, thanks to the specifically designed current collector texture.

7.4.1.2 Carbon Nanomaterial Coating

Carbon nanomaterial with high corrosion resistance, good conductivity, low cost and low weight has been widely used to modify the current collectors. Portet et al. [137] reported that the synthesis of CNFs prepared by a catalytic CVD process on a pretreated Al current collector, and it can be found that the CNFs surface treatment is efficient to improve the Al current collector/active material interface properties in terms of conductivity and contact surface. Subsequently, Wu et al. [138] compared the charge/discharge capacity and cycle stability at high rate of $LiFePO_4$ electrodes using three types of Al current collectors, including smooth un-etched Al foil, anodization-etched Al foil, and the etched Al foil covered with a conformal C coating. The results unequivocally demonstrated the strong effects exerted by the surface structure and composition of the Al current collectors on the power performance. In particular, the use of the C coated current collector not only remarkably increases the power-delivering capability, but also greatly enhances its cycle stability under high rate. The enhancements are consistent with observed reduction in overall charge-transfer resistance, which can be attributed to the removal of the native insulating oxide surface layer of the current collector and to the improved adhesion at the active layer/current collector interface.

Moreover, Xiao et al. [140] used the multilayer graphene films as electrically conductive coating layers and interfacial barrier layers to enhance the anticorrosion performances of Al foil. In detail, the multilayer graphene films, which are directly and conformably grown on industrial Al foil via plasma-enhanced CVD method, served as extraordinary conductive shield that could efficiently thwart the attacks of the anions or other coordinating species to Al^{3+}. As a consequence, the possibility of the coordination between Al^{3+} and other species would be drastically minimized or even disabled thanks to the impenetrable signature of graphene. $LiMn_2O_4$ cells using graphene-armored Al foil as current collectors would demonstrate superior electrochemical performance, including better long-term cycling as well as rate performance and significantly ameliorated self-discharging property.

Apart from in LIBs, carbon nanomaterials coated the Al current collector was also proposed to enhance the redox reactions of polysulfides in Li–S batteries. Based on this consideration, Huang et al. [141] incorporated a graphene-carbon nanotube hybrid (GNH) coating layer onto the Al foil to construct an artificial interface between the current collector and electrode materials (Fig. 7.7b). The GNH materials are highly conductive and mechanically stable. The hybrid structure also renders GNH materials self-dispersible properties to maintain the highly porous structures. The working Li–S batteries with artificial interfaces between the Al foil and the electrode layers demonstrate not only high sulfur utilization and long cycling life,

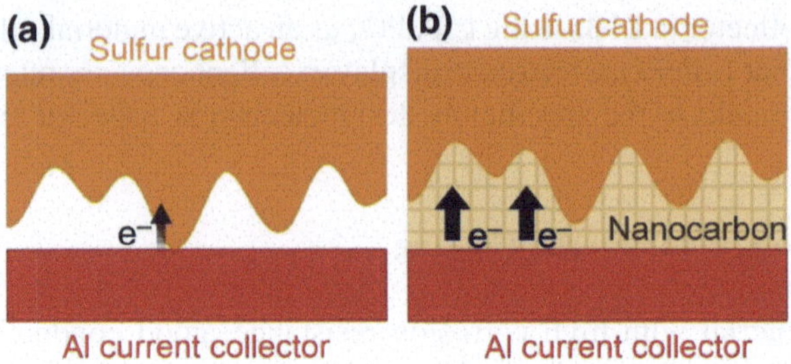

Fig. 7.7 Schematic of the GNH coating layer as an artificial interface between the Al foil and sulfur cathodes. Reproduced with the permission from Ref. [141], Copyright © 2017 Elsevier B.V. and Science China Press

but also unexpected low over-potential for rapid redox reaction of polysulfides and high energy efficiency of a Li–S battery.

7.4.1.3 Al Nanorod Electrodepositing

Electrodeposition is an effective way to prepare a nanoarchitectured Al current collector to host positive electrode materials. Perre et al. [136] reported the mechanism for electrodeposition of Al on a planar Al substrate and how this can be extended to the deposition of free standing Al nanostructured rods using porous alumina as template with the goal of producing current collectors for the positive side in 3D microbatteries. Lecoeur et al. [135] reported a template-free method to electrodeposit structured aluminum onto aluminum substrates via the use of four different imidazolium-based ionic liquids. The pulsed electrodeposition method utilizes an electrolytic bath composed of the ionic liquids and $AlCl_3$ as the source of aluminum. These deposits were used as current collectors in LIBs using $LiFePO_4$ as an active material. Electrochemical tests confirmed the well-known capacity retention for $LiFePO_4$ electrodes and reveal the feasibility to achieve high rate capability.

7.4.2 Surface Modification of Cu Current Collector

Similarly, the optimized Cu surface structure could effectively reduce the additional charge transfer and transport resistances at the interface between current collector and electrode materials. In this section, we discuss the modifying methods of Cu current collector, including coating oxide and carbon nanomaterial and electrodepositing Cu nanorods.

7.4.2.1 Metal Oxide Nanomaterial Coating

The surface structure and material composition of current collectors significantly affect the electrochemical performances of LIBs. Yuan et al. [142] prepared a high-performance CuO/Cu composite current collector with an array-pattern porous structure using a special chemical etching method (Fig. 7.8a). With this method, array-pattern blind holes were formed on the current collectors and synchronously coat their surfaces with a layer of oatmeal-like CuO, which is significantly different from the traditional methods based on chemical depositing or etching. The prepared samples inherited the properties of porous current collectors and benefited from the morphological advantages of CuO with nanostructures. The new CuO/Cu composite current collector with array-pattern porous structures significantly enhance the electrochemical performances of batteries in terms of reversible capacity, cycling stability, electrical conductivity and coulombic efficiency.

7.4.2.2 Carbon Nanomaterial Coating

Similarly, the carbon nanomaterial was also used to modify the Cu current collectors. For instance, Ke et al. [35] developed a facile and mass-producible strategy named electric discharge method to successfully improve the surface properties of Cu foils with rough carbon layer. This developed carbon coated Cu foil is not only beneficial to enhance the performances of LIBs when used as anode current collector, but also mass-producible due to the facility of the present modification method. The studies in both half and full LIBs indicate that the presence of robust and coarse carbon layer on Cu anode current collector can effectively improve the utilization efficiency of graphite anode materials and reduce the total polarization resistance, which can enhance the discharge capacity, rate capability, cycle life as well as low-temperature performance of LIBs.

CNFs, as a carbon-based composite material matrix, have attracted tremendous attention in the field of energy storage devices, due to their large specific surface area, high electron conductivity, and good chemical and mechanical stability [143]. He et al. [144] reported that CNFs were deposited on Cu foam by a floating catalyst method and a Mn_3O_4 layer was then coated onto the deposited CNFs via a hydrothermal process based on the redox reaction of carbon and potassium permanganate. The obtained architecture of Mn_3O_4-coated CNFs on Cu foam was directly used as an anode for LIBs without using any binder or conducting additive (Fig. 7.8b).

7.4.2.3 Cu Nanorod Electrodepositing

To construct nanoscale electrodes, nanoarchitectured current collector as a substrate is an effective way. Taberna et al. [145] reported that a two-step electrode design consisting of the electrochemically assisted template growth of Cu nanorods onto a current collector followed by electrochemical plating of Fe_3O_4. The nanoarchitectured Cu current collector was made by direct electrodeposition of Cu into the pores of an anodized alumina oxide membrane placed on top of a commercial Cu foil which was polished and cleaned in ethanol before use. This electrode demonstrate a factor of six improvement in power density over planar electrodes while maintaining the same total discharge time, thanks to their unique nanostructure (Fig. 7.8c).

Furthermore, Bai and Liu [146] reported that Cu-coated carbon cloth (Cu/CC) fabrics with protuberance surface have been prepared in the work and evaluated as anode current collectors in LIBs. Firstly, the copper layers are formed on the carbon cloth surface by electrodepositing method. The Cu/CC electrode has a good electrical conductivity and flexibility, which are expected to be used in flexible batteries and silicon based LIBs (Fig. 7.8d).

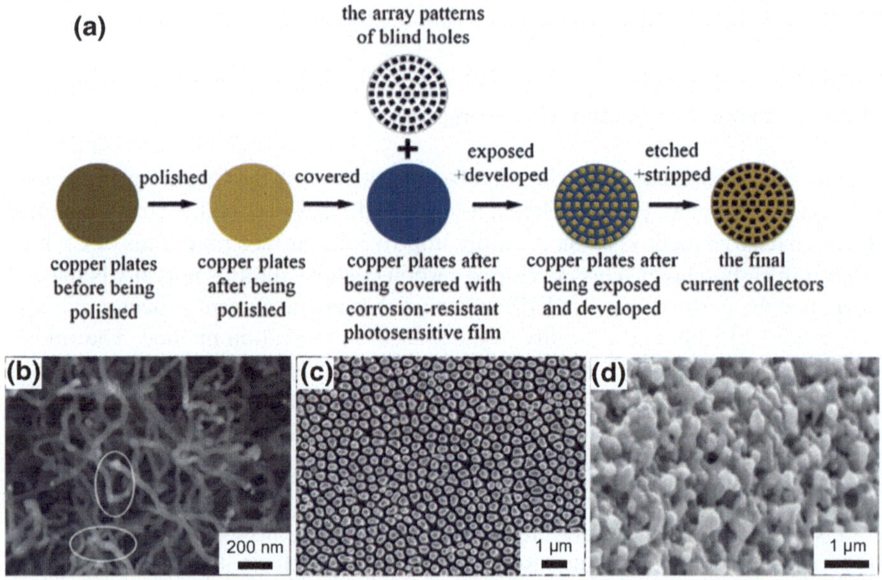

Fig. 7.8 **a** Procedure to prepare the CuO/Cu composite current collector with array-pattern porous structures; SEM images of **b** the current collector of CNFs deposited on Cu foam; **c** top view of the Cu current collector obtained after electrolysis and membrane removal, and **d** Cu/CC current collector. **a** Was reproduced with the permission from Ref. [142], Copyright © 2016 Elsevier Ltd; **b** was reproduced with the permission from Ref. [144], Copyright © 2014 Royal Society of Chemistry; **c** was reproduced with the permission from Ref. [145], Copyright © 2006, Springer Nature; **d** was reproduced with the permission from Ref. [146], Copyright © 2017, Springer Science Business Media, LLC

7.4.3 *Nanostructured Current Collector*

Some novel current collectors, such as 3D porous current collector and carbon-based nanomaterial films, can effectively improve the electrochemical performance of batteries.

7.4.3.1 3D Porous Current Collector

Current collectors based on a completely new design of 3D multifunctional architectures and micrometer-scale roughness that can simultaneously improve the essential reaction interphases for the fundamental energy storage processes, including electronic and ionic conductivity, mass and charge transport, and alleviation of the volume variation, are thus urgently required while still remaining very challenging [147–150] 3D current electrode can ensure a short electron transport length for all the redox-active material. In addition, the nano/microscale current collector is beneficial for uniform current distribution and hence reduces mechanical stresses during repeated charge/discharge processes. Among the 3D materials, porous metals have been considered as promising candidates for the construction of 3D current collectors because of their high surface area and superior electronic conductivity. Up to now, chemical deposition [151], electrodeposition [152, 153], de-alloying [154], and magnetron sputtering/e-beam evaporation [147, 148] are the main methods that have been employed to prepare 3D porous metal current collectors. Moreover, 3D porous polymer current collectors also drawn researchers' attention due to their properties of lightweight and flexibility [155].

As a commercial material with a porous structure, nickel foam provides a large inner space and specific surface area [156]. Although nickel foam is usually used as a negative current collector for LIBs, it is seldom selected as a positive current collector due to the severe corrosion of the $LiPF_6$ salt in organic electrolytes [157–159]. Liu et al. [160] introduced the electroless Ni–P alloy layer to protect the positive current collector in LIBs. The crystal Ni–P alloy coating was prepared by a facile electroless plating technique to cover the surface of nickel foam and subsequently annealed under an Ar atmosphere. The nickel foam, coated with the crystal Ni–P alloy, exhibited excellent stability and cycle life in LIBs, which was attributed to the outstanding corrosion resistance of the crystal Ni–P alloy film in the electrolyte containing the $LiPF_6$ salt. While, Huang et al. [161] reported a dendritic Ni–P-coated melamine foam, this 3D porous dendritic current collector (3DDC) with light-weight, low-cost, and robust has high electronic conductivity, abundant interconnected ion/electron and mass transport pathways, and superior electrochemical stability over a wide voltage range (0–5 V). Furthermore, the 3DDC is dendritic and amphipathic, which enables the easy fabrication of a robust, binder-free electrode. As a proof-of-concept application, a binder-free 3DDC/GeO_2 electrode has been successfully obtained that shows superior electrochemical performance even compared to binder-containing conventional electrodes in terms of high specific capacity, cycling stability, and rate capability.

Yang et al. [151] reported that a 3D porous current collector with a submicron skeleton and high electroactive surface area can significantly improve the electrochemical performance of electrode material (Fig. 7.9). Poetz et al. [162] developed a 3D current collector for the positive electrode based on a non-woven PET plated with Al by physical vapour deposition (PVD). The advantages of this new current collector are: reduced weight of inactive components, possibility of higher mass loadings without delamination of active material and reduced effort of metal, which is a critical point concerning sustainability. The advantages of our approach compared to the proposed methods in literature are an easy and economically feasible preparation, an excellent mechanical stability even during physical pressing (calendaring) of the electrode, excellent electrochemical cycling behavior and rate capability of electrodes using this 3D current collector. Furthermore, this new 3D current collector has a smooth surface, and therefore, the problem concerning safety features through separator damaging is negligible.

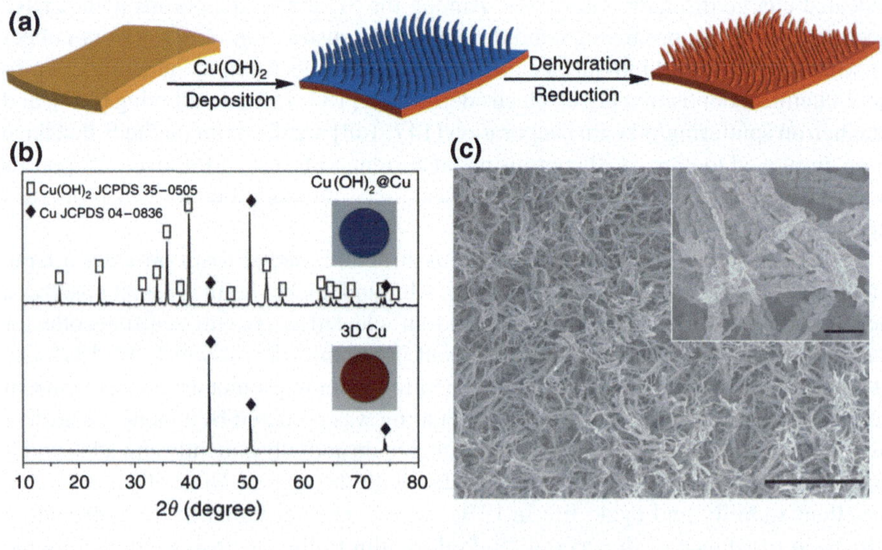

Fig. 7.9 Preparation and characterization for 3D porous Cu foil. **a** Schematic presentation of the procedures to prepare a 3D porous Cu foil from a planar Cu foil. **b** X-ray diffraction profiles of $Cu(OH)_2$ on the Cu foil and the final 3D Cu foil. The insets show digital images of the corresponding samples. **c** SEM images of the porous Cu (scale bar, 50 mm). The inset shows the high-magnification image (scale bar, 2 mm). Reproduced with the permission from Ref. [151], Copyright 2015, Nature Publishing Group

7.4.3.2 Nanocarbon-Based Film Current Collector

Carbon-based nanomaterials show much lower density, better chemical stability and well known electronic conductivity. A variety of nanocarbon current collectors such as CNTs, graphite sheets, graphene, and carbon paper, with high specific surface areas, have been developed to improve the cycle stability and gravimetric energy density [163–168]. Flexible graphite, Grafoil®, was investigated as a replacement for Cu current collectors in flexible battery applications [169]. Aliahmad et al. [170] developed the paper-based LIBs using CNT-coated wood microfibers. The current collectors were fabricated from wood microfibers that were coated with CNT through an electrostatic layer-by-layer nano-assembly process. The microfibers are individually coated with CNT and transformed into a paper which preserves the important characteristics of the paper, such as texture and porous structure allowing high surface area and stronger binding with electrode materials. The fabrication and use of CNT-microfiber paper as current collectors in LIBs have a wide range of electronic applications as they provide flexibility, lightweight, and high capacity features. Wang et al. [171] reported a lightweight, thin, and flexible current collector for LIBs based on super-aligned carbon nanotube (SACNT) films drawn from SACNT arrays. Flexible and free-standing electrodes on CNT current collectors were fabricated by coating cross-stacked CNT films with the electrode slurry. The CNT current collectors with tunable thickness functioned as a mechanical support for the structural integrity and enabled excellent mechanical durability and efficient electron transfer at the electrode/CNT interface. As a result, the graphite-CNT electrodes displayed outstanding cycling stability and rate capability. By comparing the *IR* drop data of electrodes with CNT and Cu current collectors at various rates, it can be concluded that the graphite/CNT interface displayed lower contact resistance compared with the graphite/Cu interface, leading to a much better performance upon long term cycling or at large currents. Zhong et al. [172] reported that a CNT film was used as the current collector and $LiCoO_2$ as the cathode for LIBs. The battery showed a high initial discharge capacity of 132.8 mA h g^{-1}, a high retention rate of over 80% after 500 cycles under 1 C and a low self-discharge rate of 1.5%. When the areal density of the CNT film is 16 mg cm^{-2}, the energy density of the $LiCoO_2$-CNT battery is 25% larger than that of a $LiCoO_2$-Al battery. The superior properties of the battery may be attributed to the high electrochemical stability of the graphitic structure of CNTs and the porous surface of the CNT film that ensures a high contact surface area between the current collector and the cathode material. The CNT film is promising as a new current collector for LIBs. Furthermore, in 2017, Sun et al. synthesized a novel 3D porous MWCNTs conductive papers, which has excellent electrical conductivity, and electrolyte adsorption capacity, and using the MWCNTs conductive paper for SnO_2 anode material collector can effectively alleviate SnO_2 powder, agglomeration and cracking [173].

Apart from CNTs, the other carbon materials as current collectors also had been studied. Guo et al. [166, 167] reported a series of works about carbon fibers as current collectors, and experimental results demonstrate that graphitized carbon fibers electrode can serve as a multifunctional 3D current collector to enhance the

Li storage capacity and a light-weight, flexible, and free-standing 3D hollow carbon fiber current collector with high electroactive surface area can significantly improve the electrochemical performance. Qu et al. [174] reported a thin, lightweight, and flexible LIBs that uses graphite papers deposited with nano-sized metallic layers as the current collector, $LiFePO_4$ and $Li_4Ti_5O_{12}$ as the cathode and anode materials, respectively, and a PE membrane soaked in $LiPF_6$ as the separator. Using thin, flexible graphite paper as a substrate for the current collector instead of a rigid, heavy metal foil demonstrates an ultra-thin LIB that also features a light weight and high flexibility. Yang et al. [175] reported that the novel surface modified carbon fiber was designed as a rough surface current collector for the growth of active material, and a method to obtain dense $Li_4Ti_5O_{12}$ coated by using a hydrothermal reaction was developed. Xiao et al. [140] reported a low density and highly porous 3D rGO/PAA nanocomposite aerogel as the scaffold for cathode materials in LIBs. The rGO/PAA is both simple and starts from readily available graphite and PAA, thereby providing a scalable fabrication procedure.

7.5 Summary and Outlook

In this chapter, we have described recent developments in traditional nanostructures and nanomaterials of batteries, including conductive additives, separators and current collectors, which are crucial components in batteries. This research continues to be a vibrant and growing in batteries field especially for the next-generation rechargeable batteries (e.g., Li-metal batteries, all-solid-state batteries). In the near future, there are a number of areas of research that are particularly promising, but concerted efforts are still required for success.

Conductive additives have the important influence on the performance of the battery, the developing direction of conductive additives will mainly focus on the development of carbon black with high ionic conductivity and multiple conductive additives. With the development of new nanostructures and nanomaterials, there will be more kinds of multiple conductive additives to fabricate rechargeable batteries with high electrochemical performance.

Engineering the separator is also a viable and simple way to improve the performance of batteries. Researches on the development of new separators started only a few years ago. Up to now, a number of new separators have been shown to possess great potential for practical battery applications. The new separators are generally developed by modification of the traditional polyolefin separators or using materials with certain functionality. The modification of traditional polyolefin separators with polymers and inorganic NPs generally improves the wettability and thermal stability of the polyolefin separator. Both polymeric separators and ceramic separators have been reported to be the potential candidates for high performance separators. For the future, the research for the development of new separators is needed to achieve advanced batteries with excellent performance in terms of power capability and safety.

Current collector plays a key role in electronically connecting electrode materials with external circuit. Its properties directly affect the performance of batteries. We summarize the surface modification methods of traditional current collectors and the development of new 3D current collectors. The optimal design of the architectures, correct selections of materials, and controllable procedures for the fabrication of advanced current collector will open a broad avenue for the development of next-generation electrochemical energy storage devices.

References

1. Wang, J., & Sun, X. (2012). Understanding and recent development of carbon coating on $LiFePO_4$ cathode materials for lithium-ion batteries. *Energy & Environmental Science, 5,* 5163–5185.
2. Qingtang, Z., Zuolong, Y., Ping, D., et al. (2010). Carbon nanomaterials used as conductive additives in lithium ion batteries. *Recent Patents on Nanotechnology, 4,* 100–110.
3. Li, X., Kang, F., & Shen, W. (2006). Multiwalled carbon nanotubes as a conducting additive in a $LiNi_{0.7}Co_{0.3}O_2$ cathode for rechargeable lithium batteries. *Carbon, 44,* 1334–1336.
4. Arora, P., & Zhang, Z. (2004). Battery separators. *Chemical Reviews, 104,* 4419–4462.
5. Lee, H., Yanilmaz, M., Toprakci, O., et al. (2014). A review of recent developments in membrane separators for rechargeable lithium-ion batteries. *Energy & Environmental Science, 7,* 3857–3886.
6. Wu, H.-C., Lin, Y.-P., Lee, E., et al. (2009). High-performance carbon-based supercapacitors using Al current-collector with conformal carbon coating. *Materials Chemistry and Physics, 117,* 294–300.
7. Jiang, J., Nie, P., Ding, B., et al. (2016). Effect of graphene modified Cu current collector on the performance of $Li_4Ti_5O_{12}$ anode for lithium-ion batteries. *ACS Applied Materials & Interfaces, 8,* 30926–30932.
8. Dudney, N. J., & Li, J. (2015). Using all energy in a battery. *Science, 347,* 131–132.
9. Park, M., Zhang, X., Chung, M., et al. (2010). A review of conduction phenomena in Li-ion batteries. *Journal of Power Sources, 195,* 7904–7929.
10. Bonaccorso, F., Colombo, L., Yu, G., et al. (2015). Graphene, related two-dimensional crystals, and hybrid systems for energy conversion and storage. *Science, 347,* 1246501.
11. Kucinskis, G., Bajars, G., & Kleperis, J. (2013). Graphene in lithium ion battery cathode materials: A review. *Journal of Power Sources, 240,* 66–79.
12. Kuroda, S., Tobori, N., Sakuraba, M., et al. (2003). Charge-discharge properties of a cathode prepared with ketjen black as the electro-conductive additive in lithium ion batteries. *Journal of Power Sources, 119–121,* 924–928.
13. Spahr, M. E., Goers, D., Leone, A., et al. (2011). Development of carbon conductive additives for advanced lithium ion batteries. *Journal of Power Sources, 196,* 3404–3413.
14. Jin, Y., Li, B., Zhang, Z., et al. (2011). Production of high purity conductive carbon black by high temperature treatment. *Nanoscience and Nanotechnology Letters, 3,* 784–787.
15. Liu, Z., Lee, J. Y., & Lindner, H. J. (2001). Effects of conducting carbon on the electrochemical performance of $LiCoO_2$ and $LiMn_2O_4$ cathodes. *Journal of Power Sources, 97–98,* 361–365.
16. Novák, P., Panitz, J. C., Joho, F., et al. (2000). Advanced in situ methods for the characterization of practical electrodes in lithium-ion batteries. *Journal of Power Sources, 90,* 52–58.
17. Imhof, R., & Novák, P. (1999). Oxidative electrolyte solvent degradation in lithium-ion batteries: An in situ differential electrochemical mass spectrometry investigation. *Journal of the Electrochemical Society, 146,* 1702–1706.
18. Jang, D. H., & Oh, S. (1998). Effects of carbon additives on spinel dissolution and capacity losses in 4 V $Li/Li_xMn_2O_4$ rechargeable cells. *Electrochimica Acta, 43,* 1023–1029.

19. Dominko, R., Gaberscek, M., Drofenik, J., et al. (2003). The role of carbon black distribution in cathodes for Li ion batteries. *Journal of Power Sources, 119–121,* 770–773.
20. Li, X., Kang, F., Bai, X., et al. (2007). A novel network composite cathode of $LiFePO_4$/multiwalled carbon nanotubes with high rate capability for lithium ion batteries. *Electrochemistry Communications, 9,* 663–666.
21. Zhang, H.-L., Zhang, Y., Zhang, X.-G., et al. (2006). Urchin-like nano/micro hybrid anode materials for lithium ion battery. *Carbon, 44,* 2778–2784.
22. Liu, C., Li, F., Ma, L. P., et al. (2010). Advanced materials for energy storage. *Advanced Materials, 22,* E28–E62.
23. Sakamoto, J. S., & Dunn, B. (2002). Vanadium oxide-carbon nanotube composite electrodes for use in secondary lithium batteries. *Journal of the Electrochemical Society, 149,* A26–A30.
24. Sheem, K., Lee, Y. H., & Lim, H. S. (2006). High-density positive electrodes containing carbon nanotubes for use in Li-ion cells. *Journal of Power Sources, 158,* 1425–1430.
25. Forney, M. W., Dzara, M. J., Doucett, A. L., et al. (2014). Advanced germanium nanoparticle composite anodes using single wall carbon nanotube conductive additives. *Journal of Materials Chemistry A, 2,* 14528–14535.
26. Guoping, W., Qingtang, Z., Zuolong, Y., et al. (2008). The effect of different kinds of nano-carbon conductive additives in lithium ion batteries on the resistance and electrochemical behavior of the $LiCoO_2$ composite cathodes. *Solid State Ionics, 179,* 263–268.
27. Su, F.-Y., He, Y.-B., Li, B., et al. (2012). Could graphene construct an effective conducting network in a high-power lithium ion battery? *Nano Energy, 1,* 429–439.
28. Su, F.-Y., You, C., He, Y.-B., et al. (2010). Flexible and planar graphene conductive additives for lithium-ion batteries. *Journal of Materials Chemistry, 20,* 9644–9650.
29. Zhang, B., Yu, Y., Liu, Y., et al. (2013). Percolation threshold of graphene nanosheets as conductive additives in $Li_4Ti_5O_{12}$ anodes of Li-ion batteries. *Nanoscale, 5,* 2100–2106.
30. Bi, H., Huang, F., Tang, Y., et al. (2013). Study of $LiFePO_4$ cathode modified by graphene sheets for high-performance lithium ion batteries. *Electrochimica Acta, 88,* 414–420.
31. Liu, T., Sun, S., Zang, Z., et al. (2017). Effects of graphene with different sizes as conductive additives on the electrochemical performance of a $LiFePO_4$ cathode. *RSC Advances, 7,* 20882–20887.
32. Kehrwald, D., Shearing, P. R., Brandon, N. P., et al. (2011). Local tortuosity inhomogeneities in a lithium battery composite electrode. *Journal of the Electrochemical Society, 158,* A1393–A1399.
33. Vijayaraghavan, B., Ely, D. R., Chiang, Y.-M., et al. (2012). An analytical method to determine tortuosity in rechargeable battery electrodes. *Journal of the Electrochemical Society, 159,* A548–A552.
34. Suthar, B., Northrop, P. W. C., Rife, D., et al. (2015). Effect of porosity, thickness and tortuosity on capacity fade of anode. *Journal of the Electrochemical Society, 162,* A1708–A1717.
35. Ke, L. U., Lv, W., Su, F.-Y., et al. (2015). Electrode thickness control: Precondition for quite different functions of graphene conductive additives in $LiFePO_4$ electrode. *Carbon, 92,* 311–317.
36. Ha, J., Park, S.-K., Yu, S.-H., et al. (2013). A chemically activated graphene-encapsulated $LiFePO_4$ composite for high-performance lithium ion batteries. *Nanoscale, 5,* 8647–8655.
37. Tang, R., Yun, Q., Lv, W., et al. (2016). How a very trace amount of graphene additive works for constructing an efficient conductive network in $LiCoO_2$-based lithium-ion batteries. *Carbon, 103,* 356–362.
38. Yang, J., Wang, J., Tang, Y., et al. (2013). $LiFePO_4$-graphene as a superior cathode material for rechargeable lithium batteries: impact of stacked graphene and unfolded graphene. *Energy & Environmental Science, 6,* 1521–1528.
39. Dominko, R., Gaberšček, M., Drofenik, J., et al. (2003). Influence of carbon black distribution on performance of oxide cathodes for Li ion batteries. *Electrochimica Acta, 48,* 3709–3716.
40. Chiaki, S., Gaku, O., Masataka, T., et al. (2008). The reinforcing effect of combined carbon nanotubes and acetylene blacks on the positive electrode of lithium-ion batteries. *Chemsuschem, 1,* 911–915.

41. Cheon, S. E., Kwon, C. W., Kim, D. B., et al. (2000). Effect of binary conductive agents in $LiCoO_2$ cathode on performances of lithium ion polymer battery. *Electrochimica Acta, 46,* 599–605.
42. Wang, K., Wu, Y., Luo, S., et al. (2013). Hybrid super-aligned carbon nanotube/carbon black conductive networks: A strategy to improve both electrical conductivity and capacity for lithium ion batteries. *Journal of Power Sources, 233,* 209–215.
43. Liu, X.-Y., Peng, H.-J., Zhang, Q., et al. (2014). Hierarchical carbon nanotube/carbon black scaffolds as short- and long-range electron pathways with superior li-ion storage performance. *Acs Sustainable Chemistry & Engineering, 2,* 200–206.
44. Seid, K. A., Badot, J. C., Dubrunfaut, O., et al. (2013). Multiscale electronic transport in $Li_{1+x}Ni_{1/3-u}Co_{1/3-v}Mn_{1/3-w}O_2$: A broadband dielectric study from 40 Hz to 10 GHz. *Physical Chemistry Chemical Physics, 15,* 19790–19798.
45. Seid, K. A., Badot, J. C., Dubrunfaut, O., et al. (2012). Multiscale electronic transport mechanism and true conductivities in amorphous carbon-$LiFePO_4$ nanocomposites. *Journal of Materials Chemistry, 22,* 2641–2649.
46. Hong, J. K., Lee, J. H., & Oh, S. M. (2002). Effect of carbon additive on electrochemical performance of $LiCoO_2$ composite cathodes. *Journal of Power Sources, 111,* 90–96.
47. Zhang, Q., Peng, G., Wang, G., et al. (2009). Effect of mesoporous carbon containing binary conductive additives in lithium ion batteries on the electrochemical performance of the $LiCoO_2$ composite cathodes. *Solid State Ionics, 180,* 698–702.
48. Kang, X., Utsunomiya, H., Achiha, T., et al. (2010). Effect of conductive additives and surface fluorination on the electrochemical properties of lithium titanate ($Li_{4/3}Ti_{5/3}O_4$). *Journal of the Electrochemical Society, 157,* A437–A442.
49. Utsunomiya, H., Nakajima, T., Ohzawa, Y., et al. (2010). Influence of conductive additives and surface fluorination on the charge/discharge behavior of lithium titanate ($Li_{4/3}Ti_{5/3}O_4$). *Journal of Power Sources, 195,* 6805–6810.
50. Wang, Q., Su, F.-Y., Tang, Z.-Y., et al. (2012). Synergetic effect of conductive additives on the performance of high power lithium ion batteries. *New Carbon Materials, 27,* 427–432.
51. Bian, X., Fu, Q., Qiu, C., et al. (2015). Carbon black and vapor grown carbon fibers binary conductive additive for the $Li_{1.18}Co_{0.15}Ni_{0.15}Mn_{0.52}O_2$ electrodes for Li-ion batteries. *Materials Chemistry and Physics, 156,* 69–75.
52. Wei, W., Lv, W., Wu, M.-B., et al. (2013). The effect of graphene wrapping on the performance of $LiFePO_4$ for a lithium ion battery. *Carbon, 57,* 530–533.
53. Badot, J.-C., Ligneel, E., Dubrunfaut, O., et al. (2009). A multiscale description of the electronic transport within the hierarchical architecture of a composite electrode for lithium batteries. *Advanced Functional Materials, 19,* 2749–2758.
54. Chen, Y. H., Wang, C. W., Liu, G., et al. (2007). Selection of conductive additives in Li-ion battery cathodes-A numerical study. *Journal of the Electrochemical Society, 154,* A978–A986.
55. Bauer, W., Nötzel, D., Wenzel, V., et al. (2015). Influence of dry mixing and distribution of conductive additives in cathodes for lithium ion batteries. *Journal of Power Sources, 288,* 359–367.
56. Bockholt, H., Haselrieder, W., & Kwade, A. (2016). Intensive powder mixing for dry dispersing of carbon black and its relevance for lithium-ion battery cathodes. *Powder Technology, 297,* 266–274.
57. Jiang, R., Cui, C., & Ma, H. (2013). Using graphene nanosheets as a conductive additive to enhance the rate performance of spinel $LiMn_2O_4$ cathode material. *Physical Chemistry Chemical Physics, 15,* 6406–6415.
58. Abraham, K. M. (1993). Directions in secondary lithium battery research and development. *Electrochimica Acta, 38,* 1233–1248.
59. Deimede, V., & Elmasides, C. (2015). Separators for lithium-on batteries: A review on the production processes and recent developments. *Energy Technology, 3,* 453–468.
60. Yang, M., & Hou, J. (2012). Membranes in lithium ion batteries. *Membranes, 2,* 367–383.
61. Djian, D., Alloin, F., Martinet, S., et al. (2007). Lithium-ion batteries with high charge rate capacity: Influence of the porous separator. *Journal of Power Sources, 172,* 416–421.

62. Wu, M.-S., Chiang, P.-C. J., Lin, J.-C., et al. (2004). Correlation between electrochemical characteristics and thermal stability of advanced lithium-ion batteries in abuse tests-short-circuit tests. *Electrochimica Acta, 49,* 1803–1812.
63. Love, C. T. (2011). Thermomechanical analysis and durability of commercial micro-porous polymer Li-ion battery separators. *Journal of Power Sources, 196,* 2905–2912.
64. Huang, X. (2014). Performance evaluation of a non-woven lithium ion battery separator prepared through a paper-making process. *Journal of Power Sources, 256,* 96–101.
65. Tarascon, J. M., & Armand, M. (2001). Issues and challenges facing rechargeable lithium batteries. *Nature, 414,* 359–367.
66. Yuan, F., Chen, H. Z., Yang, H. Y., et al. (2005). PAN-PRO solid polymer electrolytes with high ionic conductivity. *Materials Chemistry and Physics, 89,* 390–394.
67. Pu, W., He, X., Wang, L., et al. (2006). Preparation of PVDF-HFP microporous membrane for Li-ion batteries by phase inversion. *Journal of Membrane Science, 272,* 11–14.
68. Chiu, C. Y., Yen, Y. J., Kuo, S. W., et al. (2007). Complicated phase behavior and ionic conductivities of PVP-co-PMMA-based polymer electrolytes. *Polymer, 48,* 1329–1342.
69. Kim, D. W., Oh, B., Park, J. H., et al. (2000). Gel-coated membranes for lithium-ion polymer batteries. *Solid State Ionics, 138,* 41–49.
70. Park, J.-H., Park, W., Kim, J. H., et al. (2011). Close-packed poly(methyl methacrylate) nanoparticle arrays-coated polyethylene separators for high-power lithium-ion polymer batteries. *Journal of Power Sources, 196,* 7035–7038.
71. Sohn, J.-Y., Im, J. S., Gwon, S.-J., et al. (2009). Preparation and characterization of a PVDF-HFP/PEGDMA-coated PE separator for lithium-ion polymer battery by electron beam irradiation. *Radiation Physics and Chemistry, 78,* 505–508.
72. Lee, H., Alcoutlabi, M., Toprakci, O., et al. (2014). Preparation and characterization of electrospun nanofiber-coated membrane separators for lithium-ion batteries. *Journal of Solid State Electrochemistry, 18,* 2451–2458.
73. Liu, Q., Xia, M., Chen, J., et al. (2015). High performance hybrid Al_2O_3/poly(vinyl alcohol-co-ethylene) nanofibrous membrane for lithium-ion battery separator. *Electrochimica Acta, 176,* 949–955.
74. Liang, X., Yang, Y., Jin, X., et al. (2015). The high performances of SiO_2/Al_2O_3-coated electrospun polyimide fibrous separator for lithium-ion battery. *Journal of Membrane Science, 493,* 1–7.
75. Jeong, H.-S., & Lee, S.-Y. (2011). Closely packed SiO_2 nanoparticles/poly(vinylidene fluoride-hexafluoropropylene) layers-coated polyethylene separators for lithium-ion batteries. *Journal of Power Sources, 196,* 6716–6722.
76. Shi, C., Zhang, P., Chen, L., et al. (2014). Effect of a thin ceramic-coating layer on thermal and electrochemical properties of polyethylene separator for lithium-ion batteries. *Journal of Power Sources, 270,* 547–553.
77. Kim, K. J., Kwon, H. K., Park, M.-S., et al. (2014). Ceramic composite separators coated with moisturized ZrO_2 nanoparticles for improving the electrochemical performance and thermal stability of lithium ion batteries. *Physical Chemistry Chemical Physics, 16,* 9337–9343.
78. Zhu, X., Jiang, X., Ai, X., et al. (2015). A highly thermostable ceramic-grafted microporous polyethylene separator for safer lithium-ion batteries. *ACS Applied Materials & Interfaces, 7,* 24119–24126.
79. Zhu, X., Jiang, X., Ai, X., et al. (2016). TiO_2 ceramic-grafted polyethylene separators for enhanced thermostability and electrochemical performance of lithium-ion batteries. *Journal of Membrane Science, 504,* 97–103.
80. Yang, P., Zhang, P., Li, C., et al. (2015). The functional separator coated with core-shell structured silica-poly(methyl methacrylate) sub-microspheres for lithium-ion batteries. *Journal of Membrane Science, 474,* 148–155.
81. Xu, W., Wang, Z., Shi, L., et al. (2015). Layer-by-layer deposition of organic-inorganic hybrid multilayer on microporous polyethylene separator to enhance the electrochemical performance of lithium-ion battery. *ACS Applied Materials & Interfaces, 7,* 20678–20686.

82. Shin, W.-K., Yoo, J.-H., & Kim, D.-W. (2015). Surface-modified separators prepared with conductive polymer and aluminum fluoride for lithium-ion batteries. *Journal of Power Sources, 279,* 737–744.
83. Hu, S., Lin, S., Tu, Y., et al. (2016). Novel aramid nanofiber-coated polypropylene separators for lithium ion batteries. *Journal of Materials Chemistry A, 4,* 3513–3526.
84. Ye, W., Zhu, J., Liao, X., et al. (2015). Hierarchical three-dimensional micro/nano-architecture of polyaniline nanowires wrapped-on polyimide nanofibers for high performance lithium-ion battery separators. *Journal of Power Sources, 299,* 417–424.
85. Amatucci, G. G., Blyr, A., Sigala, C., et al. (1997). Surface treatments of $Li_{1+x}Mn_{2-x}O_4$ spinels for improved elevated temperature performance. *Solid State Ionics, 104,* 13–25.
86. Qiao, R., Wang, Y., Olalde-Velasco, P., et al. (2015). Direct evidence of gradient Mn(II) evolution at charged states in $LiNi_{0.5}Mn_{1.5}O_4$ electrodes with capacity fading. *Journal of Power Sources, 273,* 1120–1126.
87. Man, C., Jiang, P., Wong, K.-W., et al. (2014). Enhanced wetting properties of a polypropylene separator for a lithium-ion battery by hyperthermal hydrogen induced cross-linking of poly(ethylene oxide). *Journal of Materials Chemistry A, 2,* 11980–11986.
88. Li, Z., Pauric, A. D., Goward, G. R., et al. (2014). Manganese sequestration and improved high-temperature cycling of Li-ion batteries by polymeric aza-15-crown-5. *Journal of Power Sources, 272,* 1134–1141.
89. Lee, Y., Lee, H., Lee, T., et al. (2015). Synergistic thermal stabilization of ceramic/co-polyimide coated polypropylene separators for lithium-ion batteries. *Journal of Power Sources, 294,* 537–544.
90. Wang, J., Hu, Z., Yin, X., et al. (2015). Alumina/Phenolphthalein polyetherketone ceramic composite polypropylene separator film for lithium ion power batteries. *Electrochimica Acta, 159,* 61–65.
91. Zhang, F., Ma, X., Cao, C., et al. (2014). Poly(vinylidene fluoride)/SiO_2 composite membranes prepared by electrospinning and their excellent properties for nonwoven separators for lithium-ion batteries. *Journal of Power Sources, 251,* 423–431.
92. Sheng-Heng, C., & Arumugam, M. (2014). Bifunctional separator with a light-weight carbon-coating for dynamically and statically stable lithium-sulfur batteries. *Advanced Functional Materials, 24,* 5299–5306.
93. Zhao, D., Qian, X., Jin, L., et al. (2016). Separator modified by Ketjen black for enhanced electrochemical performance of lithium-sulfur batteries. *RSC Advances, 6,* 13680–13685.
94. Juan, B., Tony, J., Markus, K., et al. (2015). Functional mesoporous carbon-coated separator for long-life, high-energy lithium-sulfur batteries. *Advanced Functional Materials, 25,* 5285–5291.
95. Stoeck, U., Balach, J., Klose, M., et al. (2016). Reconfiguration of lithium sulphur batteries: "Enhancement of Li–S cell performance by employing a highly porous conductive separator coating". *Journal of Power Sources, 309,* 76–81.
96. Zhang, Z., Wang, G., Lai, Y., et al. (2015). Nitrogen-doped porous hollow carbon sphere-decorated separators for advanced lithium-sulfur batteries. *Journal of Power Sources, 300,* 157–163.
97. Balach, J., Jaumann, T., Klose, M., et al. (2016). Improved cycling stability of lithium-sulfur batteries using a polypropylene-supported nitrogen-doped mesoporous carbon hybrid separator as polysulfide adsorbent. *Journal of Power Sources, 303,* 317–324.
98. Guangmin, Z., Lu, L., Da-Wei, W., et al. (2015). A flexible sulfur-graphene-polypropylene separator integrated electrode for advanced Li–S batteries. *Advanced Materials, 27,* 641–647.
99. Ting-Zhou, Z., Jia-Qi, H., Hong-Jie, P., et al. (2016). Rational integration of polypropylene/graphene oxide/nafion as ternary-layered separator to retard the shuttle of polysulfides for lithium-sulfur batteries. *Small (Weinheim an der Bergstrasse, Germany), 12,* 381–389.
100. Chi-Hao, C., Sheng-Heng, C., & Arumugam, M. (2016). Effective stabilization of a high-loading sulfur cathode and a lithium-metal anode in Li–S Batteries Utilizing SWCNT-modulated separators. *Small (Weinheim an der Bergstrasse, Germany), 12,* 174–179.

101. Kang, G.-D., & Cao, Y.-M. (2014). Application and modification of poly(vinylidene fluoride) (PVDF) membranes-A review. *Journal of Membrane Science, 463,* 145–165.
102. Liang, Y., Cheng, S., Zhao, J., et al. (2013). Heat treatment of electrospun polyvinylidene fluoride fibrous membrane separators for rechargeable lithium-ion batteries. *Journal of Power Sources, 240,* 204–211.
103. Kim, J.-H., Kim, J.-H., Choi, E.-S., et al. (2014). Nanoporous polymer scaffold-embedded nonwoven composite separator membranes for high-rate lithium-ion batteries. *RSC Advances, 4,* 54312–54321.
104. Wu, D., Shi, C., Huang, S., et al. (2015). Electrospun nanofibers for sandwiched polyimide/poly (vinylidene fluoride)/polyimide separators with the thermal shutdown function. *Electrochimica Acta, 176,* 727–734.
105. Zhai, Y., Wang, N., Mao, X., et al. (2014). Sandwich-structured PVdF/PMIA/PVdF nanofibrous separators with robust mechanical strength and thermal stability for lithium ion batteries. *Journal of Materials Chemistry A, 2,* 14511–14518.
106. Banerjee, A., Ziv, B., Shilina, Y., et al. (2016). Improving stability of Li-ion batteries by means of transition metal ions trapping separators. *Journal of the Electrochemical Society, 163,* A1083–A1094.
107. Xiao, W., Zhao, L., Gong, Y., et al. (2015). Preparation and performance of poly(vinyl alcohol) porous separator for lithium-ion batteries. *Journal of Membrane Science, 487,* 221–228.
108. Lu, C., Qi, W., Li, L., et al. (2013). Electrochemical performance and thermal property of electrospun PPESK/PVDF/PPESK composite separator for lithium-ion battery. *Journal of Applied Electrochemistry, 43,* 711–720.
109. Miao, Y.-E., Zhu, G.-N., Hou, H., et al. (2013). Electrospun polyimide nanofiber-based nonwoven separators for lithium-ion batteries. *Journal of Power Sources, 226,* 82–86.
110. Zhihong, L., Wen, J., Qingshan, K., et al. (2013). A Core@sheath nanofibrous separator for lithium ion batteries obtained by coaxial electrospinning. *Macromolecular Materials and Engineering, 298,* 806–813.
111. Le Mong, A., & Kim, D. (2016). Tailor-made pore controlled poly (arylene ether ketone) membranes as a lithium-ion battery separator. *Journal of Power Sources, 304,* 301–310.
112. Chun, S.-J., Choi, E.-S., Lee, E.-H., et al. (2012). Eco-friendly cellulose nanofiber paper-derived separator membranes featuring tunable nanoporous network channels for lithium-ion batteries. *Journal of Materials Chemistry, 22,* 16618–16626.
113. Weng, B., Xu, F., Alcoutlabi, M., et al. (2015). Fibrous cellulose membrane mass produced via forcespinning® for lithium-ion battery separators. *Cellulose, 22,* 1311–1320.
114. Liao, H., Hong, H., Zhang, H., et al. (2016). Preparation of hydrophilic polyethylene/methylcellulose blend microporous membranes for separator of lithium-ion batteries. *Journal of Membrane Science, 498,* 147–157.
115. Kim, J.-M., Kim, C., Yoo, S., et al. (2015). Agarose-biofunctionalized, dual-electrospun heteronanofiber mats: Toward metal-ion chelating battery separator membranes. *Journal of Materials Chemistry A, 3,* 10687–10692.
116. Jiang, F., Yin, L., Yu, Q., et al. (2015). Bacterial cellulose nanofibrous membrane as thermal stable separator for lithium-ion batteries. *Journal of Power Sources, 279,* 21–27.
117. Kim, J.-H., Kim, J.-H., Choi, K.-H., et al. (2014). Inverse opal-inspired, nanoscaffold battery separators: A new membrane opportunity for high-performance energy storage systems. *Nano Letters, 14,* 4438–4448.
118. Luo, X., Pan, W., Liu, H., et al. (2015). Glass fiber fabric mat as the separator for lithium-ion battery with high safety performance. *Ionics, 21,* 3135–3139.
119. Zhang, B., Wang, Q., Zhang, J., et al. (2014). A superior thermostable and nonflammable composite membrane towards high power battery separator. *Nano Energy, 10,* 277–287.
120. He, M., Zhang, X., Jiang, K., et al. (2015). Pure inorganic separator for lithium ion batteries. *ACS Applied Materials & Interfaces, 7,* 738–742.
121. Holtmann, J., Schäfer, M., Niemöller, A., et al. (2016). Boehmite-based ceramic separator for lithium-ion batteries. *Journal of Applied Electrochemistry, 46,* 69–76.

122. Raja, M., Angulakshmi, N., Thomas, S., et al. (2014). Thin, flexible and thermally stable ceramic membranes as separator for lithium-ion batteries. *Journal of Membrane Science, 471,* 103–109.
123. Padmaraj, O., Nageswara Rao, B., Jena, P., et al. (2014). Electrochemical studies of electrospun organic/inorganic hybrid nanocomposite fibrous polymer electrolyte for lithium battery. *Polymer, 55,* 1136–1142.
124. Raja, M., Kumar, T. P., Sanjeev, G., et al. (2014). Montmorillonite-based ceramic membranes as novel lithium-ion battery separators. *Ionics, 20,* 943–948.
125. Raja, M., Sanjeev, G., Prem Kumar, T., et al. (2015). Lithium aluminate-based ceramic membranes as separators for lithium-ion batteries. *Ceramics International, 41,* 3045–3050.
126. Nunes-Pereira, J., Costa, C. M., Sousa, R. E., et al. (2014). Li-ion battery separator membranes based on barium titanate and poly(vinylidene fluoride-co-trifluoroethylene): Filler size and concentration effects. *Electrochimica Acta, 117,* 276–284.
127. Goodenough, J. B., & Park, K.-S. (2013). The Li-ion rechargeable battery: A perspective. *Journal of the American Chemical Society, 135,* 1167–1176.
128. Goodenough, J. B. (2013). Evolution of strategies for modern rechargeable batteries. *Accounts of Chemical Research, 46,* 1053–1061.
129. Myung, S.-T., Hitoshi, Y., & Sun, Y.-K. (2011). Electrochemical behavior and passivation of current collectors in lithium-ion batteries. *Journal of Materials Chemistry, 21,* 9891–9911.
130. Zhang, S. S., & Jow, T. R. (2002). Aluminum corrosion in electrolyte of Li-ion battery. *Journal of Power Sources, 109,* 458–464.
131. Zhang, X., Winget, B., Doeff, M., et al. (2005). Corrosion of aluminum current collectors in lithium-ion batteries with electrolytes containing $LiPF_6$. *Journal of the Electrochemical Society, 152,* B448–B454.
132. Yang, H., Kwon, K., Devine, T. M., et al. (2000). Aluminum corrosion in lithium batteries an investigation using the electrochemical quartz crystal microbalance. *Journal of the Electrochemical Society, 147,* 4399–4407.
133. Wen, J., Yu, Y., & Chen, C. (2012). A review on lithium-ion batteries safety issues: Existing problems and possible solutions. *Materials Express, 2,* 197–212.
134. Zhang, X., Ross, P. N., Kostecki, R., et al. (2001). Diagnostic characterization of high power lithium-ion batteries for use in hybrid electric vehicles. *Journal of the Electrochemical Society, 148,* A463–A470.
135. Lecoeur, C., Tarascon, J.-M., & Guery, C. (2010). Al current collectors for Li-ion batteries made via a template-free electrodeposition process in ionic liquids. *Journal of the Electrochemical Society, 157,* A641–A646.
136. Perre, E., Nyholm, L., Gustafsson, T., et al. (2008). Direct electrodeposition of aluminium nano-rods. *Electrochemistry Communications, 10,* 1467–1470.
137. Portet, C., Taberna, P. L., Simon, P., et al. (2006). Modification of Al current collector/active material interface for power improvement of electrochemical capacitor electrodes. *Journal of the Electrochemical Society, 153,* A649–A653.
138. Wu, H.-C., Wu, H.-C., Lee, E., et al. (2010). High-temperature carbon-coated aluminum current collector for enhanced power performance of $LiFePO_4$ electrode of Li-ion batteries. *Electrochemistry Communications, 12,* 488–491.
139. Lecoeur, C., Tarascon, J.-M., & Guery, C. (2011). Al current collectors for Li-ion batteries made via an oxidation process in ionic liquids. *Electrochemical and Solid-State Letters, 14,* A6–A9.
140. Xiao, H., Pender, J. P., Meece-Rayle, M. A., et al. (2017). Reduced-graphene oxide/poly(acrylic acid) aerogels as a three-dimensional replacement for metal-foil current collectors in lithium-ion batteries. *ACS Applied Materials & Interfaces, 9,* 22641–22651.
141. Huang, J.-Q., Zhai, P.-Y., Peng, H.-J., et al. (2017). Metal/nanocarbon layer current collectors enhanced energy efficiency in lithium-sulfur batteries. *Science Bulletin, 62,* 1267–1274.
142. Yuan, W., Luo, J., Yan, Z., et al. (2017). High-performance CuO/Cu composite current collectors with array-pattern porous structures for lithium-ion batteries. *Electrochimica Acta, 226,* 89–97.

143. Wu, Y. P., Rahm, E., & Holze, R. (2003). Carbon anode materials for lithium ion batteries. *Journal of Power Sources, 114,* 228–236.
144. Yue, H., Li, F., Yang, Z., et al. (2014). Facile preparation of Mn_3O_4-coated carbon nanofibers on copper foam as a high-capacity and long-life anode for lithium-ion batteries. *Journal of Materials Chemistry A, 2,* 17352–17358.
145. Taberna, P. L., Mitra, S., Poizot, P., et al. (2006). High rate capabilities Fe_3O_4-based Cu nano-architectured electrodes for lithium-ion battery applications. *Nature Materials, 5,* 567–573.
146. Bai, B., & Liu, Q. (2017). Enhanced cycle performance of silicon-based anode by annealing Cu-coated carbon cloth current collector for flexible lithium-ion battery. *Catalysis Letters, 147,* 2962–2966.
147. Woo, K. S., Ho, Y. J., Bongki, S., et al. (2014). Graphite/silicon hybrid electrodes using a 3D current collector for flexible batteries. *Advanced Materials, 26,* 2977–2982.
148. Kim, G., Jeong, S., Shin, J.-H., et al. (2014). 3D amorphous silicon on nanopillar copper electrodes as anodes for high-rate lithium-ion batteries. *ACS Nano, 8,* 1907–1912.
149. Liu, B., Soares, P., Checkles, C., et al. (2013). Three-dimensional hierarchical ternary nanostructures for high-performance Li-Ion battery anodes. *Nano Letters, 13,* 3414–3419.
150. Sakineh, C., Chuang, P., Di, H., et al. (2014). Ideal three-dimensional electrode structures for electrochemical energy storage. *Advanced Materials, 26,* 2440–2445.
151. Yang, C.-P., Yin, Y.-X., Zhang, S.-F., et al. (2015). Accommodating lithium into 3D current collectors with a submicron skeleton towards long-life lithium metal anodes. *Nature Communications, 6,* 8058.
152. Zhang, H., Yu, X., & Braun, P. V. (2011). Three-dimensional bicontinuous ultrafast-charge and -discharge bulk battery electrodes. *Nature Nanotechnology, 6,* 277–281.
153. Kim, J.-H., Kang, S. H., Zhu, K., et al. (2011). Ni–NiO core-shell inverse opal electrodes for supercapacitors. *Chemical Communications, 47,* 5214–5216.
154. Lang, X., Hirata, A., Fujita, T., et al. (2011). Nanoporous metal/oxide hybrid electrodes for electrochemical supercapacitors. *Nature Nanotechnology, 6,* 232–236.
155. Chen, Z., Ren, W., Gao, L., et al. (2011). Three-dimensional flexible and conductive interconnected graphene networks grown by chemical vapour deposition. *Nature Materials, 10,* 424–428.
156. Soria, M. L., Chacón, J., & Hernández, J. C. (2001). Metal hydride electrodes and Ni/MH batteries for automotive high power applications. *Journal of Power Sources, 102,* 97–104.
157. Kim, T. K., Chen, W., & Wang, C. (2011). Heat treatment effect of the Ni foam current collector in lithium ion batteries. *Journal of Power Sources, 196,* 8742–8746.
158. Du, N., Fan, X., Yu, J., et al. (2011). Ni_3Si_2-Si nanowires on Ni foam as a high-performance anode of Li-ion batteries. *Electrochemistry Communications, 13,* 1443–1446.
159. Yang, C., Zhang, D., Zhao, Y., et al. (2011). Nickel foam supported Sn-Co alloy film as anode for lithium ion batteries. *Journal of Power Sources, 196,* 10673–10678.
160. Liu, T., Zhao, L., Wang, D., et al. (2013). Corrosion resistance of nickel foam modified with electroless Ni–P alloy as positive current collector in a lithium ion battery. *RSC Advances, 3,* 25648–25651.
161. Huang, X.-L., Xu, D., Yuan, S., et al. (2014). Dendritic Ni–P-coated melamine foam for a lightweight, low-cost, and amphipathic three-dimensional current collector for binder-free electrodes. *Advanced Materials, 26,* 7264–7270.
162. Poetz, S., Fuchsbichler, B., Schmuck, M., et al. (2014). Development of a 3D current collector for the positive electrode in lithium-ion batteries. *Journal of Applied Electrochemistry, 44,* 989–994.
163. Hu, L., Wu, H., La Mantia, F., et al. (2010). Thin, flexible secondary Li-ion paper batteries. *ACS Nano, 4,* 5843–5848.
164. Lytle, J. C., Wallace, J. M., Sassin, M. B., et al. (2011). The right kind of interior for multifunctional electrode architectures: Carbon nanofoam papers with aperiodic submicrometre pore networks interconnected in 3D. *Energy & Environmental Science, 4,* 1913–1925.
165. Guo, J., Sun, A., & Wang, C. (2010). A porous silicon-carbon anode with high overall capacity on carbon fiber current collector. *Electrochemistry Communications, 12,* 981–984.

166. Tong-Tong, Z., Xiong-Wei, W., Chun-Peng, Y., et al. (2017). Graphitized carbon fibers as multifunctional 3D current collectors for high areal capacity Li anodes. *Advanced Materials, 29,* 1700389.
167. Liu, L., Yin, Y.-X., Li, J.-Y., et al. (2017). Free-standing hollow carbon fibers as high-capacity containers for stable lithium metal anodes. *Joule, 1,* 563–575.
168. Huan, Y., Sen, X., Ya-Xia, Y., et al. (2017). Advanced porous carbon materials for high-efficient lithium metal anodes. *Advanced Energy Materials, 7,* 1700530.
169. Yazici, M. S., Krassowski, D., & Prakash, J. (2005). Flexible graphite as battery anode and current collector. *Journal of Power Sources, 141,* 171–176.
170. Aliahmad, N., Agarwal, M., Shrestha, S., et al. (2013). Paper-based lithium-ion batteries using carbon nanotube-coated wood microfibers. *IEEE Transactions on Nanotechnology, 12,* 408–412.
171. Wang, K., Luo, S., Wu, Y., et al. (2013). Super-aligned carbon nanotube films as current collectors for lightweight and flexible lithium ion batteries. *Advanced Functional Materials, 23,* 846–853.
172. Zhong, S.-W., Hu, J.-W., Wu, Z.-P., et al. (2015). Performance of lithium ion batteries using a carbon nanotube film as a cathode current collector. *Carbon, 81,* 850–852.
173. Sun, X., Qiu, Z., Chen, L., et al. (2017). Three-dimensional porous carbon nanotube papers as current collector and buffer for SnO_2 anodes. *NANO, 12,* 1750141.
174. Qu, H., Hou, J., Tang, Y., et al. (2016). Thin flexible lithium-ion battery featuring graphite paper based current collectors with enhanced conductivity. *Canadian Journal of Chemistry, 95,* 169–173.
175. Yang, L. Y., Li, H. Z., Cheng, L. Z., et al. (2017). A three-dimensional surface modified carbon cloth designed as flexible current collector for high-performance lithium and sodium batteries. *Journal of Alloys and Compounds, 726,* 837–845.

Chapter 8
Conclusions and Perspectives on New Opportunities of Nanostrucutres and Nanomaterials in Batteries

Ya You and Yu-Guo Guo

Abstract Nanostructured materials and nanotechnologies show great potential in improving the electrochemical performance of rechargeable batteries to meet the commercial requirements in terms of energy, power, service life, cost, and safety. In this section, we summarize the advantages and challenges of some emerging advanced rechargeable batteries, discuss where the battery technology is heading both in short-term and long-term, and give perspectives on the opportunities of nanostructures and nanomaterials in these technologies and how they can benefit electrochemical energy storage systems.

Abbreviation

EES	Electrochemical energy storage
LIB	Li-ion battery
EV	Electric vehicle
NCA	$LiNi_{0.8}Co_{0.15}Al_{0.05}O_2$
SEI	Solid electrolyte interface
TM	Transition metal
CVD	Chemical vapor deposition
ALD	Atomic layer deposition
TEM	Transmission electron microscopy
SEM	Scanning electron microscopy
AFM	Atomic force microscopy
XPS	X-ray photoelectron spectroscopy
SSE	Solid-state electrolyte
ORR	Oxygen reduction reaction

Y. You
University of Texas at Austin, Austin, TX 78712, USA

Y.-G. Guo (✉)
Institute of Chemistry, Chinese Academy of Sciences, Beijing 100190, People's Republic of China
e-mail: ygguo@iccas.ac.cn

Y.-G. Guo, *Nanostructures and Nanomaterials for Batteries*,
https://doi.org/10.1007/978-981-13-6233-0_8

OER Oxygen evolution reaction

Rechargeable batteries have become one of the most efficient electrochemical energy storage (EES) technologies and our modern civilizations are more dependable on them than at any time before [1–3]. Throughout the history of EES systems development (Fig. 8.1), one can conclude that the desire for improving energy density drives the technology progress over the past 150 years, from Pb-acid batteries (1850s), Ni–Cd batteries (1890s) and Ni–MH cells (1960s) to, finally, Li-ion batteries (LIBs) at present [4]. The commercial LIBs based on "intercalation chemistry" and non-aqueous liquid electrolytes offer high gravimetric energy densities up to around 250 Wh kg^{-1} and volumetric energy densities of <650 Wh L^{-1} [5], which is almost five times higher than that of the Pb-acid batteries. Nevertheless, there is still an ever-growing demand for pushing the energy densities to a higher level for powering portable electronic devices and extending the driving range of electric vehicles. For example, to achieve a 300-mile driving range for electric vehicles (EVs) and plug-in hybrid EVs, a target energy density of at least 300 Wh kg^{-1} is desired in cells level in the near future [6]. Developing advanced technologies with even higher energy densities (e.g., 500 Wh kg^{-1} and 1000 Wh L^{-1}) will be a long-term goal. In addition to the energy density, other parameters such as power, cost, cycle life, safety, sustainability and temperature tolerance should also be taken into consideration for the evaluation of a rechargeable battery. Basically, EES systems with high energy-densities, high power, low-cost, long-term cycle stability, high safety and abundant resources are highly desired. The above-mentioned parameters largely depend on the intrinsic properties of active cathode and anode materials used in cells as well as the engineering techniques employed in fabricating the cells. Since the energy density of an EES system is mainly determined by the specific capacities and working potentials of the cathode and anode, active electrode materials have been extensively investigated during the past decades [7]. A significant boost in the energy density requires new redox chemistry beyond the limitation of the currently "intercalation mechanism" [8]. There is always a trad-off between high energy density and high cycle stability; as a result, a reliable long-term operation needs exquisite design in the materials structures, compositions and interphases. The repaid development of nanotechnology and nanostructured materials, which has aroused intensively attention during the past decades in various fields, offers new opportunities in addressing the challenges of advanced EES systems due to the unique physical and chemical properties of nanomaterials [9–11]. In this section, we will discuss where the battery technology is heading in both the short-term and long-term and how nanotechnologies (including nanostructures and nanomaterials) can benefit these EES systems.

Fig. 8.1 Potential battery technologies approached and developed in recent years. Reproduced with the permission from Ref. [12]. Copyright © Springer-Verlag Berlin Heidelberg 2017

8.1 Near-Term Technologies

Near-term technologies are those techniques which have already been partially employed in commercial products or are expected to mature within the next 5 years. LIBs have dominated the battery market in applications of portable electronic devices and EVs due to the high gravimetric and volumetric energy densities compared to other EES systems. Undoubtedly, optimizing the commercial LIBs in terms of the active component materials used in cells as well as the engineering techniques employed in fabricating the cells is desired to realize the >300 Wh kg^{-1} goal. Since Sony Corporation has announced the first commercial Li-ion technology in 1991 [12], its energy density shows incremental improvement during the past decades, which is largely resulted from the progress in engineering because the component electrode materials still remain the same with minor modifications. As a result, it is desired to develop advanced cathode and anode materials with higher capacity and lower cost relative to the currently employed $LiCoO_2$ and graphite [13, 14]. Constructing LIBs composed of Ni-rich cathode and silicon anodes represents one of the most efficient strategy to achieve this short-term goal since they have been developed to a level that enables their partial use in the electrodes [4].

8.1.1 Layered Nickel-Rich Cathode Materials

From a cost and sustainable perspective, transition-metal substituted layered oxides, such as $LiNi_{1/3}Mn_{1/3}Co_{1/3}O_2$ were developed to reduce the content of the toxic, rare and expensive Co elements. Ni-rich transition metal (TM) layered oxides ($LiNi_xTM_{1-x}O_2$, $x \geq 0.7$; TM = Mn, Co, Al and their combinations, etc.) are considered as promising cathode materials for near-term LIBs due to its high capacity and low cost [15–18]. In fact, it has already been partially included in commercial products, with $LiNi_{0.8}Co_{0.15}Al_{0.05}O_2$ (NCA) being a representative example. The discharge capacity of commercial NCA is limited to 180 mA h g^{-1} due to the incorporation of 5% Al [15, 19]. To meet the 300 Wh kg^{-1} goal, a fine-tuning of the synthesis procedure, particle morphology, compositions, microstructures and surface properties is necessary [20–22]. Morphology control is expected to affect cycle life, for example, the detachment of primary particles and particle cracks at prolonged cycling counts one of the most severe causes of capacity loss [18]. Decreasing the particle sizes to nanometers could alleviate the particles collapse and facilitates the ionic transport, however, nanosized cathode particles are not preferred in practical cells due to the severe unfavored electrode-electrolyte reactions and low electrode density [20]. Therefore, micro-sized secondary particles composed of nanometer-sized primary particles are expected to ensure efficient ionic transportation as well as a high tap density and mitigate surface degradation process [21]. In addition, coating inert materials, such as Al_2O_3, AlF_3, and Li_3PO_4, etc. are also desired to reduce the cathode surface reactivity toward electrolyte resulting from the highly oxidizing and reactive Ni^{4+} and diminish the formation of rock-salt NiO phase on the surface [6, 23]. The thickness of these surface coatings should be limited to <10 nm to ensure a good ionic and electronic transport at the electrode|electrolyte interphase [24]. The quality of surface coating, including the uniformity and coverage percentage, plays a critical role on the electrode|electrolyte interphase stability [25]. Typical coating procedure was carried out in liquid solutions followed by heat treatment. A precise control of the nucleation and growth of coating materials on the surface of active cathode particles to form core-shell structures instead of forming separate phases remains a significant challenge. In addition, the adoption of compositional gradient particles with a Mn-rich shell and a Ni-rich core is emerging as a promising technical solution to stabilize the surface [26].

Cationic substitution has shown great potential in stabilizing the bulk structure at deep charge, such as suppress the phase transitions, reduce the anisotropic stress, and mitigate the thermal runaway. Al^{3+} ions appear at the moment as the most promising dopants [27]. A small dose of 1% Al has drastically reduced the transition-metal ions dissolution from the cathode and lithium metal plating/dendrite on the graphite anode. Besides, the poor chemical stability against water/carbon dioxide in air, which is another prominent issue for high-Ni oxide cathodes [28–30], has been proved to be alleviated by Al doping [31]. The future research direction could be developing advanced doping nanotechnologies to ensure efficient and uniform cation substitution among the micro-sized particles. Electrolyte compositions, including the salt, solvents and additives also have great impact on surface reactivity, solid electrolyte interface (SEI) property, temperature tolerance, cycle life and rate capability of cathodes [32]. As a result, the development of high-voltage resistant electrolytes or electrolytes that can form stable SEI films on the cathode surface would be beneficial for the adoption of nickel-rich cathodes for automobile applications [33]. The ionic transport among the SEI film and the dynamic evolution of SEI also need investigation for further fundamental understanding [34].

8.1.2 Silicon Anodes

To fulfill the EVs near-term goal, it is desired to use high-specific-capacity alternative relative to the traditional graphite anode [35]. Si has shown great potential as an alternative anode because of its high theoretical capacity (>4,000 mA h g^{-1}) and feasible operating voltage (~0.3 V vs. Li^{+}/Li) [36]. The major challenge associated with Si-based anode is the huge volume change during lithiation and delithiation, which leads to pulverizations of active particles, severe surface passivation and poor Coulombic efficiency. Incorporating Si nanoparticles into conductive, nanoporous and robust matrix to buffer the volume change has been widely employed to improve the cycle stability of Si anodes [35, 37–41]. Considerable efforts have been devoted into incorporating Si nanoparticles into nanostructured carbon materials including porous carbon, carbon nanotube [42] and graphene, because of their low cost, facile synthesis and high conductivity [43–48]. Despite the great improvement in cycling performance with these designed composite structures, the practical industry employed SiO_x as the first-generation Si-based anodes because they are easy to produce in large quantities and available at a reasonable price [49]. The silicon oxide is blended with graphite with limited content (<5 wt%) in practical batteries, indicating the infancy of silicon technologies. The silicon oxide could offer a buffer matrix for the volume change, nevertheless, the low initial Coulombic efficiency (50–60% without carbon coating layer) still remains a big challenge. A pre-lithiation process by heating or pressing lithium metal with silicon-based anode is preferred to improve the initial Coulombic efficiency without sacrificing the long-term cycling. Future research directions could include a further refining of the composite structure including the surface area, pore sizes and electronic conductivities, to optimize the trade-off between cycle stability and Coulombic efficiency. Innovations in developing novel binders and smart electrode structure design are also desired to avoid the peeling off of active electrode materials. With the help of efficient electrolyte additives, a robust and stable SEI would be beneficial for the stable extensively cycling.

8.2 Long-Term Technologies

The long-term technologies have the potential to deliver significantly higher gravimetric (volumetric) energy densities or show obvious cost advantage relative to the state-of-the-art LIBs. However, the commercial success of these techniques is still not guaranteed due to the concern of their long-term operation, cost and safety. Active materials, such as metal anodes and conversion cathodes with new redox chemistry bypassing the limitation of intercalation mechanism of the current LIBs, bring new opportunities to boost the energy density, but severe challenges as well [11, 50]. Basically, these emerging post-LIBs show two major problems: irreversible phase transitions of active materials and/or unstable electrode|electrolyte interphases. As a result, innovations in structure and interphase stabilization strategy to improve the operation stability of component materials should be the focus of future research. Besides, the working principles and cell construction of many emerging post-LIBs techniques are quite different from what is typically employed in commercial LIBs, which requires creative engineering techniques for fabricating cells to facilitate their practical use [51, 52]. In spite of the electrode materials, other cell components, such as current collectors, binders, conductive additives, electrolytes, and separators, also require optimization for various techniques. The ultimate goal of the post-LIBs is to use Li metal anode accompanied with solid-state electrolytes. In this section, we will give a general outlook on the fundamental issues and show how to gain better understanding on the behavior of the electrodes and electrode/electrolyte interphases with nanotechnologies. Some potential solutions enabled by nanostructures and nanomaterials will be discussed too.

8.2.1 Lithium-Rich Cathodes

Li-rich cathodes $x\mathrm{Li_2MnO_3}\ (1-x)\ \mathrm{LiMO_2}$ have the potential to deliver a capacity as high as >250 mA h g^{-1} [53, 54]. Notably, Li- and Mn-rich $\mathrm{Li_{1+x}M_{1-x}O_2}$ offers the highest volumetric energy densities paired with graphite-silicon anodes among all the developed Li-ion based batteries. The commercial perspective, however, still remains an open question because of the various disadvantages, including low Coulombic efficiency, poor rate capability, TM ions dissolution and voltage decay [55]. Many of the above issues, such as poor high-rate performance and TM ions dissolution, could be addressed by bulk and surface modifications [53, 56], however, research is expected to continue in the pursuit of more robust solutions. Moreover, the long-standing voltage decay problem is quite challenging because it originates from thermodynamically flavored phase transitions from layered structure to spinel phase [57]. The realistic solutions based on cost-effective approaches are less available for Li- and Mn-rich layered materials relative to Ni-rich oxides, leading to a relatively slower progress towards commercialization. The future research should focus on the strategies to improve the bulk and surface electrochemical stability, which includes:

(i) dopant substitution to mitigate TM ions migration, (ii) uniform surface nanocoating; (iii) electrolytes that are high-voltage durable and enables the formation of stable, robust SEI film on the electrode surface; Note that the density of both Ni-rich and Li- and Mn-rich compounds are lower compared with the $LiCoO_2$, increasing the tap and press density of these materials active powder and the electrodes contribute to another useful research direction.

8.2.2 Metal Anodes

To further boost the energy densities of post-LIB batteries, an ultimate goal would be using metal anodes due to the high specific capacities and suitable redox potentials, such as Li (3862 mA h g^{-1}), Na (1166 mA h g^{-1}), Mg (2206 mA h g^{-1}), and Al (2981 mA h g^{-1}) etc. However, there are still many challenges to overcome before the practical utilization of the metal anodes, which includes: (i) the inevitable growth of metal dendrite; (ii) poor Coulombic efficiency; (iii) infinite volume changes, and (iv) corrosion and passivation of metal surfaces. The redox potential of the above-mentioned metal is beyond the electrochemical stable window of the currently employed non-aqueous electrolyte (include organic solvent and salt) and an SEI would instantaneously form at the metal-electrolyte interphase [58, 59]. Therefore, the property of SEI plays a critical role on the metal plating and stripping behavior and cycle life of metal anodes. An ideal SEI should be a thin, homogeneous, and compact structure with high ionic conductivity and high elastic strength [58, 60]. Various "nano" strategies have been developed to modify the metal-electrolyte interphase and improve the metal plating/stripping efficiencies [61, 62]. Despite substantial progress achieved at limited current densities and cycle life, we believe that great advances still await discovery. Several possible research directions are provided below, which may lead to pathways for possible solutions.

1. ***Advanced SEI engineering strategies and techniques***. Research into artificial SEI and metal surface modification is making important progress and should continue. Advanced engineering strategies and techniques, such as chemical vapor deposition (CVD), atomic layer deposition (ALD) are required to fabricate homogeneous surface layers in large scale. In addition to surface coating, a controllable in situ growth of a robust and stable SEI layer by means of reacting with metal should also be explored. Typically, surface layer with single composition could not address all the issues of metal anodes. Therefore, lamination of pre-fabricated layers or construction multi-layers on the metal anode contribute to another possible research direction. For example, it could be possible to construct a multilayer composed of a mechanically strong bottom layer, a highly Li-ion-conductive medium layer, and a robust surface protective layer.
2. ***3D nanostructured host***. Embedding metal anodes into 3D connected host with nanostructures would be an effective way to maintain the integrity of the metal and minimize the "dead metal" upon cycling. The nanostructured host could

effectively accommodate the volume changes and suppress the formation of dendrites. Various nanostructured host, such as carbon nanotube, graphene, copper nanowire and porous metal foil have been developed and certain progress have been achieved [63]. A careful control of the nanostructure, wettability, pore volume, and pore sizes are required to compromise the energy density and cycle life [64–66]. Note that an unstable SEI and low Coulombic efficiency is still possible by incorporation metal into a matrix. Therefore, combining the nanostructured host and SEI engineering strategies would be a feasible way to synergistically improve the stability and electrochemical performances of metal anodes.

3. ***Safety issue***. The use of flammable metal, such as Li and Na, in rechargeable batteries brings another concern of the safety of metal batteries. In addition to use flame-retardant additives and nonflammable solid electrolytes, it is also important to integrate smart detection function into metal batteries. Previously, modification of separators and current collectors were employed to demonstrate the Li dendrite detection and battery shut-down mechanisms. Because temperature variates significantly in the battery thermal-abuse process, temperature-sensitive materials, such as thermo-responsive polymers and shape-memory alloys could be integrated into the battery configuration to detect the failure. It would be also worthwhile to explore how to predict the internal short-circuit by real-time monitoring the gas or voltage fluctuation. We also expect that the design of Li-metal anodes compatible with manufacturing outside the glove box will emerge as an area of high priority.
4. ***Advanced characterization techniques on metals***. A clear understanding of metal dendrite nucleation, growth and stripping is desired for developing feasible strategies. Despite some ex situ characterization techniques give much information, they may destroy the original morphology and composition of dendrite and SEI [67]. It has been challenging to reveal the microstructure of metal and SEI in atomic-scale resolution upon cycling due to their instability during transfer or characterization process. As a result, in situ or operando techniques, including operando transmission electron microscopy (TEM), scanning electron microscopy (SEM), atomic force microscopy (AFM), cryo-TEM, X-ray tomographic microscopy, X-ray photoelectron spectroscopy (XPS), and time-of-flights secondary ion mass spectroscopy, should be developed to provide more real-time information on this process, such as the critical stable metal nuclei size.

8.2.3 All-Solid-State Batteries

The attempt to build all-solid-state batteries contributes to a feasible option to enable the employment of Li metal anodes. A typical solid-state battery is consisted of an intercalation cathode, a Li anode and a solid-state electrolyte that that blocks electric contact yet enables Li^+ transport between both electrodes [51]. The use of a nonflammable solid-state electrolyte (SSE) with wide electrochemical stable window is expected to improve the energy output, safety, and suppress the short circuit caused by dendrite formation [68]. However, the all-solid-state batteries are encountered with various challenges, such as poor thermodynamic stability at the electrode|electrolyte interphase [69, 70], high dependence on temperature, low power density, and fabrication and cost concerns. As the key component, solid-state electrolytes have been revisited during the past five years. In addition to certain success in improving bulk conductivity, the ion transports at the electrode|electrolyte interphase and the chemical/electrochemical stability of these interphases are the research focus to make a solid-state battery works its way into commercial success [71, 72]. The eventual commercialization of this technology is going to be an enormous challenge and long-term interdisciplinary research and development in science and engineering will be required. The future development directions of all-solid-state batteries are provided below.

1. ***Enable high power density at room temperature***. Researches have been focused on the enhancement of ionic transport in solid electrolytes since the 20th century, because solid-state batteries suffer from low power densities due to the low ionic conductivities of solid electrolytes, especially at room and low temperatures. Although a high bulk ionic conductivity up to 10^{-3} S cm^{-1} at room temperature has been realized on some oxide and sulfide electrolytes, the power density is quite limited due to the high SSE thickness [52, 73]. For example, a thickness in the <100 μm range are necessary for SSE with ionic conductivity of ~5×10^{-3} S cm^{-1} to enable the cell energy densities and power densities competitive with the state-of-the-art Li-ion technology [73]. Thick SSE can be employed to gain understanding on solid-state systems and their fundamental limitations; however, a thickness of <100 μm is ultimately required for technologically relevant devices. Scalable and cost-effective manufacturing techniques to construct thin and robust SSE may be one of the most serious challenges facing solid-state battery technology. Many of the ceramic SSE can be made using vapor deposition processes, such as pulsed laser deposition (PLD), physical vapor deposition (PVD), CVD, and ALD, although aspects of cost and scalability using such methods need further consideration. The solid polymer electrolytes are easy to fabricate as soft and flexible films with adjustable thickness; however, they suffer from low ionic conductivity at room temperature. Presently, embedding nanosized ceramic particles into the polymer matrix to form a polymer-ceramic composite electrolyte is effective to combine the advantage of high ionic conductivity and high flexibility [74]. The Li^+ ionic conductivity and diffusion pathways can be altered by designing the morphology and geometry of the inorganic fillers, such as nanoparticles, nanowires and vertical aligned nanochannels [75–77].
2. ***Construct chemically/electrochemically stable electrode|electrolyte interphase***. The total SSE resistance consists the bulk resistance, the SSE-anode interphase resistance, and the SSE-cathode interphase resistance. Whereas some electrolytes, such as garnet, NASICON and sulfide Li^+ electrolytes can give a high bulk conductivity of >1 mS cm^{-1} at 25 °C, a large resistance to Li^+ transfer between the electrode and electrolyte dominates the total resistance of the electrolyte. The huge ionic transfer resistance at the electrode|electrolyte interphase

is attributed to a poor wetting ability of the SSE, a high insulating surface layer forms upon exposure to moist air, and narrow electrochemical window of SSE. Nanocomposite materials could provide a sufficiently large contact area at the interphase. As a result, achievement of fast mass transportation is possible by rational design of the structure of electrode composites and SSE materials. Moreover, nanocoating is another way to inhibit the unfavorable reaction between electrode and electrolyte, alleviate the volume variations, and combine the electrochemical advantages of different SSEs. The quality of the surface coating is of vital importance on the electrode|electrolyte interphase and advanced surface engineering nanotechnologies show great promise to construct high-quality coating layers with good ionic and electronic transportation, good uniformity and high coverage percentage. Polymer electrolyte coating on both sides of ceramic SSE improved the wetting on both electrodes and resulted in a significantly reduced resistance at the electrode|SSE interface [78].

3. ***Suppress the dendrite formation***. Solid electrolyte is expected to physically block the Li dendrite penetration from anode to cathode and prevent the short circuit [79]. Unfortunately, it has been observed that the Li dendrite can grow along the defects and grain boundaries of SSE and bring safety issues. The preferential Li plating is attributed to the poor wettability between solid–solid Li|SSE interfaces and induces an inhomogeneous distribution of Li^+ flux [78]. Proper nanoengineering techniques, such as depositing artificial SEI layers between Li metal and SSE should be a feasible way to wet both the Li anode and the SSE. Sulfide and LISICON electrolytes can be reduced by Li to generate an in situ formed Li-conductive passivation nanolayer at the Li|SSE interface, making them promising to demonstrate a good wettability with Li metal. A fine control of the microstructures of SSE, such as defects and porosity, would also be a feasible strategy to suppress the Li dendrite growth.

8.2.4 Sodium-Ion Batteries

From a cost and sustainable perspective, it is more appealing to use Na-ion rather than Li-ion based batteries for energy storage in grid-scale due to the rich abundance and low-cost of sodium precursor materials. Taking carbonates as an example, Na_2CO_3 are about 25–30 times cheaper than Li_2CO_3. Since the cost of the carrier-ions-precursors only counts for a small percentage of the overall material cost, the use of Al foil as anode current collector contributes to a greater reduction in cost [80]. In addition, the use of low-cost and non-toxic elements such as Fe, Mn, Cu, and Ti instead of Ni and Co should further reduce the cost [81]. The energy storage chemistry and electrochemical properties of Na-ion batteries are similar with that of Li-ion batteries in many aspects, nevertheless, there exist significant differences between them [82]. As a result, besides the well-known component materials working well in Li-ion based batteries, the development of advanced electrode materials with novel structures and compositions is still the research focus [83]. The future development direction of Na-ion batteries should focus on a few aspects: (i) improve the energy density to decrease the normalized cost per kWh; (ii) develop advanced synthesis technology to reduce the cost of hard carbon anodes; (iii) optimize the electrolyte compositions and the purification technology of $NaPF_6$ salt [83, 84]; (iv) enhance the long-term operation stability; (v) improve the chemical stability of electrode materials against moisture and carbon dioxide in air; (vi) more efforts in the investigation of safety issues such as thermal abuse.

Innovations in the synthesis procedures, stabilization strategies, and the characterization techniques of the electrode materials are expected to address the issues listed above [85]. For cathode materials, advanced synthesis techniques, such as co-precipitation and sol-gel method should be applied to obtain cathode particles with high tap density, low surface area and good kinetics. Intensive investigation has shown that element substitution stabilizes the bulk structure by minimizing the phase transitions and volume changes upon Na^+ ions insertion/extraction [86, 87]. More efforts should be devoted into the research of the surface/interphase of active particles [88], such as the composition, property, structure and evolution of surface SEI film and how these properties affect the diffusion of Na^+ ions in nanoscale and the interphase resistance [69, 89]. In addition, fundamental understanding of the degradation mechanism and the dynamic evolution of chemical species at the cathode-air interphase of air-sensitive compounds, such as transition-metal layered oxides, should provide a vital guide on the feasible strategies for further optimization [81, 90]. Nanosized materials are expected to alleviate the severe particle pulverization caused by huge volume change during the cycling process of high-capacity anode materials, such as P, Sb, transition-metal sulfides and sodium metal [91–93]. Distributing the nanoparticles uniformly among a conductive matrix is required to minimize the undesired surface reactions and improve the electronic conductivity [94]. Decreasing the particles size to nanoscale plus surface carbon nanocoating is an effective strategy to enhance the kinetics for ions and electrons transportation, improving the high-rate capabilities [95]. It is worth noting that there is a trad-off between the cycling performance, rate capability and the energy density with the introduction of nanostructure and nanoparticles. Therefore, an exquisite control of the morphology, particle size, and surface area of the electrodes is required.

8.2.5 Conversion-Type Batteries

In the state-of-the-art LIBs, both anode and cathodes operate on intercalation chemistry and their capacities are being pushed to the theoretical limitations. To further boost the energy density, electrode materials based on conversion reactions, such as binary transitional metal oxides and sulfides as proposed by Armand et al., have attracted great attention [96]. Nowadays, the concept of conversion-type materials has been extended to some newly emerging systems, such as sulfur and oxygen cathodes [11]. Extensive efforts have been devoted in improving the feasibility of conversion-type batteries and certain progress have been achieved to address their potential in surpassing the state-of-the-art LIBs. However, challenges still exist in the science and engineering aspects of these newly emerging EES systems. Some typical conversion-type electrode materials are highlighted in the following.

1. ***Sulfur cathodes***. Low electronic/ionic conductivity of sulfur and high solubility of the polysulfide species produced during the cycling are still two major challenges for sulfur cathodes [97]. Distributing active sulfur into a nanoporous and conductive matrix has been regarded as one of the most promising strategies to address the above issues [98]. The morphology, microstructure, pore sizes and mass concentration of the matrix have significant impact on the final performance [99, 100]. However, a trade-off must be made between energy density and cycling life because the introduction of inactive matrix, such as carbon and TiO_2, leads to a diminished energy output [99–102]. Free-standing sulfur composite cathodes without current collectors turns out to be an effective strategy to further improve the capacity calculated based on the electrode mass. Despite the superior gravimetric energy density of sulfur cathodes, their volumetric energy densities are often neglected. Unfortunately, the volumetric energy density is normally lower than expectation, which are attributed to the low electrode density of most plausible sulfur composite electrodes. It is estimated that a S cathode should have an areal sulfur loading of $\geq$6 mg cm^{-2} and >70 wt%, an electrode porosity of <70%, a specific capacity of 1000 mA h g^{-1}, and an anode excess of <50 wt%, to make a Li–S cell comparable to the commercial LIBs (~600 Wh L^{-1}) in volumetric densities [103]. In the future, research emphasis should be placed on electrode design to improve the packing density and simultaneously alleviate polysulfide dissolution. Fundamental understandings on the conversion electrochemistry may help to reveal the operational feasibilities. In addition, to intrinsically inhibit the capacity loss caused by dissolution of active materials, advanced sulfur cathodes based on novel electrochemistry without forming dissolvable polysulfide intermediates is desirable [104]. Apart from smart design on the cathode, modifications on the separators and passivation of metal anodes, such as nanocoating and artificial SEI layers, are also necessary for satisfactory advancements.
2. ***Oxygen cathode***. The active material of oxygen cathodes is O_2 gas from the air, which avoids the use of solid active materials and promises a significant increase in the energy density [105]. The poor reversibility, low energy efficiency and long-term cycle stability remain the major problems in most metal-oxygen systems. Li–O_2 cells can operate in both aqueous and non-aqueous electrolytes with LiOH and Li_2O_2, respectively, as reaction products [106, 107]. The improvement in the reversibility of the electrochemical reaction lies in improving the decomposition efficiency of both LiOH and Li_2O_2 during charging and avoiding unfavorable side reactions among electrolyte, carbon substrate and O_2 [108, 109]. The future development direction of O_2 cathodes should focus on a few aspects: (i) a well control of the nucleation, growth and distribution of discharge products; (ii) construct efficient catalyst cathodes with highly dispersive nanostructured oxygen reduction reaction (ORR) and oxygen evolution reaction (OER) catalysts among conductive and porous substrate; (iii) non-carbonaceous electrode materials to

avoid the side reactions with superoxide and peroxide intermediates; (iv) stable electrolyte against O_2 attack. Compared with Li–O_2 batteries, Zn–O_2 batteries show lower gravimetric energy densities due to the smaller gravimetric capacity of Zn relative to Li anode, but have the advantage of high volumetric energy density, long durability and high safety [110]. Primary Zn–O_2 battery has been successfully employed in audiphones, navigation lights, telecommunication systems, and railway signal amplifiers for a long time [111]. Nevertheless, secondary Zn–O_2 battery is still in the research phase since many problems remain to be solved, which includes: (i) Zn metal passivation and corrosion; (ii) poor oxygen reduction and evolution efficiencies; (iii) low reversibility of oxygen cathode caused by the formation of carbonate; (iv) cathode degradation upon extensively cycling [112–114]. To meet the commercial standards in terms of cycling stability and rate capability, the research focus should be improving the efficiency, reversibility and stability of both Zn and O_2 electrodes. Possible strategies may involve with developing advanced non-precious metal ORR and OER catalysts; smart design of the structure, pore sizes, hydrophilic property and corrosion-resistivity of air electrodes; surface modifications of zinc anode and scavenger additives to suppress the formation of ZnO passivation layers.

3. ***Transition metal oxide, sulfide and fluoride***. The electrochemical-driven conversion reactions of transition-metal oxide, sulfide and fluoride cathodes/anodes are typically evolved with breaking and forming chemical bonds and generating multiple reaction products [115, 116]. These reaction features are responsible for the huge volume variations, high overpotential, low energy efficiency and insufficient cycle life. Well-designed nanostructured architectures could effectively alleviate the capacity fading caused by powder pulverizations and decreasing the pronounced voltage hysteresis [117]. Nevertheless, the low Coulombic efficiency, which is aroused by the intrinsic reaction property, remains a huge challenge. Advanced lithiation process in ambient environment and ultrathin artificial SEI layer would be two feasible strategies to improve the energy efficiency. The operation voltages of oxides or sulfides are either too low as cathodes or too high as anodes. Transition metal fluorides, on the contrary, are more promising relative to the oxides and sulfides because of the higher average operating voltages [118]. In addition, low-cost, abundant and non-toxic transition metals, such as Fe, Mn, Cu and Ti, should be preferentially chosen. The strategies to facilitate the development of transition-metal fluorides materials may include: (i) construction of three-dimensional efficient Li^+/e^- conductive networks with nanoscale wiring to achieve fast electrons/mass transport; (ii) utilize solid–liquid reaction interface instead of solid–solid interface to improve the conversion efficiency; (iii) operando characterization techniques to reveal the conversion mechanisms in real time.

8.2.6 Other Opportunities

The properties of a battery in pack or cell level do not only depend on the active materials, but also on other components, including the carbon additive, current collector, electrolyte, separator, and cell and battery design. To improve the energy density of a battery cell/pack, it is desired to further increase the active materials ratio and decrease the amount of inactive components since the active electrode materials in the state-of-the-art LIBs only counts 50 wt% by weight and as little as 25 wt% in the pack used in EVs. Therefore, a near-future strategy should also focus on increasing the thickness of electrodes, which is challenged by the deteriorated mechanical properties and sluggish Li^+/e^- transportation kinetics among the electrodes. Besides moderate innovations in engineering to prevent the electrode crack and control the porosity of electrodes, a well design in the content and microstructure of conductive additives, binder, and current collectors would also be a viable strategy. For conductive additives, the dispersity, conductivity, surface area, morphology, and electrolyte wettability are important factors needs comprehensive evaluation. A combination of carbon additives with various dimensions, such as 2D graphene and 0D carbon black, would be a feasible way to construct three-dimensional conductive networks and maximize the contact area with active materials. Developing light-weight and low-cost current collectors, such as graphene or carbon nanotube films [119–121] would give a push in the energy density. In addition, 3D porous current collectors with well-aligned Li diffusion channels are expected to ensure efficient electrolyte penetration and good Li^+/e^- transport kinetics in thick electrodes [122]. The structure of the electrode film requires smart design too. For example, to achieve a trade-off between electrode density and powder density, the active particles could be deposited layer by layer to form an asymmetric porous structure with a gradient distribution of pores; macrospores near the separator and nanopores near the current collectors.

The chemical and electrochemical stability of inactive components is one of the important factors affecting the long-term cycle stability of batteries. The metal current collectors would be corroded at extreme conditions, such as high voltages and under the attack of acid, base or strong oxidizing/reducing intermediates, leading to internal resistance increment and performance loss of the battery [123, 124]. Therefore, efficient passivation or protection strategy should be applied. Surface modification with nanomaterials with high conductivity and corrosive-resistance would be the most efficient strategy. The physical contact between current collector and coating layer should be sufficient and stable to eliminate the interphase resistance.

Separators, which are used to avoid short circuit caused by electric contact between cathode and anode electrodes, are also a key component in battery configuration since it plays a vital role in determining the internal resistance and safety of the batteries [125]. Both polymeric and ceramic separators are considered as potential high-performance separators [126–128]. A future research direction could be developing ceramic/polymer composite separators, such as surface modification of polymer separator with ceramic nanolayers and use nanosized ceramic particles as filler in the polymer substrate. It is worth noting that the commercial Celgard separators could not meet the requirement of all kinds of rechargeable batteries due to the various battery configurations, working principles and electrolyte compositions. Therefore, the development of advanced separators with high-safety, low-cost, good electrolyte wettability, and mechanical property for long-term post-LIB technologies are highly demanded. In addition, functional separators, which have the abilities to inhibit the shuttle process of polysulfides, selective conductivity of various metal ions, or detect/prevent the inner short circuit would be a promising research direction in the future.

References

1. Dunn, B., Kamath, H., & Tarascon, J.-M. (2011). Electrical energy storage for the grid: A battery of choices. *Science, 334,* 928–935.
2. Armand, M., & Tarascon, J. M. (2008). Building better batteries. *Nature, 451,* 652–657.
3. Tarascon, J. M., & Armand, M. (2001). Issues and challenges facing rechargeable lithium batteries. *Nature, 414,* 359–367.
4. Choi, J. W., & Aurbach, D. (2016). Promise and reality of post-lithium-ion batteries with high energy densities. *Nature Reviews Materials, 1,* 16013.
5. Manthiram, A. (2017). An outlook on lithium ion battery technology. *ACS Central Science, 3,* 1063–1069.
6. Myung, S.-T., Maglia, F., Park, K.-J., et al. (2017). Nickel-rich layered cathode materials for automotive lithium-ion batteries: Achievements and perspectives. *ACS Energy Letters, 2,* 196–223.
7. Nitta, N., Wu, F., Lee, J. T., et al. (2015). Li-ion battery materials: Present and future. *Materials Today, 18,* 252–264.
8. Whittingham, M.-S. (1976). Electrical energy storage and intercalation chemistry. *Science, 192,* 1126–1127.
9. Lu, J., Chen, Z., Ma, Z., et al. (2016). The role of nanotechnology in the development of battery materials for electric vehicles. *Nature Nanotechnology, 11,* 1031–1038.
10. Zhu, C., Usiskin, R. E., Yu, Y., et al. (2017). The nanoscale circuitry of battery electrodes. *Science 358,* eaao2808.
11. Xin, S., Chang, Z., Zhang, X., et al. (2017). Progress of rechargeable lithium metal batteries based on conversion reactions. *National Science Review, 4,* 54–70.
12. Placke, T., Kloepsch, R., Dühnen, S., et al. (2017). Lithium ion, lithium metal, and alternative rechargeable battery technologies: The odyssey for high energy density. *Journal of Solid State Electrochemistry, 21,* 1939–1964.
13. Wood, D. L., Li, J., & Daniel, C. (2015). Prospects for reducing the processing cost of lithium ion batteries. *Journal of Power Sources, 275,* 234–242.
14. Nykvist, B., & Nilsson, M. (2015). Rapidly falling costs of battery packs for electric vehicles. *Nature Climate Change, 5,* 329–332.
15. Liu, W., Oh, P., Liu, X., et al. (2015). Nickel-rich layered lithium transition-metal oxide for high-energy lithium-ion batteries. *Angewandte Chemie International Edition, 54,* 4440–4457.
16. Manthiram, A., Knight, J. C., Myung, S.-T., et al. (2016). Nickel-rich and lithium-rich layered oxide cathodes: Progress and perspectives. *Advanced Energy Materials, 6,* 1501010.
17. Li, W., Song, B., & Manthiram, A. (2017). High-voltage positive electrode materials for lithium-ion batteries. *Chemical Society Reviews, 46,* 3006–3059.
18. Sun, Y.-K., Myung, S.-T., Park, B.-C., et al. (2009). High-energy cathode material for long-life and safe lithium batteries. *Nature Materials, 8,* 320–324.
19. Grenier, A., Liu, H., Wiaderek, K. M., et al. (2017). Reaction heterogeneity in $LiNi_{0.8}Co_{0.15}Al_{0.05}O_2$ induced by surface layer. *Chemistry of Materials, 29,* 7345–7352.

20. Ryu, H.-H., Park, K.-J., Yoon, C. S., et al. (2018). Capacity fading of Ni-rich $Li[Ni_xCo_yMn_{1-x-y}]O_2$ ($0.6 \leq x \leq 0.95$) cathodes for high-energy-density lithium-ion batteries: Bulk or surface degradation? *Chemistry of Materials, 30,* 1155–1163.
21. Zheng, J., Kan, W. H., & Manthiram, A. (2015). Role of Mn content on the electrochemical properties of nickel-rich layered $LiNi_{0.8-x}Co_{0.1}Mn_{0.1+x}O_2$ ($0.0 \leq x \leq 0.08$) cathodes for lithium-ion batteries. *ACS Applied Materials & Interfaces, 7,* 6926–6934.
22. Yoon, C. S., Jun, D.-W., Myung, S.-T., et al. (2017). Structural stability of $LiNiO_2$ cycled above 4.2 V. *ACS Energy Letters, 2,* 1150–1155.
23. Jo, C.-H., Cho, D.-H., Noh, H.-J., et al. (2015). An effective method to reduce residual lithium compounds on Ni-rich $Li[Ni_{0.6}Co_{0.2}Mn_{0.2}]O_2$ active material using a phosphoric acid derived Li_3PO_4 nanolayer. *Nano Research, 8,* 1464–1479.
24. Li, L., Xu, M., Yao, Q., et al. (2016). Alleviating surface degradation of nickel-rich layered oxide cathode material by encapsulating with nanoscale Li-ions/electrons superionic conductors hybrid membrane for advanced Li-ion batteries. *ACS Applied Materials & Interfaces, 8,* 30879–30889.
25. Cho, Y., Oh, P., & Cho, J. (2013). A new type of protective surface layer for high-capacity Ni-based cathode materials: Nanoscaled surface pillaring layer. *Nano Letters, 13,* 1145–1152.
26. Kim, U.-H., Myung, S.-T., Yoon, C. S., et al. (2017). Extending the battery life using an Al-doped $Li[Ni_{0.76}Co_{0.09}Mn_{0.15}]O_2$ cathode with concentration gradients for lithium ion batteries. *ACS Energy Letters, 2,* 1848–1854.
27. Conry, T. E., Mehta, A., Cabana, J., et al. (2012). Structural underpinnings of the enhanced cycling stability upon Al-substitution in $LiNi_{0.45}Mn_{0.45}Co_{0.1-y}Al_yO_2$ positive electrode materials for Li-ion batteries. *Chemistry of Materials, 24,* 3307–3317.
28. Cho, D.-H., Jo, C.-H., Cho, W., et al. (2014). Effect of residual lithium compounds on layer Ni-rich $Li[Ni_{0.7}Mn_{0.3}]O_2$. *Journal of the Electrochemical Society, 161,* A920–A926.
29. Eom, J., Kim, M. G., & Cho, J. (2008). Storage characteristics of $LiNi_{0.8}Co_{0.1+x}Mn_{0.1-x}O_2$ ($x = 0$, 0.03, and 0.06) cathode materials for lithium batteries. *Journal of the Electrochemical Society, 155,* A239–A245.
30. Liu, H., Yang, Y., & Zhang, J. (2007). Reaction mechanism and kinetics of lithium ion battery cathode material $LiNiO_2$ with CO_2. *Journal of Power Sources, 173,* 556–561.
31. You, Y., Celio, H., Li, J., et al. (2018). Modified high-nickel cathodes with stable surface chemistry against ambient air for lithium-ion batteries. *Angewandte Chemie International Edition, 57,* 6480–6485.
32. An, S. J., Li, J., Daniel, C., et al. (2016). The state of understanding of the lithium-ion-battery graphite solid electrolyte interphase (SEI) and its relationship to formation cycling. *Carbon, 105,* 52–76.
33. Son, I. H., Park, J. H., Kwon, S., et al. (2015). Self-terminated artificial SEI layer for nickel-rich layered cathode material via mixed gas chemical vapor deposition. *Chemistry of Materials, 27,* 7370–7379.
34. Li, W., Dolocan, A., Oh, P., et al. (2017). Dynamic behaviour of interphases and its implication on high-energy-density cathode materials in lithium-ion batteries. *Nature Communications, 8,* 14589.
35. Franco Gonzalez, A., Yang, N.-H., & Liu, R.-S. (2017). Silicon anode design for lithium-ion batteries: Progress and perspectives. *The Journal of Physical Chemistry C, 121,* 27775–27787.
36. Jin, Y., Zhu, B., Lu, Z., et al. (2017). Challenges and recent progress in the development of Si anodes for lithium-ion battery. *Advanced Energy Materials, 7,* 1700715.
37. Feng, K., Li, M., Liu, W., et al. (2018). Silicon-based anodes for lithium-ion batteries: From fundamentals to practical applications. *Small (Weinheim an der Bergstrasse, Germany), 14,* 1702737.
38. Wu, H., Chan, G., Choi, J. W., et al. (2012). Stable cycling of double-walled silicon nanotube battery anodes through solid-electrolyte interphase control. *Nature Nanotechnology, 7,* 310–315.
39. Teki, R., Datta, M. K., Krishnan, R., et al. (2009). Nanostructured silicon anodes for lithium ion rechargeable batteries. *Small (Weinheim an der Bergstrasse, Germany), 5,* 2236–2242.

40. Ji, L., Lin, Z., Alcoutlabi, M., et al. (2011). Recent developments in nanostructured anode materials for rechargeable lithium-ion batteries. *Energy & Environmental Science, 4,* 2682–2699.
41. Wu, H., Yu, G., Pan, L., et al. (2013). Stable Li-ion battery anodes by in-situ polymerization of conducting hydrogel to conformally coat silicon nanoparticles. *Nature Communications, 4,* 1943.
42. Zhang, Y.-C., You, Y., Xin, S., et al. (2016). Rice husk-derived hierarchical silicon/nitrogen-doped carbon/carbon nanotube spheres as low-cost and high-capacity anodes for lithium-ion batteries. *Nano Energy, 25,* 120–127.
43. Casimir, A., Zhang, H., Ogoke, O., et al. (2016). Silicon-based anodes for lithium-ion batteries: Effectiveness of materials synthesis and electrode preparation. *Nano Energy, 27,* 359–376.
44. Wu, H., & Cui, Y. (2012). Designing nanostructured Si anodes for high energy lithium ion batteries. *Nano Today, 7,* 414–429.
45. Ng, S.-H., Wang, J., Wexler, D., et al. (2006). Highly reversible lithium storage in spheroidal carbon-coated silicon nanocomposites as anodes for lithium-ion batteries. *Angewandte Chemie International Edition, 45,* 6896–6899.
46. Mokkelbost, T., Fossdal, A., Dahl, Ø., et al. (2011). Preparation of silicon nanostructures for lithium ion battery anodes. *ECS Transactions, 35,* 149–158.
47. Wang, C.-M., Li, X., Wang, Z., et al. (2012). In situ TEM investigation of congruent phase transition and structural evolution of nanostructured silicon/carbon anode for lithium ion batteries. *Nano Letters, 12,* 1624–1632.
48. Fang, S., Tong, Z., Nie, P., et al. (2017). Raspberry-like nanostructured silicon composite anode for high-performance lithium-ion batteries. *ACS Applied Materials & Interfaces, 9,* 18766–18773.
49. Szczech, J. R., & Jin, S. (2011). Nanostructured silicon for high capacity lithium battery anodes. *Energy & Environmental Science, 4,* 56–72.
50. Lin, D., Liu, Y., & Cui, Y. (2017). Reviving the lithium metal anode for high-energy batteries. *Nature Nanotechnology, 12,* 194–206.
51. Manthiram, A., Yu, X., & Wang, S. (2017). Lithium battery chemistries enabled by solid-state electrolytes. *Nature Reviews Materials, 2,* 16103.
52. Kerman, K., Luntz, A., Viswanathan, V., et al. (2017). Review—Practical challenges hindering the development of solid state Li ion batteries. *Journal of the Electrochemical Society, 164,* A1731–A1744.
53. Jafta, C. J., Ozoemena, K. I., Mathe, M. K., et al. (2012). Synthesis, characterisation and electrochemical intercalation kinetics of nanostructured aluminium-doped $Li[Li_{0.2}Mn_{0.54}Ni_{0.13}Co_{0.13}]O_2$ cathode material for lithium ion battery. *Electrochimica Acta, 85,* 411–422.
54. Li, H., Wang, Z., Chen, L., et al. (2009). Research on advanced materials for Li-ion batteries. *Advanced Materials, 21,* 4593–4607.
55. Jarvis, K. A., Deng, Z., Allard, L. F., et al. (2011). Atomic structure of a lithium-rich layered oxide material for lithium-ion batteries: Evidence of a solid solution. *Chemistry of Materials, 23,* 3614–3621.
56. He, X., Wang, J., Kloepsch, R., et al. (2014). Enhanced electrochemical performance in lithium ion batteries of a hollow spherical lithium-rich cathode material synthesized by a molten salt method. *Nano Research, 7,* 110–118.
57. Chen, L., Su, Y., Chen, S., et al. (2014). Hierarchical $Li_{1.2}Ni_{0.2}Mn_{0.6}O_2$ nanoplates with exposed {010} planes as high-performance cathode material for lithium-ion batteries. *Advanced Materials, 26,* 6756–6760.
58. Tikekar, M. D., Choudhury, S., Tu, Z., et al. (2016). Design principles for electrolytes and interfaces for stable lithium-metal batteries. *Nature Energy, 1,* 16114.
59. Suo, L., Borodin, O., Gao, T., et al. (2015). "Water-in-salt" electrolyte enables high-voltage aqueous lithium-ion chemistries. *Science, 350,* 938–943.
60. Zhamu, A., Chen, G., Liu, C., et al. (2012). Reviving rechargeable lithium metal batteries: Enabling next-generation high-energy and high-power cells. *Energy & Environmental Science, 5,* 5701–5707.

61. Lin, D., Liu, Y., Liang, Z., et al. (2016). Layered reduced graphene oxide with nanoscale interlayer gaps as a stable host for lithium metal anodes. *Nature Nanotechnology, 11,* 626–632.
62. Sun, Y., Liu, N., & Cui, Y. (2016). Promises and challenges of nanomaterials for lithium-based rechargeable batteries. *Nature Energy, 1,* 16071.
63. Mukherjee, R., Thomas, A. V., Datta, D., et al. (2014). Defect-induced plating of lithium metal within porous graphene networks. *Nature Communications, 5,* 3710.
64. Kazyak, E., Wood, K. N., & Dasgupta, N. P. (2015). Improved cycle life and stability of lithium metal anodes through ultrathin atomic layer deposition surface treatments. *Chemistry of Materials, 27,* 6457–6462.
65. Liu, W., Lin, D., Pei, A., et al. (2016). Stabilizing lithium metal anodes by uniform Li-ion flux distribution in nanochannel confinement. *Journal of the American Chemical Society, 138,* 15443–15450.
66. Yan, K., Lu, Z., Lee, H.-W., et al. (2016). Selective deposition and stable encapsulation of lithium through heterogeneous seeded growth. *Nature Energy, 1,* 16010.
67. Andersson, A. M., Abraham, D. P., Haasch, R., et al. (2002). Surface characterization of electrodes from high power lithium-ion batteries. *Journal of the Electrochemical Society, 149,* A1358–A1369.
68. Huang, B., Xu, B., Li, Y., et al. (2016). Li-ion conduction and stability of perovskite $Li_{3/8}Sr_{7/16}Hf_{1/4}Ta_{3/4}O_3$. *ACS Applied Materials & Interfaces, 8,* 14552–14557.
69. Lu, Y., Li, L., Zhang, Q., et al. (2018). Electrolyte and interface engineering for solid-state sodium batteries. *Joule, 2,* 1747–1770.
70. Li, Y., Zhou, W., Chen, X., et al. (2016). Mastering the interface for advanced all-solid-state lithium rechargeable batteries. *Proceedings of the National Academy of Sciences of the United States of America, 113,* 13313–13317.
71. Xin, S., You, Y., Wang, S., et al. (2017). Solid-state lithium metal batteries promoted by nanotechnology: Progress and prospects. *ACS Energy Letters, 2,* 1385–1394.
72. Truong, L., & Thangadurai, V. (2012). First total H^+/Li^+ ion exchange in garnet-type $Li_5La_3Nb_2O_{12}$ using organic acids and studies on the effect of Li stuffing. *Inorganic Chemistry, 51,* 1222–1224.
73. Zheng, F., Kotobuki, M., Song, S., et al. (2018). Review on solid electrolytes for all-solid-state lithium-ion batteries. *Journal of Power Sources, 389,* 198–213.
74. Zhang, X., Xie, J., Shi, F., et al. (2018). Vertically aligned and continuous nanoscale ceramic-polymer interfaces in composite solid polymer electrolytes for enhanced ionic conductivity. *Nano Letters, 18,* 3829–3838.
75. Chinnam, P. R., & Wunder, S. L. (2017). Engineered interfaces in hybrid ceramic-polymer electrolytes for use in all-solid-state Li batteries. *ACS Energy Letters, 2,* 134–138.
76. Zekoll, S., Marriner-Edwards, C., Hekselman, A. K. O., et al. (2018). Hybrid electrolytes with 3D bicontinuous ordered ceramic and polymer microchannels for all-solid-state batteries. *Energy & Environmental Science, 11,* 185–201.
77. Zheng, J., Tang, M., & Hu, Y.-Y. (2016). Lithium ion pathway within $Li_7La_3Zr_2O_{12}$-polyethylene oxide composite electrolytes. *Angewandte Chemie International Edition, 55,* 1–6.
78. Zhou, W., Wang, S., Li, Y., et al. (2016). Plating a dendrite-free lithium anode with a polymer/ceramic/polymer sandwich electrolyte. *Journal of the American Chemical Society, 138,* 9385–9388.
79. Liu, K., Liu, Y., Lin, D., et al. (2018). Materials for lithium-ion battery safety. *Science Advances 4,* eaas9820.
80. Wang, P.-F., You, Y., Yin, Y.-X., et al. (2018). Layered oxide cathodes for sodium-ion batteries: Phase transition, air stability, and performance. *Advanced Energy Materials, 8,* 1701912.
81. Deng, J., Luo, W.-B., Lu, X., et al. (2018). High energy density sodium-ion battery with industrially feasible and air-stable O3-type layered oxide cathode. *Advanced Energy Materials, 8,* 1701610.
82. You, Y., & Manthiram, A. (2018). Progress in high-voltage cathode materials for rechargeable sodium-ion batteries. *Advanced Energy Materials, 8,* 1701785.

83. Kubota, K., & Komaba, S. (2015). Review—Practical issues and future perspective for Na-ion batteries. *Journal of the Electrochemical Society, 162,* A2538–A2550.
84. Mortemard de Boisse, B., Cheng, J. H., Carlier, D., et al. (2015). O3-$NaxMn_{1/3}Fe_{2/3}O_2$ as a positive electrode material for Na-ion batteries: Structural evolutions and redox mechanisms upon Na^+ (de)intercalation. *Journal of Materials Chemistry A, 3,* 10976–10989.
85. You, Y., Wu, X.-L., Yin, Y.-X., et al. (2013). A zero-strain insertion cathode material of nickel ferricyanide for sodium-ion batteries. *Journal of Materials Chemistry A, 1,* 14061–14065.
86. Wang, P.-F., You, Y., Yin, Y.-X., et al. (2016). Suppressing the P2–O2 phase transition of $Na_{0.67}Mn_{0.67}Ni_{0.33}O_2$ by magnesium substitution for improved sodium-ion batteries. *Angewandte Chemie International Edition, 55,* 7445–7449.
87. You, Y., Kim, S. O., & Manthiram, A. (2017). A honeycomb-layered oxide cathode for sodium-ion batteries with suppressed P3–O1 phase transition. *Advanced Energy Materials, 7,* 1601698.
88. You, Y., Xin, S., Asl, H. Y., et al. (2018). Insights into the improved high-voltage performance of Li-incorporated layered oxide cathodes for sodium-ion batteries. *Chem, 4,* 2124–2139.
89. Komaba, S., Ishikawa, T., Yabuuchi, N., et al. (2011). Fluorinated ethylene carbonate as electrolyte additive for rechargeable Na batteries. *ACS Applied Materials & Interfaces, 3,* 4165–4168.
90. Yao, H.-R., Wang, P.-F., Gong, Y., et al. (2017). Designing air-stable O3-type cathode materials by combined structure modulation for Na-ion batteries. *Journal of the American Chemical Society, 139,* 8440–8443.
91. Bayley, P. M., Trease, N. M., & Grey, C. P. (2016). Insights into electrochemical sodium metal deposition as probed with in situ ^{23}Na NMR. *Journal of the American Chemical Society, 138,* 1955–1961.
92. Liu, J., Kopold, P., Wu, C., et al. (2015). Uniform yolk-shell Sn_4P_3@C nanospheres as high-capacity and cycle-stable anode materials for sodium-ion batteries. *Energy & Environmental Science, 8,* 3531–3538.
93. Song, J., Yu, Z., Gordin, M. L., et al. (2015). Advanced sodium ion battery anode constructed via chemical bonding between phosphorus, carbon nanotube, and cross-linked polymer binder. *ACS Nano, 9,* 11933–11941.
94. Zhu, C., Mu, X., van Aken, P. A., et al. (2014). Single-layered ultrasmall nanoplates of MoS_2 embedded in carbon nanofibers with excellent electrochemical performance for lithium and sodium storage. *Angewandte Chemie International Edition, 53,* 2152–2156.
95. Hu, Z., Wang, L., Zhang, K., et al. (2014). MoS_2 nanoflowers with expanded interlayers as high-performance anodes for sodium-ion batteries. *Angewandte Chemie International Edition, 53,* 12794–12798.
96. Yu, S.-H., Feng, X., Zhang, N., et al. (2018). Understanding conversion-type electrodes for lithium rechargeable batteries. *Accounts of Chemical Research, 51,* 273–281.
97. Song, J., Yu, Z., Gordin, M. L., et al. (2016). Advanced sulfur cathode enabled by highly crumpled nitrogen-doped graphene sheets for high-energy-density lithium-sulfur batteries. *Nano Letters, 16,* 864–870.
98. Wang, H., Yang, Y., Liang, Y., et al. (2011). Graphene-wrapped sulfur particles as a rechargeable lithium-sulfur battery cathode material with high capacity and cycling stability. *Nano Letters, 11,* 2644–2647.
99. Xin, S., You, Y., Li, H.-Q., et al. (2016). Graphene sandwiched by sulfur-confined mesoporous carbon nanosheets: A kinetically stable cathode for Li–S batteries. *ACS Applied Materials & Interfaces, 8,* 33704–33711.
100. Wei Seh, Z., Li, W., Cha, J. J., et al. (2013). Sulphur–TiO_2 yolk-shell nanoarchitecture with internal void space for long-cycle lithium–sulphur batteries. *Nature Communications, 4,* 1331.
101. Xu, D.-W., Xin, S., You, Y., et al. (2016). Built-in carbon nanotube network inside a biomass-derived hierarchically porous carbon to enhance the performance of the sulfur cathode in a Li–S battery. *ChemNanoMat, 2,* 712–718.
102. You, Y., Zeng, W., Yin, Y.-X., et al. (2015). Hierarchically micro/mesoporous activated graphene with a large surface area for high sulfur loading in Li–S batteries. *Journal of Materials Chemistry A, 3,* 4799–4802.

103. Xue, W., Miao, L., Qie, L., et al. (2017). Gravimetric and volumetric energy densities of lithium-sulfur batteries. *Current Opinion in Electrochemistry, 6,* 92–99.
104. Xin, S., Yu, L., You, Y., et al. (2016). The electrochemistry with lithium versus sodium of selenium confined to slit micropores in carbon. *Nano Letters, 16,* 4560–4568.
105. Wang, Z., You, Y., Yuan, J., et al. (2016). Nickel-doped $La_{0.8}Sr_{0.2}Mn_{1-x}Ni_xO_3$ nanoparticles containing abundant oxygen vacancies as an optimized bifunctional catalyst for oxygen cathode in rechargeable lithium-air batteries. *ACS Applied Materials & Interfaces, 8,* 6520–6528.
106. Liu, T., Leskes, M., Yu, W., et al. (2015). Cycling Li-O_2 batteries via LiOH formation and decomposition. *Science, 350,* 530–533.
107. Lim, H., Yilmaz, E., & Byon, H. R. (2012). Real-time XRD studies of Li–O_2 electrochemical reaction in nonaqueous lithium-oxygen battery. *The Journal of Physical Chemistry Letters, 3,* 3210–3215.
108. Chang, Z.-W., Xu, J.-J., Liu, Q.-C., et al. (2015). Recent progress on stability enhancement for cathode in rechargeable non-aqueous lithium-oxygen battery. *Advanced Energy Materials, 5,* 1500633.
109. Xu, Y.-F., Chen, Y., Xu, G.-L., et al. (2016). RuO_2 nanoparticles supported on MnO_2 nanorods as high efficient bifunctional electrocatalyst of lithium-oxygen battery. *Nano Energy, 28,* 63–70.
110. Lee, J.-S., Tai Kim, S., Cao, R., et al. (2011). Metal-air batteries with high energy density: Li–air versus Zn–air. *Advanced Energy Materials, 1,* 34–50.
111. Cheng, F., & Chen, J. (2012). Metal–air batteries: From oxygen reduction electrochemistry to cathode catalysts. *Chemical Society Reviews, 41,* 2172–2192.
112. Dhavale, V. M., & Kurungot, S. (2015). Cu–Pt nanocage with 3-D electrocatalytic surface as an efficient oxygen reduction electrocatalyst for a primary Zn–air battery. *ACS Catalysis, 5,* 1445–1452.
113. Shinde, S. S., Lee, C.-H., Sami, A., et al. (2017). Scalable 3-D carbon nitride sponge as an efficient metal-free bifunctional oxygen electrocatalyst for rechargeable Zn–air batteries. *ACS Nano, 11,* 347–357.
114. Müller, S., Holzer, F., & Haas, O. (1998). Optimized zinc electrode for the rechargeable zinc–air battery. *Journal of Applied Electrochemistry, 28,* 895–898.
115. Shan, T.-T., Xin, S., You, Y., et al. (2016). Combining nitrogen-doped graphene sheets and MoS_2: A unique film–foam–film structure for enhanced lithium storage. *Angewandte Chemie International Edition, 55,* 12783–12788.
116. Zhang, W.-M., Wu, X.-L., Hu, J.-S., et al. (2008). Carbon coated Fe_3O_4 nanospindles as a superior anode material for lithium-ion batteries. *Advanced Functional Materials, 18,* 3941–3946.
117. Poizot, P., Laruelle, S., Grugeon, S., et al. (2000). Nano-sized transition-metal oxides as negative-electrode materials for lithium-ion batteries. *Nature, 407,* 496–499.
118. Li, C., Chen, K., Zhou, X., et al. (2018). Electrochemically driven conversion reaction in fluoride electrodes for energy storage devices. *npj Computational Materials, 4,* 22.
119. Wang, K., Luo, S., Wu, Y., et al. (2013). Super-aligned carbon nanotube films as current collectors for lightweight and flexible lithium ion batteries. *Advanced Functional Materials, 23,* 846–853.
120. Li, X., Yang, J., Hu, Y., et al. (2012). Novel approach toward a binder-free and current collector-free anode configuration: Highly flexible nanoporous carbon nanotube electrodes with strong mechanical strength harvesting improved lithium storage. *Journal of Materials Chemistry, 22,* 18847–18853.
121. Rana, K., Singh, J., Lee, J.-T., et al. (2014). Highly conductive freestanding graphene films as anode current collectors for flexible lithium-ion batteries. *ACS Applied Materials & Interfaces, 6,* 11158–11166.
122. Nara, H., Yokoshima, T., Mikuriya, H., et al. (2017). The potential for the creation of a high areal capacity lithium-sulfur battery using a metal foam current collector. *Journal of the Electrochemical Society, 164,* A5026–A5030.
123. Kühnel, R.-S., Lübke, M., Winter, M., et al. (2012). Suppression of aluminum current collector corrosion in ionic liquid containing electrolytes. *Journal of Power Sources, 214,* 178–184.

124. Mun, J., Yim, T., Choi, C. Y., et al. (2010). Linear-sweep thermammetry study on corrosion behavior of Al current collector in ionic liquid solvent. *Electrochemical and Solid-State Letters, 13,* A109–A111.
125. Miao, Y.-E., Zhu, G.-N., Hou, H., et al. (2013). Electrospun polyimide nanofiber-based nonwoven separators for lithium-ion batteries. *Journal of Power Sources, 226,* 82–86.
126. Ryou, M.-H., Lee, Y. M., Park, J.-K., et al. (2011). Mussel-inspired polydopamine-treated polyethylene separators for high-power Li-ion batteries. *Advanced Materials, 23,* 3066–3070.
127. Jung, Y. S., Cavanagh, A. S., Gedvilas, L., et al. (2012). Improved functionality of lithium-ion batteries enabled by atomic layer deposition on the porous microstructure of polymer separators and coating electrodes. *Advanced Energy Materials, 2,* 1022–1027.
128. Zhai, Y., Xiao, K., Yu, J., et al. (2015). Fabrication of hierarchical structured SiO_2/polyetherimide-polyurethane nanofibrous separators with high performance for lithium ion batteries. *Electrochimica Acta, 154,* 219–226.

The manufacturer's authorised representative in the EU is Springer Nature Customer Service Centre GmbH, Europaplatz 3, 69115 Heidelberg, Germany. If you have any concerns regarding our products, please contact ProductSafety@springernature.com

Printed and bound by CPI Group (UK) Ltd, Croydon, CR0 4YY
25/09/2026
02222884-0004